ASTROPHYSICAL AND LABORATORY PLASMAS

ASTROPHYSICAL AND LABORATORY PLASMAS

A Festschrift for Professor Sir Robert Wilson

Edited by

A. J. WILLIS

University College London,
London, United Kindom

and

T.W. HARTQUIST

Max-Planck-Institut für extraterrestrische Physik,
Garching, Germany

Reprinted from *Astrophysics and Space Science*
Volume 237, Nos. 1–2, 1996

KLUWER ACADEMIC PUBLISHERS
DORDRECHT / BOSTON / LONDON

A C.I.P. Catalogue record for this book is available from the Library of Congress.

ISBN 0-7923-4151-1

Published by Kluwer Academic Publishers,
P.O. Box 17, 3300 AA Dordrecht, The Netherlands.

Kluwer Academic Publishers incorporates
the publishing programmes of
D. Reidel, Martinus Nijhoff, Dr W. Junk and MTP Press.

Sold and distributed in the U.S.A. and Canada
by Kluwer Academic Publishers,
101 Philip Drive, Norwell, MA 02061, U.S.A.

In all other countries, sold and distributed
by Kluwer Academic Publishers Group,
P.O. Box 322, 3300 AH Dordrecht, The Netherlands.

Cover photo: copyright Anglo-Australian Observatory, photography by David
Malin

Printed on acid-free paper

TABLE OF CONTENTS

SIR ROBERT WILSON – A DEDICATION

SIR ROBERT WILSON, C.B.E., FRS

ALLAN J. WILLIS

Department of Physics & Astronomy, University College London, Gower Street, London WC1E 6BT, UK

and

THOMAS W. HARTQUIST

Max-Planck-Institut fuer extraterrestrische Physik, D-85740 Garching, Germany

1. DEDICATION

On 1 October 1994, Sir Robert Wilson FRS, CBE became Professor Emeritus in University College London. To commemorate this occasion we decided, with the kind cooperation of Professor J E Dyson in his role as editor of Astrophysics and Space Science, to invite some of Sir Robert's friends to prepare a volume in his honour.

Robert began his research career in 1948 as a postgraduate student in astronomy in the University of Edinburgh. He wrote his doctoral thesis in 1952 on optical spectroscopy of hot O stars and reported the detection of lines in the Pfund series of ionized helium. His formal supervisor was Professor W M H Greaves, FRS.

However, one suspects that Robert's need for guidance, if any, was short lived. Already in September 1950 his independence drew a reprimand from Dr A E Baker. On the 26th of that month the sun in Edinburgh appeared to be blue, a condition which promoted locals to overtax the telephone lines and personnel of Edinburgh's observatory. Spotting Robert after the crisis had passed, Dr Baker expressed displeasure and inquired about the absence and whereabouts of a certain young student whose help in dealing with the public would have lightened others' burdens. Robert replied that he had rushed up to the dome and worked continuously to perform optical spectroscopy of the sun. The instinctive dash to the telescope resulted in the publication of 'The Blue Sun of 1950 September' in Monthly Notices (Wilson, 1951); this paper has appeared in at least one collection of great contributions in meteorological research.

After completing his doctoral research Robert served on the staff of the Royal Observatory, Edinburgh from 1952 to 1957. During this period he performed the analysis of observations that led him to the conclusion that O stars have high velocity winds with mechanical luminosities which correlate with the stellar radiative luminosities, a result which suggests a mechanism for the

Astrophysics and Space Science 237: 3–9, 1996.
© 1996 *Kluwer Academic Publishers. Printed in Belgium.*

winds' acceleration. Robert's important paper on this subject (Wilson 1957) greatly predated those based on rocket ultraviolet data and was submitted from Victoria, Canada where Robert was a visitor for a year at the invitation of Professor B Petrie, who was the director there.

In Victoria, Robert developed his long standing interest in interstellar dust and extinction. He made observations of the 4430 Å absorption feature and discovered other diffuse bands (Wilson, 1958). His attempts to look for polarization in the 4430 Å band were foiled by the insufficiencies of the available polarimeter.

Robert was called back to England in 1958 when he received a letter from Sir John Crockcroft stating that his team in Harwell was on the verge of solving the world's energy problem and would like him to join in the effort to achieve fusion in the Zeta device. Having worked in astronomy until that time Robert felt totally inadequate during his first year as a laboratory physicist leading the plasma spectroscopy group supporting the Zeta device team. Though all of his previous work had been optical spectroscopy, he quickly concluded that ultraviolet spectroscopy must be implemented to diagnose the plasma. He stated that, 'At that time in Harwell one could just ask for anything, and it was done'. 'Robert asked' ! In addition, he developed an analytically tractable model, including plasma losses, of the time dependant ionization structure in a rapidly heated plasma as well as decided that argon (because a variety of its ions abundant over a wide range of temperatures in equilibrium plasma possess ultraviolet resonance lines) be introduced into the Zeta plasma. These efforts culminated in an important paper (Burton and Wilson, 1961) in which the authors demonstrated that Sir John Crockcroft had been wrong. The plasma in Zeta was not fully confined but was escaping to the walls on a timescale of 100 μ–sec; neither had it the previously supposed temperature of 10^7 K. Using high dispersion spectroscopy, Robert measured the linewidths of ultraviolet features of ions of neon, argon and krypton to deduce turbulent and thermal contributions to the linewidths. He concluded that the Zeta plasma reached temperatures of the order of only 10^6 K (Wilson, 1961).

Amongst Robert's other contributions in laboratory plasma spectroscopy was his realization that coherent scattering of laser radiation by fluctuations would permit diagnosis. Accordingly, he established an effort in his group to pursue this direction.

After the demonstration that the Zeta device did not produce a fusion environment the problem of what to do with Robert arose. He solved it by proposing the study of natural plasma (i.e. the sun). The approval of a rocket programme to pursue this study was given in 1962 when the group was still at Harwell. Shortly thereafter the entire fusion team was moved to

Culham where laboratory work continued in the plasma spectroscopy group and Robert's solar physics initiative was realized.

Robert's first achievement in solar physics was the discovery that the Zeta device was an important astrophysical facility. He led an effort that showed that many unidentified extreme ultraviolet solar lines could be matched with features seen in the Zeta spectrum and that the lines were formed by various ions of iron (Wilson, 1964). The very first spectrum of the solar limb was obtained shortly thereafter with a rocket-borne normal incidence ultraviolet spectrograph (Burton, Ridgeley and Wilson, 1967).

As controlled fusion's short term unattainability became clearer, support for the entire Culham Laboratory diminished. In 1968 Brian Flowers in his role as chairman of the Science Research Council (SRC) approached Robert about the possibility that his group's potential funding problems be solved by the SRC taking over the financial support. Thus, the Astrophysics Research Unit of the SRC was created ; while it initially remained in Culham where Robert was its director from 1968 to 1972, the Unit eventually moved to Rutherford Appleton Laboratory where its arrival marked the beginnings of that laboratory's work in space research.

The Unit published data obtained with two extreme ultraviolet and soft X-ray grazing incidence spectrographs that revealed satellite and forbidden lines of many Helium-like ions (Jones, Freeman and Wilson, 1968). The powerful diagnostic utility of the ratios of observed line strengths were exploited in the initial report and in later work. An ultraviolet echelle spectrograph was flown to obtain high spectral resolution observations (Jones et al, 1970); further analysis of them demonstrated that the lines were broadened by turbulence, and Robert was involved in efforts extending for a number of years in using these data to understand heating of the solar corona. He played a leading role in the development of a rocket project to obtain an ultraviolet chromospheric flash spectrum of the Sun during an eclipse (Gabriel et al 1971); the results provided considerable new insight into the chromospheric and coronal structures.

During the Culham years Robert was involved in a number of projects other than directing the laboratory spectroscopy effort, overseeing the development of theoretical analysis of plasma spectra and the solar ultraviolet and X-ray programme. He also conducted a stellar rocket ultraviolet spectroscopy program (Burton et al, 1973) and led the British team in the joint UK/Belgium ultraviolet sky-survey experiment (S2/68) launched in the ESRO satellite TD-1. The TD-1 work resulted in the production of three major UV spectrophotometric stellar catalogues which have been extensively used by the community. The data were analyzed in a number of detailed studies yielding important results on interstellar extinction (Nandy et al, 1976), the interstellar radiation background (Gondhalekhar and Wilson, 1975) and chemical abundances of Wolf-Rayet stars (Willis and Wilson, 1978) as well as in other areas.

Despite all of his involvement with rocket-borne astrophysics, Robert did not see a rocket launch until January, 1978. He claims that during the Culham years he was so busy and that his presence at no launch site was essential; he simply did not make time. He now regrets that he did not go to the eclipse launch.

The story that includes Robert's first viewing of a rocket launch goes back at least as far as 1962. During that year, he met Sir Harrie Massie who was then Chairman of the British National Committee of Space Research, a position in which Robert succeeded Sir Harrie upon his retirement from it in 1983. In the early autumn of 1964 Sir Harrie convened a meeting in University College London (UCL) in response to ESRO's call for proposals for its first large space mission. The participants were Sir Harrie, John Adams (the director of Culham Laboratory), Ieun Maddick (the deputy director of Aldermaston), Robert Boyd (the head of the space research group at UCL) and Robert. (All of Sir Harrie's guests were eventually knighted for their services to science.) With immediate approval from the meeting's other participants Sir Harrie invited Robert to undertake the development of a proposal for the ESRO mission and to lead a team drawing upon the staff and resources of Aldermaston, Culham and UCL. The series of encouraging reviews and setbacks that occurred as the British proposal for a large ultraviolet observatory with an echelle spectrograph was considered by ESRO is detailed elsewhere (Boggess and Wilson, 1987). After considerable frustration, Robert obtained permission from the UK Atomic Energy Authority (who, as joint owners with ESRO of the proposal, could give him the authority to act) to give the proposal which ESRO did not accept to another agency. Robert sent the design report with no request for his own involvement to Leo Goldberg, to act upon it as his saw fit in his capacity as chairman of NASA's Astronomy Missions Board. This outright gift of the concept to NASA led, as described more thoroughly by Boggess and Wilson (1987), to the effort resulting in the launch of the International Ultraviolet Explorer (IUE) satellite.

Robert directed the British effort on IUE first from Culham and then from UCL where he was appointed to the Perren Professorship in Astronomy in 1972. The main contribution of the British hardware part of the IUE team was the development of the on–board ultraviolet cameras (SEC Vidicons) and associated software. These cameras lie at the heart of the overall IUE scientific instrument and their development was for a long period of time at the core of the critical path for the whole IUE programme. Consequently, Robert as UK Project Director, together with Mr Peter Barker (the UK

IUE Project Manager) shouldered the ultimate management responsibility for ensuring the timely success of the satellite programme, and for convincing NASA, in particular, that the cameras would be delivered punctually and to specification. After several, often frought years, this was achieved, at which point the critical path switched across the Altantic to NASA. IUE was launched on 26 January 1978 and, after a successful orbital switch–on, optimisation and calibration, and a post–launch commisioning phase, was opened up to Guest Observers worldwide in April of that year. Although initially with a design lifetime of 3 years, the IUE satellite has far–outlived this timeframe, and is currently still fully operational some 18 years later. The satellite has become undoubtedly the most successful and productive astronomical space facility ever, inspiring 10 international conferences, involved many thousands of guest observer astronomers and resulted in over 3000 papers published in conferences and the major scientific journals.

After arriving at UCL, Robert quickly embarked on setting up a new astrophysics research group, initially emphasising studies of hot, massive stars and the interstellar medium. Programmes in these areas, and in stellar chromospheres that had been started with TD-1 and rocket data, were to be greatly expanded with the IUE ultraviolet spectra being generated. Highlights included the determination of the UV interstellar extinction curves for the Large and Small Magellanic Clouds (Nandy *et al.* 1981, 1882), the discovery of mass loss from hot subdwarf stars (Darius, Giddings & Wilson, 1979), the discovery of selective line eclipses in the WC8+O9I binary system γ Velorum (Willis & Wilson 1976), and the detection of emission in the Lyman band of molecular hydrogen in the pre–main sequence star T Tauri (Brown *et al*, 1981). In addition Robert extended his research interests to UV studies of X-ray binary systems, leading an international effort to use IUE continuously for 2 full weeks for a coordinated multi–wavelength programme of observations involving several ground–based observatories and other satellites. Both massive and low–mass X–ray binary systems were observed, delineating the effects of interaction of the primary and compact object. In HZ Her, the effect of the re–processed X–rays on the inside hemisphere of the primary star was clearly seen in the UV spectra, as was its accretion disk (Gursky *et al*, 1980). In the high mass systems, like Cyg X–1, observational confirmation of the phase–dependent Hatchett & McCray effect (X–ray ionisation induced changes in the UV P–Cygni resonance line profiles) was confirmed for the first time.

Robert also initiated several ultraviolet studies of active galaxies with IUE, in particular the first UV variability studies of Seyfert galaxies (Barr, Willis & Wilson 1983) and quasars (Ulrich *et al*, 1980). Robert pioneered techniques for pushing IUE to the limit of its sensitivity for the observation of faint extragalactic sources, leading, *inter alia*, to the detection of the UV spectrum of the quasar Q2204–408 (z = 3.2, m = 17.5) down to a rest frame wavelength of 300 Å. Further, Robert used IUE to obtain UV spectra of the twin quasar Q0957+561 A,B, covering the Lyman emission and absorption systems. These unique data showed that the UV ratio of the two objects was the same as observed at radio wavelengths, confirming a fundamental prediction of the theory of graviational lensing – no chromatic aberration (Gondhalekar & Wilson, 1980). Further IUE monitoring of the twin quasar, revealed a variation in image A detected in image B with a time delay of 1.8 years, yielding on simple analysis a value for Hubble's constant in the frame of the quasar. The result demonstrated that the universe is expanding in a region at a cosmological distance scale (z = 1.4) in a manner compatible with the expansion detected locally and thus supported the validity of important assumptions of homogeneity underlying cosmological theory (Gondhalekar & Wilson, 1982). The measured value of Hubble's constant in the quasar frame corresponds to Hubble's constant of 67 km^{-1} Mpc^{-1} at the current epoch, a result which is in excellent agreement with, and considerably predated the values inferred following studies of Cepheid variables in other galaxies made with the Hubble Space Telescope.

In addition to his research activities at UCL, Robert also undertook heavy responsibilities in the administration and policy–making activities at university, national and international levels. As well as leading the astronomy programme at UCL, as Perren Professor (1972–1994), he acted as Dean of the Science Faculty at UCL during a particularly difficult financial period (1982–85) for the College, and as Head of the Department of Physics & Astronomy (1987–93). At the national level, Robert has served with distinction on many boards and committees, including the Councils of the SERC and NERC. Robert was chairman of the British National Committee for Space Research (1983–88), of the Anglo–Australian Telescope Board (1986–89), of the James Clerk Maxwell Telescope Board (1987–91), and of the Council of the Royal Institution (1992–93). He has also served on many committees of the Royal Society. On the international scene Robert was a Vice–President of the International Astronomical Union (1979–85), and a member of the COSPAR Bureau (1986–90). He is a Trustee of the UK Space Education Trust (1985–present) and a trustee of the International Academy of Astronautics (1987–present).

Sir Robert has been honoured a number of times including on the occasions of his election to fellowship of the Institute of Physics in 1968 and the Royal Society in 1975; since 1976 he has been a Foreign Member of the Royal Society of Liege. He was made a Commander of the British Empire in 1978 and a Knight Batchelor in 1989. He received the Royal Astronomical Society's Herschel Medal in 1986 and the Science Award of the International Astronautics Academy in 1987. In 1988 he became the first non-citizen of the US to receive its Presidential Award for achievements in technological excellence. He was elected an Honorary Fellow of University College London in 1989. Particularly given his association with and respect for Sir Harrie Massey, the conferment upon Sir Robert in 1994 of the Royal Society/COSPAR Massey Award was fitting. In 1995 Queens University Belfast presented him with an honorary Doctorate of Science.

Those who know Sir Robert well recognize in him a total absence of pettiness and value his generosity of spirit and gentleness. His hospitality contributes with his sincere interest in others and tendency in the direction of being a bon vivant to his success as a truly great host. His insistence upon identifying the fundamental question underlying a research project and his intrinsically encouraging nature have influenced greatly those scientists around him on many particular instances and provided a standard which arises frequently in ones memory and serves as a more general guide in ones further research.

References

Barr, P., Willis, A.J., & Wilson, R., 1983. *MNRAS*, **202**, 453

Boggess, A., & Wilson, R., 1987. *in Exploring the Universe with the IUE Satellite*, (ed. Y Kondo), Kluwer Acad. Press., p 3

Brown, A., Jordan, C., Millar, T.J., Gondhalekar, P.M., & Wilson, R., 1981. *Nature*, **290**, 34

Burton, W.M., & Wilson, R., 1961. *Proc. Phys. Soc.*, **78**, 1416

Burton, W.M., Ridgeley, A., & Wilson, R., 1967. *MNRAS*, **135**, 207

Burton, W.M., *et al.*, 1973. *Nature*, **246**, 37

Darius, J., Giddings, J.R., & Wilson, R., 1979. *in The First Year of IUE* (ed A J Willis), UCL, p 363

Gabriel, A., *et al.*, 1971. *Astrophys. J.*, **169**, 595

Gondhalekar, P.M., & Wilson, R., 1975. *Astron. Astrophys.*, **38**, 329

Gondhalekar, P.M., & Wilson, R., 1980. *Nature*, **285**, 461

Gondhalekar, P.M., & Wilson, R., 1982. *Nature*, **296**, 415

Gursky, H., *et al*, 1980. *Astrophys. J.*, **237**, 163

Jones, B.B., Freeman, F.F., & Wilson, R., 1968. *Nature*, **219**, 252

Jones, B.B. *et al.*, 1970. *Nature*, **226**, 249

Nandy, K., Thompson, G.I., Jamar, C., Monfils, A., & Wilson, R., 1976. . Astrophys., **51**, 63

Nandy, K., Morgan, D.H., Willis, A.J., Gondhalekar, P.M., & Wilson, R., 1981. *MNRAS*, **196**, 955

Nandy, K., McLachlan, A., Thompson, G.I., Morgan, D.H., Willis, A.J., Gondhalekar, P.M., Houziaux, L., & Wilson, R., 1982. *MNRAS*, **201**, 1p.

Ulrich, M.H., *et al*, 1980. *MNRAS*, **192**, 561

Willis, A.J., & Wilson, R., 1976. *Astron. Astrophys.*, **47**, 429

Willis, A.J., & Wilson, R., 1978. *MNRAS*, **182**, 559.

Wilson, R., 1951. *MNRAS*, **111**, 478

Wilson, R., 1957. *Mem. Soc. Roy. Sci. Liege*, **20**, 85.

Wilson, R., 1958. *Astrophys. J.*, **128**, 57

Wilson, R., 1961. *Proc. Phys. Soc.*, **78**, 1223

Wilson, R., 1964. *Ann. d'Astrophys.*, **27**, 771

SOLAR PHYSICS

THE CHROMOSPHERE-CORONA TRANSITION REGION
IN LATE-TYPE STARS

C. JORDAN

Department of Physics (Theoretical Physics), University of Oxford, 1 Keble Road, Oxford, OX1 3NP, UK

Abstract. Observations of the Sun show that the chromosphere-corona transition region has a complex geometry and dynamic nature. In spite of this, observations of stellar transition regions show common behaviour as well as systematic trends. The basic methods used in making models of the transition region are set out. Observations relating to inhomogeneities in the solar transition region are summarized. The structure and energy balance of stellar transition regions and the trends emerging are discussed.

1. Introduction

Observations of the Sun show the great complexity of the transition region. There is spatial structure on the scale of the supergranulation network and down to the currently observable limit of about 1 arcsec. There is also indirect evidence for structure on even smaller scales. The conditions in the supergranulation network differ from those in the cell interiors and a transition region also occurs in dynamic structures such as spicules. Red-shifted lines indicating downflows are also observed. It is therefore not surprising that the detailed physical processes operating are poorly understood. However, it *is* remarkable that spatially integrated observations of transition regions in a wide variety of stars, show behaviour in common and systematic trends. Either the complexities are unimportant compared with the average overall control by the energy input and losses, or the processes producing the complexities behave systematically from star to star.

The transition region can be defined to havetwo parts: the lower transition region between 2×10^4 K and 2×10^5 K and the upper transition region between 2×10^5 K and T_c, the average coronal temperature. The division is near the temperature at which the emission measure distribution ($\int N_e N_H dh$ versus T_e), passes through a minimum value in many stars. Only the lower transition region is observable with the International Ultraviolet Explorer (IUE) and the Goddard High Resolution Spectrograph (GHRS) on the Hubble Space Telescope (HST). Broad band X-ray observations with the *Einstein* Observatory and ROSAT yield average coronal temperatures and emission measures, and a few stellar spectra were obtained with EXOSAT. Observations with the Extreme Ultraviolet Explorer (EUVE), are now allowing the upper transition region to be studied in many nearby stars. The ASCA satellite is also providing spectra of hot sources.

Astrophysics and Space Science **237**: 13–32, 1996.

Section 2 summarizes methods of analysis in the solar context, some historical material, and solar observations relating to spatial inhomogeneities. Observations and models of stellar transition regions and some trends emerging are summarized in Section 3. The energy balance within the transition region is discussed in Section 4, and Section 5 draws attention to further observations required to resolve outstanding issues.

2. The Solar Transition Region

Athay (1976) covers much of the early work on the solar transition region. Mariska (1992) has given a comprehensive review of observations and theoretical developments since the time of the Skylab missions in 1973 to 1974. Here I can select only a few topics for discussion.

2.1. The Emission Measure Distribution

Although Allen (1961) and Ivanov-Kholodnyi & Nikol'skii (1962) made early contributions, the formulation used by Pottasch (1963) has become widely adopted, with various refinements. The flux at the Earth in an effectively optically thin emission line is given by

$$\dot{F}_\oplus = hc \int N_2 A_{21} dV / \lambda 4\pi d^2 \qquad (1)$$

where N_2 is the population density of the excited level, A_{21} is the spontaneous transition probability. For a collisionally excited line, and the assumption of a *spatially uniform plane parallel atmosphere*, the flux at the Sun is

$$F_\odot = const. \times \int N_e N_H g(T_e) dh \qquad (2)$$

where N_e and N_H are the electron and hydrogen number densities. The constant includes a factor of 0.5 (only half the photons created are observed), and depends on the atomic data and, here, the element abundance. The function $g(T_e)$ is given by

$$g(T_e) = T_e^{-1/2} e^{-W_{12}/kT_e} N_{ion}/N_E \qquad (3)$$

where W_{12}/kT_e is the excitation energy of the transition and N_{ion}/N_E is the relative ion population.

Since each ion has a high population over only a limited range of temperature, Pottasch (1963) put

$$< g(T_e) >= 0.7 g(T_{max}) \qquad (4)$$

where T_{max} is the temperature at which $g(T_e)$ has its maximum value. Removing $0.7g(T_{max})$ from the integral then allows the quantity $\int_{\Delta H} N_e N_H dh$

to be found, where ΔH corresponds to the temperature range ΔT_e over which the line is predominantly formed.

In deriving relative abundances of different elements it is important to use a *fixed* temperature range for each line, e.g. $\Delta logT_e = 0.3$, and find the individual normalization constants (see Jordan & Wilson 1971; Jordan & Brown 1981), since two lines with the same T_{max}, may have different shaped $g(T_e)$ functions. In regions where the emission measure distribution is varying rapidly it is important to iterate the solution, re-calculating the optimum temperature of line formation (Burton *et al.* 1971; Withbroe 1975). It is useful to plot the locus of solutions such that the total line flux would be produced at each temperature in turn (see e.g. Jordan *et al.* 1987), since these give upper limits to the acceptable emission measure distribution as a function of temperature.

2.2. Electron Densities

In modern work the electron pressure in the transition region is measured from density sensitive line ratios. Ideally, lines from the same element and stage of ionization, not greatly separated in wavelength should be used (see Mason & Monsignori Fossi 1995 for suitable transition region and coronal lines). The volume edited by Lang (1994) contains assessments of calculations available. For the important lines of C III and Si III see the reviews by Berrington (1994) and Dufton & Kingston (1994). Cook *et al.* (1995) have recently recomputed the theoretical line ratios in O IV] using new excitation rates by Zhang, Graziani & Pradhan (1994). They find that the different pairs of O IV] transitions in the Sun, and in Capella, yield significantly different electron densities and suggest that the contribution to the line at 1404.8 Å from S IV] may be greater than predicted by the calculations by Dufton *et al.* (1982). This issue needs to be resolved urgently given the importance of the O IV] lines. Current measurements for the quiet Sun give electron pressures in the range 6×10^{14} to 10^{15} cm^{-3} K.

2.3. Modelling from the Emission Measure Distribution

Once the electron pressure has been found at some temperature, the value of P_e elsewhere can be found from the equation of hydrostatic equilibrium (on the assumption that regions at different temperatures are physically connected). This approach has been used in its simplest form for many years (see Jordan & Wilson 1971). Harper (1992) gives the formulation in spherical symmetry in the context of the hybrid bright giant stars, allowing for the radial extent of the atmosphere. In this case an *apparent* emission measure can be defined by

$$Em(0.3)_{\text{app}} = \int N_e N_H G(r) f(r) dr \tag{5}$$

where $G(r)$ is the fraction of photons not intercepted by the star, given by

$$G(r) = 0.5(1 + (1 - (\frac{R*}{r})^2)^{\frac{1}{2}})$$

(6)

and

$$f(r) = \left(\frac{r}{R*}\right)^2.$$

(7)

Note that with $G(r) = 0.5$, $Em(0.3)_{app}$ is *half* the value of the emission measure found using a plane parallel approximation. Initially all variables except T_e are taken as constant over the region of line formation, so that

$$\frac{dr}{dT_e} = \frac{\sqrt{2}Em(0.3)_{app}T_e k^2}{P_e P_H G(r) f(r)}.$$

(8)

Observations of solar and stellar line widths show the presence of non-thermal broadening which may give a turbulent support term. The equation of hydrostatic equilibrium is then given by

$$\frac{dP_{Tot}}{dT_e} = \frac{d(P_g + P_T)}{dT_e} = \frac{-\mu m_H}{kT_e f(r)} g_* P_g \left(\frac{dr}{dT_e}\right)$$

(9)

where P_g and P_T are the gas and turbulent pressures, respectively, and μ is the mean molecular weight. P_T is taken as $0.5\rho\xi^2$, where ρ is the density and ξ is the most probable turbulent velocity, determined from a measured line width. In the Sun, the widths of *optically thin* lines formed in the lower transition region can be fitted by $\xi^2 \propto T_e^{1/2}$ (Jordan 1991), but there are few measurements in the upper transition region, and some approximation, such as $\xi = $ constant above 2×10^5 K, must be made. With

$$x = \frac{N_H}{N_e}, \quad P_g = P_e(1 + 1.1x) = P_H \frac{(1 + 1.1x)}{x}$$

$$\mu = \frac{1.31x}{(1 + 1.1x)}$$

(10)

and

$$Y(T_e) = P_T/P_g$$

(11)

equations (8) to (11) can be combined and the pressure at any T_e can be related to the pressure at some reference temperature, so that

$$P_{Tot}^2 = P_{Tot}^2(ref) \pm 2\sqrt{2} \, 1.31 m_H g_* k \int \frac{Em(0.3)(1 + 1.1x)(1 + Y(T_e))}{G(r)f(r)^2} dT_e$$

(12)

One can then find the variation of the pressures and temperature with height above a chosen base (chromospheric) height, starting with $G(r) = 0.5$ and $f(r) = 1$, and iterating until the solution converges. In the lower transition region of the Sun and other main-sequence stars the radial extent is not significant, but in the presence of turbulent support the radial terms can become important in stars with low surface gravity. Other geometries can be treated in a similar manner.

2.4. SPATIAL INHOMOGENEITIES AND THE HEIGHT OF THE TRANSITION REGION

Solar observations provide estimates of the effects of spatial inhomogeneities on the integrated stellar line fluxes. The area covered by the supergranulation network boundaries and the boundary/cell interior contrast was studied by Reeves (1976) using observations made with the Harvard instrument on the Apollo Telescope Mount (ATM) on Skylab. (See Mariska 1992 for a discussion of other work). Reeves (1976) found that the width of the network was essentially constant for lines formed between 10^4 K and 3×10^5 K, but increased in lines of Ne VII and Ne VIII, formed above 5×10^5 K. The network structure was not discernible in lines of Mg X and Si XII. Although small loops can be observed in images illustrated by Reeves (1976), there does also appear to be a component that extends through the transition region to the corona (see also Dowdy, Emslie & Moore 1987). The network/cell interior contrast was found to have a maximum value of about 6 at 2×10^5 K. However, at 2×10^4 K the contrast was between a factor of one and two, and by 6×10^5 K it had decreased to about a factor of three. Reeves (1976) used these observations to estimate the contribution of the network and cell interiors to the intensity of the average quiet Sun. If the network and cell interior areas were taken to be equal, his results show that the use of spatially integrated observations to make a mean transition region model would give emission measures only a factor of 1.7 lower than those appropriate to the network at 2×10^5 K (observed with 5 arcsec resolution), with a smaller factor at lower and higher temperatures.

However, there is evidence that structure exists on a smaller scale, for example, within spicules. Several sets of authors have combined *spatially resolved* observations above the limb, where spicules will dominate, and observations on the disc, to derive model dependent 'filling factors' (see Mariska 1992). The spicule emission in lines such as those of C IV has been observed directly, e.g. by Dere, Bartoe & Brueckner (1986), and through the limb/disc ratio measurements discussed below. Spicules observed in the $H\alpha$ line can extend to heights of around 10^4 km above the limb. Because the Sun is the only star for which any small scale structure can be investigated this issue is now discussed in some detail. The Naval Research Laboratory's (NRL) High Resolution Telescope and Spectrograph (HRTS) has achieved a resolution of about 1 arcsec. Dere *et al.* (1987) combined the intensities in the C IV resonance lines, observed in network elements assumed to be spicules, with the spatial extent of resolved elements along the slit, and the electron pressure measured from the O IV lines. Following Dere *et al.* (1987) (see their Figure 8), let the spicule diameter be D = 2400 km, and N the number of filaments within the spicule. The slit width, s, is 0.5 arcsec. The *volume* emission measure derived for a particular feature from the flux observed at the Earth is invariant, so that, for the same N_e,

$$\Delta h D s = N \pi r^2 s \tag{13}$$

where Δh is the thickness of the equivalent uniform layer and r is the filament radius. Using the most frequent values found by Dere *et al.* (1987), $\Delta h = 1.8$ km, $Nr^2 = 1.4 \times 10^{13}$ cm and with $N = 10$, $r = 12$ km. On a line of sight down the *end* of such a spicule, the fraction of the area filled by the filaments is

$$f_A = \frac{4N\pi r^2}{\pi D^2} = 9.4 \times 10^{-4}. \tag{14}$$

However, for a spicule lying along the solar surface, provided the spicule is optically thin along a line of sight normal to the Sun's surface , the emitting area of the filaments as a fraction of the total observed area is approximately

$$f_A = \frac{N 2r}{D} = 0.031\sqrt{N} \tag{15}$$

which gives the substantially larger value of $f_A = 0.1$, if $N = 10$. In practice spicules will lie at a range of angles to the surface, so the filling factor given by equation (14) should be regarded as a lower limit.

Dere *et al.* (1987) did not discuss the length, L, of the C IV emitting material within the spicule, which can be related to the optical depth, τ, in the C IV line at 1548 Å, by

$$\tau = 5.3 \times 10^{-18} \int N_e dl. \tag{16}$$

If there are $\sqrt{N}$ filaments along the line of sight when the spicule is seen *side on*, each with path-length 2r, and $N_e = 10^{10}$ cm^{-3}, the optical depth is 0.39, whihc corresponds to it being more or less optically thin. But if the spicule is viewed *end on*, then

$$\tau = 5.3 \times 10^{-8} L(cm). \tag{17}$$

Thus if $L \geq 190$ km, τ will be ≥ 1.0. Dere *et al.* (1986) quote a mean height of 3900 km for the C IV emission at the limb, so if this originates mainly from spicules they must have τ about 20 giving obvious effects in the CIV line ratios and profiles when observations are made on the disc. Even if the majority of spicules lie at some intermediate angle to the solar surface, the above limiting cases will lead to the majority having $\tau \geq 1$. Montesinos & Jordan (in preparation) have found that the Si IV resonance lines are optically thick at some locations in the supergranulation network. Scaling the Si IV opacity to that of C IV, using mean emission measures and a constant pressure, one finds τ to be about 10 for C IV. It is important to measure *simultaneously* the absolute line intensities, the opacities from line ratios and profiles, and the electron densities from the O IV] lines, but because of the film dynamic range this is difficult to do.

In the 1960's, when there were doubts about the existence of the steep transition region temperature gradient, Bob Wilson proposed an elegant method of measuring the height of the emission in transition region lines (Burton, Ridgeley & Wilson 1967) from the ratio of line intensities measured on the disc and at the limb, with a projected spectrograph slit length of about 0.4 $R_\odot$ and a width of about 8 arcsec. (The limb spectra obtained were also used to make the first identifications in the Sun of the important intersystem lines of C III, N IV, O IV, O V and Si II, and the Fe XII forbidden lines). Provided the emission is formed in a thin layer compared with the slit function width, in practice greater than 8 arcsec, the line emissivities cancel and the limb to disc ratio becomes

$$\frac{I(limb)}{I(disc)} = 2\pi \int_{-\infty}^{h} f(x - x_o)dx \qquad (18)$$

where $f(x - x_o)$ is the combined image and slit function. Thus the limb/disc ratio yields the *average* height of the emission above the lower slit jaw. This average includes contributions from the supergranulation cell interiors and boundaries, and emission from spicules, since the effective slit width is *larger* than the vertical scale of the inhomogenities at the limb. The observations gave the first direct evidence for a steep temperature gradient in the transition region up to $T_e = 2 \times 10^5$ K. In later work the absolute mean height of the optically thin (mainly intersystem lines) was measured as 1700 km $\pm$ 700 km (Burton *et al.* 1973). Greater mean heights were found for lines, such as those of C IV, that become optically thick at the limb. The low average height found for all the optically thin transition region lines suggests that the transition region in or around spicules with a significant vertical extent, cannot be the major contributor to the spatially averaged flux on the disc. Using a *narrow* (2 arcsec) slit at the limb, Mariska *et al.* (1978) and other authors measured the relative positions of optically thin lines at the limb, and found a height of between 1 and 5 arcsec above the white light limb. The measurement by the Culham group remains the most accurate to date; the method that does not attempt to directly resolve the regions, but relies only on the scale of any height fluctuations being less than the slit function effective width gives paradoxically the best data.

3. Observations and Models of Stellar Transition Regions

Although the first evidence for transition regions around other stars came
from observations with the Copernicus satellite of Procyon (F5 IV-V) (Evans,
Jordan & Wilson 1975) and Capella (G1 III + G8 III) (Dupree 1975) only
since the advent of IUE have systematic studies of a wide range of stars been
possible. The results of observations with IUE up to about 1987 have been
reviewed by Jordan & Linsky (1989). More recent reviews may be found
in Rolfe (1990), Ulmschneider, Priest & Rosner (1991), and the proceedings of the Cambridge series of workshops (Wallerstein 1990; Giampapa &
Bookbinder 1992; Caillault 1994).

Fewer lines are observable with IUE than in the solar spectrum, but they
are sufficient to establish the emission measure distribution. Observations
with the GHRS on the HST have greater sensitivity and allow more lines to
be observed. Emission measure distributions have been found for a number
of pre-main sequence, main sequence, giant and bright giant stars, and a few
supergiants (see Jordan & Linsky 1989). The K-type hybrid bright giants
continue to be of interest (see Harper 1992). Extensive new observations
of Capella have been made with the GHRS by Linsky *et al.* (1995), giving
improved measurements of line profiles and density sensitive line ratios. The
line profiles appear to have both broad and narrow components.

One of the important results that has emerged from observations with
IUE is that the *shape* of the emission measure distribution between $T_e =
2 \times 10^4$ and 10^5 K is very similar in all G/K stars with hot coronae. Figure
1 shows the emission measure distribution for a typical main-sequence star,
ξ Boo A (G8 V), from a re-analysis of IUE data by Philippides (1995) based
on updated atomic data and solar photospheric abundances. The emission
measure loci and the mean distribution that give the optimum fit to all line
fluxes are shown. The electron pressure derived from the Si III] line in comparison to the mean distribution is $P_e = 1.0 \times 10^{16}$ cm^{-3} K. Above 10^5 K the
distributions may differ, but this region is difficult to study from observations with IUE alone. The similarity in the emission measure distributions is
also implied by the almost linear correlation between the fluxes in transition
region lines, for example those of C IV and C II (Oranje 1986; Capelli *et al.*
1989; Rutten *et al.* 1991) (see also Section 3.1). Given the influence of area
factors on the spatially averaged fluxes this suggests that the area of the
dominant feature (in the Sun, the supergranulation network and structures
within it) remains constant through the lower transition region in all these
stars, although the area may not be the same.

Observations from the *Einstein* Observatory (see Schmitt *et al.* 1990)
and ROSAT provide mean coronal emission measures and temperatures,

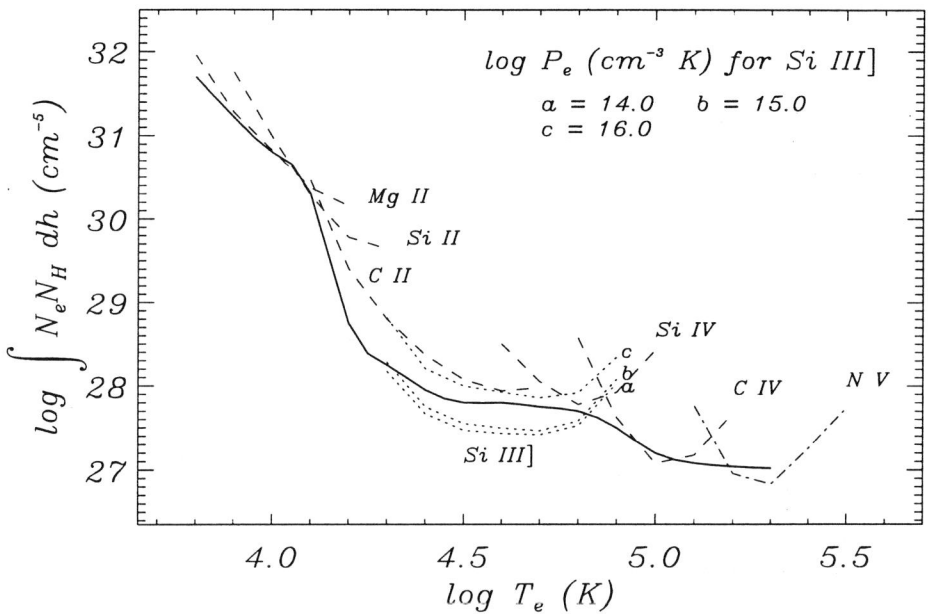

Fig. 1.   The emission measure distribution for ξ Boo A (Philippides 1995). The limiting loci and mean iterated distribution are shown. The best fit electron pressure is 1.0×10^{16} cm^{-3} K.

subject to various approximations in the spectral fitting process (e.g. one temperature, two temperature and continuous emission measure distribution fits). The moderate resolution spectra being obtained with the ASCA satellite are giving information on high temperature plasma (e.g. the study of β Cet (K0 III) by Drake *et al.* 1994). Although line spectra of three stars were obtained with EXOSAT (Lemen *et al.* 1989; Schrijver 1985) the spectra being obtained with the EUVE are giving information on the upper transition region and corona in a wide variety of nearby stars, e.g. Procyon (Drake, Laming & Widing 1995), Capella (Dupree *et al.* 1993; Jordan 1995), ξ Boo A+B (G8V + K2 V) (Jordan 1995), χ^1 Ori (Haisch, Drake & Schmitt 1994), α Cen (A + B) (G2 V + K0 V) (Schrijver *et al.* 1995), the M dwarfs AU Mic (Monsignori Fossi & Landini 1994) and AT Mic (Monsignori Fossi *et al.* 1994), the rapidly rotating K star, AB Doradus (Rucinski *et al.* 1994) and a number of RS CVn binary stars. Because different spectral modelling techniques and emissivity codes have been used it is difficult to make intercomparisons. From the data available so far it appears that in the G/K main-sequence stars the apparent emission measure distribution increases immediately above 2×10^5 K, as in the Sun. However, in the giant star Capella, and to a lesser extent in Procyon, the emission measure con-

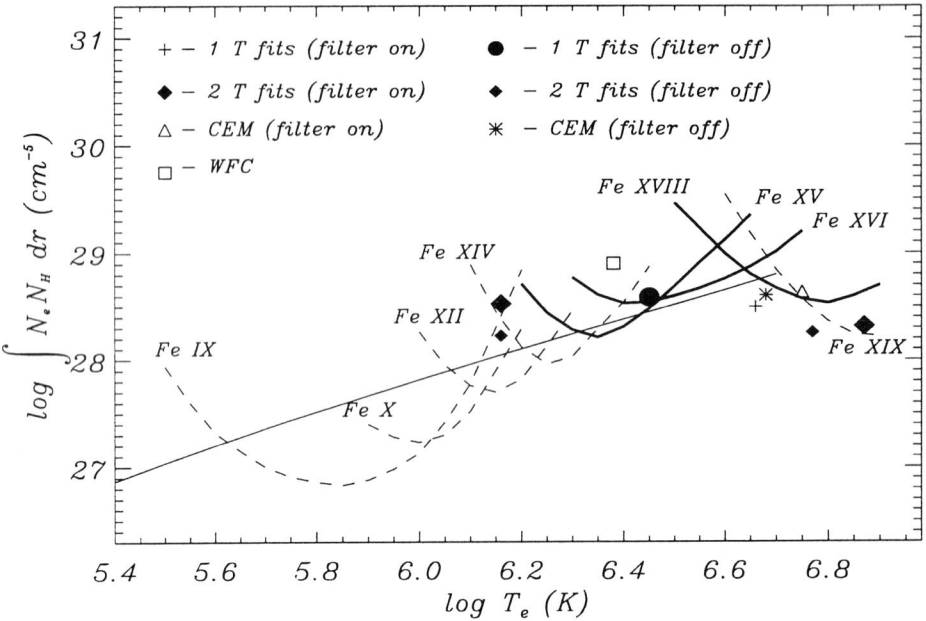

Fig. 2. The emission measure distribution for ξ Boo A. The limiting loci are from EUVE observations; those derived from upper limits to line fluxes are shown as dashed lines. Several alternative fits to ROSAT spectra are also shown. The line indicates the predictions of a spherically symmetric energy balance solution.

tinues to decrease and passes through a minimum value above 2×10^5 K. These stars, like some other evolved stars, have a maximum coronal emission measure that is lower than in the main-sequence stars, relative to the region below 2×10^5 K (see Brown *et al.* 1984; Brown *et al.* 1991; Dupree *et al.* 1993 and Section 3.1). Figure 2 shows the emission measure distribution derived for ξ Boo A (ξ Boo B is not expected to make a significant contribution), from EUVE spectra and observations with ROSAT (Jordan 1995; Philippides 1995), compared with the results of a spherically symmetric energy balance model.

Observations with IUE and ROSAT are also useful in studying variations in the transition region line and X-ray fluxes, which show the presence of active regions. Ayres *et al.* (1995) find variations of up to 29 % in the C IV flux from main-sequence and giant stars. Haisch & Schmitt (1994) detect variations in the X-ray fluxes from giant stars observed with ROSAT. If variations in the EUVE line fluxes could be observed, the contribution from active regions could be modelled.

In modelling from the emission measure distribution there are fewer density sensitive lines available in stellar spectra. The only suitable lines in *low* resolution observations made with IUE are those of Si III] and C III] at 1892 Å and 1909 Å, respectively. However, because of the continuum present it is often necessary to observe these lines at *high* resolution, so the measurements are limited to a relatively small number of stars. The lines of O IV] around 1400 Å may then just be observable. Observations with the GHRS on the HST give a significant improvement over IUE for these density sensitive lines (see Linsky *et al.* 1995 for observations of Capella). In main-sequence stars where the density is similar to, or an order of magnitude or so greater than in the Sun, the C III] line is weak and the line of Si III] is the best density indicator. In giant stars with hot coronae the electron pressure is similar to the Sun at G1 III (e.g. in Capella, see Linsky *et al.* 1995), but by K0 III is up to an order of magnitude lower than in the Sun (e.g. in β Cet, see Eriksson *et al.* 1983). The C III] line may then be marginally useful. However, the chromospheric densities in these giants are too high to make use of the C II] lines around 2335 Å. These lines provide the best way of measuring N_e in non-coronal giants and supergiants, and also in the K-type bright giants which have coronae (see Harper 1992). To determine N_e from individual lines, such as those of Si III] and C III], the mean emission measure distribution found from a range of permitted lines is used together with the level populations as a function of P_e and T_e, to compute the predicted fluxes. The pressure is then determined from the best agreement between the observed and computed line fluxes.

Models of the average transition region can then be made as discussed in Section 2.3. The common shape of the emission measure distribution between 2×10^4 and 10^5 K implies that all the stars with hot coronae have a common variation of dT_e/dh with T_e, and detailed modelling shows that at a given temperature the gradient increases almost linearly with the transition region pressure.

3.1. TRENDS IN LINE FLUXES AND OTHER PARAMETERS

The near constant ratio of the C IV to C II line fluxes found by Rutten *et al.* (1991) and others (see above) has been confirmed by Ayres *et al.* (1995). In the data sets including luminosity classes IV - V, and III - I, the ratios differ only by about a factor of two, with those in the evolved stars being marginally larger. Small differences are expected in the ion populations from the density dependence of di-electronic recombination. They also find that the ratio of the Si IV to C IV line fluxes in class IV -V stars, where O IV] does not contribute significantly to the total Si IV flux, remains constant over three orders of magnitude in the ROSAT X-ray flux, with a mean value of about 0.5. This appears to conflict with the work of Doschek, Dere & Lund (1992) who find ratios of 0.22 to 1.4 in the solar transition region, which they attribute to abundance variations. If this were so one would expect higher ratios in the more active stars.

The trends in coronal properties are relevant to the upper transition region and the region around 1 to 2×10^5 K. Scalings between X-ray fluxes and C IV lines fluxes have been studied by a number of authors since around 1981. Ayres, Marstad & Linsky (1981) found the relation

$$\frac{F_X}{F_{bol}} \propto \left(\frac{F_{CIV}}{F_{bol}}\right)^{3/2} \tag{19}$$

while Montesinos & Jordan (1988), from a sample of G-K stars only, and Rutten *et al.* (1991), from a wider sample, found relations of the form

$$F_X \propto F_{CIV}^x \tag{20}$$

with x = 1.7 and 1.3, respectively. For G-K dwarfs, Ayres *et al.* (1995), find a mean power of 1.53 in a relation of the form given by (19), but a higher power for F9-G2 dwarfs (1.72) compared with G8-K5 dwarfs (1.32). The G5-K0 giants have a similar power to that of the G8-K5 dwarfs. Thus the range of powers found by different authors using different samples seems understandable. However, earlier type F IV-V stars and Hertzsprung gap giants are relatively deficient in their X-ray flux (see also Simon & Drake 1989), and Procyon (F5 IV-V) and Capella (G1 III), share this behaviour.

Correlations with stellar rotation rates are not reviewed here, but we note that Capella has a low coronal temperature for its rotation rate (see Jordan & Montesinos 1991). That such correlations exist suggests that in most stars with coronae the underlying heating processes and energy balance are related to the strength and configuration of the magnetic fields, generated by dynamo action in the stellar convection zones.

4. The Energy Balance in the Transition Region

The emission measure distribution is determined by the balance between the divergence of the non-thermal energy flux, $\mathbf{div}\, F_M$, the radiation losses, dF_R/dr, and the net energy transferred by conduction, $\mathbf{div}\, F_C$. Energy transferred by flows and by non-classical thermal conduction are not explicitly considered here, but would form part of the empirical energy input found from the emission measure distribution. Thus in a spherically symmetric geometry, where there is a variable surface area, $A(r)$, involved,

$$\frac{d(A(r)F_M)}{A(r)dr} = \frac{-dF_R}{dr} - \frac{d(A(r)F_C)}{A(r)dr} \tag{21}$$

where

$$F_C = -\kappa T_e^{5/2} dT_e/dr \tag{22}$$

and

$$\frac{dF_R}{dr} = N_e N_H P_{\text{rad}}(T_e) \tag{23}$$

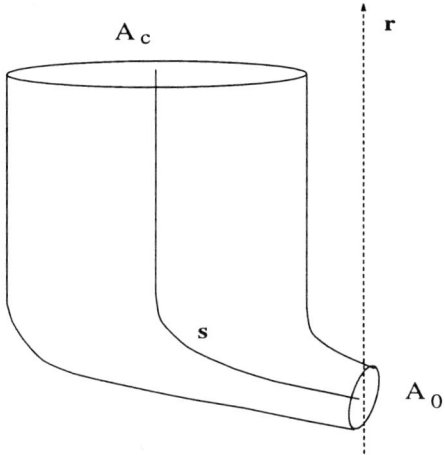

Fig. 3. A schematic diagram of an elementary flux tube, with area A_c in the low corona and area A_0 at around 2×10^5 K.

where $P_{rad}(T_e)$ is the radiative power loss function. We adopt $P_{rad}(T_e) = 1.25 \times 10^{-16}/T_e$ erg cm^3 s^{-1} above 2×10^5 K, and 6.25×10^{-22} erg cm^3 s^{-1} between 10^5 K and 2×10^5 K. The regions above and below $T_e = 2 \times 10^5$ K are now considered separately.

4.1. THE REGION ABOVE $T_e = 2 \times 10^5$ K

Use of equation (23) and an intrinsic emission measure distribution that increases with temperature to a power greater than 1, shows that the radiation losses from the high transition region are relatively small compared with the *total* energy flux required to heat the corona. They can be balanced either by a small amount of heating, if F_C is constant, or by the net conductive flux, if there is no heating. The coronal heating must extend to heights far greater than the first pressure-squared scale height, which contributes most of the observed radiation. Thus, up to T_c, the observed maximum temperature, it is reasonable to balance the net conductive flux against the radiation losses. This then *determines* the emission measure distribution for a chosen geometry. In the Sun, the supergranulation boundaries provide most of the emission up to at least 5×10^5 K, but the coronal emission (ignoring active regions) is more amorphous. Thus a varying area should be included. A schematic diagram, based on the theoretical model of Gabriel (1976), of the geometry used is shown in Figure 3. Working along the path, s, the emission measure gradient can be expressed as

$$\frac{d\log Em(s)}{d\log T_e} = 3/2 + \frac{2d\log P_e}{d\log T_e} + \frac{d\log A(s)}{d\log T_e} - \frac{2P_{\mathrm{rad}}(T_e)Em(s)^2k^2}{0.8\kappa P_e^2 T_e^{3/2}}. \tag{24}$$

Using the equation of hydrostatic equilibrium, and a chosen form of A(s), and "coronal" values of $Em(T_e)$ and T_e, the solution can be found in an iterative manner. Here we will be concerned only with the energy terms at T_c, where the area is A_c, and the flux tube is radial, and at T_0, where the area is A_0.

Two approximations can be used for the conductive flux into the top of the tube. First, in equation (22) dT_e is set to the temperature "resolution" of $\sqrt{2}T_c - T_c/\sqrt{2}$ and $ds = dr$ is taken as the pressure-squared isothermal scale height, so that

$$F_C(T_c) = -\frac{\kappa T_c^{5/2} g_{eff}}{\sqrt{2}\,7.1 \times 10^7}. \tag{25}$$

where g_{eff} takes account of the radial factor f(r) and the extension by the turbulent pressure. For the Sun, with $T_c = 1.58 \times 10^6$ K and $f(r) = 1.12$, this gives 8.5×10^5 $erg\ cm^{-2}s^{-1}$. Secondly, dT_e/ds can be replaced by the emission measure, as in equation (8), so that

$$F_C(T_c) = -\frac{\kappa T_c^{3/2} 0.8 P_c^2}{\sqrt{2} Em(r_c)k^2}. \tag{26}$$

Over the *whole* corona, made up of N such structures, each with area A_c, the mean *apparent* emission measure, $Em(r_c)_{app}$, is given by

$$Em(r_c)_{app} = \frac{Em(r_c)N A_c G(r_c)}{4\pi R_*^2} = Em(r_c)f_c G(r_c) \tag{27}$$

where f_c is the area filling factor. (N.B. in the literature the coronal emission measure derived may use either G(r) = 0.5 or 1.0). The combination of equations (25) to (27) gives an expression for f_c,

$$f_c = \frac{T_c g_{eff} Em(r_c)_{app}k^2}{0.8 P_c^2 7.1 \times 10^7 G(r_c)}. \tag{28}$$

Even for the Sun there are very few simultaneous measurements of the coronal parameters, but typical values lead to $f_c \simeq 1.0$. For the main-sequence stars modelled by Jordan et al. (1987), where a range of pressures was found from the Si III] line, the mean pressures lead to values of f_c between about 0.42 and 1.7, for $G(r_c) = 0.5$ and $f(r_c) = 1.0$. In models which include a turbulent pressure the coronal electron pressure is not necessarily the same as or less than the transition region electron pressure, but modelling in progress suggests that f_c is not substantially less than 0.5.

For Procyon, the use of line fluxes from Drake *et al.* (1995), emissivities from Brickhouse *et al.* (1995), and $R_* = 2R_\odot$, gives the coronal emission measure (derived with $G(r_c) = 0.5, f(r) = 1$) as 2.9×10^{27} cm^{-5}, about a factor of two smaller than found by Drake *et al.* (1995) using different emissivities. With $T_c = 1.6 \times 10^6$ K and $g_* = 1.0 \times 10^4$ cm s^{-2}, the unknown factor is $P_c f_c^{1/2} = 9.0 \times 10^{14}$ cm^{-3} K. Schmitt, Haisch & Drake (1994) and Drake *et al.* (1995) have used line ratios in Fe X, XII and Fe XIV to estimate the coronal pressure and find about 6.4×10^{15} cm^{-3} K, which would imply $f_c = 0.02$. However, the crucial lines are very weak and the spectral fitting for Fe X shown by Drake *et al.* (1995) does not have the correct wavelength interval between the lines at 174.53 and 175.24 Å, and the flux at 219.13 Å yields only an upper limit to the Fe XIV line. For a flux in the 175.24 Å line lower by a factor of two, and the emissivities of Brickhouse *et al.* (1995), a more realistic upper limit appears to be $P_c = 7.1 \times 10^{14}$ cm^{-3} K, giving f_c around 1. The importance of accurate atomic data and line fluxes is obvious. A longer exposure of Procyon with EUVE would be worthwhile, as well as monitoring for variations due to active regions.

For Capella, using $G(r_c) = 0.5, f(r_c) = 1$, and ignoring the turbulent pressure, one finds (see Jordan 1995), $f_c^{1/2} P_c = 1.5 \times 10^{15}$ cm^{-3} K. Linsky *et al.* (1995) find a transition region pressure of about 10^{15} cm^{-3} K. From the archival EUVE spectrum and emissivities of Brickhouse *et al.* (1995) we find that the lines of Fe XXI are in their low density limit ($N_e \leq 5 \times 10^{10}$ cm^{-3}). Spherically symmetric models with turbulent pressure support suggest that the additional factors are unlikely to reduce the coronal pressure by more than a factor of two, so f_c is again close to 1. Ayres *et al.* (1995) and Haisch & Schmitt (1994) find only small variations in the C IV line and ROSAT PSPC X-ray fluxes, respectively. Thus it appears that active regions are not the *dominant* cause of the EUVE emission.

The application of equations (23) and (25) shows that in the main-sequence stars the conductive flux from the corona exceeds the coronal radiation losses by an order of magnitude. However, in Capella and Procyon, the conductive flux is lower and the ratio $F_C(T_c)/F_R(T_c)$ is less than a factor of two.

4.2. THE REGION AT AND BELOW THE MINIMUM EMISSION MEASURE

In the transition region it is usual to take $G(r) = 0.5, f(r) = 1$ and give $2Em(r)_{app}$ as the observed emission measure. Around the temperature of the minimum emission measure, T_0,

$$Em(r_0)_{app} = \frac{Em(s_0)A_0 f_c}{2A_c} \tag{29}$$

and the conductive flux is given by

$$F_C(T_0) = -\frac{\kappa T_0^{3/2} 0.8 P_0^2 f_c A_0}{2\sqrt{2} k^2 Em(r_0)_{app} A_c}.$$ (30)

From the form of equation (24) it can be seen that with energy balance between conduction and radiation, and when P_e and $A(s)$ are constant, $dlog Em(s)/dlog T_e$ is positive provided the last term does not exceed $3/2$. The condition for a *minimum* in the emission measure distribution gives

$$Em(r_0)_{app} = \frac{4.1 \times 10^{-4} (P_0/k) T_0^{0.75} A_0 f_c}{P_{rad}^{0.5}}.$$ (31)

In the Sun, the area occupied by the supergranulation network appears to be constant in the region of the minimum emission measure Em_{min}. In the presence of a turbulent pressure P_e will not be constant, and numerical calculations are required to find Em_{min} at a chosen temperature. In the Sun and main-sequence stars Em_{min} occurs at around 2×10^5 K. Use of parameters from Jordan et al. (1987), and their mean pressures, gives values predicted by equation (31), with the area factors set to 1, significantly larger than those observed, as in the Sun. In other words, the observed value of Em_{min}, if area factors are neglected, cannot satisfy energy balance between radiation and conduction. This type of problem was recognized early on by Giovanelli (1949), in models with a constant conductive flux from the corona, and a variety of solutions have been proposed (see e.g Kuperus & Athay 1967; Kopp & Kuperus 1968). The problem can be solved if the area factors are invoked. Kopp (1972) used the spatially integrated observations then available to explore the range of pressures and area factors that would satisfy energy balance between radiation and conduction. He concluded that the transition region emission must be concentrated in area, as later confirmed with the observations from Skylab, and also that the low transtion region required another source of direct energy deposition. Gabriel (1976) made a network model that removed most of the problem around T_0, and again pointed out the need for additional heating below T_0 (see below also). In the Sun and the main-sequence stars discussed above, equating the observed values of Em_{min} to those predicted by equation (31) gives values of $A_0 f_c/A_c$ that lie between 0.03 and 0.09 for the Sun, to 0.35 for χ^1 Ori. The solar value is within the range discussed in Section 2.4.

In Procyon the minimum emission measure may occur at a slightly higher temperature (around 3×10^5 K, see Jordan et al. 1986; Drake et al. 1995). Because the transition region pressure in Procyon is found from opacity arguments (Brown & Jordan 1981), the quantity derived is $P_0 A_0 f_c/A_c$, and has a value 2.3×10^{14} cm^{-3} K. Using equation (31), this predicts that the observed emission measure should be $2Em_{app} = 1.2 \times 10^{26}$ cm^{-5}, comparable with the value found by Jordan *et al.* (1986), but *lower* than the value proposed by Drake *et al.* (1995). If the observed emission measure is *higher* than that predicted, then heating other than by conduction is implied.

In Capella, Em_{min} appears to occur between $4 - 8 \times 10^5$ K (see Linsky *et al.* 1995; Brickhouse *et al.* 1995; Jordan 1995). With $P_0 = 10^{15}$ cm^{-3} K, from Linsky *et al.* (1995), the predicted minimum value is comparable with that observed if $T_0 = 4 \times 10^5$ K, but gives $A_o f_c / A_c = 0.04 - 0.14$ if $T_0 = 8 \times 10^5$ K. It is important to determine more accurately the temperature and value of the minimum emission measure, as well as the electron pressure.

As discussed above, the inclusion of area factors allows solutions where the radiation is balanced by the net conductive flux, down to the region of Em_{min}. At lower temperatures (below 10^5 K in the main-sequence stars), even without the area factors, modelling from the observed emission measure distribution shows that the energy deposited by thermal conduction is considerably less than that lost by radiation, so some additional source of heating is required. The situation is complicated by the apparent *imbalance* at temperatures between 10^5 K and 2×10^5 K. *Without the area factors* the conductive flux at 2×10^5 K is comparable with the radiation losses summed down to about 3×10^4 K, and this has led to the suggestion that processes other than classical thermal conduction could be carrying energy down from the region above 2×10^5 K (e.g. diffusion, see Fontenla, Avrett & Loeser 1990, 1991; turbulent conductivity, see Cally 1991). However, in evolved stars, including Capella, Procyon, β Dra (G2 Ib-II) (Brown *et al.* 1984, and the hybrid bright giants (Hartmann *et al.* 1985, Brown *et al.* 1991; Harper 1992), the thermal conduction back at the temperature of Em_{min} does not appear to be sufficient to account for the radiation losses down to 3×10^4 K. The inclusion of the area factors reduces $F_C(T_e)$ and increases $F_R(T_e)$, so that there is a clear need for additional heating.

The energy required to heat the lower transition region compared with the corona also increases when the area factors are included. In the spatially averaged models, in the main-sequence stars the losses from the lower transition region above about 2×10^4 K are small compared with the losses from the corona by conduction and radiation, but this is *not* the case in the evolved stars mentioned above. With the inclusion of the area factors the energy required to heat the lower transition region increases to about 25 % to 50 % of the total. Jordan *et al.* (1987) suggested that if the heating of the lower transition region came from the same waves that heated the corona, the *relatively* small amount of heating in the lower transition region might be a cause of the ubiquitous shape of the emission measure distribution, but this seems unlikely with the larger radiation losses found with restricted areas. However, the *form* of the heating function is still

$$\frac{dF_M}{dlogT_e} = -\sqrt{2}Em(T_e)P_{rad}(T_e) \simeq constant. \tag{32}$$

Thus the conclusion that the shape of the emission measure distribution in the lower transition region is the inverse of the power loss distribution is unchanged, but the underlying reason must be the particular form of the heating function. Böhm Vitense (1987) has also discussed the relation between the emission measure distribution and wave heating. Simon & Drake (1989) suggested that acoustic heating may be important in Procyon and Capella, and Mullan & Cheng (1994) have proposed that this could account for the form of the emission measure distribution in Procyon. To match the form of equation (32) with a wave-flux requires $\xi^2 \propto T_e^{1/2} ln T_e \beta^{1/2}$, for waves travelling at the Alfvén velocity ($\beta = P_g/8\pi B^2$), and $\xi^2 \propto T_e^{1/2} ln T_e$, for acoustic waves. In the Sun the variation is close to $T_e^{1/2}$.

The observed scaling between the transition region and X-ray fluxes may be accounted for in the following way. If one balances a magnetic mode of heating against the thermal conduction in the corona, then if a common dissipation process is assumed, a scaling such that $P_c \propto T_c^2 g_*$ results (see Montesinos & Jordan 1993, for details). The coronal emission measure (and the *Einstein* Observatory X-ray flux) then scales as $P_c^{3/2}/g_*^{1/2}$. The minimum emission measure and any wave dissipation to balance the radiation below Em_{min} scale as P_0. Thus, apart from the area factors one expects, at least for the main-sequence stars, $F_X \propto F_{CIV}^{3/2}/g_*^{1/2}$, similar to the scalings discussed in Section 3.1.

5. Further Observations Required

At present many of the numerical results above must be considered simply illustrative. Apart from the general need to obtain good uv, EUVE and X-ray spectra, to improve our understanding of the energy balance we will need specifically to improve measurements of: (i) the temperature and value of the minimum emission measure, since this may provide an estimate of the area coverage; (ii) the electron pressure in both the lower transition region and corona, since modelling cannot even begin without these; (iii) the widths/profiles of *optically thin* emission lines at all temperatures, although this may be possible above 2×10^5 K only in the Sun; (iv) variations in EUVE and/or X-ray fluxes, since this is the only way of determining the contribution of active regions. Detailed studies of a few well chosen stars are important in understanding the physics underlying the broader trends.

References

Allen, C.W. 1961, *Mem. Soc. Roy. Liege*, **4**, 241.
Athay, R.G. 1976, "The Solar Chromosphere and Corona: Quiet Sun", D. Reidel, Dordrecht.
Ayres, T.R., Marstad, N.C. & Linsky, J.L. 1981, *Ap.J.*, **247**, 545.

Ayres, T.R. *et al.* 1995, *Ap.J.Suppl.*, **96**, 223.

Berrington, K.A. 1994, *Atomic Data Nucl. Data*, **57**, 71.

Böhm-Vitense, E. 1087, *Ap.J.*, **317**, 750.

Brickhouse, N.S., Raymond, J.C. & Smith, B.W. 1995, *Ap.J.Suppl.*, **97**, 551.

Brown, A. & Jordan, C. 1981, *M.N.R.A.S.*, **196**, 757.

Brown, A., Jordan, C., Stencel, R.E., Linsky, J.L & Ayres, T.R. 1984, *Ap.J.*, **283**, 731.

Brown, A., Drake, S.A., Van Steenberg, M.E. & Linsky, J.L. 1991, *Ap.J.*, **373**, 614.

Burton, W.M., Ridgeley, A. & Wilson, R. 1967, *M.N.R.A.S.*, **135**, 207.

Burton, W.M., Jordan, C., Ridgeley, A. & Wilson, R. 1971, *Phil. Trans. Roy. Soc. Lond. A.*, **270**, 81.

Burton, W.M., Jordan, C., Ridgeley, A. & Wilson, R. 1973, *Astron. & Astrophys.*, **27**, 101.

Caillault, J.-P. 1994, ed. "Cool Stars, Stellar Systems, and the Sun. Eighth Cambridge Workshop", *ASP Conf. Ser.*, **64**, San Francisco.

Cally, P.S. 1991, in "Mechanisms of Chromospheric and Coronal Heating", P. Ulmschneider, E.R. Priest & R. Rosner, eds., Springer-Verlag, Berlin, p. 103.

Capelli, A., Cerruti-Sola, M., Cheng, C.C. & Pallavicini, R. 1989, *Astron. Astrophys.*, **213**, 226.

Cook, J.W., Keenan, F.P., Dufton, P.L., Kingston, A.E., Pradhan, A.K., Zhang, H.L., Doyle, J.G. & Hayes, M.A. 1995, *Ap.J.*, **444**, 936.

Dere, K.P., Bartoe, J.-D.F. & Brueckner, G.E. 1986, *Ap.J.*, **305**, 947.

Dere, K.P., Bartoe, J.-D.F., Brueckner, G.E., Cook, J.W. & Socker, D.G. 1987, *Sol. Phys.*, **114**, 223.

Doschek, G.A., Dere, K.P. & Lund, P.A. 1992, *Ap.J.*, **381**, 583.

Dowdy, J.F., Emslie, A.G. & Moore, R.L. 1987, *Sol. Phys.*, **112**, 255.

Drake, J.J., Laming, J.M. & Widing, K.G. 1995, *Ap.J.*, **443**, 393.

Drake, S.A., Singh, K.P., White, N.E. & Simon, T. 1994, *Ap.J.Letts.*, **436**, L87.

Dufton, P.L. & Kingston, A.E. 1994, *Atomic Data Nucl. Data*, **57**, 273.

Dufton, P.L., Hibbert, A., Kingston, A.E. & Doschek, G.A. 1982, *Ap.J.*, **247**, 338.

Dupree, A.K. 1975, *Ap.J. Letts.*, **200**, L27.

Dupree, A.K., Brickhouse, N.S., Green, J.C. & Raymond, J.C. 1993, *Ap.J.Letts.*, **418**, L41.

Eriksson, K., Linsky, J.L. & Simon, T. 1983, *Ap.J.*, **272**, 665.

Evans, R., Jordan, C. & Wilson, R. 1975, *M.N.R.A.S.*, **172**, 585.

Fontenla, J.M., Avrett, E.H. & Loeser, R. 1990, *Ap.J.*, **355**, 700.

Fontenla, J.M., Avrett, E.H. & Loeser, R. 1991, *Ap.J.*, **377**, 712.

Gabriel, A.H. 1976, *Phil. Trans. Roy. Soc. Lond. A.*, **281**, 339.

Giampapa, M. & Bookbinder, J.A. 1992, eds. "Cool Stars, Stellar Systems, and the Sun. Seventh Cambridge Workshop", *ASP Conf. Ser.*, **26**, San Francisco.

Giovanelli, R.G. 1949, *M.N.R.A.S.*, **109**, 372.

Haisch, B. & Schmitt, J.H.M.M. 1994, *Ap.J.*, **426**, 716.

Haisch, B., Drake, J. & Schmitt, J.H.M.M. 1994, *Ap.J.Letts.*, **421**, L39.

Harper, G.M. 1992, *M.N.R.A.S.*, **256**, 37.

Ivanov-Kholodnyi, G.S. & Nikol'skii, G.M. 1962, *Astron. Zh.*, **39**, 777.

Jordan, C. 1991, in "Mechanisms of Chromospheric and Coronal Heating", P. Ulmschneider, E.R. Priest & R. Rosner, eds., Springer- Verlag, Berlin, p. 300.

Jordan, C. 1995, in "Astrophysics in the Extreme Ultraviolet", IAU Coll. 152, S. Bowyer & R. Malina, eds., CUP, Cambridge, In Press.

Jordan, C. & Brown, A. 1981, in "Solar Phenomena in Stars and Stellar Systems", R.M. Bonnet & A.K. Dupree, eds., D. Reidel, Dordrecht, p. 199.

Jordan, C. & Linsky, J.L. 1989, in "Exploring the Universe with the IUE Satellite", Y. Kondo, ed. in chief, Kluwer, Dordrecht, p. 259.

Jordan, C. & Montesinos, B.M. 1991, *M.N.R.A.S.*, **252**, 25P.

Jordan, C. & Wilson, R. 1971, in "Physics of the Solar Corona", C.J. Macris, ed., D. Reidel, Dordrecht, p. 219.

Jordan, C., Brown, A., Walter, F.M. & Linsky, J.L. 1986, *M.N.R.A.S.*, **218**, 465.
Jordan, C., Ayres, T.R., Brown, A., Linsky, J.L. & Simon, T. 1987, *M.N.R.A.S.*, **225**, 903.
Kopp, R.A. 1972, *Sol. Phys.*, **27**, 373.
Kopp, R.A. & Kuperus, M. 1968, *Sol. Phys.*, **4**, 212.
Kuperus, M. & Athay, R.G. 1967, *Sol. Phys.*, **1**, 361.
Lang, J. 1994, Special ed., *Atomic Data Nucl. Data*, **57**.
Lemen, J.R., Mewe, R., Schrijver, C.J. & Fludra, A. 1989, *Ap.J.*, **341**, 474.
Linsky, J.L., Wood, B.E., Judge, P., Brown, A., Andrulis, C. & Ayres, T.R. 1995, *Ap.J.*, **442**, 381.
Mariska, J.T. 1992, "The Solar Transition Region", CUP, Cambridge.
Mariska, J.T., Feldman, U. & Doschek, G.A. 1978, *Ap.J.*, **226**, 698.
Mason, H.E. & Monsignori Fossi, B.C. 1995, *Astron. & Astrophys. Rev.*, In press.
Monsignori Fossi, B.C. & Landini, M. 1994, *Astron. & Astrophys.*, **284**, 900.
Monsignori Fossi, B.C., Landini, M., Del Zanna, G. & Drake, J.J. 1994, in "Cool Stars, Stellar Systems, and the Sun. Eighth Cabridge Workshop", J.-P. Caillault, ed., ASP Conf. Ser., **64**, San Francisco, p. 44.
Montesinos, B.M. & Jordan, C. 1988, in "A Decade of UV Astronomy with IUE", ESA SP-281, Vol. 1, p.283.
Montesinos, B.M. & Jordan, C. 1993, *M.N.R.A.S.*, **264**, 900.
Mullan, D.J. & Cheng, Q.Q. 1994, *Ap.J.*, **435**, 435.
Oranje, B.J. 1986, *Astron. & Astrophys.*, **154**, 185.
Philippides, D. 1995, D.Phil.Thesis, University of Oxford, In preparation.
Pottasch, S.R. 1963, *Ap.J.*, **137**, 945.
Reeves, E.M. 1976, *Sol. Phys.*, **46**, 53.
Rolfe, E.J. 1990, ed. "Evolution in Astrophysics: IUE Astronomy in the era of New Space Mission", ESA SP-310.
Rucinski, S.M., Mewe, R., Kaastra, J.S., Vilhu, O. & White, S.M. 1994, preprint.
Rutten, R.G.M., Schrijver, C.J., Lemmens, A.F.P. & Zwaan, C. 1991, *Astron. & Astrophys.*, **252**, 203.
Schmitt, J.H.M.M., Haisch, B. & Drake, J. 1994, *Science*, **265**, 1420.
Schmitt, J.H.M.M., Collura, A., Sciortino, S., Vaiana, G.S., Harnden, F.R. & Rosner, R. 1990, *Ap.J.*, **365**, 704.
Schrijver, C.J. 1985, *Space Sci. Rev.*, **40**, 3.
Schrijver, C.J., Mewe, R., van den Oord, G.H.J. & Kaastra, J.S. 1995, *Astron. & Astrophys.*, In press.
Simon, T. & Drake, S.A. 1989, *Ap.J.*, **293**, 551.
Ulmschneider, P., Priest, E.R. & Rosner, R. 1991, eds. "Mechanisms of Chromospheric and Coronal Heating", Springer-Verlag, Berlin.
Wallerstein, G. 1990, ed. "Cool Stars Stellar Systems, and the Sun. Sixth Cambridge Workshop", *ASP Conf. Ser.*, **64**, San Francisco.
Withbroe, G.L. 1975, *Sol. Phys.*, **45**, 301.
Zhang, H.L., Graziani, M. & Pradhan, A.K. 1994, *Astron. & Astrophys.*, **283**, 319.

THE SOLAR X-RAY CORONA

L. GOLUB

Smithsonian Astrophysical Observatory
60 Garden St., Cambridge MA.

Abstract. The solar corona, and the coronae of solar-type stars, consist of a low-density magnetized plasma at temperatures exceeding 10^6 K. The primary coronal emission is therefore in the UV and soft x-ray range. The observed close connection between solar magnetic fields and the physical parameters of the corona implies a fundamental role for the magnetic field in coronal structuring and dynamics. Variability of the corona occurs on all temporal and spatial scales – at one extreme, as the result of plasma instabilities, and at the other extreme driven by the global magnetic flux emergence patterns of the solar cycle.

1. Introduction

The corona is a portion of the Sun's outer atmosphere beginning slightly above the visible surface and extending many solar radii out. A precise definition of the term "corona" is to some extent dependent on one's theoretical bias, and one may choose to think in terms of a modified plane-parallel model, or in terms of a composite, multi-component model made up of relatively isolated individual structures.

In either case, the most important physical fact about the corona is that it reaches very high temperatures, more than 10^6 K. Moreover, this temperature increase is found to occur over very short distances, with the rise from $< 10^4$ K to $> 10^6$ K occurring within less than a thousandth of the solar radius. If we pick a temperature well above that of the photosphere, such as 10^5 K, then we may define any portion of the atmosphere above this temperature as *corona*. Because the rise in temperature is so dramatically steep, this choice is adequate for many purposes, since a large change in this cutoff value will correspond to only a very small change in actual physical location.

At visible wavelengths, the corona is extremely faint relative to the disk, having a maximum brightness ratio of $\approx 10^{-6}$, decreasing to $\approx 10^{-9}$ within a single solar diameter away from the visible limb. However, at UV and soft x-ray wavelengths, the situation is reversed. Because of the high temperature of the coronal gas, its primary emission is in the UV and soft x-ray portion of the spectrum. Therefore, an instrument in which the visible light is blocked while the short wavelengths are transmitted permits viewing of the coronal emission on the disk and out to several solar radii above the limb.

Figure 1 shows a superposition of both on-disk and limb observations. It was obtained during a total solar eclipse in 1991, using ground-based data from the CFH-T in Hawaii and x-ray data from the NIXT sounding

Astrophysics and Space Science **237**: 33–48, 1996.

rocket (Golub *et al.* 1990). The ground-based eclipse permitted the white-light photo of the outer corona to be obtained, while the *uneclipsed* sun was viewed at the same time from above White Sands, New Mexico, where the eclipse had not yet started. The combination of the two observations shows that the streamer structures originate at the solar surface, typically in the brighter places called "active regions." This type of comparison brings home clearly the point that the corona is three-dimensional, with its roots at or below the solar photosphere and outer extension far into interplanetary space.

Observations of the high-temperature solar emission were first carried out from sounding rockets (Baum *et al.* 1946) and techniques for high resolution x-ray imaging were developed under NASA's Suborbital program (Vaiana, Krieger & Timothy 1973) during the late 60's and early 70's. The high temperature corona emits predominantly in isolated spectral lines which fall in the XUV and soft x-ray spectral regions, and many of the important lines were observed and identified by Sir Robert and co-workers in the 60's (Wilson 1964; Jones, Freeman & Wilson 1968). These studies and technological development efforts led to the first series of high resolution studies with extended temporal coverage, carried out from *Skylab* (*viz.* Orrall 1981); these will be discussed in the next section. Most recently, the *Yohkoh* satellite has significantly advanced the study of coronal activity and variability, using a combination of soft x-ray and hard x-ray imaging to study coronal activity, and a major new solar observatory – the Solar and Heliospheric Observatory (SoHO) – carries a large complement of instruments which are expected to provide a comprehensive view of the sun from its interior out to the solar wind.

The solar-stellar connection

If our sun, which is a typical middle-aged low-mass star, has a corona and is a source of x-ray emission, then it is reasonable to ask whether other stars also have coronae and emit x-rays. Within the past two decades this question has been answered in the affirmative: not only do other stars emit x-rays, but the sun is rather below average in activity level. Stars of nearly all spectral types are found to emit UV and x-rays and to display tracers of activity which are detectable in ground-based observations.

Ground-based observations may be used to determine the level of activity on stars, using methods which range from detection of chromospheric lines (Wilson 1963) to direct detection of magnetic fields (Robinson, Worden & Harvey 1980). Surveys of all spectral types, but especially of solar-type stars, have been carried out, most notably at Mt. Wilson (Vaughan 1980).

The direct detection of material at transition region and coronal temperatures had to await observations from space: the International Ultraviolet

Fig. 1.   Composite photo showing the white-light corona seen from the CFH-T in Hawaii at the 11 July 1991 eclipse and the on-disk x-ray corona observed from the NIXT sounding rocket at the same time.

Explorer (IUE) was launched into a quasi-geosynchronous * orbit on 26 January 1978. The satellite provided ultraviolet spectra of astronomical objects ranging from comets and planets to active galactic nuclei and quasars. For stellar studies, the spectra extended the ground-based observations to more highly ionized species, such as Si IV, C IV and N V, thus permitting the extension of ground-based chromospheric studies into what would appear to

---

* The satellite circulates over Central and South America in a pattern which allows access by ground stations feeding both Europe and North America.

be temperatures more characteristic of the chromosphere-corona transition region.

Early IUE surveys extended our knowledge of high-temperature atmospheres on solar-type stars, and also showed a cutoff in coronal emission for late-type giants and supergiants (Linsky & Haisch, 1979). Detailed analysis of emission from late-type stars shows that in general they are solar-like in their properties, as was shown by data from both the IUE (Linsky 1980) and the *Einstein* Observatory, launched on 13 November 1978.

The *Einstein* Observatory had higher sensitivity to x-rays than previous experiments and it also had the ability to produce high resolution images, which yields a high signal-to-noise ratio. As it turns out, flare stars and RS CVn stars are only a few orders of magnitude brighter than "normal" stars, so that the increase in sensitivity of the *Einstein* observations was more than enough to allow the less active solar-type stars to be seen. A survey published after the first year of observation (Vaiana *et al.*, 1981), showed an "x-ray H-R diagram" nearly indistinguishable from its optical counterpart.

2. Magnetic Fields and X-ray Emission

In seeking to explain the existence of a corona on the sun, the major questions to be answered concern:

- coronal heating: the high temperature seems to compel the need to invoke some non-thermal, i.e., mechanical, source of energy. What is that source and how does it transfer energy to the coronal plasma?
- coronal structure: in addition to the gross correlation between magnetic fields and coronal heating, there is fine structure in the corona. What determines the scale size of the "loops"?
- stability: the overall appearance of the corona is stable on several days' timescale, but instabilities and rapid energy release occur on timescales of minutes and seconds.
- currents: how is energy stored in the corona and what causes its sudden release?

The key to answering these questions seems to be in the close connection between the presence, at and above the photosphere, of strong magnetic fields and the locations of the brightest, hottest regions in the corona. Magnetic flux is seen to emerge from the solar interior, rising and breaking through the surface in the form of bipolar regions. Here 'bipolar' indicates that the magnetic field is re-entrant to the solar surface – field of one polarity emerges and field of the opposite polarity re-enters, usually at a nearby location. The overall appearance is roughly that of the field from a magnetic dipole lying horizontally just below the surface. Corresponding to this magnetic structure, the hot coronal plasma is seen to form loop-shaped structures which appear to trace the magnetic topology.

The life of such a region is divided into two main stages, the emergence of magnetic flux and then the subsequent diffusion of that flux across the solar surface. The x-ray loop structures are seen to emerge and grow in accordance with the evolution of the magnetic field.

These processes are directly observed in the corona by x-ray imaging techniques, and at the photospheric level by magnetic field maps, or magnetograms, which can measure the field strength[*] directly. The regions of emerged flux are seen in magnetograms as neighboring patches of opposite polarity fiel. The bottom panel of Fig. 2 shows such a magnetogram.

For large regions containing more than 10^{20} Mx [1 Mx $\equiv$ 1 gauss-cm^2] total flux, the field is seen to emerge, grow in size and then gradually spread out across the solar surface. The calculated timescale for ohmic diffusion of the magnetic field in the photosphere by classical collisional electric resistivity is too slow by many orders of magnitude. It appears necessary to invoke turbulent diffusion, in which the field is moved about by the convective motions at or below the photosphere, in order to account for the rapid spreading of emerged magnetic flux. This process is also, in some theories, closely connected with the heating of the corona, by one of any number of proposed mechanisms whereby the convective motions feed energy into the coronal plasma via the magnetic field (for a recent review, see Narain & Ulmschneider 1990).

Fig. 2 also shows a near-simultaneous coronal x-ray image, taken with the NIXT sounding rocket payload (Golub et al. 1990)[**]. The x-ray image covers temperatures from $1 - 3 \times 10^6$ K, and the hottest, brightest locations coincide with the strongest concentrations of emerged magnetic flux. This can be seen by locating the x-ray bright regions in the top photo and comparing to the black-and-white bipolar magnetic areas in the bottom photo.

Note also that there are extensive regions of weaker magnetic flux in the magnetogram. However, the measured values of magnetic field strength are now known to be deceptively low. The actual photospheric strength of the magnetic field is found to be well over 10^3 G, (Frazier & Stenflo 1972) so that the appearance of weaker average field means that the B is concentrated into small magnetic elements with a low filling factor: the photospheric field is intermittent. In the corona, the field cannot be measured directly, but appears to be space-filling, as expected from the low plasma β. Above the weak-field regions the loop structures are larger and weaker than in active regions; such locations have sometimes been called "quiet corona", but time resolved observations show that it is not at all quiet, as we describe in the next section.

---

[*] Either just the line of sight component, or more recently the full vector field.
[**] The data could not be taken exactly simultaneously because of a total eclipse; note the shadow of the moon entering the x-ray f.o.v. from the west.

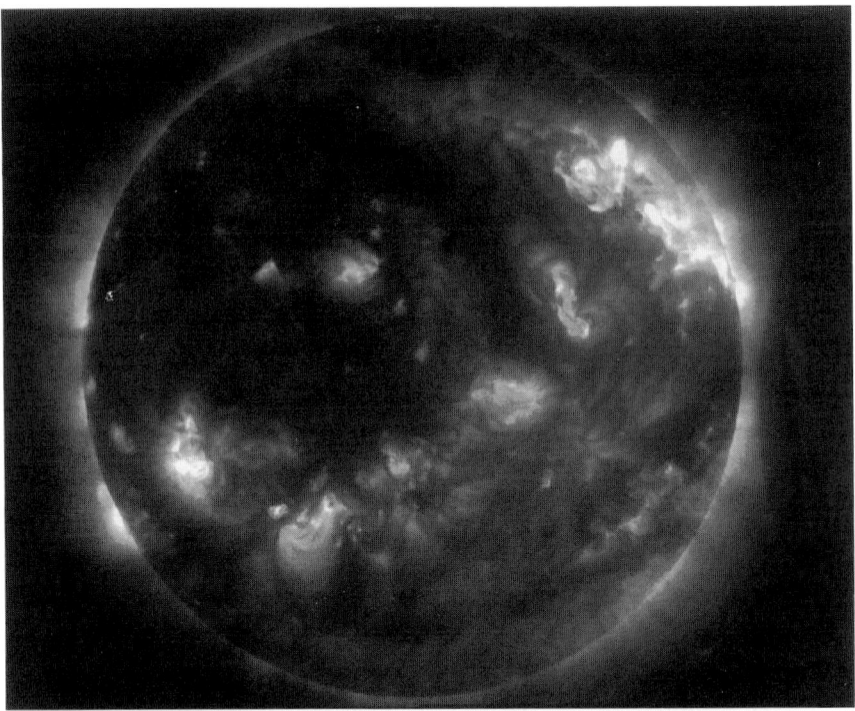

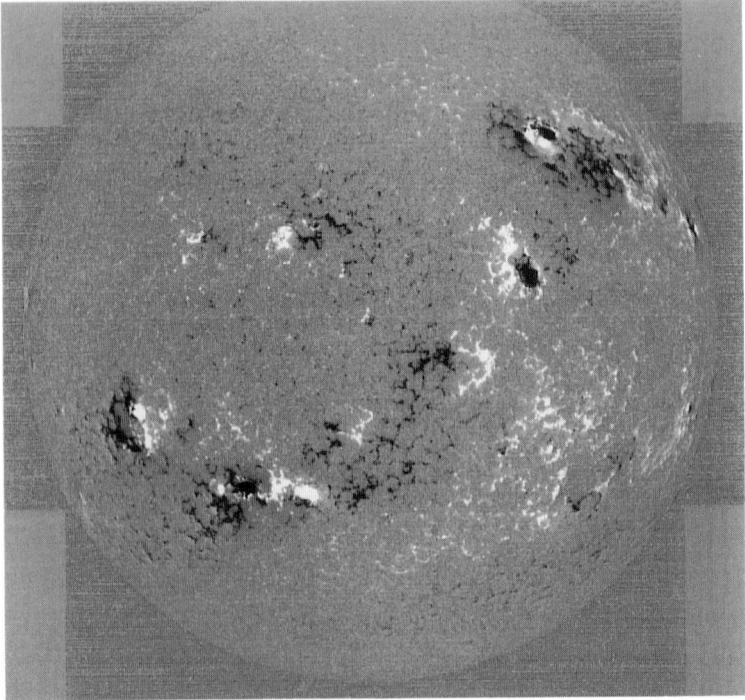

Fig. 2.   Near-simultaneous longitudinal magnetic field map (bottom, from NSO/Kitt Peak), and x-ray images of the sun (top, from the NIXT rocket), 11 July 1991.

3. Short-term Variability

X-ray variability.

Prior to *Skylab* it was generally expected that the x-ray corona would show variations on time scales governed by the emergence and diffusion of magnetic flux. Although flares certainly were known, the general view concerning rapid events was that "Coronal events are rare." (Dunn, 1971). This view was completely reversed by the *Skylab* data, leading to the realization that all parts of the corona are varying on nearly every possible timescale (Vaiana & Rosner 1978) and that future instruments should be designed with the capability to obtain both high spatial and temporal resolution coronal imaging (Golub 1991).

An example of the dynamic changes seen in active regions is shown in Fig. 3. The four panels cover about 1-1/2 days and show the changes in the corona induced by the emergence of new magnetic flux near a pre-existing region. The older region is seen to consist of closed (re-entrant) loops and follows the magnetic field in being spread-out and fairly diffuse. To the west (right in this image) a newly emerging region is compact and very bright, with correspondingly strong emerging magnetic field. The dynamic and highly variable x-ray emission in the newly emerging region is evident, as is also the flare-like activity associated with the formation of interconnecting loops between the two regions. The complexity of the coronal structures, i.e. of the magnetized plasma loops, is also quite evident.

Transient loop brightenings.

A study carried out by Sheeley & Golub (1979) using *Skylab* data from the NRL S-082 and AS&E S-054 instruments provided one of the only studies of coronal variability at high spatial and temporal resolution. The study consisted of a set of nested exposures with time resolution down to two minutes at the center of the set, and focussed on two x-ray bright points (XBPs). These are defined as small, short-lived, magnetically bipolar regions of enhanced x-ray emission in the low corona (Golub *et al.* 1974). Active regions (ARs) were also seen in the data, but were excluded from the study because the large number of loops in the ARs made tracking of individual structures difficult.

The XBPs were seen to consist of a small number of loops, the number varying from two to six from one observation to another. Individual loops were seen to brighten and fade rapidly on timescales of about six minutes, which corresponds to the timescale for cooling by radiation and conduction in a plasma with the observed temperature and density. At any given moment, the "bright point" is seen to consist of several "independently-evolving miniature loops". This study concluded that the coronal structures

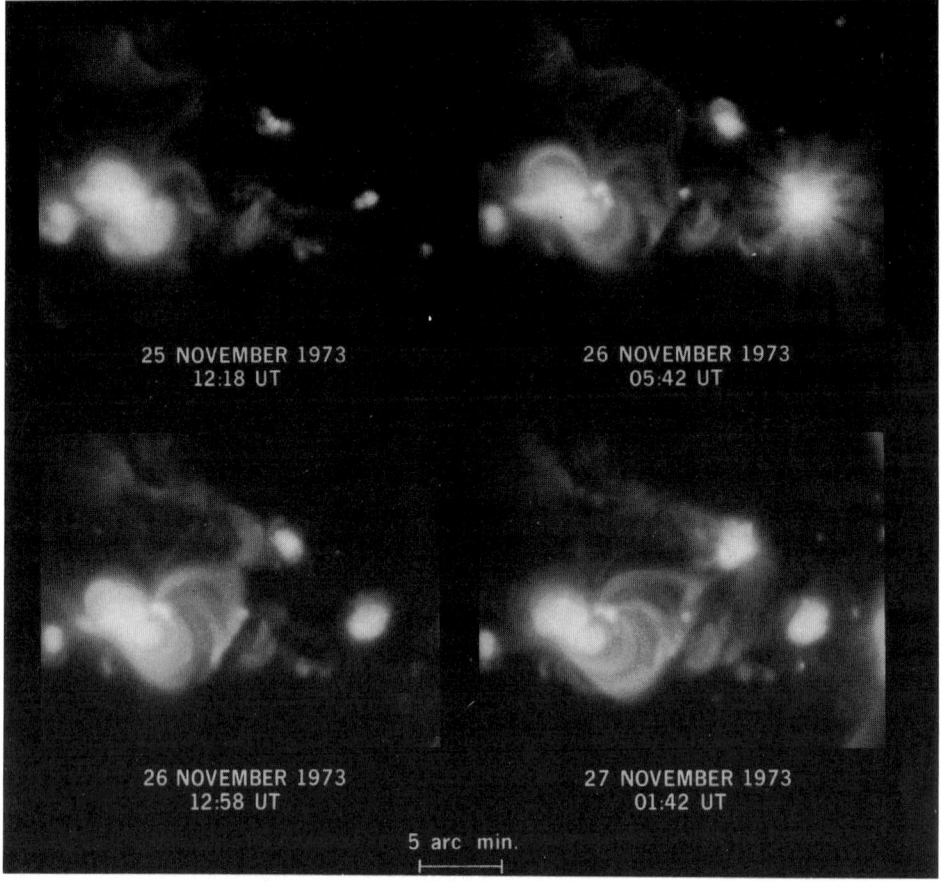

Fig. 3. The rapid reorganization of coronal loop structures in response to newly emerging magnetic flux.

are brightening and fading as fast as they can, and that the observed variability is consistent with an intermittent delta-function energy input, followed by energy redistribution by conduction and radiative losses without significant additional energy input.

The Yohkoh soft x-ray telescope (SXT) has provided the clearest images to date of the extremely dynamic and variable nature of the solar corona. While originally designed primarily as a flare mission, this satellite has proven to be extremely useful for studying the activity of the corona on spatial scales consistent with the resolution (2.5 arcsec pixels) and on temporal scales from seconds to months, and even years.

One of the finest examples of this variability are the so-called "transient brightenings" seen in active regions (Shimizu *et al.* 1992). Some (but not all) active regions are particularly dynamic in showing repeated, small flare-like brightenings, which are clearly seen to take place in closed magnetic structures, i.e., in "loops". An example of this phenomenon is shown in **Fig. 4, from Shimizu et al. (1992).**

The brightenings have a power-law spectrum of energy, from 10^{29} erg down to the instrumental threshold at $\approx 10^{25}$ erg, with a slope $\alpha = 1.7$. Thus, the larger events blend into the distribution of events which one would normally call "flares", but the slope is not consistent with the suggestion that these brightenings make a significant contribution to the heating of active region coronae (Shimizu 1994).

Flares in x-rays.

Solar flares emit high levels of radiation at nearly all wavelengths, from the radio to x-rays and even gamma rays. Flares and the large-scale magnetic rearrangements in the corona associated with flares, eject relativistic electrons, protons, heavy ions and perhaps neutrinos. They produce microwave radio bursts with timescales of milliseconds and long wavelength radio noise storms lasting days. Coronal soft x-ray emission may increase temporarily by a factor of 1000 in a flare, and enhancements may last several minutes up to tens of hours. At the earth, upper atmospheric disturbances, auroral displays, ground-level particle events and a host of related phenomena are produced.

The definition of a flare is somewhat controvertial. An often-used classification defines a flare as 'a rapid temporary heating of a restricted part of the solar corona and chromosphere.' However, 'rapid' might mean a few seconds or it might mean several hours. 'Restricted' may mean a volume so small that it is below our ability to resolve, or it may mean a volume nearly as large as the sun itself. If we ask, how much heating must occur for an event to be called a flare, the possible answers cover a range of six to nine orders of magnitude.

Of course, some things are considered nearly certain. There is nearly universal agreement that magnetic fields play a crucial role in controlling solar activity in general, and flares in particular. We cannot do justice to the extensive range of ideas in the literature on this subject; for details of present flare theory the reader is referred to Tandberg-Hanssen & Emslie (1988) and to the article by Priest in this volume.

As an indication of the possible upper end of the flare size scale in the **solar corona, Fig. 5 shows the coronal brightening associated with a large** filament eruption. This event evolves to a size greater than a solar radius in extent, and such events often are responsible for major readjustments of the large-scale coronal structure. The timescale of this event is long, taking nearly an hour to brighten and several hours to fade away. The latter timescale typically does not agree with the somewhat shorter times calculated for radiative and conductive cooling, so that continued energy input

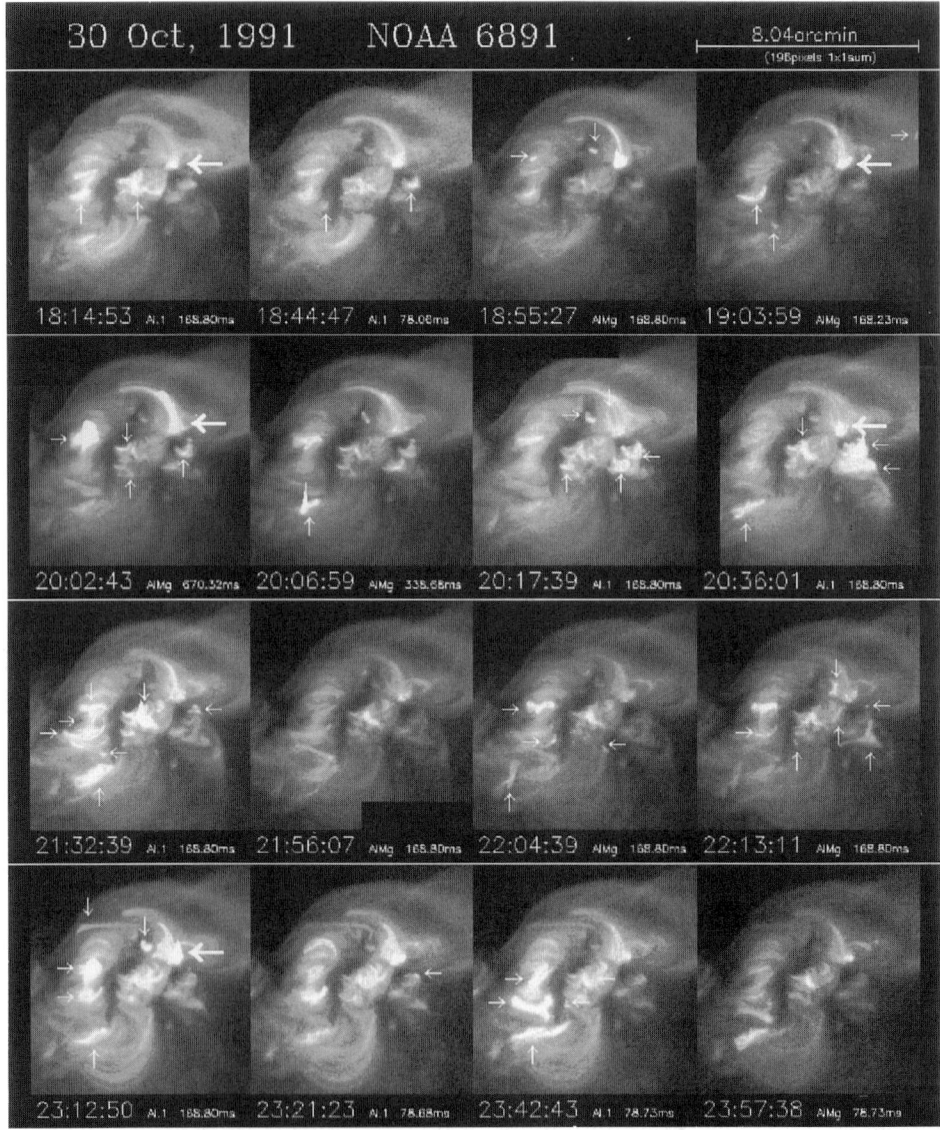

Fig. 4.   A time sequence of x-ray images from *Yohkoh*, showing successive transient brightenings in active region loops.

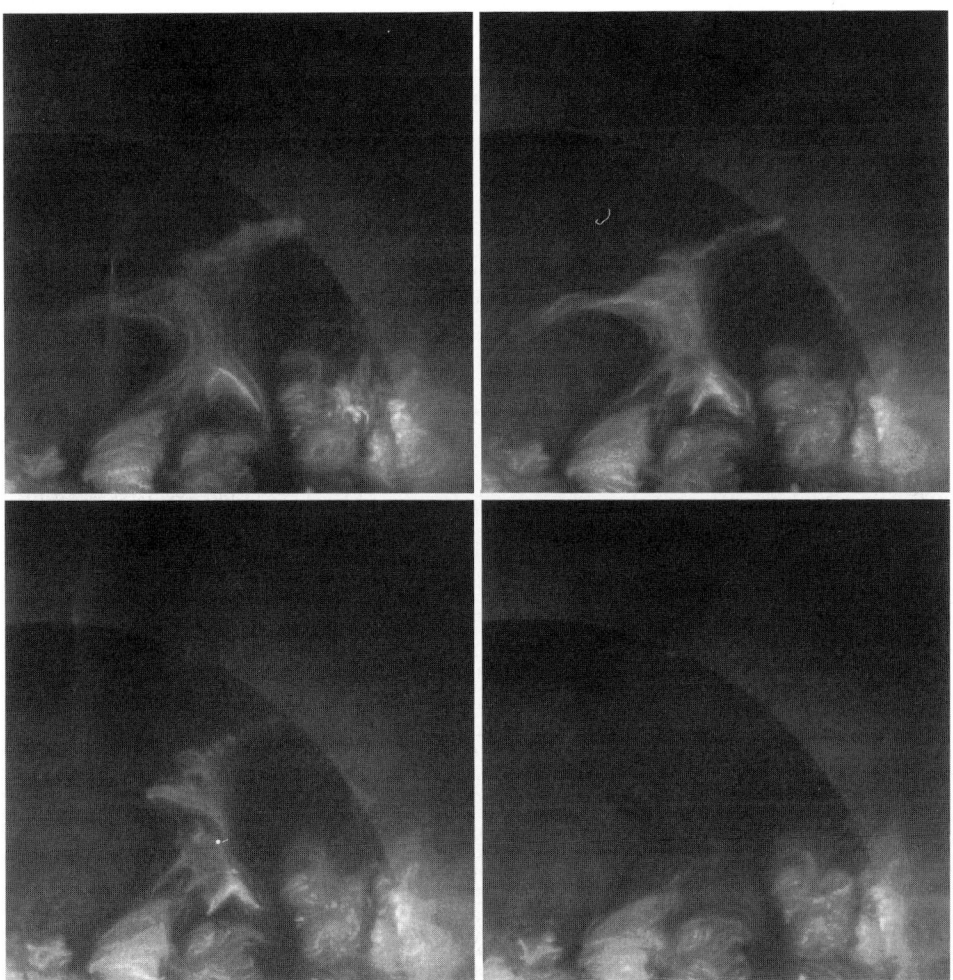

·Fig. 5. The 'scorpion' flare, a large filament eruption event observed by the *Yohkoh* SXT on 26 Feb. 1992.

during the flare decay stage is often invoked.

4. Long-Term & Solar Cycle Variations

The most dramatic difference seen in coronal imaging data between solar minimum and other phases of the solar cycle is the vast increase in the number of x-ray bright points (XBP) seen at minimum (Golub 1980). There is considerable controversy concerning the nature of XBP, i.e. whether they represent new magnetic flux reaching the surface, or reprocessed magnetic flux from previous active region emergence (see Harvey-Angle 1993 for a thorough review). However, the obervational fact is undeniable that at solar minimum the low corona is dominated by these small-scale features.

As an example, Fig. 6 shows a comparison between two x-ray images taken a few years apart. The top image was taken in 1973, during the declining phase of that cycle and the bottom image was taken in 1976, at solar minimum. The data were selected at times when the averaged sunspot numbers were nearly identical, so that the instantaneous level of flux emergence was not a factor. Nevertheless, there is a clear difference between the two observations in the number of small features seen: the solar minimum corona is seen to have triple the number of small-scale features. It even appears that there is very little structure in the corona other than that associated with the XBP.

Finally, the changes in x-ray luminosity and in structural composition of the corona as a function of phase in the magnetic cycle are again shown with dramatic clarity by the Yohkoh SXT. Figure 7 shows how the corona has evolved during four years of observations by the SXT. The overall soft x-ray luminosity has decreased by a factor of twenty in that time and the large-scale structure, representing the evolved magnetic field of large active regions, has been replaced by the weaker, less organized field structure of the numerous x-ray bright points.

5. Future Observations

With the launch of the Solar and Heliospheric Observatory (SoHO) by ESA and NASA at the end of 1995, a major new observational capability is added to the solar arsenal. Combined with the *Yohkoh* satellite and with the Transition Region and Coronal Explorer (TRACE) to be launched at the end of 1997, it would appear that solar physics is in a healthy state and that it can look forward to a decade of progress as this millenium closes.

While all of this is true, there are still many areas of solar and solar-terrestrial research which urgently require further attention, both on strictly scientific grounds and because of their direct importance to society. It is now becoming apparent that relatively small changes in the solar output can lead to major changes in the earth's climate, even if the direct cause is not yet clearly understood. Short-term variations, particularly mass ejections and high-speed solar wind streams, are now known to cause damaging effects at ground level, in the upper atmosphere, and in the near-earth space environment. Thus, the activity of the corona has direct consequences for power distribution networks, for the survival of satellites and astronauts in space, for long-range communications, and possibly even for long-term global climate changes.

A wide range of instrumentation will be used in attacking these prob-

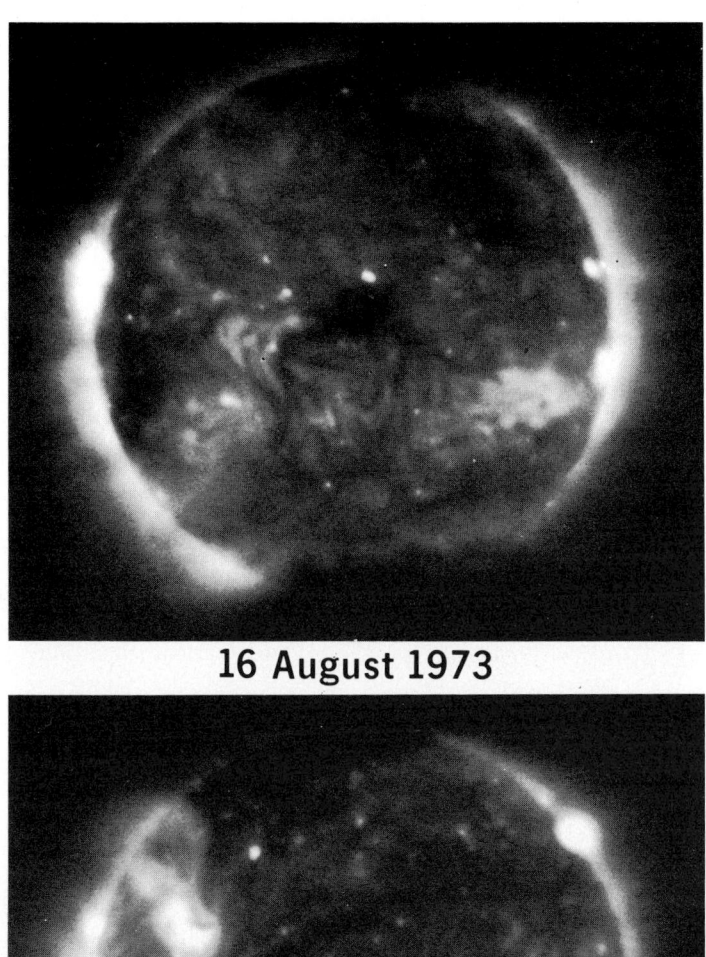

16 August 1973

17 November 1976

Fig. 6.   Comparison of the XBP number density during the declining phase of the solar cycle (top) and solar minimum (bottom).

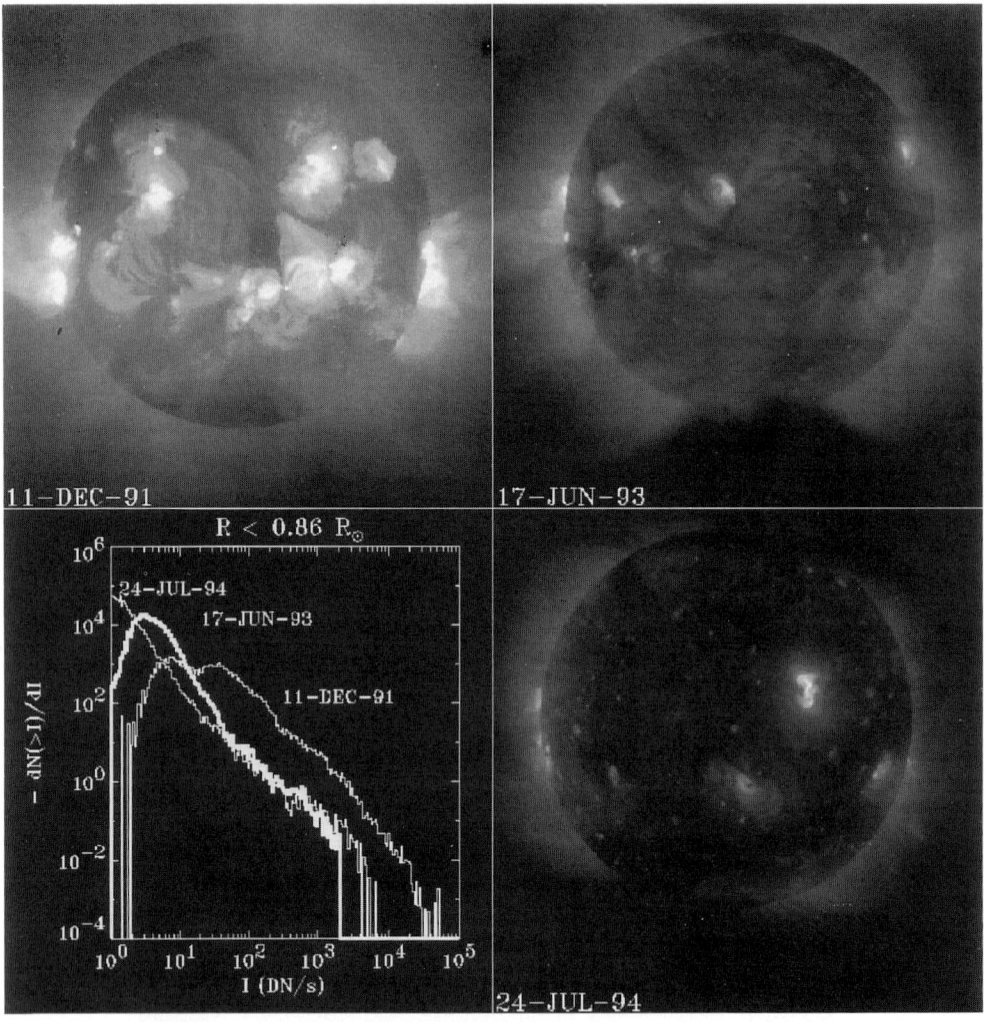

Fig. 7. Change in the coronal x-ray emission as a function of phase in the solar cycle, as seen by the Yohkoh SXT (photo courtesy H. Hara, NAOJ).

lems, including ground-based observing networks, NASA sounding rockets and other rapid, low-cost methods for putting instruments in space, and also larger missions such as the planned Solar-B observatory. A Solar Probe mission would allow direct *in situ* measurements relevant to coronal heating and solar wind acceleration, as well as offering the possibility of obtaining observations of the coronal structure with unprecedented effective spatial resolution. It is, of course, impossible to predict how much of this will actually happen.

Acknowledgements

I would like to thank Dr. E. DeLuca for helpful discussions and Dr. T. Hartquist for a critical reading of the manuscript. I am also grateful to Drs. L. Acton, H. Hara and T. Shimizu for providing illustrations used in this chapter.

References

Baum, W.A., Johnson, F.S., Oberly, J.J., Rockwood, C.C., Strain, C.V. and Tousey, R.: 1946, 'Solar Ultraviolet Spectrum to 88 km', *Phys. Rev.* **70**, 781.

Dunn, R.B.: 1971, 'Coronal Events Observed in 5303Å, in *Physics of the Solar Corona*, ed. C.J. Macris, D. Reidel Publ. Co., Dordrecht-Holland.

Frazier, E.N. and Stenflo, J.O.: 1972, 'On the Small-Scale Structure of Solar Magnetic Fields', *Sol. Phys*, **27**, 330.

Golub, L.: 1991, 'X-ray Observations of Global Solar Activity', in *Flare Physics in Solar Activity Maximum 22*, ed. Y. Uchida, *et al.*, Springer-Verlag, Berlin.

Golub, L.: 1980, 'Solar X-ray Bright Points', *Phil. Trans. Royal Soc. London A* **297**, 595.

Golub, L., Herant, M., Kalata, K., Lovas, S., Nystrom, G., Pardo, F., Spiller, E. and Wilczynski, J.: 1990, 'Sub-arcsecond Observations of the Solar X-ray Corona', *Nature*, No. 6269, 842.

Golub, L., Krieger, A.S., Silk, J.K., Timothy, A.F. and Vaiana, G.S.: 1974, 'Solar X-ray Bright Points', *Astrophys. J. (Letters)* **189**, L93.

Harvey-Angle, K.L.: 1993, 'Magnetic Bipoles on the Sun', Ph.D. Thesis, U. of Utrecht.

Jones, B.B., Freeman, F.F. and Wilson, R.: 1968, 'XUV and Soft X-ray Spectra of the Sun', *Nature* **219**, 252.

Linsky, J.L.: 1980, 'Stellar Chromospheres', *Ann. Rev. Astron. & Astrophys.* **18**, 439.

Linsky, J.L. and Haisch, B.M.: 1979, 'Outer Atmospheres of Cool Stars, I"', *Astrophys. J. (Letters)* **229**, L27.

Narain, U. and Ulmschneider, P.: 1990, 'Chromospheric and Coronal Heating Mechanisms', *Space Sci. Rev.*, **54**, 377.

Orrall, F.Q. (ed.) 1981. *Solar Active Regions*, Colorado Assoc. Univ. Press, Boulder, Col.

Robinson, R.D., Worden, S.P. and Harvey, J.W.: 1980, 'Observations of Magnetic Fields on Two Late-Type Dwarf Stars', *Astrophys. J. Lett.* **23**, L155.

Sheeley, N.R., Jr. and Golub, L.: 1979, 'Rapid Changes in the Fine Structure of a Coronal Bright Point and a Small Coronal Active Region', *Sol. Phys.* **63**, 119.

Shimizu, T., Tsuneta, S., Acton, L.W., Lemen, J.R., and Uchida, Y.: 1992, 'Transient Brightenings in Soft X-rays Observed by the Soft X-ray Telescope on Yohkoh', *Publ. Astronom. Soc. Japan*, **44**, L147.

Shimizu, T.: 1994, 'Active Region Transient Brightenings', in *X-ray Solar Physics from Yohkoh*, eds. Y. Uchida, T. Watanabe, K. Shibata and H.S. Hudson, Universal Academy Press, Tokyo, Japan.

Tandberg-Hanssen, E. and Emslie, A.G. 1988. *The Physics of Solar FLares*, Cambridge University Press UK.

Vaiana, G.S., Krieger, A.S. and Timothy, A.F.: 1973, 'Identification and Analysis of Structures in the Corona', *Sol. Phys.* **32**, 81.

Vaiana, G.S., *et al.*: 1981, 'Results from an Extensive EINSTEIN Stellar Survey', *Astrophys. J.* **245**, 163.

Vaiana, G.S. and Rosner, R.: (1978), 'Recent Advances in Coronal Physics', *Ann. Rev. Astron. Astrophys.* **16**, 393.

Vaughan, A.H.: 1980, 'Comparison of Activity Cycles in Old and Young Main-Sequence Stars', *Publ. Astron. Soc. Pac.* **92**, 392.

Wilson, O.C.: 1963, 'A Probable Correlation Between Chromospheric Activity and Age in Main-Sequence Stars', *Astrophys. J.* **138**, 832.

Wilson, R.: 1964, 'The Zeta/Solar lines Between 170 Å and 220 Å; IAU Symp. No. 23, Liege, Belgium, ed. J-L Steinberg.

CORONAL HEATING BY MAGNETIC RECONNECTION

ERIC. R. PRIEST

Mathematical and Computational Sciences Dept, The University, St Andrews KY16 9SS, Scotland

Abstract. The theory of magnetic reconnection has advanced substantially over the past few years. There now exists a new generation of fast two-dimensional models known as almost-uniform reconnection and nonuniform reconnection, depending on the boundary conditions. Also, we are beginning to explore the uncharted region of three-dimensional reconnection, where regimes of "spine reconnection" and "fan reconnection" have been discovered. Furthermore, part of the coronal heating problem appears to have been solved with recent observational support for the Converging Flux Model in which heating is produced by coronal reconnection driven by footpoint motions.

1. Introduction

As well as being an outstanding project manager and group leader, Bob Wilson was a scientist of unusual vision and inspiration. It was he who realised the need to observe the Sun in X-rays and the ultra-violet in order to study the properties of the corona. It is only now that we are reaping the full benefit of the early rocket flight programs that he led and are beginning to understand just how the corona is heated to enormous temperatures.

In the present paper I would like to describe some recent ideas on the theory of magnetic reconnection and on how part of the corona is heated to a few million degrees.

Soft X-ray images from rockets and Skylab of the corona revealed a three-fold structure of coronal hole, coronal loops and X-ray bright points, all dominated by the magnetic field. The very best pictures now (Figure 13) from the Normal Incidence X-ray Telescope (NIXT) of Leon Golub have a five-times better spatial resolution and reveal extremely fine-scale structure.

The interaction of a magnetic field and plasma in the corona are modelled with magnetohydrodynamics (MHD), whose equations are a unification of the equations of electromagnetism and fluid mechanics. In MHD we add an electric field $\mathbf{v} \times \mathbf{B}$ on moving plasma to the normal version of Ohm's Law to give

$$\mathbf{E} + \mathbf{v} \times \mathbf{B} = \frac{\mathbf{j}}{\sigma}. \tag{1.1}$$

In addition we neglect the displacement current in Ampère's Law so that

$$\nabla \times \mathbf{B}/\mu = \mathbf{j}, \tag{1.2}$$

the remaining equations of Maxwell that we require being

$$\nabla \times \mathbf{E} = -\frac{\partial \mathbf{B}}{\partial t} \tag{1.3}$$

Astrophysics and Space Science **237**: 49–73, 1996.
© 1996 *Kluwer Academic Publishers. Printed in Belgium.*

and

$$\nabla \times \mathbf{B} = 0. \tag{1.4}$$

Finally, we add a magnetic force $(\mathbf{j} \times \mathbf{B})$ to the equation of motion to give

$$\rho \frac{d\mathbf{v}}{dt} = -\nabla p + \mathbf{j} \times \mathbf{B}, \tag{1.5}$$

which is the first of the two main equations of MHD and expresses the fact that plasma is accelerated by pressure gradients and magnetic forces, where $\mathbf{j}$ is given by (1.2).

The second main equation is the *induction equation*, obtained by eliminating $\mathbf{j}$ and $\mathbf{E}$ between (1.1), (1.2) and (1.3), to give

$$\frac{\partial \mathbf{B}}{\partial t} = \nabla \times (\mathbf{v} \times \mathbf{B}) + \eta \nabla^2 \mathbf{B}, \tag{1.6}$$

where $\eta = (\mu\sigma)^{-1}$ is the *magnetic diffusion coefficient*. It states that magnetic fields change in time due to two effects (on the right) namely transport of the magnetic field with the plasma and diffusion through it. The ratio of these two effects is the magnetic Reynolds number

$$R_{\mathrm{m}} = \frac{VL}{\eta},$$

which is enormous in most of the universe (typically $10^6 - 10^{12}$ in the Sun), so that the magnetic field is frozen to the plasma to a very high approximation. The exception is in singularities, where, as we shall see below, magnetic field lines may break and reconnect and magnetic energy is released.

2. Two-Dimensional Magnetic Reconnection Theory

In the classical regime of Sweet and Parker there is a simple diffusion region between opposing field lines, but the resulting reconnection rate as measured by the Alvenic Mach number of the flow towards the region of field dissipation

$$M = \frac{1}{R_m^{1/2}} \tag{2.1}$$

is much too slow to explain flare energy release since R_m is so large. This led Petschek to propose the first fast mechanism, in which two pairs of slow-mode MHD shock waves stand in the flow and radiate from a tiny central diffusion region (Figure 1). Incoming magnetic energy is converted by the

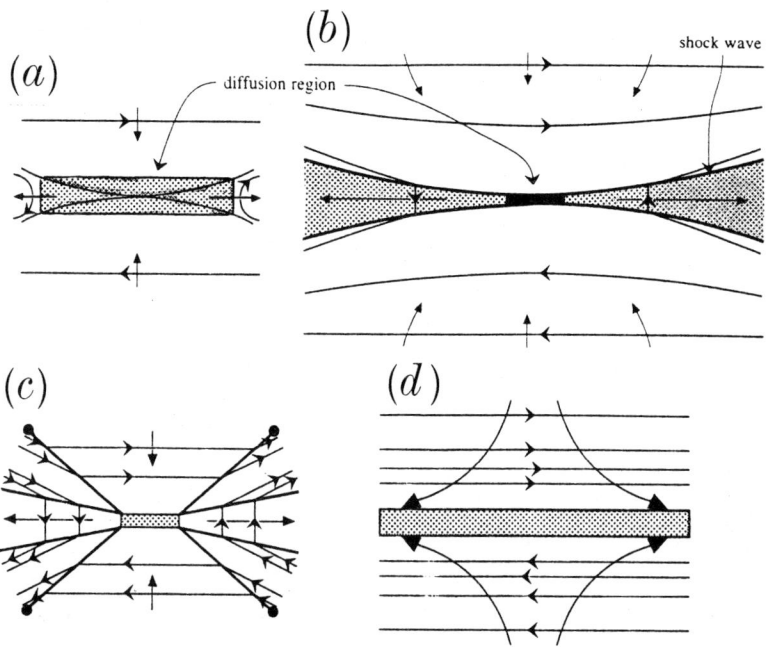

Fig. 1. Classical regimes of reconnection: (a) Sweet-Parker, (b) Petschek, (c) Sonnerup, (d) Sonnerup-Priest.

shocks into the kinetic and thermal energy of two hot fast streams of plasma moving to left and right. The resulting reconnection rate is much faster and possesses a maximum of

$$M = \frac{\pi}{8 \log R_m},$$ (2.2)

which is typically 0.01.

The classical Petschek mechanism has recently been generalised in two ways to produce a new generation of reconnection mechanisms. First of all, the simple Petschek analysis has been generalised to give an *Almost-Uniform Reconnection* family (Priest and Forbes, 1986), the validity of which has been confirmed numerically (Yan, Lee and Priest, 1992). Figure 2 sketches several members of the family including (a) a slow compressional regime, (b) a Petschek-like regime, (c) a Sonnerup-like regime and (d) a flux pile-up regime. The procedure is simply to linearise about a uniform field in the inflow region and to impose boundary conditions on the inflow boundary - both the magnitude and direction of the velocity - so that as the direction is changed different regimes are produced. As one progresses through the different regimes, (a) to (d), the maximum reconnection rate increases and the diffusion region lengthens - when it becomes too long it goes unstable

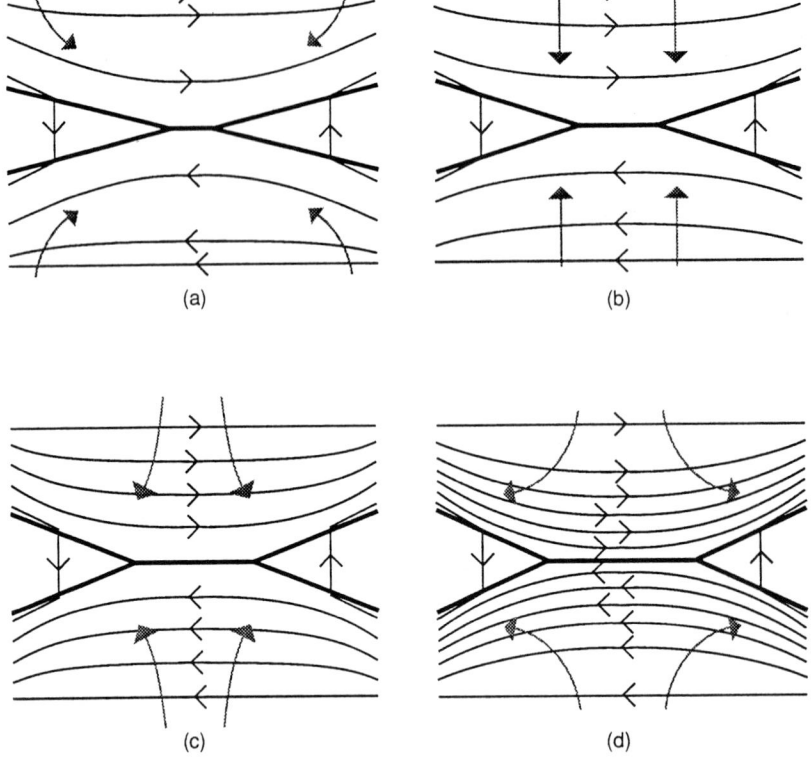

Fig. 2.   Almost-uniform regimes of reconnection

and a new regime of *impulsive bursty reconnection* results (Priest, 1986; Lee and Fu, 1986).

Secondly, a family of *Non-Uniform Reconnection* models (Figure 3) has been developed by Priest and Lee (1990) and Strachan and Priest (1993) and their validity has been confirmed numerically by Yan, Lee and Priest (1993). Here in the simplest case, one imposes $B_y + iB_x = B_i(z^2/L^2 - 1)^{\frac{1}{2}}$ in the inflow region and deduces the flow perpendicular to the field from $\mathbf{E} + \mathbf{v} \times \mathbf{B} = 0$. It possesses highly curved inflow field-lines, jets of plasma escaping along the separatrices (see also Soward and Priest, 1986), and reversed current spikes. These features are present in the numerical experiments of Biskamp (1986). In addition, numerical solutions show that a steady state tends to be maintained only when the magnetic diffusivity is enhanced in the diffusion region, as it is likely to be by micro-instabilities (Scholer, 1989). Indeed, the analytical theories now explain Biskamp's scalings and demonstrate that fast reconnection does indeed exist (Priest and Forbes, 1992).

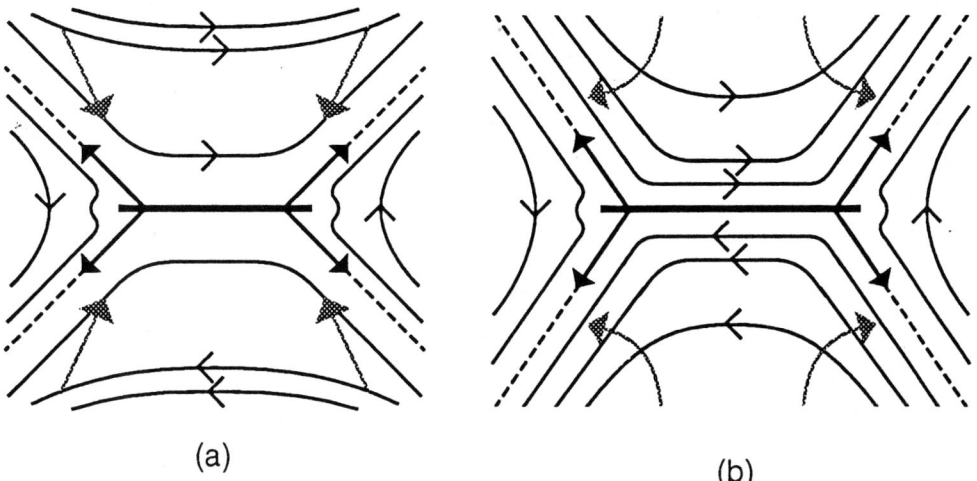

(a)

(b)

Fig. 3.   Nonuniform regimes of reconnection

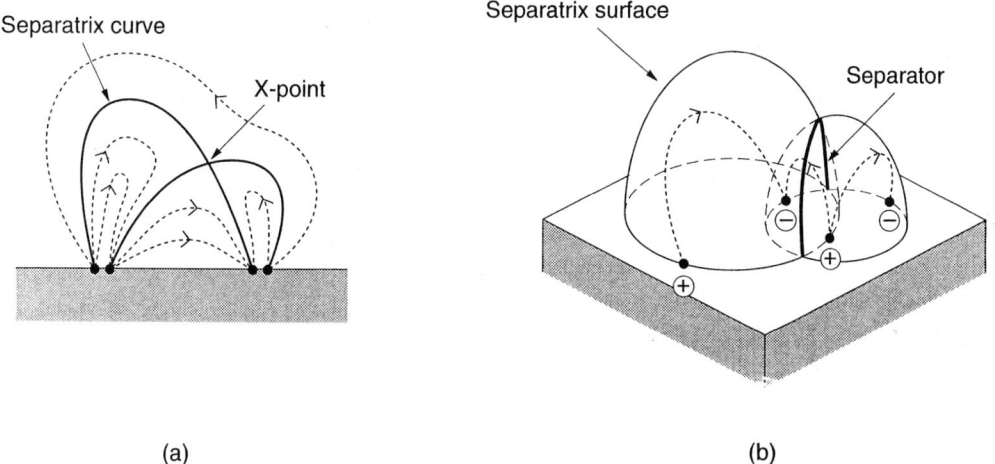

(a)

(b)

Fig. 4.   (a) Separatrix curves in 2D. (b) Separatrix surfaces in 3D.

3. Three-Dimensional Magnetic Reconnection

A general magnetic configuration in two dimensions (Figure 4) contains *separatrix curves*, which separate the plane into topologically distinct regions, in the sense that all the field lines in one region start at a particular source and end at a particular sink. The separatrices intersect in an *X-point* where the field vanishes and the field is locally hyperbolic. Reconnection then occurs by the breaking and rejoining of field lines at the X-point and a transfer of flux across the separatrices from one topological region to another. In three dimensions, some configurations have similar properties, with *separa-*

trix surfaces separating the volume into topologically different regions. They intersect each other in a *separator*, a field line which ends at null points or on the boundary.

A start has recently been made on trying to understand 3D reconnection; for example, Schindler et al (1988) have proposed a general theory, while Priest and Forbes (1989), Lau and Finn (1990) have focussed on null points, and Priest and Forbes (1992) have set up a theory for magnetic flipping in the absence of null points. Here we outline some progress that Priest and Titov (1995) and Priest and Demoulin (1995) have been making on reconnection at nulls and in the absence of nulls, respectively.

3.1. AT NULL POINTS

The simplest null point has field components

$$(B_x, B_y, B_z) = (x, y, -2z) \tag{3.1}$$

or, in cylindrical polars

$$(B_r, B_\phi, B_z) = (R, 0, -2z).$$

Two distinct families of field lines pass through the null point (Figure 5). The *null spine* is an isolated field line which approaches the null from above and below along the z-axis: neighbouring field lines form two bundles which spread out as they approach the xy-plane. The *null fan* is a surface of field lines (the xy-plane) which spread out from the null. Thus the spine and the fan form the skeleton of the field lines near the null.

The kinematics of steady reconnection may be studied by solving

$$\nabla \times \mathbf{E} = 0 \tag{3.2}$$

and

$$\mathbf{E} + \mathbf{v} \times \mathbf{B} = 0. \tag{3.3}$$

Equation (3.2) implies that the electric field can be written as $\mathbf{E} = -\nabla \Phi$ in terms of a potential (Φ) and then (3.3) implies that $(\mathbf{B} \cdot \nabla)\Phi = 0$ or

$$\Phi = \Phi(c, k), \tag{3.4}$$

where c and k are constants describing a field line. Also (3.3) implies that the velocity normal to the magnetic field is

$$\mathbf{v}_\perp = \frac{\mathbf{E} \times \mathbf{B}}{B^2}. \tag{3.5}$$

Thus the boundary conditions determine $\Phi(c, k)$ and so $\mathbf{E}$ and $\mathbf{v}_\perp$ follow. This approach has been applied to a single null, a pair of nulls and fields without nulls, as follows.

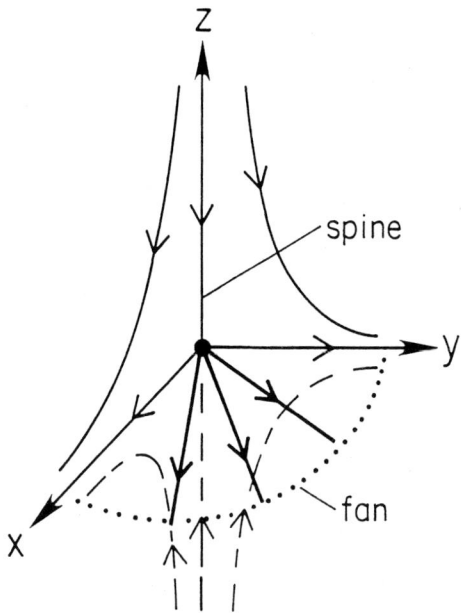

Fig. 5.   The structure of a 3D null point

First of all, let us consider a 3D null point and ask what is the nature of reconnection near such a point? How do magnetic flux surfaces reconnect? If you impose continuous footpoint motions on any cylindrical surface (such as $R = 1$) surrounding the spine and crossing the fan, then a singular motion is driven along the spine axis, so we refer to such a process as *spine reconnection*. Consider field lines in any vertical plane and suppose the footpoints on $R = 1$ move down on the right and up on the left (Figure 6). Then the field lines approach the null, break and reconnect, while the other ends move across the top and bottom of the cylinder through the spine. Suppose the same process occurs in all other vertical planes but is modulated in ϕ: then what are the implications for flux surfaces formed from footpoints that lie initially on the top edge and then move down? The other ends of the field lines (on the top or bottom surface) move in and through the spine. The flux surface therefore moves in and touches the null, forming a fold along the spine. It then reconnects with an oppositely placed flux surface and unfurls from the spine like a cylindrical bubble. A simple solution for the equations of the flux surfaces is

$$R^2 z = \pm 1 - t \sin \phi \qquad (3.6)$$

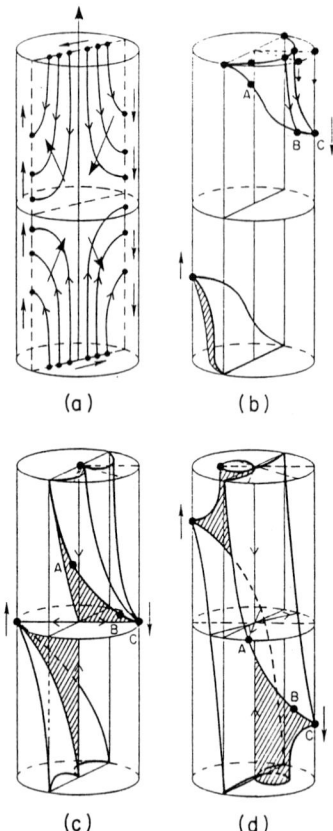

Fig. 6.   Motion of field lines and flux surfaces in spine reconnection

and a simple solution for the field line velocity is

$$(v_{\perp R}, v_{\perp \phi}, v_{\perp z}) = \frac{E_0(\phi)}{R(R^2 + 4z^2)}(2z, 0, R). \tag{3.7}$$

It can be seen that a singularity exists all along the spine $(R = 0)$ and so a key question is whether it can be resolved by diffusion.

   If instead you impose continuous motions on surfaces (such as $z = \pm 1$) that cross the spine, then singular behaviour is driven at the fan surface. Suppose the footpoints on the top move in a straight line from right to left (Figure 7). Then the other ends twirl around the z-axis like a swirling skirt. Consider a flux surface made of field lines whose footpoints march across the top in a straight line: the flux surface distorts and becomes a vertical surface plus a semicircle. It then breaks and reconnects with a similar flux surface on the opposite side of the null point. During this process, magnetic field lines rotate rapidly in one direction above the fan and in the opposite

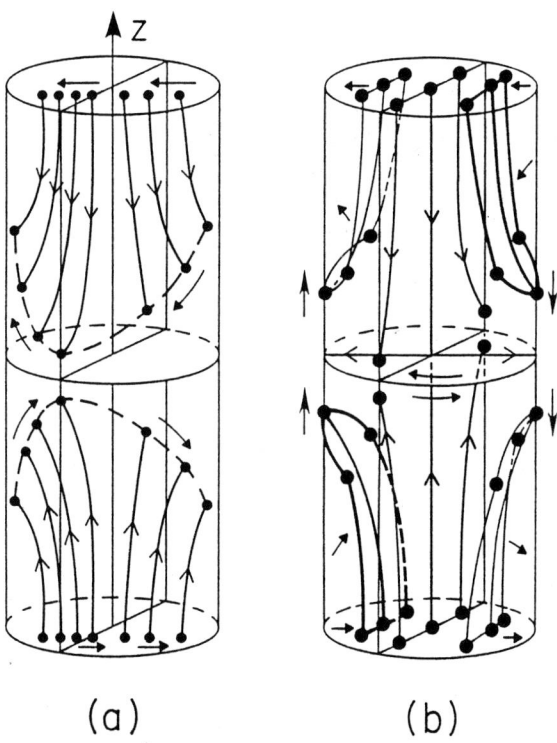

Fig. 7. Motion of (a) field lines and (b) flux surfaces in fan reconnection

direction below it. A simple solution has field line velocity

$$(v_{\perp x}, v_{\perp y}, v_{\perp z}) = \frac{1}{(x^2 + y^2 + z^2)(4 + y^2 z)^{3/2}} \tag{3.8}$$

$$\left(\frac{2xy(z^3 - 1)}{z^{1/2}}, \frac{2(x^2 + 4z^2 + y^2 z^3)}{z^{1/2}}, (4 + y^2 z + x^2 z)yz^{1/2}\right)$$

from which we note that there is a singularity at the fan $(z = 0)$, and so again the question is: can it be resolved by diffusion?

In the simplest case, namely linear reconnection, the answer is no, and it leads to an

Antireconnection Theorem:

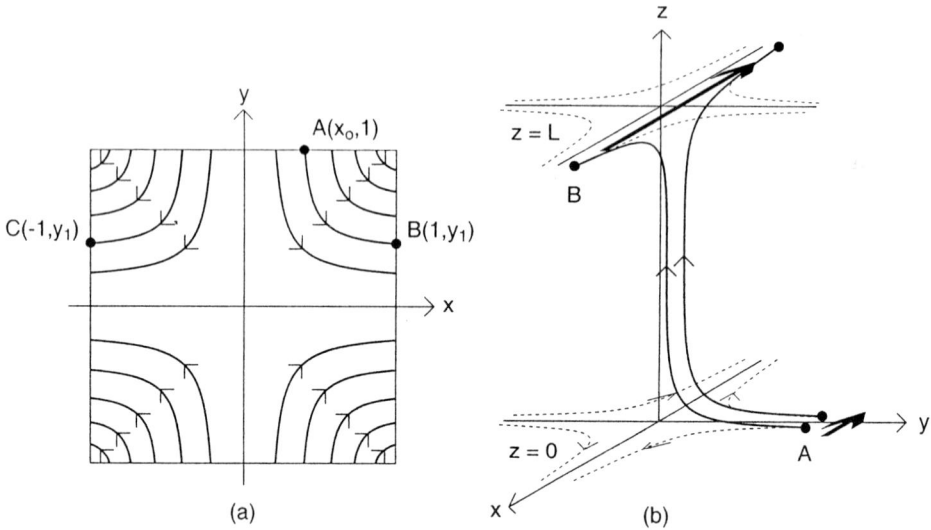

Fig. 8.   The mapping of footpoints for (a) a 2D X-field and (b) a 3D sheared X-field

Steady MHD reconnection in 3D with convective plasma flow across the spine or fan of a radial null point is impossible in an inviscid plasma with a highly subAlvénic flow and a uniform magnetic diffusivity.

The implication of the theorem is that probably nonlinearity is required for such reconnection.

3.2. IN THE ABSENCE OF NULL POINTS

When there is a null point, two-dimensional reconnection is associated with the fact that the mapping of field lines from one footpoint to another is discontinuous. For example, with the simple X-point field

$$B_x = x, \quad B_y = -y \tag{3.9}$$

the point (x_0, y_0) on a boundary will map to (x_1, y_1) in such a way that, when (x_0, y_0) crosses a separatrix, the point (x_1, y_1) suddenly jumps in location (Figure 8). In a classic paper Schindler et al (1988) realised that, when no nulls (or bald patches, Titov et al, 1993) are present, the mapping is continuous, so that separatrices do not exist and the 2D concept of reconnection based on flux transfer across separatrices is no longer applicable. They proposed instead a concept of "general reconnection" to include all effects of local nonidealness that produce a component ($E_{||}$) of electric field parallel to the magnetic field. Furthermore, Priest and Forbes (1989) suggested a local definition of reconnection in terms of potential singular lines, while Priest and Forbes (1992) proposed a concept of *magnetic flipping* and Lau and Finn (1990) analysed kinematic reconnection.

Here I shall report on a new theory for reconnection without nulls (Priest and Demoulin, 1995) which builds on the above ideas. It has four steps for investigating reconnection in a given 3D configuration:

(i) first, the volume of consideration is surrounded by a closed surface S;

(ii) then the mapping of footpoints of field lines from one part of S to another is calculated; for instance, if a small component ($B_z = l \leq 1$) is added to (3.1) to create a sheared X-field the mapping becomes continuous, so that, as the point (x_0, y_0) crosses the y-axis in the plane $z = 0$, say, the other end (x_1, y_1) in the plane $z = 1$ moves continuously (Figure 8);

(iii) next, so-called *quasi-separatrix layers* (QSL) are identified where the gradients of the mapping are very large;

(iv) finally, reconnection occurs when there is a breakdown of ideal MHD and a change of connectivity of plasma elements: this takes place in quasi-separatrix layers where the field line velocity greatly exceeds the plasma velocity - it may be driven by slow regular boundary motions of footpoints (across a QSL).

It may be noted that these definitions of a QSL and magnetic reconnection involve a mapping to a boundary and therefore refer to global properties of a configuration. The concept of a quasi-separatrix layer may be defined formally as follows. Split the surface into parts S_0 and S_1 where the field lines enter and leave the volume, respectively, and set up orthogonal coordinates (u, v) in S and w normal to S. Then field lines map (u_0, v_0) in S_0 to (u_1, v_1) in S_1. Next, form the *displacement gradient tensor*

$$F = \begin{pmatrix} \partial u_1/\partial u_0 & \partial u_1/\partial v_0 \\ \partial v_1/\partial u_0 & \partial v_1/\partial v_0 \end{pmatrix} \tag{3.10}$$

from the gradients of the mapping functions $u_1(u_0, v_0)$ and $v_1(u_0, v_0)$ and evaluate the norm

$$N = \sqrt{\left[\left(\frac{\partial u_1}{\partial u_0} \right)^2 + \left(\frac{\partial u_1}{\partial v_0} \right)^2 + \left(\frac{\partial v_1}{\partial u_0} \right)^2 + \left(\frac{\partial v_1}{\partial v_0} \right)^2 \right]}. \tag{3.11}$$

Finally, define a quasi-separatrix layer as the region where

$$N \gg 1. \tag{3.12}$$

The properties of F are as follows. First of all, a difference δu_0 and δv_0 in footpoint positions maps to

$$\begin{pmatrix} \delta u_1 \\ \delta v_1 \end{pmatrix} = F \begin{pmatrix} \delta u_0 \\ \delta v_0 \end{pmatrix}. \tag{3.13}$$

Secondly, a surface element dS_0 transforms to

$$dS_1 = J dS_0,$$

where $J = (\partial u_1/\partial u_0)(\partial v_1/\partial v_0) - (\partial u_1/\partial v_0)(\partial v_1/\partial u_0)$ is the Jacobian. Thus flux conservation ($B_1 dS_1 = B_0 dS_0$) implies that

$$B_1 = \frac{B_0}{J}, \tag{3.14}$$

where J is finite and nonzero if the field has no nulls or singularities. Thirdly, the displacement gradient tensor may be written as the product

$$F = RU \tag{3.15}$$

of one matrix (R) representing a rotation through ϕ and another (U) representing a stretching by λ_+ (an eigenvalue) along $\mathbf{e}_+$ (an eigenvector) together with a compression by λ_- along $\mathbf{e}_-$. Thus a quasi-separatrix layer (where $N \gg 1$) is associated with a large expansion along one direction and a large compression along the other, such that N is approximately equal to the largest eigenvalue

$$N \approx \lambda_{\max}. \tag{3.16}$$

Consider as an example the sheared X-field

$$(B_x, B_y, B_z) = (x, -y, l) \tag{3.17}$$

inside a cube with $l \ll 1$. The mapping from the base S_0 to the top and sides (S_1) is given by

$$x_1 = x_0 e^{z_1/l} \quad , \quad y_1 = y_0 e^{-z_1/l}. \tag{3.18}$$

Thus, when the point $A(x_0, y_0, 0)$ on S_0 is so close to the y-axis that $2x_0 < \epsilon$, A maps to a point B on the top ($z_1 = 1$) and

$$F = \begin{pmatrix} \epsilon^{-1} & 0 \\ 0 & \epsilon \end{pmatrix}, \tag{3.19}$$

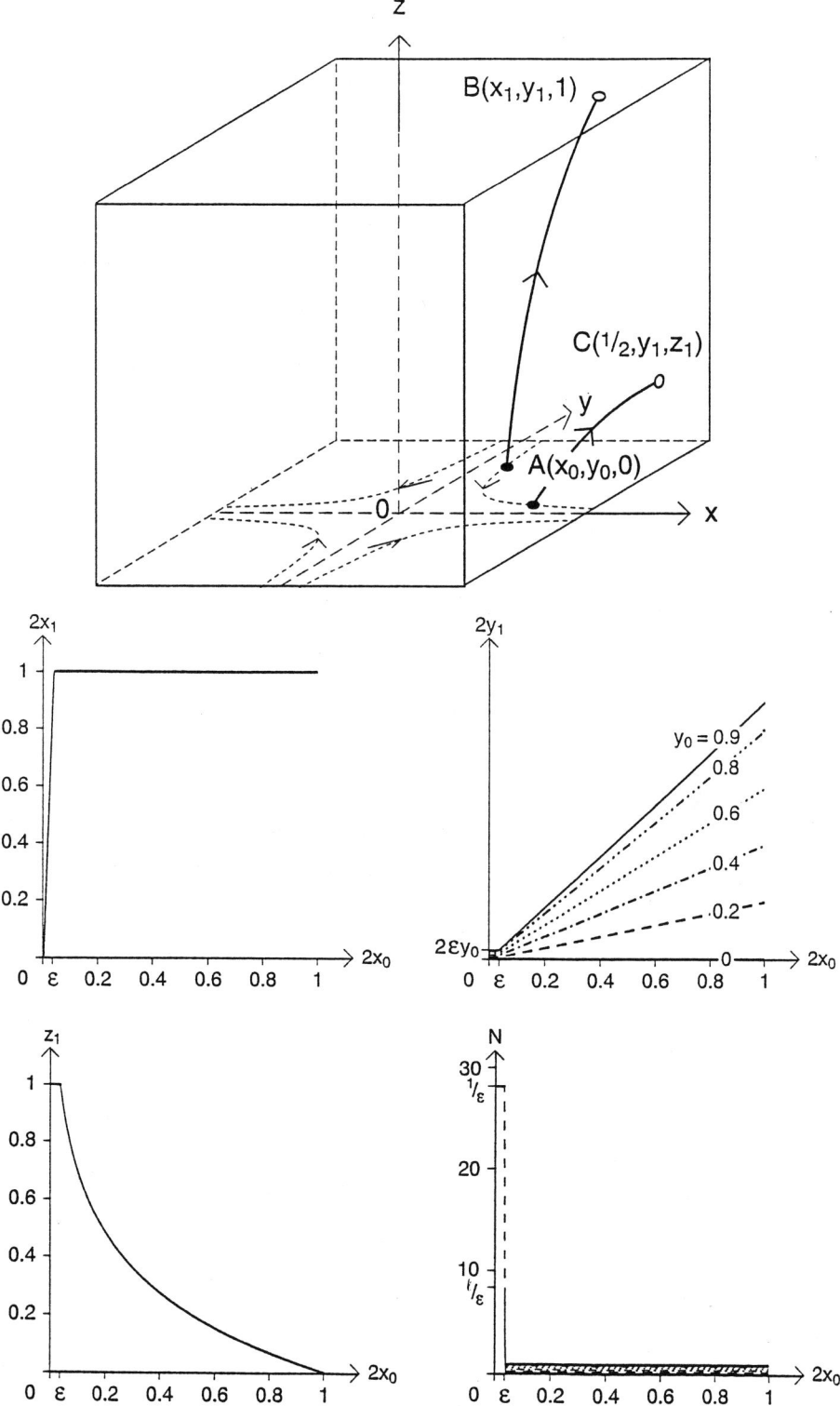

Fig. 9. Sheared X-field in a cube together with the variations of x_1, y_1, z_1 and N with x_0 and y_0

while

$$N \approx \frac{1}{\epsilon},$$ (3.20)

where

$$\epsilon = e^{-1/l} \ll 1.$$ (3.21)

On the other hand, when $\epsilon < 2x_0 < 1$, A maps to C on the side $(x_1 = \frac{1}{2})$ and the elements of F and the value of N are of order unity. The resulting variations of x_1, y_1, z_1, N with x_0 are shown in Figure 9, which reveals the quasi-separatrix layer as a very narrow region of width ϵ where $N \gg 1$. When $l = 0.1$ the value of N in the QSL is 10^4, and even when l is as large as 0.3, N is about 28 in the QSL. If the cube is replaced by a hemisphere or sphere similar forms are produced but the functions become continuous and differentiable.

Having located a quasi-separatrix layer, we may now consider kinematic reconnection satisfying (3.2) and (3.3) and producing a potential (3.4) and field line velocity (3.5). Thus suppose we impose the field line velocity components $v_{\perp 1x}$ and $v_{\perp 1y}$ on the top side $(z = 1)$ of the cube and deduce the function $\Phi(x_1, y_1)$ together with $\mathbf{E}$ and $\mathbf{v}_\perp$ throughout the cube. The resulting electric field on the base $(z = 0)$ of the cube has components

$$
\begin{aligned}
E_{x0} &= \frac{\partial \Phi}{\partial x_1}\frac{\partial x_1}{\partial x_0} + \frac{\partial \Phi}{\partial y_1}\frac{\partial y_1}{\partial x_0}, \\
E_{y0} &= \frac{\partial \Phi}{\partial x_1}\frac{\partial x_1}{\partial y_0} + \frac{\partial \Phi}{\partial y_1}\frac{\partial y_1}{\partial y_0},
\end{aligned}
$$ (3.22)

which depend partly on the electric field components on the top $(E_{x1} = \partial \Phi/\partial x_1,\ E_{y1} = \partial \Phi/\partial y_1)$ and partly on the gradients of the mapping functions $(x_1(x_0, y_0))$ and $y_1(x_0, y_0))$. Thus $\mathbf{E}_0$ is large where the gradients of the mapping are large, namely in the quasi-separatrix layer. For example, if on the top $(z = 1)$ and side $(x = \frac{1}{2})$ of the cube we impose

$$v_{\perp 1x} = 0, \quad v_{\perp 1y} = v_0 x_1$$

and

$$v_{\perp 1x} = 0, \quad v_{\perp 1y} = \tfrac{1}{2}v_0,$$ (3.23)

respectively, then the resulting velocity on the base $(z = 0)$ along the x-axis $(y = 0)$ is

$$
v_{\perp y0} = \begin{cases} \frac{v_0 x_0}{\epsilon^2} & \text{if} \quad |x_0| < \frac{1}{2}\epsilon \\ \frac{v_0}{4 x_0} & \text{if} \quad x_0 > \frac{1}{2}\epsilon \end{cases}.
$$ (3.24)

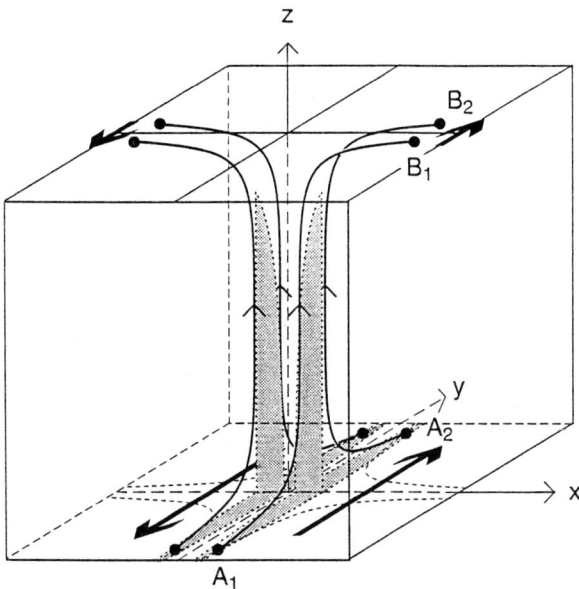

Fig. 10.   Quasi-separatrix layers (shaded) produced by footpoint motion on the top side of a cube.

This peaks at $x_0 = \frac{1}{2}\epsilon$ with a value of $v_0/(2\epsilon)$ and so, if this peak value exceeds the Alfvén speed, there will exist two diffusive layers centred on $x_0 = \pm\frac{1}{2}\epsilon$ where the field lines are unfrozen and flip rapidly through the plasma (Figure 10). In other words the field lines move more quickly than the plasma and become disconnected from it.

4. Reconnection and Coronal Heating

4.1. INTRODUCTION

A major problem in solar physics is: how is the corona heated to a few million degrees, presumably by the magnetic field? The corona has a three-fold structure of coronal holes, coronal loops and X-ray bright points, which was revealed by soft X-ray images from early rockets and Skylab. In these pictures, the active regions above sunspot groups are rather fuzzy and the bright points are mainly just unresolved points of emission. However, recently, the remarkable NIXT photographs from rocket flights (at a temperature of 2-3 x 10^6K and with a five-times better spatial resolution of 500 km) have shown that active regions consist of many fine-scale loops and that bright points appear to include several interacting loops (Figure 14).

X-ray bright points (XBP) were discovered in 1970 from rocket images and were studied with Skylab by Golub et al (1974, 1976). They show up as diffuse clouds typically 20 Mm across with a bright central core. They are

uniformly distributed over the solar surface, with about 200 being present
at one time and 1500 being born each day. Their lifetimes vary between 2
and 48 hours, with a mean value of 8 hours, and they are situated above
pairs of opposite polarity magnetic fragments in the photosphere.

It was natural to assume that the magnetic fragments represent emerging
flux and this became the standard explanation for bright points (Heyvaerts
et al, 1977; Forbes and Priest, 1984). However, Harvey (1984) showed that
two-thirds of the bright points instead lie above so-called "cancelling magnetic features" (CMF), where pairs of photospheric magnetic fragments are
approaching and cancelling (Martin, 1984).

So what is happening in an XBP/CMF event? It cannot be simple sub-
mergence of magnetic flux since this would not explain the brightening
(which starts well before cancellation). Also, no chromospheric fibrils join
the fragments and the fragments are initially widely separated (and therefore unconnected). We need to include the effect of the ambient magnetic
field to which the fragments are initially connected.

4.2. CONVERGING FLUX MODEL

Together with Clare Parnell and Sara Martin, we are proposing a Converging
Flux Model (Priest et al, 1994), which has three phases (Figure 11). In the
Pre-interaction Phase a pair of oppositely directed magnetic fragments in the
photosphere are unconnected and approach one another. They are separated
by a channel of overlying flux, which is squeezed by the approach until a
null point forms in the photosphere (Figure 11(ii)). In the Interaction Phase
the null point moves upwards and coronal reconnection creates an X-ray
bright point, whose structure consists of two newly reconnected and heated
flux tubes, one a small loop linking the fragments and the other a large
loop (as seen in NIXT and Yohkoh images) linking to distant locations.
In the Cancellation Phase the fragments come into contact and cancel by
photospheric reconnection. In the special case when the initial fragments are
equal in magnitude the final state consists of two disconnected fields, one
above the photosphere and one below.

A simple way of modelling the above processes is to represent the sources
by poles of flux $\pm f$ at locations $z = \pm a$, where $z = x + iy$. The field
components (B_x, B_y) due to such sources together with a uniform horizontal
ambient field (B_o) may be written

$$B_y + iB_x = \frac{if/\pi}{z - a} - \frac{if/\pi}{z + a} + iB_o = iB_o \frac{z^2 - b^2}{z^2 - a^2} \qquad (4.1)$$

where

$$b = (a^2 - ad)^{\frac{1}{2}} \qquad (4.2)$$

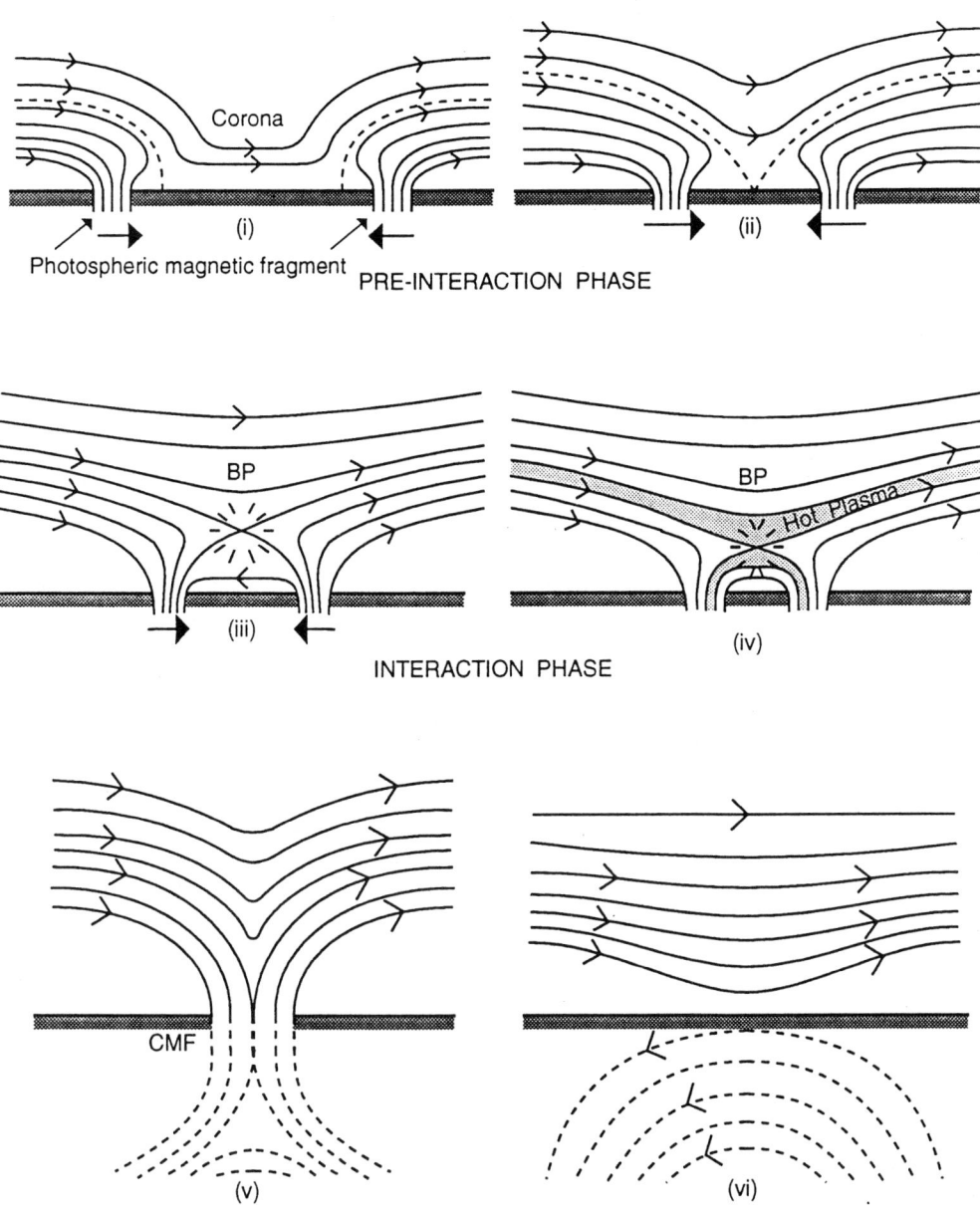

Fig. 11. The magnetic field lines for the Converging Flux Model, showing: the Pre-interaction, Interaction and Cancellation Phases.

is the half-width of the channel and $d = 2f/(\pi B_o)$ we refer to as the "inter-action distance".

In the Pre-interaction Phase, the poles are assumed to approach at a speed much slower than the Alfven speed and to make the overlying field evolve through a series of potential states given by (4.1). As the source position (a) decreases, the half-width (b) of the channel decreases while its flux is conserved, until at the interaction distance (a=d) the null point forms at the origin. It is a second-order null point with $B_y + iB_x \approx -iB_o\, z^2/d^2$ so that the field components are

$$(B_x, B_y) = \frac{B_o}{d^2}(-x^2 + y^2, 2xy) \tag{4.3}$$

When a<d, we have the Interaction Phase with reconnection driven at the null point by the motion of the sources. Such reconnection would probably be in the flux pile-up regime (Priest and Forbes, 1986) and may be impulsive and bursty (Priest, 1986; Lee and Fu, 1986) as the diffusion region goes unstable to secondary tearing, which could explain the rapid time-variations observed in bright points (Sheeley and Golub, 1979). As a preparation for performing a numerical experiment on such reconnection, a simple model may be set up for continuing evolution through potential states. In equation (2.1) b^2 now becomes negative and so the field vanishes at an X-point on the y-axis at a height $\mid b \mid = (ad - a^2)$, which rises as a decreases to a maximum of $\frac{1}{2}d$ when $a = \frac{1}{2}d$ and then decreases to zero as the sources approach the origin.

If instead there is no reconnection, the topology is preserved and a current sheet forms so that the magnetic energy (W) exceeds the energy (W_o) of the potential state by an amount that can be released by reconnection to give the bright point. The field is given by

$$B_y + iB_x = \frac{iB_o(z^2 + h^2)^{\frac{1}{2}}z}{z^2 - a^2}, \tag{4.4}$$

which tends to a uniform field iB_o at infinity and has a cut (the current sheet) stretching along the y-axis from the origin to $z = ih$. The condition that the flux above the sheet is preserved gives the length of the sheet as $h = (d^2 - a^2)^{\frac{1}{2}}$, which increases from zero to d as a decreases from d to zero.

The energy (W) is given by

$$2\mu W = \int B^2 dV = \int \mathbf{B}.\nabla \times \mathbf{A}\ dV = \int \mathbf{A}.\nabla \times \mathbf{B} + \nabla.(\mathbf{A} \times \mathbf{B})\ dV \tag{4.5}$$

since $\mathbf{B} = \nabla \times \mathbf{A}$. We assume $\mathbf{A}$ vanishes on the current sheet, the only place where $\nabla \times \mathbf{B} = \mu \mathbf{j}$ is nonzero, and so the first term vanishes. The second term

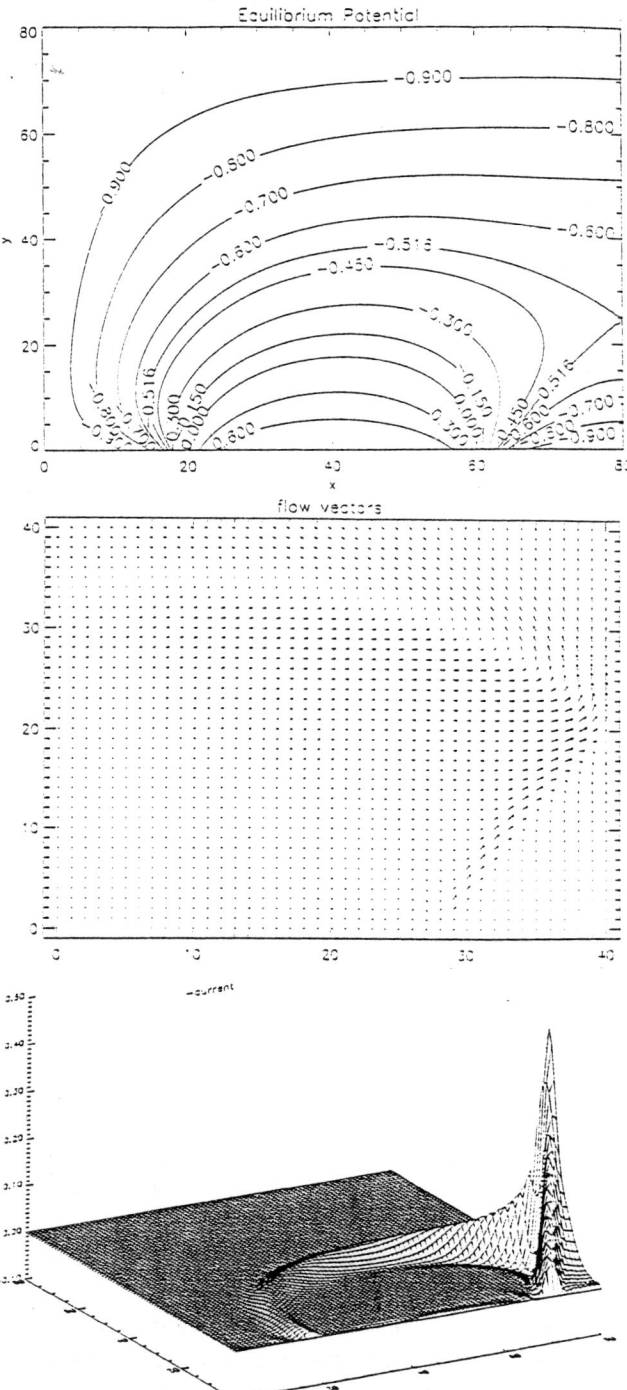

Fig. 12.   A numerical experiment on converging flux showing the initial magnetic field (top), the resulting velocity vectors in one quarter of the domain (middle), the current density (bottom).

may be transformed by the divergence theorem to give $2\mu W = \int \mathbf{A} \times \mathbf{B}.d\mathbf{S}$, or, in our two-dimensional geometry,

$$2\mu W = -2 \int_o^\infty (AB_x)_{y=o} \ dx. \tag{4.6}$$

The potential field ($\mathbf{B}_o$) with the same normal field (B_y) and therefore flux function on the base ($y = 0$) has energy

$$2\mu W = -2 \int_o^\infty (AB_{ox})_{y=o} \ dx, \tag{4.7}$$

and so by subtracting from (2.6) we obtain the stored energy in excess of potential as

$$2\mu W_f = 4 \int_o^\infty (AB_{sx})_{y=o} \ dx. \tag{4.8}$$

Here $\mathbf{B} - \mathbf{B}_o = \mathbf{B}_s + \mathbf{B}_{si}$, where $\mathbf{B}_s$ is the field due to the sheet and $\mathbf{B}_{si}$ due to its image, so that $B_{sx} = B_{six}$ on $y = 0$.

However, the contribution due to the sheet is

$$B_{sx} = \int_o^h \frac{\mu I(y)}{2\pi(x^2 + y^2)^{\frac{1}{2}}} \frac{y}{(x^2 + y^2)^{\frac{1}{2}}} \ dy \tag{4.9}$$

where

$$I(y) = \frac{2B_{ys}}{\mu} = \frac{2B_o}{\mu} \frac{(h^2 - y^2)^{\frac{1}{2}} y}{y^2 + a^2} \tag{4.10}$$

is the current in the sheet. Thus, after integrating over x, 4.8) reduces to

$$W_f = \frac{B_o^2 d^3}{2\mu} \int_o^h \frac{2(h^2 - y^2)^{\frac{1}{2}} y}{d(y^2 + a^2)} \left(\frac{\pi}{2} - tan^{-1} \frac{a}{y} \right) dy \tag{4.11}$$

The factor outside the integral represents the energy in a cube of side d, while the integral is a dimensionless factor that depends on a/d. The resulting W_f represents an estimate (an upper limit) of the energy released by the reconnection. Now, the energy released in a bright point is between 3.10^{20} and 3.10^{21}J (3.10^{27}-3.10^{28} erg) and so we find W_f is of this order for $d = 5$-10 Mm.

Furthermore, the duration of the bright point is roughly the duration of the interaction phase, namely d/v_o, where v_o is the speed of approach. Putting $d = 7.5$ Mm and $v_o = 0.5$ km s^{-1}, we obtain a duration of 3.10^4sec or 8 hours, as required.

We have just started a numerical experiment on slow reconnection in the Converging Flux Model. Figure 12 shows the left half of the initial magnetic field due to four sources. The right-hand boundary is an axis of symmetry with an X-point about one-third of the way up it. The left-hand source on the base is held fixed and the compressible, resistive MHD evolution is studied in response to the motion of the right-hand source to the right, so driving reconnection at the X-point. The resulting flow vectors in one-quarter of the domain are shown in Figure 13. There are strong inflows and outflows and a vortex layer along the separatrix. The current density shows an intense spike at the X-point and a strong current everywhere along the separatrix.

4.3. APPLICATION TO SPECIFIC BRIGHT POINTS

We have considered particular bright points that were observed on the NIXT flight of July 11, 1991 in Fe XVI at 2-3 x 10^6K, as shown in Figure 13.

In the full disc image there is an active region surrounded by a collection of 5 bright points, labelled A-D. The photospheric magnetogram below bright point C shows four discrete sources of flux which we have represented by poles, two being of positive polarity and two negative. The resulting field lines in the photospheric plane (Figure 15a) show that the plane is split into four topologically different regions by four separatrix field lines which intersect at two X-type neutral points. Each region contains only field lines that link two particular poles. When the central source moves to the right, flux is transferred between the regions and the field lines that have reconnected brighten to give shapes that compare well with bright point C. In three dimensions the separatrix field lines become dome-like surfaces (Figure 14b) and these surfaces intersect in a separator field line that links the two photospheric neutral points. Two field lines before and after reconnection show that, during the process of reconnection in three dimensions, the field lines approach the separator and then reconnect at the two null points before moving away from it. Bright point D in Figure 13 has an attractive seagull-like shape; below it in the photosphere there is one positive source of flux and three negative ones. The resulting field lines are shown in Figure 15b with three separate lobes of flux going from the central source to the three sinks. As the central pole moves to the left, so flux is transferred from the right-hand region to the upper and lower lobes, creating hot reconnected loops on the edges of the lobes which outline the wings of the seagull.

5. Conclusion

Understanding a complex phenomenon like the solar corona demands the complemetary skills of a wide range of scientists including instrumentalists, specroscopists, plasma theorists and computational experts. It is gratifying that so much progress is being made from the early beginnings initiated by Bob Wilson.

Here I have been focussing on only part of the story. Magnetic reconnec-

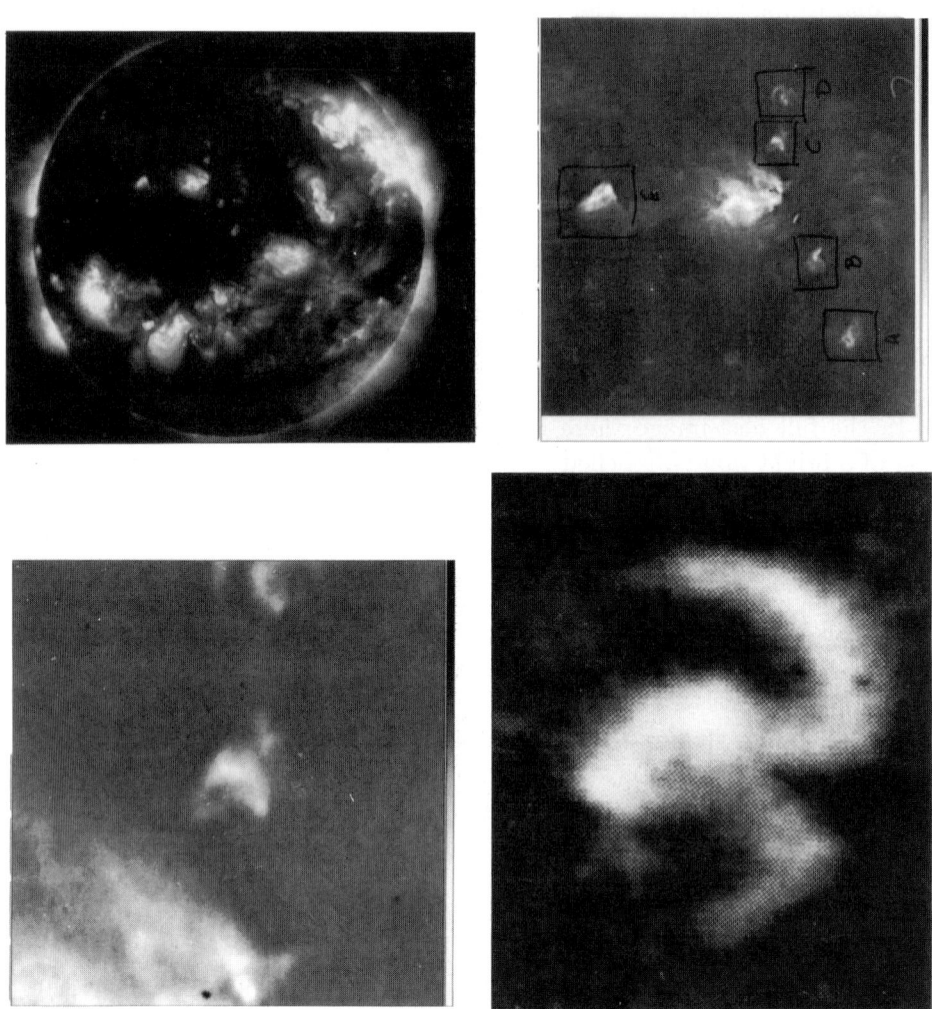

Fig. 13.   NIXT images from 11 July, 1991, showing full disc, close-up of collection of bright points, close-ups of bright points C (bottom left) and D (bottom right). (Courtesy L. Golub).

tion theory is being developed in several ways: there is a new generation of fast reconnection models; and the elements of a three-dimensional model are emerging. In particular, reconnection may take place at 3D null points by *spine reconnection* or *fan reconnection,* in which the *antireconnection theorem* implies that such reconnection probably needs nonlinear effects. Furthermore, reconnection in the absence of null points may sometimes occur within *quasi-separatrix layers,* where the norm of the displacement gradient tensor is large and an electric field component parallel to the magnetic field is produced.

Evidence for 3D reconnection in the Sun's corona is beginning to appear.

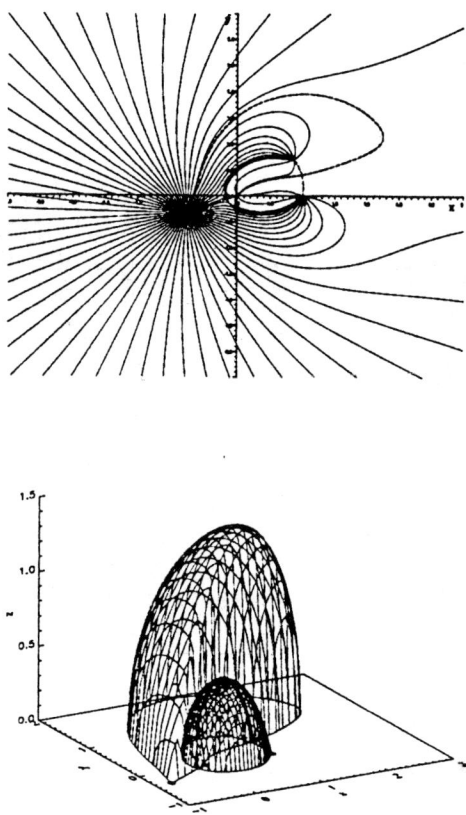

Fig. 14. (a) Field lines in the photosphere due to four flux sources below bright point C. (b) Resulting separatrix surfaces above the photosphere.

Reconnection is thought to be driven in large solar flares by an eruption that occurs due to an MHD catastrophe (Priest and Forbes, 1990). Other potential locations are coronal heating regions, and recently a solution has been found to part of the long-standing coronal heating problem, namely how x-ray bright points are heated. A new *Converging Flux Model* agrees in a natural way with many observations and suggests that converging motions of magnetic sources in the photosphere drive reconnection in the overlying corona. Furthermore, a detailed comparison with X-ray images of the highest spatial resolution leads to the explanation of the internal structure of bright points. Indeed, coronal reconnection driven by footpoint motion may represent an elementary heating mechanism that is heating all coronal loops and even coronal holes and not just bright points.

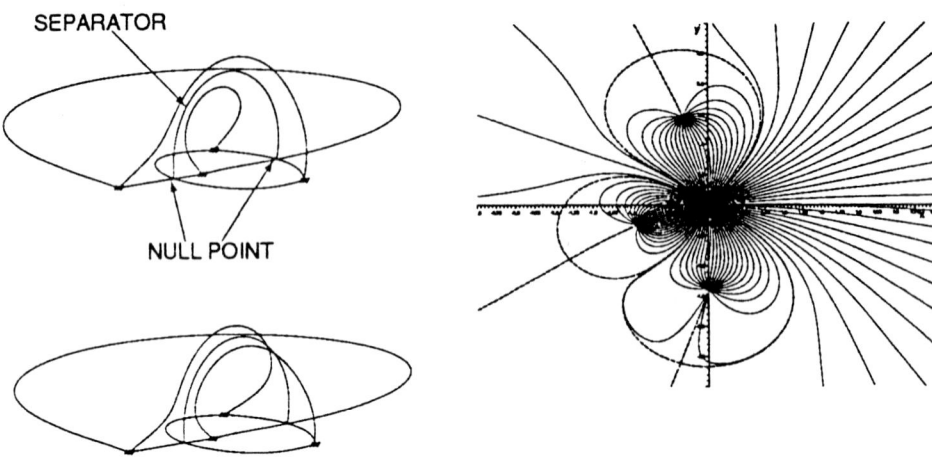

Fig. 15.   (a) Field lines before (above) and after reconnection (below). (b) Photospheric field lines below bright point D.

References

Biskamp,D.: (1986) 'Magnetic reconnection via current sheets', *Phys. Fluids* **29** 1520–1531.

Forbes, T.G. and Priest, E.R.: (1984) 'Numerical simulation of reconnection in an emerging flux region', *Solar Phys.* **94**, 315-340.

Golub, L., Krieger, A.S., Silk, J., Timothy, A. and Vaiana G.: (1974) 'Solar x-ray bright points', *Astrophys. J.* **189**, L93 - L97.

Golub, L., Krieger, A.S. and Vaiana, G.S.: (1976a) 'Distribution of lifetimes for coronal soft x-ray bright points', *Solar Phys.* **49**, 79-116.

Harvey, K.L.: (1984) 'Solar cycle variation of ephemeral active regions', *Proc. 4th European Meeting on Solar Phys.* ESA SP 220, 235-236.

Heyvaerts, J., Priest, E.R. and Rust, D.M.: (1977) 'An emerging flux model for the solar flare phenomenon', *Astrophys. J.* **216**, 123-137.

Lau,Y.T. and Finn, J.M.: (1990) '3D kinematic reconnection in the presence of field nulls and closed field lines', *Astrophys. J.* **350** 672–691

Lee, L.C. and Fu, Z.F.: (1986) 'Multiple x-line reconnection, 1, A criterion for the transition from a single x-line to a multiple x-line reconnection', *J Geophys Res.* **91**, 6807-6815.

Martin, S.F.: (1984) 'Dynamical signatures of quiet-Sun magnetic fields' in *Small-scale Dynamical Processes in Stellar Atmospheres* (ed S. Keil) Sac. Peak Observatory, p30.

Parnell, C.E., Golub, L. and Priest, E.R.: (1993) 'The 3D structure of x-ray bright points', *Solar Phys.* **151**, 57–74.

Priest, E.R.: (1986) 'Magnetic reconnection on the Sun', *Mit. Astron. Ges.* **65**, 41-51.

Priest, E.R. and Demoulin, P.: (1995) '3D reconnection without null points'.

Priest, E.R. and Forbes, T.G.: (1986) 'New models for fast steady state reconnection', *J. Geophys. Res.* **91** 5579-5588

Priest, E.R. and Forbes, T.G.: (1989) 'Steady reconnection in three dimensions', *Solar*

Phys. **119** 211–214.

Priest, E.R. and Forbes, T.G.: (1992a) 'Magnetic flipping – reconnection in three dimensions without null points', *J. Geophys. Res.* **97** 1521–1531.

Priest, E.R. and Forbes, T.G.: (1992b) 'Does fast magnetic reconnection exist?', *J. Geophys. Res.* **97**, 16757-16772.

Priest, E.R. and Lee, L.C.: (1990) 'Nonlinear magnetic reconnection models with separatrix jets', *J Plasma Phys.* **44**, 337-360.

Priest, E.R., Parnell, C.E. and Martin, S.F.: (1994) ' A converging flux model of an X-ray bright point and an associated cancelling magnetic feature', *Astrophys. J.* **427**, 459-474.

Priest, E.R. and Titov, V.S.: (1995) 'Magnetic reconnection at 3D null points'.

Schindler, K., Hesse, M. and Birn, J.: (1988) 'General magnetic reconnection, parallel electric fields and helicity', *J. Geophys. Res.* **93**, 5547-5557.

Scholer, M.: (1989) 'Undriven magnetic reconnection in an isolated current sheet' *J Geophys Res.* **94**, 8805-8812.

Sheeley, N.R. and Golub, L.: (1979) 'Rapid changes in the fine structure of a coronal bright point and a small coronal active region,' *Solar Phys.* **63**, 119-126.

Strachan, N.R. and Priest, E.R.: (1994) 'A general family of nonuniform reconnection models with separatrix jets', *Geophys. Astrophys. Fluid Dynamics* **74**, 245-274.

Titov, V.S., Priest, E.R. and Demoulin, P.: (1993) 'Conditions for the appearance of 'bald patches' at the solar surface' *Astron. Astrophys.*, **276**, 564-570.

Yan, M., Lee, L.C. and Priest, E.R.: (1992) 'Fast magnetic reconnection with small shock angles', *J. Geophys. Res.* **97**, 8277-8293.

Yan, M., Lee, L.C. and Priest, E.R.: (1993) 'Magnetic reconnection with large separatrix angles', *J. Geophys. Res.* **98**, 7593-7602.

HOT STARS AND STELLAR X-RAY SOURCES

CHEMICALLY PECULIAR HOT STARS

KEITH C. SMITH
Department of Physics & Astronomy
University College London

Abstract. The upper main sequence is home to a diverse family of chemically peculiar stars, including λ Boo, Am–Fm, Bp–Ap, HgMn, He-weak, and He-rich varieties. This paper presents an informal review of the physical properties of these objects, including their location in the H–R diagram, frequency of incidence, rotation, binarity, magnetic fields, and variability. Part of the discussion is devoted to describing the bizarre surface compositions encountered in chemically peculiar stars, with an emphasis on insights provided by a generation of ultraviolet observations obtained with the *IUE* satellite. The paper concludes with an overview of the radiative diffusion mechanism which has been proposed as an explanation of the chemically peculiar phenomenon.

1. Introduction

The upper main sequence of the H–R diagram is populated by a 'zoo' of chemically peculiar (CP) stars, the diversity and complexity of which has fascinated stellar spectroscopists for decades. Several quite distinct families of CP stars are now recognized, encompassing spectral types from early F through early B, and characterized variously by intense magnetic fields, unusually slow rotation, rapid oscillations, photometric and spectroscopic variability, and – most spectacularly – surface abundances which can depart from the solar (cosmic) pattern by many orders of magnitude.

This paper is intended as an introduction to the chemically peculiar zoo, rather than a formal review. The emphasis is on the impact that space-based observations have had on the subject, in particular the rôle of *IUE*, which made accessible the rich ultraviolet spectra of these objects for the first time. For the reader who is interested in pursuing the subject at greater depth, there is no better starting point than Wolff's (1983) monumental monograph which condenses into two hundred pages virtually everything known about these objects from some five decades of study. To bring oneself up to date, there are several conference proceedings to consult which track progress in the field over the last decade (*e.g.*, Cowley, Dworetsky & Mégessier 1986; Michaud & Tutukov 1991; Dworetsky, Castelli & Faraggiana 1993).

The physical and chemical characteristics of six major groups of CP stars are considered here; these groups are listed in Table I which gives some basic data for each. The reader should beware that the nomenclature for these stars is almost as diverse as the objects are themselves: Preston (1974) attempted to bring some order to this chaos by assigning CP stars to four major groups (CP1–4). While that system has the virtue of emphasizing the view that these stars exhibit the same phenomenon (*i.e.*, element segregation via diffusion) in different guises, it has never been extended to accom-

Astrophysics and Space Science **237**: 77–105, 1996.
© 1996 *Kluwer Academic Publishers. Printed in Belgium.*

TABLE I

Chemically peculiar stars of the upper main sequence

Classical name	Preston's group	Discovery criteria	Spectral types	Temperature domain
λ Boo	–	weak Mg II and weak metals	A0–F0	7500–9000
Am–Fm	CP1	weak Ca II and/or Sc II; enhanced metals	A0–F4	7000–10 000
Bp–Ap	CP2	enhanced Sr, Cr, Eu, and/or Si	B6–F4	7000–16 000
HgMn	CP3	enhanced Hg II and/or Mn II	B6–A0	10 500–16 000
He-weak	CP4	weak He I compared with colours	B2–B8	14 000–20 000
He-rich	–	enhanced He I compared with colours	B2	20 000–25 000

modate the λ Boo or He-rich varieties, and therefore the more descriptive classical names are retained here.

Broadly speaking, CP stars form magnetic and non-magnetic sequences which, though quite distinct at the cool extreme, merge at higher effective temperatures. The magnetic sequence is defined by what are usually referred to as Bp–Ap stars, comprising the cool SrCrEu and hot Si sub-types over-lapping at around 10 000 K. The non-magnetic sequence is populated by Am–Fm and λ Boo stars below 10 000 K, and by HgMn stars above 10 500 K. Above approximately 16 000 K, the magnetic and non-magnetic sequences become less distinct. The HgMn stars merge with a rather heterogeneous collection of He-weak stars, some members of which share the properties of weak or non-existent magnetic fields and enhanced P, Xe, and Ga lines observed in the hotter HgMn stars, while others are distinctly magnetic and exhibit enhanced lines of Sr, Ti, and Si as seen in the Bp–Ap stars. At yet higher effective temperatures, the He-rich stars are encountered, many of which have strong magnetic fields and are spectrum variables.

The CP phenomenon seems to terminate at the hottest He-rich stars (around 25 000 K) with the onset of substantial mass loss. Of course, stars with anomalous surface compositions are observed at higher effective tem-peratures and luminosities; for example the OBC/OBN stars and the Wolf–Rayet stars. However, unlike the CP stars considered here, they are thought to exhibit the products of nuclear burning at their surfaces – either via mass transfer in binaries, through mixing of material from the core, or as a result of extensive mass loss in the late stages of post–main-sequence evolution (see, *e.g.*, Willis 1991).

1.1. CLASSIFICATION AND SOME HISTORICAL REMARKS

λ Boo Stars. These comprise what was until relatively recently, a highly heterogeneous and ill-defined group. The spectrum of λ Boo itself was first described by Morgan, Keenan & Kellman (1943). Over two decades later, only seven objects had been designated as being like λ Boo in character (Sargent 1967), of which only three were recognized as forming a chemical-ly distinct group by Baschek & Searle (1969). The λ Boo stars were then neglected until the 1980s when the launch of *IUE* prompted the discovery of several unique characteristics in their ultraviolet spectra – notably strong C I lines and the presence of broad, unidentified absorption features at 1600 and 3040 Å. A precise working definition of λ Boo stars in the visual domain has now been proposed (Gray 1988, 1991): they are characterized by a weak Mg II λ4481 line; a Ca II K-line type of A0 or slightly later; a hydrogen-line type between A0 and F0; and a weak metallic-line spectrum.

Am–Fm Stars. The metallic-lined stars were first recognized as a distinct class by Titus & Morgan (1940), and were introduced into the MK classifica-tion system by Roman, Morgan & Eggen (1948). They can be recognized at classification dispersions by the fact that the Ca II K line is relatively weak, and the metallic lines relatively strong, as compared with the spectral type indicated by the Balmer lines. This leads naturally to a classification system based on three spectral criteria: for example, the 'classic' Am star 63 Tau is classified kA2hF0mF3. Conti's (1970) revision of the Am definition stipulat-ed that only one of these anomalies need be present, and thereby extended the class to include the so-called 'hot' Am stars, the best-known example of which is Sirius (α CMa). Some late A- and early F-type giants and sub-giants also exhibit disparate K-line and metal-line spectral types – these are the δ Del stars. Similarities with the surface abundances of Am stars have prompted the suggestion that the δ Del stars are evolved metallic-line stars.

Bp–Ap Stars. From the perspective of classification, this group compris-es an amalgamation of several spectroscopically defined peculiarity types: the λ4200-Si, Si, Si-Cr-Eu, Sr-Cr-Eu, and Sr stars (Jaschek & Jaschek 1958; Osawa 1965). Nowadays, in recognition of the fact that these groups actually form a continuous sequence in effective temperature, they tend to be identi-fied as the single generic magnetic Bp–Ap type, although it is still useful to distinguish between the cooler SrCrEu and hotter Si stars. It is interesting to note that the appellation *p* was, historically, applied to any star which did not fit into the Harvard sequence (to imply peculiar, of course). Even now, some authors use the Bp–Ap label for non-magnetic HgMn stars – a source of potential confusion to newcomers in the field! As suggested by the descriptive peculiarity types, Bp–Ap stars are identified at classification dispersions by the presence of strong lines of Si, Sr, Cr, and Eu. Possibly the most fascinating subset of objects within this group are the rapidly oscillating Ap (or roAp) stars which exhibit short-period ($\sim$few min), small-amplitude ($\sim$few mmag) oscillations (Kurtz & Martinez 1993). About 23 roAp stars are currently known; they are high-overtone ($n \gg l$), low-degree ($l \leq 3$) p-mode pulsators in which the oscillation axis is aligned with the magnetic axis, both of which are oblique to the rotation axis.

HgMn Stars. Manganese stars were first recognized as a distinct class by Morgan (1931) when he identified thirteen stars with spectra similar to the (former) prototype α And. Earlier, Lockyer & Baxendall (1906) had called attention to unidentified lines in this star at 3944, 3984, 4137, 4206, and 4282 Å; these were all attributed to manganese by Baxendall (1914), *except* λ3984, which remained a mystery for nearly fifty years. Its identification was announced by W.P. Bidelman in December 1961 at the 109th meeting of the American Astronomical Society, when he showed that λ3984 is due to singly-ionized mercury (Bidelman 1962a,b).* The presence of a strong Hg II λ3984 line and/or Mn II spectrum are now the defining characteristics of the HgMn class (*e.g.*, Jaschek & Jaschek 1987).

He-weak Stars. For many years it seemed that the CP phenomenon did not extend beyond the hot HgMn and magnetic Bp types until Bidelman described the peculiar spectrum of 3 Cen A in 1960. This B5 star is character-ized by weak He I and many sharp lines of P II, Ga II, and noble-gas elements. It was soon joined by the B4 star α Scl which also exhibits anomalously weak He I, but which has a metallic spectrum replete with lines of Cr II, Ti II, and even Sr II, the latter being unprecedented for a star of such early spectral type (Jugaku & Sargent 1961). These unusual objects heralded the discovery of the new, He-weak class of CP stars, recognizable at classification disper-sions by a He I spectrum which is too weak for the spectral type implied by the photometric colours or hydrogen lines. Numerous examples of these stars turned up in early classification studies of the Orion and Scorpius-Centaurus associations (Sharpless 1952; Garrison 1967). At least three sub-classes of He-weak stars are now recognized: (a) the P-Ga type, historically designated the "phosphorus stars", of which 3 Cen A is the prototype; (b) the Sr-Ti type or "α Scl stars"; and (c) the Si type or "blue helium-weak stars" (Jaschek & Jaschek 1974). To add confusion to an already complex morphology, there exists a small subset of He-weak stars exhibiting enhanced ^{3}He abundances. For a long time 3 Cen A was considered to be unique in this respect, with

---

* In retrospect, this minor historical episode offers a salutary lesson: Hg II was one of many 'exotic' species (such as Ga II, Pt II, Au II, and Bi II) not included in Charlotte Moore's *Multiplet Table* by reason of not being of astrophysical interest and therefore 'lost' to a generation of astronomers. Lines of all of these species have now been identified in the spectra of (some) HgMn stars.

a ^{3}He/^{4}He ratio of approximately 5 (Sargent & Jugaku 1961), although a whole cohort of such objects turned up in a careful spectroscopic search made by Hartoog & Cowley (1979). They found that the ^{3}He stars occupy a narrow strip in the ($T_{\rm eff}$, $\log g$) plane between the He-rich B stars, and He-weak B stars with normal He isotope ratios.

He-rich Stars. There are two distinct categories of main-sequence early-type stars with anomalously strong helium lines: the *extreme* He stars which are low mass, highly evolved objects in which helium is enriched throughout the star, and *intermediate* He stars with essentially main-sequence gravities and moderate helium enhancements confined to the surface layers. Only the intermediate helium stars are considered here. The first He-rich star of this type to be studied in detail was σ Ori E (Greenstein & Wallerstein 1958). Many have since been identified in the Orion OB1 association. As a group they can be distinguished from the other varieties of He-strong stars by the presence of essentially normal hydrogen lines and Population I space velocities. Only about 24 He-rich stars are currently known.

2. Physical characteristics

2.1. LOCATION IN H–R DIAGRAM

The distribution of CP stars in the Hertzsprung–Russell (H–R) diagram is illustrated in Fig. 1 using data for open clusters from North (1993). Note that the Bp–Ap class has been split into its Si and SrCrEu sub-types. The He-weak stars are difficult to distinguish from the Bp Si types so they are treated together for the purposes of this plot. Evolutionary tracks and the zero-age main sequence for solar-composition stars are superposed (Schaller *et al.* 1992). As can be seen, North's data show that essentially all CP stars lie on the main-sequence band (a few stars fall below the ZAMS because of observational errors). In fact most groups of CP stars fill this band, and North concludes that there is therefore no reason to suspect that the frequency of CP stars changes during this evolutionary phase. The wedge-shaped distribution of Am–Fm stars at the cool end of the H–R diagram is a selection effect due to the finite age of the clusters in which they were observed ($t \lesssim 10^9$ yr).

2.2. FREQUENCY OF INCIDENCE

Chemically peculiar stars comprise a significant fraction of the stellar population of the upper main sequence. This fact can be appreciated by looking at the frequency of incidence of CP stars over specific intervals of spectral type in magnitude-limited samples, as illustrated in Fig. 2 (*cf.* Schneider 1993). The samples used here were taken from the 5th Revised Edition of the *Bright Star Catalogue* (Hoffleit & Warren 1991) excluding all supergiants (luminosity classes I and II), and emission-line, shell, δ Del and δ Sct type

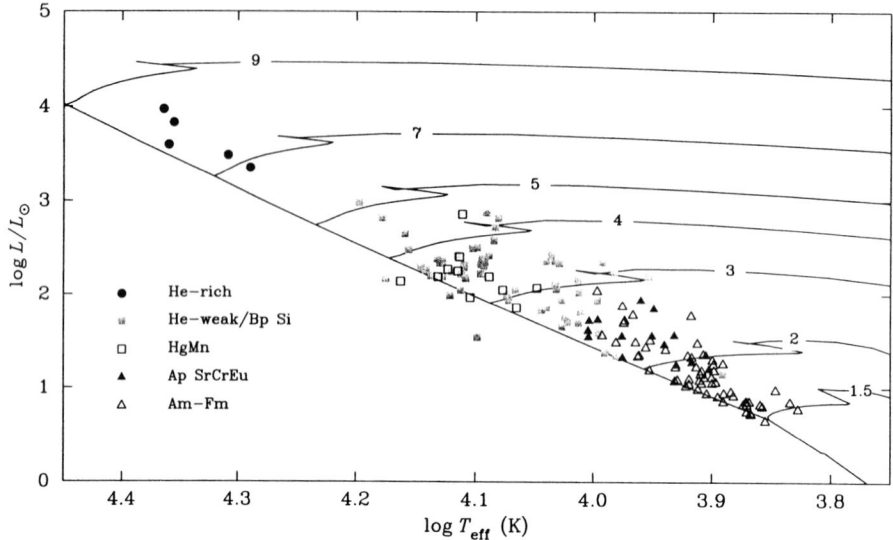

Fig. 1. The Hertzsprung–Russell diagram for several classes of chemically peculiar stars in open clusters. Magnetic classes are indicated by filled black symbols, mixed magnetic/non-magnetic by filled gray symbols, and non-magnetic by open symbols. Data from North (1993). Evolutionary tracks for solar-composition stars of several initial masses (as indicated) are superposed (Schaller *et al.* 1992).

stars; giants *are* included because many CP stars are designated luminosity class III due to abnormally weak He I lines. The fraction of CP stars of each peculiarity type relative to the total number of stars was determined in each spectral subinterval by reference to the *General Catalogue of Am and Ap Stars* (Renson, Gerbaldi & Catalano 1991). The λ Boo and He-rich stars are not considered here because they have rather low frequencies: although some 18% of field A stars have a weak Mg II λ4481 line characteristic of the λ Boo class (Abt & Morrel 1995), λ Boo stars proper comprise only 1% of the A star population (Gray 1988). One should also bear in mind that the spectral classifications in the BSC can be in error by a few sub-types, so the tails of the frequency distributions may be somewhat smeared, although the overall impression should be essentially unaffected. Note that the larger $V \leq 6^{m}\!.5$ sample (shaded) exhibits slightly lower frequencies than the $V \leq 6^{m}\!.0$ sample because of incompleteness related to detection difficulties in fainter stars. The summed sample demonstrates that CP phenomena are observed in more than 25 % of upper main-sequence stars over a fairly large range in spectral type, peaking at over 50 % near A6.

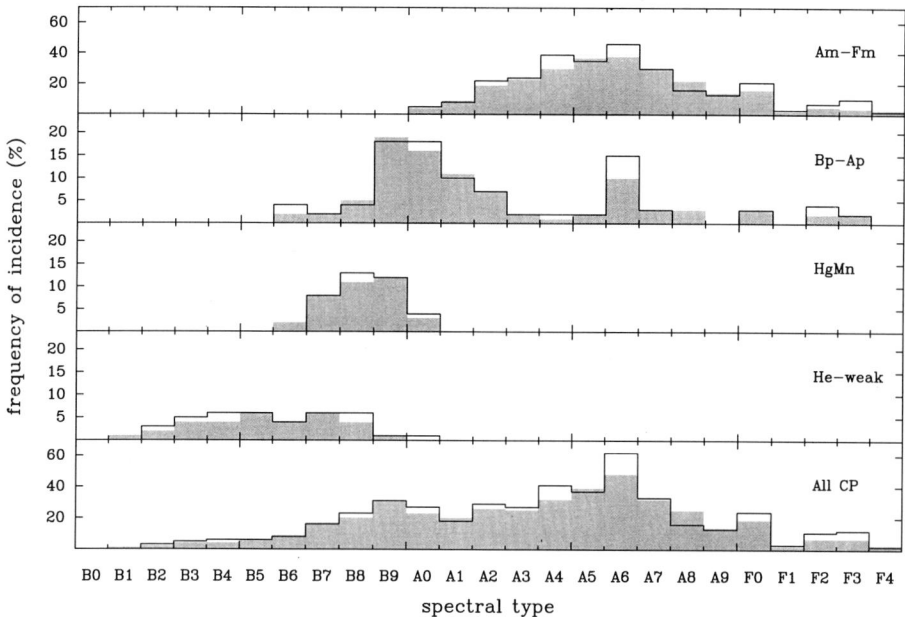

Fig. 2.   Frequency of incidence of chemically peculiar stars as a function of spectral type. The frequencies are calculated with respect to all (*i.e.*, normal + CP) stars in magnitude-limited samples of $V \leq 6\overset{m}{.}0$ (outline) and $V \leq 6\overset{m}{.}5$ (shaded) taken from the *Bright Star Catalogue*. The summed frequencies for all chemically peculiar types is shown in the lowest panel.

2.3. ROTATION

The rotational velocities of CP stars (excluding the λ Boo and He-rich types) are unusually low when compared with apparently normal stars of the same spectral type. This is illustrated in Fig. 3, which compares the observed rotational velocity distributions of Am, HgMn, and magnetic Bp stars with those of normal stars of similar spectral type, using data taken from Wolff & Wolff (1976) and Wolff & Preston (1978). The distribution of rotational velocities among magnetic Bp–Ap stars of different subclasses has been examined by Wolff (1983) who finds that the Si group includes many rapid rotators ($v \sin i \geq 100 \, \mathrm{km \, s^{-1}}$) whereas the SrCrEu group does not. There is some evidence for an overall decrease in $\langle v \sin i \rangle$ with decreasing mass in the Bp–Ap class as a whole. Although relatively few He-rich stars are known, their distribution of rotational velocities is thought to be consistent with that of normal B stars of similar spectral type. The rotational velocities of the λ Boo stars also appear to be comparable with normal A stars (Abt & Morrel 1995).

    A long-standing question concerns the rôle of rotation in inhibiting the

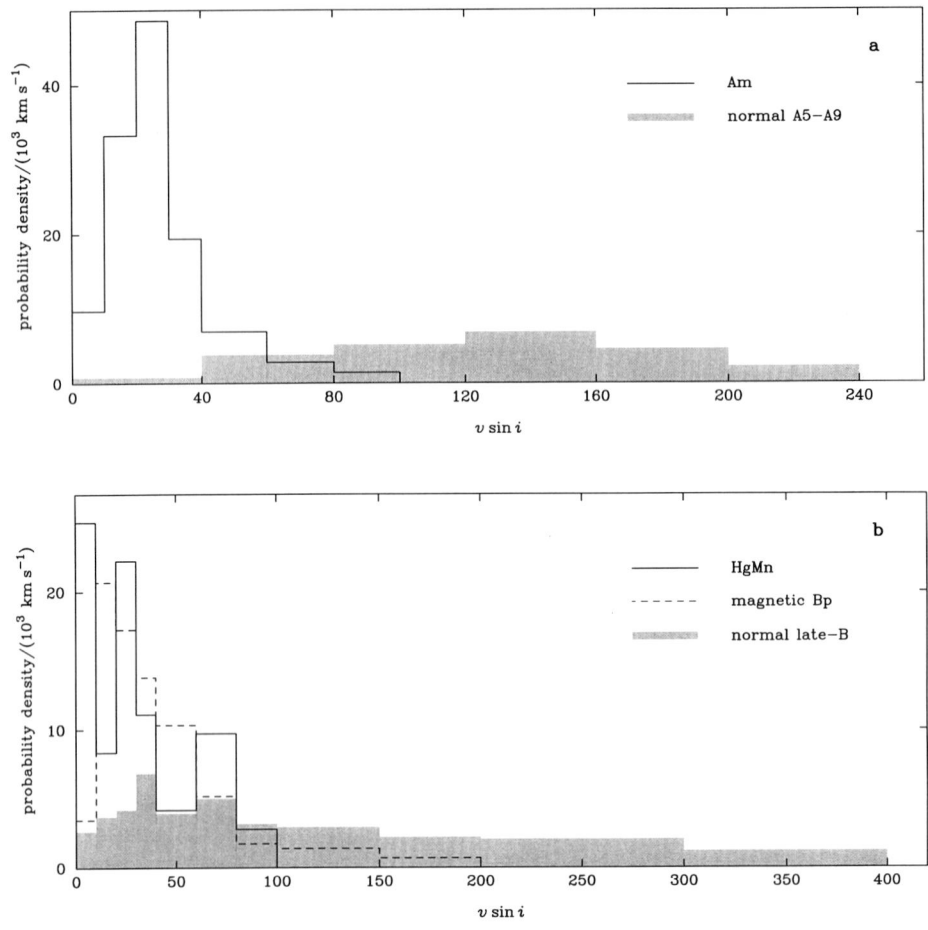

Fig. 3. Apparent rotational velocity distributions for **a** Am and normal A5-A9 stars and
b HgMn, magnetic Bp, and normal late-B stars. Data from Wolff & Wolff (1976) and
Wolff & Preston (1978).

development of abundance anomalies in CP stars. Notice in particular that
there is very little overlap of the Am and normal-A star distributions at
$v \sin i \leq 40 \, \mathrm{km\,s^{-1}}$ in Fig. 3a. This observation is reinforced when the
smearing effect of $\sin i$ on the *apparent* velocity distributions is removed
by assuming a random distribution of inclinations for the rotation axes and
using Lucy's (1974) iterative inversion procedure (Wolff & Wolff 1974). Thus
nearly all slowly rotating A stars are chemically peculiar. Unlike Abt &
Moyd (1973) before them, the Wolffs found that there is considerable overlap between the velocity distributions of Am and normal A stars at intermediate velocities; so is slow rotation a sufficient condition for the appearance
of Am characteristics? Very recently, Abt & Morrel (1995) made a defini-

tive contribution to this argument: their 'epic' survey of over 1700 A stars has shown that the velocity distributions of normal A and Am/Ap types exhibit a very small degree of overlap at intermediate velocities (between 7 and 10% depending on spectral type). They convincingly argue that this overlap can attributed to misclassification of some normal stars due to the use of abnormal MK standards such as α Lyr and α Dra, and they conclude that essentially all A stars with rotational velocities below 120 km s^{-1} are chemically peculiar.

The late-B stars present a somewhat different picture. Wolff & Preston's (1978) survey of rotational velocities in HgMn stars found no examples with $v \sin i \geq 100$ km s^{-1}, but did reveal an apparent surfeit of slowly rotating late-B stars ($v \sin i \leq 40$ km s^{-1}) which do *not* exhibit HgMn characteristics (see Fig. 3b). They concluded that slow rotation is a necessary but *insufficient* condition for the appearance of HgMn characteristics in a star. Wolff & Preston took special care to ensure a high detection efficiency for HgMn stars in their survey, but it is nonetheless interesting to note that numerous apparently normal, slowly rotating late-B stars observed at intermediate dispersions during the 1970s (Dworetsky 1976) have since been whittled down to nothing by studies at coudé dispersions (Cowley 1980; Cowley *et al.* 1982; Holweger, Gigas & Steffen 1986; Ramella *et al.* 1989; Lemke 1989, 1990).

A second issue concerns the influence of magnetic braking on the rotational velocities of magnetic Bp–Ap stars. The discussion has centered on the apparent decline in $\langle v \sin i \rangle$ of Ap stars in clusters of increasing age, a proposed explanation for which is spin-down during the main-sequence lifetime (Stift 1976; Wolff 1975; Abt 1979; Wolff 1981). Recent and more numerous data on the rotation periods of Ap stars in clusters show that such systematics can in fact be entirely ascribed to conservation of angular momentum as those stars evolve to larger radii on the main sequence (North 1984, 1987; Borra *et al.* 1985). Thus differences in the rotational velocity distributions of normal and Bp–Ap stars must be acquired during the pre–main-sequence evolution. Very little is known of the pre-MS evolution of progenitor CP stars, although some angular-momentum loss mechanisms have been discussed by Stępień (1994).

2.4. BINARITY

The statistics of chemically peculiar stars in spectroscopic binary (SB) systems have been evaluated in several studies (Abt & Snowden 1973; Aikman 1976; Gerbaldi, Floquet & Hauck 1985). This problem has been revisited recently by Seggewiss (1993) who used as principal data sources the *Eighth Catalogue of Orbital Elements of Spectroscopic Binary Systems* (Batten, Fletcher & MacCarthy 1989) and the *General Catalogue of Ap and Am stars* (Renson, Gerbaldi & Catalano 1991). His results are summarized in Fig. 4. The relative frequency of Am–Fm, Bp–Ap and HgMn stars in spec-

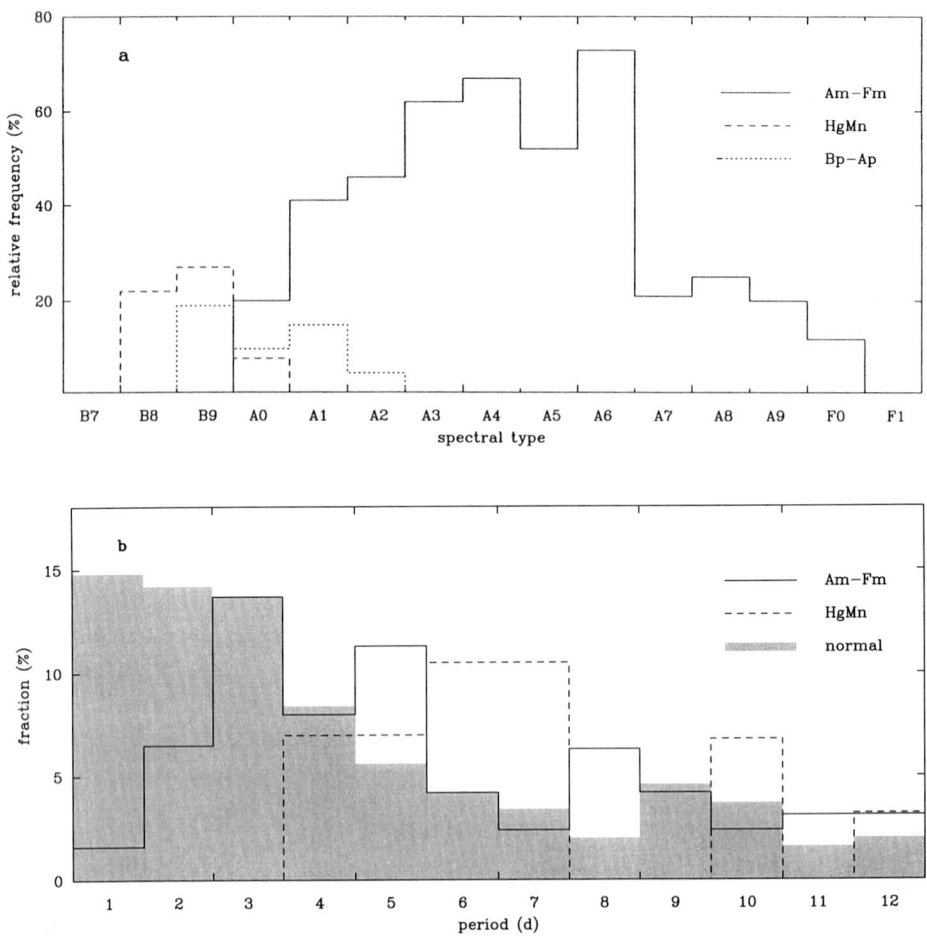

Fig. 4. **a** Relative frequency of chemically peculiar stars in spectroscopic binary systems as a function of spectral type. **b** Fraction of normal and chemically peculiar spectroscopic binaries as a function of period in bins of 1 day. Data from Seggewiss (1993).

troscopic binaries as a function of spectral type is shown in Fig. 4a. [Note that the relative frequencies are calculated with respect to the total number of stars (normal and CP) in spectroscopic binaries of a given spectral type.] Comparing these statistics with the frequency of CP stars in general (Fig. 2) leads to the conclusion that Am and HgMn stars may be somewhat over-represented in SB systems whereas Bp–Ap stars are under-represented.

Even more interesting is the fraction of normal and CP stars in SB systems of a given period, as illustrated in Fig. 4b. There is a dramatic difference in the frequency of long-period systems when comparing normal with Am, HgMn and Bp–Ap type stars. The statistics for systems with periods of more than 12 days (which are not plotted) are particularly instructive: about one third of Am binaries, one half of HgMn binaries, and two-thirds of Bp–Ap binaries have such long periods compared with only one quarter of normal binary systems. Short period systems ($P < 3$ days) are quite common among normal stars, but are non-existent in the HgMn and Bp–Ap types.

2.5. MAGNETIC FIELDS

The first evidence for a magnetic field in a star other than the Sun was reported more than forty years ago by Babcock (1947) who detected a strong and variable field in the Ap star 78 Vir. In fact it now seems likely that *all* magnetic stars on the upper main sequence are chemically peculiar since no fields have been detected in 'normal' A and B stars (Landstreet 1982). Some CP stars have sufficiently strong fields and sharp lines that magnetically split Zeeman components can be resolved (Mathys & Lanz 1992); however, the 'rule-of-thumb' condition that B in kG needs to be numerically larger than $v \sin i$ in km s^{-1} in order to resolve the components is rarely attained. So, such stars are scarce and more usually Zeeman polarimetry of the metal lines or one of the hydrogen Balmer lines is employed as a diagnostic of the mean longitudinal field strength (this approach exploits a wavelength displacement between σ components recorded in left and right circular polarization).

The longitudinal magnetic fields of magnetic CP stars frequently exhibit predominantly sinusoidal variations matched to the stellar rotation period. Their RMS amplitudes are typically a few hundred gauss, although a small but conspicuous fraction have fields of several kG and above. The magnetic variations are now understood in terms of the rigid rotator or oblique-dipole rotator model, first developed by Stibbs (1950). The basic premise of this model is that the stellar magnetic field, usually assumed dipolar, is frozen into the surface of the star with an axis of symmetry inclined at some angle (β) with respect to the rotation axis. As the star rotates, so does the aspect of the magnetic field with respect to the distant observer, and thus the apparent field intensity. This hypothesis explains quite naturally the phased magnetic field, line-strength, and radial-velocity variations observed in Bp–Ap stars with the assumption that the variable elements are concentrated in patches on the stellar surface.

In recent surveys, magnetic fields above a 100–400 G detection threshold have been discovered in about 50 % of Bp–Ap and He-weak stars (Borra & Landstreet 1980; Bohlender, Landstreet & Thompson 1993). It is interesting that magnetic fields (along with photometric and spectroscopic variability) are only observed among the Sr-Ti and Si subclasses of He-weak stars (Borra, Landstreet & Thompson 1983; Bohlender, Landstreet & Thompson 1993). Historically, weak but detectable magnetic fields have been reported for the Am and HgMn stars too (Babcock 1958), although later work has shown that, if present, these fields must be either less than a few hundred gauss, or complex in structure (Borra & Landstreet 1980; Landstreet 1982). The

λ Boo stars also appear to be non-magnetic (Bohlender & Landstreet 1990).

The possibility of magnetic fields with complex or disordered topology has been tackled recently using a statistical approach, borrowed from solar spectroscopy, which relies on the differential broadening of lines having different Zeeman sensitivities (Stenflo & Lindegren 1977). An application of that method shows that observations of the hot Am star o Peg are consistent with a $\sim 2\,\mathrm{kG}$ field (Mathys 1988). Independent confirmation of this result has been achieved using another indirect method which exploits the anomalous magnetic intensification of one component of a close pair of Fe II lines at $\lambda\lambda 6147.7$, 6149.2 (Mathys & Lanz 1990). The same technique indicates kilogauss fields in several other Am stars, but has proved inconclusive for the HgMn stars examined thus far (Lanz & Mathys 1993). The question of magnetic fields in HgMn stars has been re-opened recently by Mathys & Hubrig (1995) who found a $3.6\,\mathrm{kG}$ field in the HgMn spectroscopic binary 74 Aqu.

The He-rich stars include amongst them the strongest magnetic fields detected in non-degenerate stars (Landstreet & Borra 1978; Borra & Landstreet 1979). All of them appear to possess comparatively dense magnetospheres – magnetically controlled regions of trapped circumstellar matter – which are observed via Balmer emission line variations and gyrosynchrotron emission in the radio waveband, as well as anomalously strong and variable ultraviolet resonance lines (Shore 1993). Some He-weak stars also exhibit evidence for co-rotating magnetospheres, but such objects are very rare (Shore *et al.* 1988).

2.6. SPECTRUM VARIABILITY

Spectrum variations in Ap stars were first noted at the turn of the century (Ludendorff 1906). About one half of Bp–Ap stars are known to be spectrum variables (Bonsack 1974). The important characteristics of spectrum variations in Ap stars were established in a pioneering study by Deutsch (1947). The variations have periods from a fraction of a day upwards, and these periods are identical to the photometric and magnetic periods. The most conspicuously variable elements are those which are characteristically anomalous in Ap stars, including Sr, Cr, Eu, and Si. Not all elements vary in phase, but all lines of a given ion do vary together. The phase relationships between elements can vary from star to star.

These characteristics can be understood in the context of the oblique rotator model described in Sect. 2.5, wherein elements are distributed inhomogeneously across the stellar surface (*e.g.*, in rings or patches). Stellar rotation carries the abundance patches across the visible hemisphere leading to line-strength and radial velocity variations. Deutsch (1958, 1970) was the first to recognise that this information could be used to derive the abundance distribution of an element across the stellar surface. In recent years, enormous progress has been made in mapping stellar surface abundances using inversion procedures (*e.g.*, maximum entropy reconstruction) applied to time-series observations of line profiles (Wehlau & Rice 1993; Hatzes 1993).

3. Surface Abundances

The literature on abundance determinations for CP stars is vast, and would demand a book-length review to cover in any detail for all six groups considered here. During the 1970s, the abundance patterns of the Am–Fm and SrCrEu Ap groups were established in comprehensive surveys which have never been superseded for completeness (Smith 1971, 1973, 1974; Adelman 1973). Rather more up-to-date abundances for Am–Fm stars have been obtained in the Crimea by Savanov (1986, 1995). Adelman (1994) has summarised abundances for HgMn stars arising from a long-term programme of analyses based on photographic spectra obtained at the Dominion Astrophysical Observatory. Smith (1992) has carried out a comprehensive abundance survey of HgMn stars using ultraviolet spectra obtained with *IUE*.

An impression of the overall abundance patterns of these three groups of CP stars can be gained from Fig. 5, which plots abundances relative to solar values (Anders & Grevesse 1989) versus atomic number (following Preston 1974). Note that the ultraviolet results for the HgMn stars are distinguished by filled symbols; *IUE* has filled in some gaps in the periodic table, notably among deficient light elements. Some of the more conspicuously anomalous elements are labelled in each case.

It is still instructive to remind ourselves of some points made by Preston (1974). First, the abundance patterns of these three groups are quite distinct; the classification scheme is evidently successful in isolating chemically similar objects. Second, the abundance patterns contain considerable dispersions, which are larger for the Ap and HgMn groups than the Am–Fm. Third, the dispersions of some elements are larger than those of others; for example compare Sr and Y in the Ap stars.

3.1. λ Boo Stars

Our knowledge of the atmospheric compositions of λ Boo stars is essentially limited to a few elements which can be observed in the optical region at the relatively high rotational velocities that characterize the class ($v \sin i \gtrsim 100 \,\mathrm{km\,s^{-1}}$). Most early analyses focussed on the prototype λ Boo itself (Baschek & Searle 1969), and the broader applicability of those results to the class as a whole rested on similarities in their spectra. The general consensus from recent studies is that the light elements CNO and S are normal, or mildly deficient, whereas Mg, Ca, Ba, and most Fe-peak elements exhibit deficiencies up to 2 dex (Venn & Lambert 1990; Holweger & Stürenburg

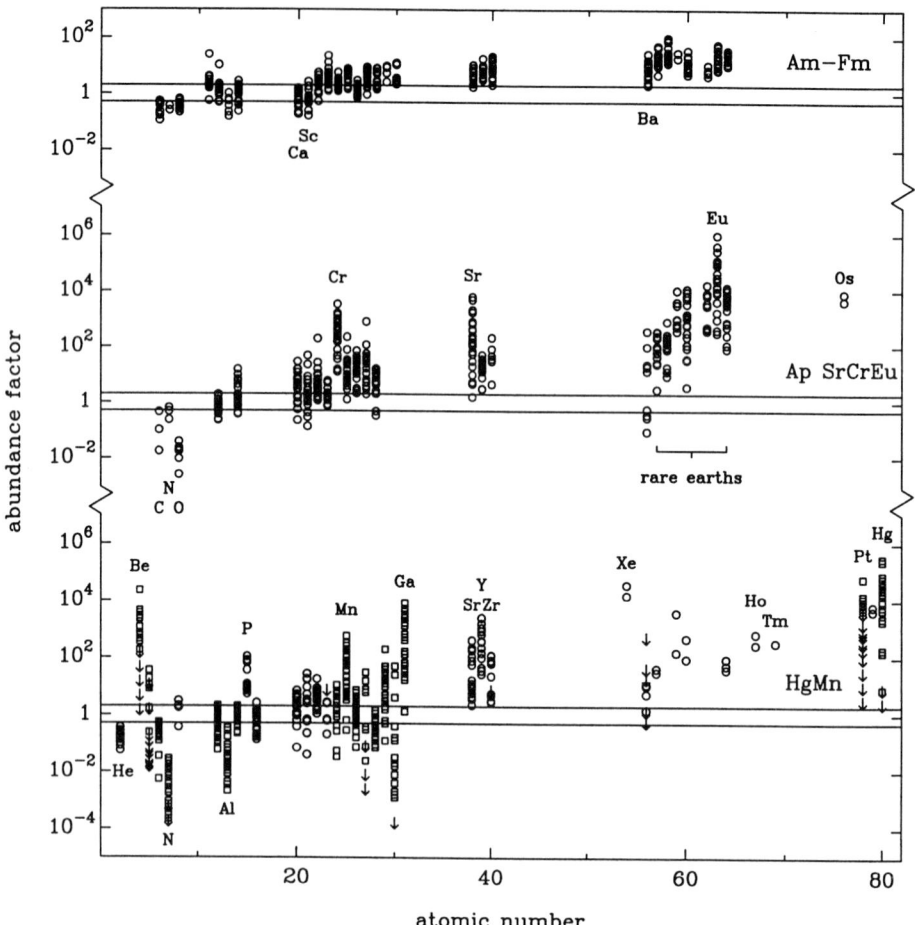

Fig. 5. Abundances of the elements (relative to the sun) for Am–Fm, cool Ap, and HgMn stars plotted versus atomic number. Pairs of horizontal lines denote a range of ±0.3 dex in which most normal stars lie. Circles denote abundances obtained from optical-region spectra; squares and arrows denote abundances and upper limits obtained from *IUE* spectra. See text for sources.

1991). The most comprehensive study to date, involving 15 λ Boo stars, shows that there is a continuous distribution in abundances in these objects, from large deficiencies in stars such as π^1 Ori and 29 Cyg, to near normal values (Stürenburg 1993).

Quantitative analysis of the relatively crowded and heavily blended UV spectra of λ Boo stars poses an even greater challenge than the optical data. To date few such studies have been carried out, although important steps in this direction have been made by Baschek & Slettebak (1988). Their analysis of *IUE* spectra of 10 λ Boo stars concentrated on the strongest spectral lines and thereby sidestepped uncertainties in the microturbulence parameters. The UV results largely confirm those obtained in the optical region, adding aluminium to a list of deficient elements in these stars.

The broad absorption feature at 1600 Å is proving a valuable tool for identifying λ Boo stars from low-dispersion *IUE* spectra (Faraggiana, Gerbaldi & Böhm 1990). Its identification remains obscure, however; C I autoionization features have been proposed by some authors, although this interpretation can be dismissed on the grounds that carbon is essentially normal in λ Boo stars (see Gerbaldi & Faraggiana 1993). It is particularly interesting that the 1600 Å feature is also observed in horizontal branch stars (Jaschek *et al.* 1985).

3.2. Am–Fm Stars

From the point of view of surface composition, Am–Fm stars are relatively innocuous compared with the extraordinary anomalies exhibited by their magnetic cousins. As one would expect from the class definition, calcium and scandium are typically deficient, by factors of 5 to 10 in most examples, but somewhat less for 'mild' Am stars, such as o Peg (*e.g.*, Guthrie 1987). Lithium is either normal or underabundant (Burkhart *et al.* 1987; Burkhart & Coupry 1988, 1989, 1991). The light elements CNO are markedly deficient in the coolest Am–Fm stars, but essentially normal in the hottest examples (Sadakane & Okyudo 1989; Roby & Lambert 1990). Zirconium is either overabundant or normal (Smith 1971, 1973). The iron-peak elements are in some cases overabundant by factors of 10 or more. Perhaps the most remarkable departures from solar composition in Am–Fm stars are to be found among the rare-earth elements which are quite typically overabundant by factors of 20, but still preserve the odd–even signature of nucleosynthetic systematics (Smith 1971; Maguzzu & Cowley 1986; van't Veer, Burkhart & Coupry 1988).

The advent of *IUE* has exposed Am–Fm stars to systematic surveys for heavy elements. Sadakane (1991) identified the Pb II λ2203 line in the UV spectrum of the hot Am star α CMa and derived a lead abundance some three orders of magnitude greater than the solar value! That star also has the distinction amongst the Am class of being the sole known member with a detectable Hg II line at λ1942, indicative of an overabundance of a factor of 20 according to Sadakane, Jugaku & Takada-Hidai (1988b). In this regard, α CMa clearly shares several properties characteristic of the HgMn class.

3.3. Bp–Ap Stars

The magnetic Bp–Ap stars exhibit perhaps the most complex abundance patterns observed among CP stars, and those patterns differ somewhat between the cool Ap (SrCrEu) and the hot Bp (Si) varieties. In some cool Ap stars, lithium has been found to be overabundant by one or two orders of magnitude, but the neutral lines used for analysis are not observed in hotter stars (Faraggiana *et al.* 1986). Unlike the HgMn stars, very few (if any) Bp–Ap types exhibit enhanced beryllium (Gerbaldi *et al.* 1986). The light elements CNO are generally deficient, to a greater extent perhaps than non-magnetic stars of similar effective temperature (Roby & Lambert 1990; Gerbaldi *et al.* 1989). Magnesium is essentially normal in the Bp–Ap types, whereas aluminium is typically deficient by an order of magnitude, with a few exceptional cases in which it appears to be normal (Sadakane, Takada & Jugaku 1983).

Silicon, which is overabundant in the hot Bp stars by definition of the class, has been studied quite extensively, both by assuming a homogeneous distribution of the element over the surface (Mégessier 1986), and by using considerably more involved surface mapping techniques which reveal ring-like distributions near the magnetic equator (Hatzes 1993). Calcium is generally normal or slightly underabundant and scandium is typically underabundant.

The iron-group elements are enhanced by quite large factors, but as Cowley (1993) has emphasized, the odd–even alternation of nucleosynthetic systematics is always preserved. Surface mapping studies show that iron can be concentrated near the magnetic pole, such as in 53 Cam and HD 215441 (Landstreet 1988; Landstreet *et al.* 1989), or primarily in the equatorial regions, such as in α^2 CVn (Khokhlova & Pavlova 1984). Whereas the horizontal gradients tend to be quite small for iron, both chromium and titanium exhibit profoundly inhomogeneous surface distributions. The titanium surface abundances appear to correlate with effective temperature and magnetic field strength according to Ryabchikova (1991), and can reach local maximum overabundances of some three orders of magnitude!

Gallium is overabundant in the hot Bp stars, although the enhancement factors are somewhat less than observed in the non-magnetic HgMn types (Takada-Hidai, Sadakane & Jugaku 1986). In the spectrum variable HD 25823 the gallium and silicon line strengths vary in anti-phase (Artru & Freire-Ferrero 1988). Strontium, enhanced by class definition in the cool Ap stars, is typically overabundant by factors of 10 to 300. Both Y and Zr are overabundant and correlate with effective temperature. Barium shows no differentiation between normal and Bp–Ap stars, although there is evidence that in both it reaches overabundances of a factor of 30 in stars hotter than 9500 K (Cowley 1976).

The cool roAp stars have been comparatively neglected in terms of abundance analyses, and the recent study by Kupka *et al.* (1994) aims to redress this with the objective of providing much-needed chemical constraints for pulsation models. Looking at three roAp stars, they find that the Fe-peak elements are normal or slightly underabundant, barium is normal (or possibly deficient), while the Sr-Y-Zr triad is overabundant. The rare-earths are enhanced, but at no point is there a violation of the odd–even systematics in their abundance pattern. In all cases examined so far, the abundance anomalies are significantly milder than observed in β CrB, a well-studied cool Ap star for which no rapid oscillations have yet been detected.

3.4. HgMn Stars

Mercury-manganese stars are characterized by a bizarre pattern of over-abundant elements including mercury and manganese (essentially by definition), phosphorus, and gallium (Heacox 1979; Guthrie 1984). Mercury is particularly unusual in that it exhibits star-to-star differences in its *isotopic* composition which are correlated with effective temperature: the cool HgMn stars (*e.g.*, χ Lup, HR 4072, and ι CrB) are dominated by the heaviest isotopes ^{202}Hg and ^{204}Hg which constitute only a small fraction of the terrestrial mixture (White *et al.* 1976). Platinum seems to exhibit analogous behaviour (Dworetsky & Vaughan 1973), but no quantitative studies have yet been published.

The launch of the *IUE* spawned a series of studies of HgMn stars devoted to surveying the prominent ultraviolet resonance lines of optically inconspicuous elements. The resonance lines are particularly powerful probes of photospheric abundances because they make observable elements which are cosmically rare or underabundant. Analyses of elements such as nitrogen, aluminium, and zinc have revealed surprisingly large deficiencies which had only been suspected from optical spectra (Sadakane, Takada & Jugaku 1983; Sadakane, Jugaku & Takada-Hidai 1988a; Smith & Dworetsky 1990). Copper was found to be overabundant in nearly all HgMn stars and joins a growing list of elements, including Ga, Al, and Mn, for which the new UV data exhibit the strong correlations with effective temperature characteristic of the HgMn class (Jacobs & Dworetsky 1981; Smith & Dworetsky 1993a; Smith 1993, 1994, 1995b).

Comprehensive analyses of the iron-group elements have painted a somewhat more complex picture: star-to-star diversity seems to be a real characteristic of the HgMn stars – not a manifestation of systematic errors between inconsistent analyses or the impact of a few pathological targets (Smith & Dworetsky 1993a). For example, iron exhibits a scatter of $\sim \pm 1$ dex which is uncorrelated with any stellar parameters (*e.g.*, T_{eff}, $\log g$, $v \sin i$). The behaviour of the light elements beryllium and boron has also proved puzzling: some HgMn stars are Be-rich, others are essentially normal; similarly, boron is overabundant in about 20 % of HgMn stars, normal in a handful, and deficient in the rest (Sadakane, Jugaku & Takada-Hidai 1985; Smith 1992). In neither case does there appear to be a single factor which distinguishes HgMn stars with overabundances from those without. However, hidden within the apparently chaotic diversity of abundances seen in some elements are star-to-star systematics revealed by correlation analysis (such as the triad Mg-Al-Cr) – these may point to a common underlying mechanism responsible for the surface 'chemistry' of these stars (Smith & Dworetsky 1993b).

Some of the most exciting work on HgMn stars has emerged in only the last few years, and is based on beautifully resolved UV spectra obtained by David Leckrone and colleagues using the Goddard High Resolution Spectrograph (GHRS) on board the *HST*. Leckrone's team have concentrated on the bright, ultra-sharp-lined HgMn star χ Lup, for which ground-based interferometric observations (White *et al.* 1976) have shown mercury to be concentrated almost entirely in its heaviest ^{204}Hg isotope (which comprises only 7 % of the terrestrial mixture!). During the early 1970s, a radiative diffusion model was proposed as an explanation for this bizarre phenomenon (Michaud, Reeves & Charland 1974). The central hypothesis of that model is that radiation pressure concentrates mercury in a very thin, high-lying layer of the atmosphere where falling electron density shifts the ionization balance from Hg^+ (which is optically opaque) in favour of Hg^{++} (which is essentially transparent). Within this 'ionization trap' a delicate balance between gravitational and radiative forces develops and drives a mass-dependent fractionation of the mercury isotopes, with the lighter isotopes rising into the Hg^{++} zone where they have been, hitherto, unobservable. The prediction that the light isotopes of mercury are 'hiding' as Hg^{++} has now been tested by Leckrone's team who obtained resolved spectra of two Hg III transitions in χ Lup (Leckrone *et al.* 1993; Wahlgren *et al.* 1995). Contrary to predictions of the theory, they found that the observed wavelengths of the Hg III features, which can be determined to within an accuracy of $\pm 2\,\mathrm{m\AA}$, are entirely consistent with the same isotopic mixture indicated by the Hg II lines.

The *HST*/GHRS data have also provided insights into the abundances of rare elements not readily observable at *IUE* resolutions: Ru, Cd, Ge, As, and Au are all overabundant in the atmosphere of χ Lup, in some cases by factors of more than 10^4. Leckrone has argued that there are superficial similarities between the heavy-element abundance pattern of this object and S stars, the atmospheres of which are replete with s-process elements, thereby re-opening the question of the rôle of nucleosynthetic processes in understanding CP stars in general.

3.5. He-weak Stars

As expected from morphological considerations, the He-weak stars represent a chemically heterogeneous group. Their helium deficiencies vary from factors of 2 to 15 (Norris 1971). A few isolated objects associated within the He-weak class have had considerable attention lavished upon them. The abundance pattern of the prototype 3 Cen A, for example, is well known following a series of detailed analyses carried during the 1960s (*e.g.*, Jugaku & Sargent 1963; Hardorp, Bidelman & Prölss 1968): it is characterized by gross excesses of P ($\sim$ 100), Kr ($\sim 10^3$), and Ga ($\sim 10^4$) coupled with deficiencies of S ($\sim$ 10) and He ($\sim$ 6); milder enhancements of Si, Mn and Fe together with moderate deficiencies of C, O, and Mg reinforce similarities with the HgMn types noted at classification dispersions. The P-Ga star ι Ori B exhibits a similar abundance pattern (Conti & Loonen 1970).

A reasonably complete sample of Sr-Ti type He-weaks was analysed by Vilhu, Tuominen & Boyarchuk (1976): they give mean abundances for 12 stars which show that C and Si are near normal, whereas Ti, Cr and Fe are enhanced by factors of 60, 4, and 6 respectively. Analyses of the ultraviolet resonance lines of light elements, observed in a selection of He-weak stars using *IUE*, demonstrate striking similarities with HgMn stars: Al is deficient by typically a factor of 10 (Sadakane, Takada & Jugaku 1983) and Ga is enhanced by up to a factor of 1000 (Takada-Hidai, Sadakane & Jugaku 1986). The He-weak star 20 Tau also exhibits detectable Be and B resonance lines (Sadakane & Jugaku 1981), albeit somewhat weaker than those observed in HgMn stars (Sadakane, Jugaku & Takada-Hidai 1985).

3.6. HE-RICH STARS

The He-rich stars are characterized by helium number fractions relative to hydrogen (ϵ_{He}) of between 0.3 and 10. In this respect they are actually quite distinct from the extreme helium stars, but appear to merge continuously with the normal B types. Their spectra are often time variable, but phase-averaged abundances tend to suggest approximately solar metallicities (Bohlender 1988). More detailed studies of stars such as σ Ori E reveal an abundance distribution across the surface comprising two, symmetric He-rich caps (radius 47°, $\epsilon_{He} = 0.8$) located at approximately 45° with respect to the intersection of the magnetic and rotational equators. The He-rich caps partially overlap metal-depleted caps centered at the magnetic poles (Hunger & Groote 1993).

Photospheric metal abundances are difficult to determine because the optical lines of interesting species, such as C II $\lambda 4267$, are weak and exhibit relatively small phase variations. The ultraviolet resonance lines of dominant species such as C IV and Si III observed using *IUE* are much more amenable to analysis, but demand the use of non-LTE techniques. Analysis of optical observations of σ Ori E data suggest near-solar metal abundances outside the polar caps, and (model-dependent) depletions of 6–10 within them (Hunger & Groote 1993).

The relationship between surface abundances and evolutionary state in He-rich stars is being explored in a new programme initiated by Zboril, Glagolevskij & North (1994); their preliminary results already support an age-dependent increase in helium abundances. The light elements CNO were found to be generally deficient, but more observations will be required before systematics in these species can be ascertained.

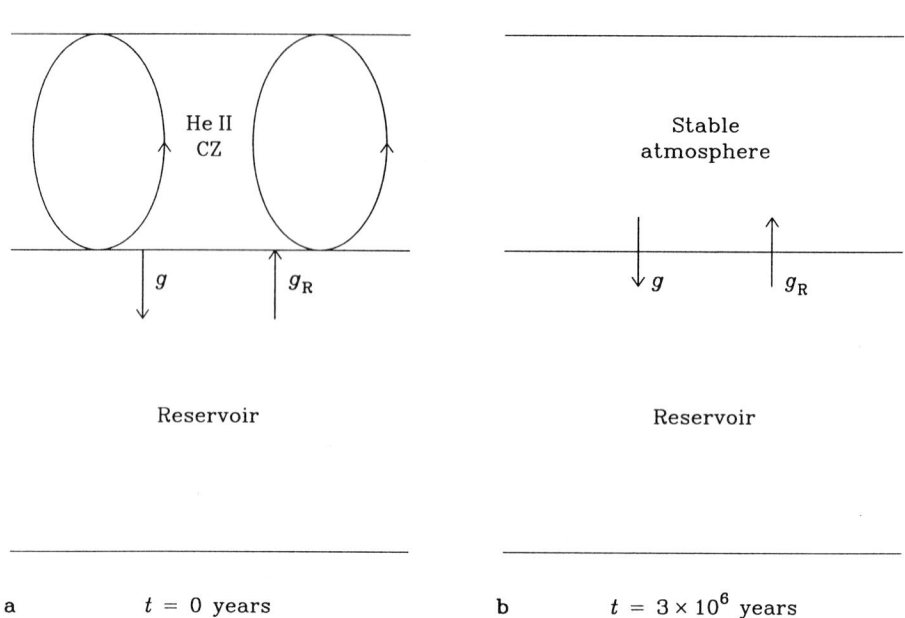

Fig. 6. Schematic diagram illustrating the parameter-free diffusion model for HgMn stars.
a A late-B star arrives on the main sequence with a helium abundance of 10 % by number
and has a superficial He II convection zone (CZ) which mixes the atmosphere. Helium sinks
through the bottom of the convection zone until its abundance has decreased by about
a factor of 3. b The He II convection zone then disappears and diffusion proceeds in the
stable atmosphere. After Michaud (1986).

4. The Radiative Diffusion Model

Since the early 1970s, an understanding of CP stars has been pursued under
the unifying umbrella of radiative diffusion theory, first advanced in this
context by Michaud (1970). The central hypothesis of the radiative diffusion
model is that sufficiently *quiescent* stellar atmospheres become chemically
differentiated via the microscopic migration of elements under the competing
influences of gravitational acceleration and radiation pressure. This simple
idea explains quite naturally why cool stars with convective envelopes and
hot stars with massive winds mark the extremes in effective temperature
between which the CP phenomenon is observed (their atmospheres are mixed
or undergo mass motion on timescales much shorter than that required for
diffusion to work its magic). Elaborate diffusion calculations are now capable
of reproducing – at least qualitatively – many of the observed chemical
characteristics of the major groups of CP stars, including the rôle of the
magnetic field topology in channelling the diffusion of charged species into
spots and rings on the stellar surface.

The diffusion model has arguably had most success when applied as a
description of HgMn stars which, uncomplicated by surface magnetic fields,

rotation or turbulence, are supposed to exhibit the end-products of diffusion processes in their 'purest form' (Vauclair & Vauclair 1982). Michaud (1981, 1986) has proposed a 'parameter-free' model for an idealized HgMn star which depends only on its mass and age (or equivalently on T_{eff} and $\log g$). The development of an HgMn star according to this model is illustrated schematically in Fig. 6; its key ingredient is that chemical differentiation of the atmosphere is triggered by the disappearance of a relatively deep He II convection zone after about 3×10^6 yr (which initially mixes the entire envelope via overshooting). That timescale agrees well with the age of the youngest open clusters in which HgMn stars are observed (Schneider 1993). Moreover, the He II convection zone can only disappear in stars with equatorial rotational velocities less than $\sim 90 \, \text{km s}^{-1}$ (Michaud 1982), which ties in nicely with the observed $v \sin i$ distribution of these objects (see §2.3).

Detailed radiative acceleration calculations have been carried out within the framework of the parameter-free model for several elements, and can be used to derive an indication of the maximum abundance that can supported by the radiation field in an atmosphere of given effective temperature and surface gravity. Large temperature-dependent overabundances are predicted for Mn and Ga (Alecian & Michaud 1981; Alecian & Artru 1987) whereas Mg and Si are expected to be supported at abundances within a factor of 2–10 of their solar values (Borsenberger, Michaud & Praderie 1984; Vauclair, Hardorp & Peterson 1979) – in good agreement with results obtained from UV resonance lines observed with *IUE* (see §3.4). The situation for Be is somewhat less satisfactory: the predicted large overabundances (factors up to 10^5) are observed in only $\sim 60\%$ of HgMn stars, the remainder exhibiting Be II resonance lines indistinguishable from normal stars. In the case of B, radiation forces are sufficient to drive the element very high into the atmosphere (unobservable at low optical depths) or out of the star entirely (Borsenberger, Michaud & Praderie 1979), and yet several HgMn stars exhibit anomalously high B abundances.

The diffusion model leads to qualitatively different solutions for the non-magnetic Am–Fm stars as compared with the hotter HgMn types because element separation occurs in the envelope, beneath a superficial H I convection zone. The convection zone carries any abundance anomalies which evolve at depth up into the photosphere where they are observed. A coherent and quantitatively accurate model for Am–Fm stars can be achieved by introducing mass loss in diffusion calculations (Michaud *et al.* 1983; Michaud & Charland 1986). The requisite mass loss rates are very low – typically less than $10^{-14} \, M_{\odot} \, \text{yr}^{-1}$ – but nonetheless have a dramatic influence on the surface abundances (Alecian 1993). Thus the diffusion model may yet prove a valuable, albeit indirect means of obtaining mass loss rates in stars where they are otherwise indeterminate. An interesting development in modelling diffusion in Am–Fm stars involves solution of the *time-dependent* problem:

Alecian (1993) has presented preliminary results which show how the surface abundance of Ca varies non-monotonically with time, passing through several phases in which it is underabundant.

For the magnetic Bp–Ap stars (and the magnetic He-weak and He-rich types too) the diffusion of charged particles is strongly influenced by the local geometry of magnetic field lines via Lorentz forces. This has several important consequences which help explain the anomalies unique to magnetic CP stars (see, *e.g.*, Mégessier 1986). Firstly, wherever the field lines are oblique to the stellar surface, diffusing ions will have both a vertical and horizontal velocity component; this can lead to the horizontal migration of some species and the development of patches and rings on the stellar surface correlated with the magnetic field geometry. Secondly, ions which are not radiatively supported can be impeded from sinking in regions where the magnetic field lines are horizontal. This occurs in the case of silicon which has a large radiative acceleration in its neutral state; horizontal field lines at the magnetic equator prevent the ionized states from sinking and overabundance factors of about 100 can thereby be supported in the atmosphere. Horizontal migration of silicon from the magnetic equator to the poles is then possible, and occurs on timescales shorter than the main-sequence lifetime of the star: magnetic Bp–Ap stars can change their stripes to spots! Other elements exhibit similar behaviour.

Diffusion calculations show that helium sinks in stable envelopes (Michaud *et al.* 1979), so how then does the diffusion model explain the existence of He-*rich* stars? The effects of mass loss coupled with a magnetic field have been identified as important factors for diffusion in hot CP stars (Vauclair 1975). The mechanism proposed by Shore (1978) posits a magnetic field which inhibits mass loss near the magnetic equator so that a balance between downward diffusion of helium and mass loss can be achieved. Observations of He-rich stars suggest that the situation is rather more complex than this simple model suggests: helium-enrichment actually occurs in caps near the intersection of the rotational and magnetic equators (see §3.6). This has prompted the suggestion that helium diffusion is influenced not only by the magnetic field but also by rotation, either via centrifugal forces, which promote the stellar wind, or via Coriolis forces, which act as angular momentum barriers that brake the wind (Hunger, Groote & Heber 1991).

Finally we should mention the λ Boo stars. Preliminary attempts to explain these objects in terms of the diffusion model required the supposition of somewhat larger mass-loss rates than for Am–Fm stars (*i.e.*, $10^{-13} M_\odot \, \mathrm{yr}^{-1}$) and indicated that the required abundance patterns would only develop after about 10^9 years (Michaud & Charland 1986); the very large light-element deficiencies derived by Venn & Lambert (1990) could not be reproduced, however. Moreover, meridional circulation currents which accompany rapid rotation in λ Boo stars are now known to disrupt the development of abundance anomalies deep beneath the convection zones of these stars (Charbonneau 1992). An alternative hypothesis has been proposed by Venn & Lambert (1990) which invokes the accretion of metal-depleted circumstellar gas onto the surface of λ Boo stars – this gas having first separated from grains which carry away refractory species. That mechanism has been used in calculations by Charbonneau (1991) and, coupled with diffusion, is capable of delivering a λ Boo abundance pattern within 10^5–10^6 yr. Unfortunately, there are some 'flies in the ointment': firstly, the accretion rates require rather delicate fine tuning, and secondly the circumstellar environment of A stars is bathed by a UV radiation field of sufficient intensity as to induce charge build-up on grains, thereby bringing into question the gas–dust separation. It remains to be seen whether these obstacles can be overcome.

5. Concluding Remarks and Prospectus

The last two decades have seen a rejuvenation in the study of chemically peculiar stars on the upper main sequence, with improving access to their ultraviolet spectra afforded by a series of space-based observatories culminating in the *HST*. As a consequence, our picture of the compositions of hot stars is much more complete. In particular, the ultraviolet window has promoted a more balanced view of the CP phenomenon by elevating the underabundant elements to the same status as the more conspicuous overabundant ones. The broad, cross-sectional surveys undertaken using the *IUE* will remain a lasting (and probably unique) legacy of the *IUE* mission. They are now being complemented by detailed studies of individual objects with the *HST*, which are adding fine detail to the broader picture.

For the future, much remains to be done in modelling the complex atmospheres of CP stars: in particular the inclusion of magnetic fields (Landstreet 1992), vertical and horizontal abundance gradients (Babel 1994; Smith 1995a), and non-LTE physics (Hubeny & Lanz 1993). Diffusion calculations, long hampered by inadequate and sparse atomic data, are now seeing the benefit of the large databases such as Kurucz's semi-empirical data for iron-group elements (Kurucz 1990), and the Opacity Project's TOPBASE (Cunto et al. 1993).

What problems remained to be solved? For the author, the compositional diversity observed among stars with essentially identical atmospheric parameters has always been puzzling, especially in the HgMn group where diffusion is unmoderated by the action of magnetic fields or surficial convection zones. How can such objects be accommodated within the parameter-free model? A possible solution to this paradox lies in the time-dependent nature of the diffusion process, which is currently being explored for iron-peak elements by Mike Seaton (this volume). Most HgMn stars analysed to date comprise a mixed population of different origins, ages, and evolutionary states: if diffusion can drive gross changes in the surface compositions of stars on relatively short timescales, then we may have a natural explanation for their apparent diversity. Observations of young CP stars in open clusters and associations are now needed to test that hypothesis, and to probe the time-dependent nature of the diffusion process in general.

Acknowledgements

I am pleased to thank my colleagues at UCL, Ian Howarth and Mike Dworetsky for their constructive comments on the manuscript, and Mike Seaton for many stimulating discussions on diffusion processes. Pierre North (University of Lausanne) and Igor Savanov (Crimean Astrophysical Observatory) generously provided data used in some of the figures. Financial support from PPARC is gratefully acknowledged.

References

Adelman, S.J.: 1973, *ApJ* **183**, 95
Adelman, S.J.: 1994, *MNRAS* **266**, 97
Abt, H.A.: 1979, *ApJ* **230**, 485
Abt, H.A. & Morrel, N.I.: 1995, *ApJS* **99**, 135
Abt, H.A. & Moyd, K.I.: 1973, *ApJ* **182**, 809
Abt, H.A. & Snowden, M.S.: 1973, *ApJS* **25**, 137
Aikman, G.C.L.: 1976, *Publs DAO* **14**, 379
Alecian, G.: 1993, 'Diffusion: Theoretical computations for non-magnetic stars' in M.M. Dworetsky, F. Castelli, R. Faraggiana, ed(s)., *Peculiar versus Normal Phenomena in A-type and Related Stars, IAU Coll. No. 138*, ASP: San Francisco, 450
Alecian, G. & Artru, M. C.: 1987, *A&A* **186**, 223
Alecian, G. & Michaud, G.: 1981, *ApJ* **245**, 226
Anders, E. & Grevesse, N.: 1989, *Geochim. Cosmochim. Acta* **53**, 197
Artru, M.-C. & Freire-Ferrero, R.: 1988, *A&A* **203**, 111
Babcock, H.W.: 1947, *ApJ* **105**, 105
Babcock, H.W.: 1958, *ApJS* **3**, 141
Babel, J.: 1994, *A&A* **283**, 189
Baschek, B. & Searle, L.: 1969, *ApJ* **155**, 537
Baschek, B. & Slettebak, A.: 1988, *A&A* **207**, 112
Batten, A.H., Fletcher, J.M. & MacCarthy, D.G.: 1989, *Publs DAO* **17**, 1
Baxendall, F.E.: 1914, *MNRAS* **74**, 250
Bidelman, W.P.: 1960, *PASP* **72**, 24
Bidelman, W.P.: 1962a, *AJ* **67**, 111
Bidelman, W.P.: 1962b, *Sky & Telescope* **23**, 140
Bohlender, D.A.: 1988, *PhD Thesis*, University of Western Ontario
Bohlender, D.A. & Landstreet, J.D: 1990, *MNRAS* **247**, 606
Bohlender, D.A., Landstreet, J.D. & Thompson, I.B.: 1993, *A&A* **269**, 355
Borra, E.F., Beaulieu, A., Brousseau, D. & Shelton, I.: 1985, *A&A* **149**, 266
Borra, E.F. & Landstreet, J.D.: 1979, *ApJ* **228**, 809
Borra, E.F. & Landstreet, J.D.: 1980, *ApJS* **42**, 421
Borra, E.F., Landstreet, J.D. & Thompson, I.B.: 1983, *ApJS* **53**, 151
Borsenberger, J., Michaud, G. & Praderie, F.: 1979, *A&A* **76**, 287
Borsenberger, J., Michaud, G. & Praderie, F.: 1984, *A&A* **139**, 147

Burkhart, C., Coupry, M.F., Lunle, M. & van't Veer, C.: 1987, A&A **172**, 257
Burkhart, C. & Coupry, M.F.: 1988, A&A **200**, 175
Burkhart, C. & Coupry, M.F.: 1989, A&A **220**, 197
Burkhart, C. & Coupry, M.F.: 1991, A&A **249**, 205
Bonsack, W.K.: 1974, PASP **86**, 408
Charbonneau, P.: 1991, ApJ **372**, L33
Charbonneau, P.: 1992, A&A **259**, 134
Conti, P.S.: 1970, PASP **82**, 781
Conti, P.S. & Loonen, J.P.: 1970, A&A **8**, 197
Cowley, C.R.: 1976, ApJS **32**, 631
Cowley, C.R.: 1980, PASP **92**, 159
Cowley, C.R.: 1993, 'Abundances in magnetic Ap stars' in M.M. Dworetsky, F. Castelli, R. Faraggiana, ed(s)., *Peculiar versus Normal Phenomena in A-type and Related Stars, IAU Coll. No. 138*, ASP: San Francisco, 18
Cowley, C.R., Dworetsky, M.M. & Mégessier, C., ed(s).: 1986, *Upper Main Sequence Stars with Anomalous Abundances, IAU Coll. No. 90*, D. Reidel: Dordercht
Cowley, C.R., Sears, R.L., Aikman, G.C.L. & Sadakane, K.: 1982, ApJ **254**, 191
Cunto, W., Mendoza, C., Ochsenbein, F. & Zeippen, C.J.: 1993, A&A **275**, L5
Deutsch, A.J.: 1947, ApJ **105**, 283
Deutsch, A.J.: 1958, in B. Lehnert, *Electromagnetic Phenomena in Cosmical Physics, IAU Symp. No. 6,* Cambridge University Press: Cambridge, 209.
Deutsch, A.J.: 1970, ApJ **159**, 985
Dworetsky, M. M.: 1976, 'A high dispersion search for peculiar A and B stars' in W.W. Weiss, H. Jenkner & H.J. Wood, ed(s)., *Physics of Ap Stars, IAU Coll. No. 32*, Universitätssternwarte Wien mit Figl-Observatorium für Astrophysik: Vienna, 549
Dworetsky, M.M., Castelli, F. & Faraggiana, R., ed(s).: 1993, *Peculiar Versus Normal Phenomena in A-Type and Related Stars, IAU Coll. No. 138*, Astronomical Society of the Pacific: San Francisco
Dworetsky, M.M. & Vaughan, A.H.: 1973, ApJ **181**, 911
Faraggianna, R., Gerbaldi, M., Castelli, F. & Floquet, M.: 1986, A&A **158**, 200
Faraggianna, R., Gerbaldi, M. & Böhm, C.: 1990, A&A **235**, 311
Garrison, R.F.: 1967, ApJ **147**, 1003
Gerbaldi, M. & Faraggiana, R.: 1993, 'The λ Boo and related stars' in M.M. Dworetsky, F. Castelli, R. Faraggiana, ed(s)., *Peculiar versus Normal Phenomena in A-type and Related Stars, IAU Coll. No. 138*, ASP: San Francisco, 368
Gerbaldi, M., Floquet, M. & Hauck, B.: 1985, A&A **146**, 341
Gerbaldi, M., Faraggiana, R., Castelli, F. & Floquet, M.: 1986, 'Be abundance in four cool CP stars' in E.J. Rolfe, ed(s)., *New Insights in Astrophysics, ESA SP-263*, ESA Publ.: Noordwijk, 49
Gerbaldi, M., Floquet, M., Faraggiana, R. & van't Veer-Menneret, C.: 1989, A&AS **81**, 127
Gray, R.O.: 1988, AJ **95**, 220
Gray, R.O.: 1991, in A.G. Davis Philip, A.R. Upgren & K.A. Janes, ed(s)., *Precision Photometry*, L. Davis Press, 309.
Greenstein, J.L. & Wallerstein, G.: 1958, ApJ **127**, 237
Guthrie, B.N.G.: 1984, MNRAS **206**, 85
Guthrie, B.N.G.: 1987, MNRAS **226**, 361
Hartoog, M.R. & Cowley, A.P.: 1979, ApJ **228**, 229
Hardorp, J., Bidelman, W.P. & Prölss: 1968, ZfA **69**, 429
Hatzes, A.P.: 1993, 'Mapping the surface distribution of elements on Ap stars using the maximum entropy method' in M.M. Dworetsky, F. Castelli, R. Faraggiana, ed(s)., *Peculiar versus Normal Phenomena in A-type and Related Stars, IAU Coll. No. 138*, ASP: San Francisco, 258
Heacox, W.D.: 1979, ApJS **41**, 675
Hoffleit, D. & Warren, W.H.: 1991, *The Bright Star Catalogue (Provisional 5th Revised*

Edition), NSSDC Astronomical Data Center: USA

Holweger, H., Gigas, D. & Steffen, M.: 1986, *A&A* **155**, 58

Holweger, H. & Stürenburg, S.: 1991, *A&A* **252**, 255

Hubeny, I. & Lanz, T.: 1993, 'Modelling A-type atmospheres: NLTE models' in M.M. Dworetsky, F. Castelli, R. Faraggiana, ed(s)., *Peculiar versus Normal Phenomena in A-type and Related Stars, IAU Coll. No. 138*, ASP: San Francisco, 98

Hunger, K., Groote, D. & Heber, U.: 1991, 'Intermediate He rich stars' in Michaud, G. & Tutukov, A., ed(s)., *Evolution of Stars: The Photospheric Abundance Connection, IAU Symp. No. 145*, Kluwer Academic: Dordrecht, 173

Hunger, K. & Groote, D.: 1993, 'He-abnormal stars' in M.M. Dworetsky, F. Castelli, R. Faraggiana, ed(s)., *Peculiar versus Normal Phenomena in A-type and Related Stars, IAU Coll. No. 138*, ASP: San Francisco, 394

Jacobs, J.M. & Dworetsky, M.M.: 1981, 'Recent results from an *IUE* survey of mercury-manganese stars' in P. Renson, ed(s)., *Upper Main Sequence Chemically Peculiar Stars, 23rd Liège Astrophys. Coll.*, Institut d'Astrophysique: Université de Liège, 153

Jaschek, M. & Jaschek, C.: 1958, *ZfA* **45**, 35

Jaschek, M. & Jaschek, C.: 1974, *ARA&A* **16**, 131

Jaschek, C. & Jaschek, M.: 1987, *Classification of Stars*, Cambridge University Press: Cambridge

Jaschek, M. Baschek, B., Jaschek, C. & Heck, A.: 1985, *A&A* **152**, 439

Jugaku, J. & Sargent, W.L.W.: 1961, *PASP* **73**, 249

Jugaku, J. & Sargent, W.L.W.: 1963, *ApJ* **138**, 90

Khokhlova, V.L. & Pavlova, V.M.: 1984, *Pis'ma Astron. Zh.* **10**, 377

Kupka, F., Ryabchikova, T., Bolgova, G., Kuschnig, R., Wiess, W.W., Mathys, G. & Le Contel, J.M.: 1994, 'Properties and position of CP2 stars on the HR diagram' in J. Zverko & J. Žižňovský, ed(s)., *Chemically Peculiar and Magnetic Stars on and Close to the Upper Main Sequence*, Astronomical Institute of Slovak Academy of Sciences: Tatranska Lomnica, 131

Kurtz, D.W. & Martinez, P.: 1993, 'Determination of luminosity, atmospheric structure, and magnetic geometry from studies of the pulsation in roAp stars' in M.M. Dworetsky, F. Castelli, R. Faraggiana, ed(s)., *Peculiar versus Normal Phenomena in A-type and Related Stars, IAU Coll. No. 138*, ASP: San Francisco, 561

Kurucz, R.L.: 1990, 'Semiempirical calculations of gf values for the iron group' in D. McNally, ed(s)., *Transactions of the IAU, Vol. 10B*, Kluwer: Dordrecht, 168

Landstreet, J.D.: 1982, *ApJ* **258**, 639

Landstreet, J.D.: 1988, *ApJ* **326**, 967

Landstreet, J.D.: 1992, *A&AR* **4**, 35

Landstreet, J.D., Barker, P.K, Bohlender, D.A. & Jewison, M.S.: 1989, *ApJ* **344**, 876

Landstreet, J.D. & Borra, E.F.: 1978, *ApJ* **224**, L5

Lanz, T. & Mathys, G.: 1993, *A&A* **280**, 486

Leckrone, D.S., Wahlgren, G.M., Johansson, S.G. & Adelman, S.J.: 1993, 'Exploring abundance and isotope anomalies in CP stars with the *HST*/GHRS: High resolution spectroscopy of χ Lup' in M.M. Dworetsky, F. Castelli, R. Faraggiana, ed(s)., *Peculiar versus Normal Phenomena in A-type and Related Stars, IAU Coll. No. 138*, ASP: San Francisco, 42

Lemke, M.: 1989, *A&A* **225**, 125

Lemke, M.: 1990, *A&A* **240**, 331

Lockyer, N. & Baxendall, F.E.: 1906, *Proc. Roy. Soc. London* **77**, 551

Lucy, L.B.: 1974, *AJ* **79**, 745

Ludendorff, H.: 1906, *Astron. Nach.* **173**, 1

Maguzzu, A. & Cowley, C.R.: 1986, *ApJ* **308**, 254

Mathys, G.: 1988, 'Stellar magnetism as a source of complication for elemental abundance determinations: a search for a magnetic field in *o* Peg' in S.J. Adelman & T. Lanz, ed(s)., *Elemental Abundance Analyses*, Institut d'Astronomie de l'Université de Lausanne: Lausanne, 101

Mathys, G. & Lanz, T.: 1990, *A&A* **230**, L21
Mathys, G. & Lanz, T.: 1992, *A&A* **256**, 169
Mathys, G. & Hubrig, S.: 1995, *A&A* **293**, 810
Mégessier, C.: 1986, 'Intermediate peculiar stars: The Bp–Ap Si Stars' in C.R. Cowley, M.M. Dworetsky & C. Mégessier, ed(s)., *Upper Main Sequence Stars with Anomalous Abundances, IAU Coll. No. 90*, D. Reidel: Dordercht, 275
Michaud, G.: 1970, *ApJ* **160**, 641
Michaud, G.: 1981, 'A parameter free model for HgMn stars' in Renson, P., ed(s)., Upper Main Sequence Chemically Peculiar Stars, 23rd Liège Astrophys. Coll., Institut d'Astrophysique: Université de Liège, 355
Michaud, G.: 1982, *ApJ* **258**, 349
Michaud, G.: 1986, 'Particle transport in non-magnetic stars' in Cowley, C. R., Dworetsky, M. M. & Mégessier, C., ed(s)., *Upper Main Sequence Stars with Anomalous Abundances, IAU Coll. No. 90*, D. Reidel: Dordrecht, 459
Michaud, G. & Charland, Y.: 1986, *ApJ* **311**, 326
Michaud, G., Montmerle, T., Cox, A.N., Magee, N.H., Hodson, S.W. & Martel, A.: 1979, *ApJ* **234**, 206
Michaud, G., Reeves, H. & Charland, Y.: 1974, *A&A* **37**, 313
Michaud, G., Tarasick, D., Charland, Y. & Pelletier, C.: 1983, *ApJ* **269**, 239
Michaud, G. & Tutukov, A., ed(s).: 1991, *Evolution of Stars: The Photospheric Abundance Connection, IAU Symp. No. 145*, Kluwer Academic: Dordrecht
Morgan, W.W.: 1931, *ApJ* **73**, 104
Morgan, W.W., Keenan, P.C. & Kellman, E.: 1943, *An Atlas of Stellar Spectra*, University of Chicago Press: Chicago
Norris, J.: 1971, *ApJS* **23**, 213
North, P.: 1984, *A&A* **141**, 328
North, P.: 1987, *A&AS* **69**, 371
North, P.: 1993, 'Chemically peculiar stars in clusters: upper and lower age limits of CP stars' in M.M. Dworetsky, F. Castelli, R. Faraggiana, ed(s)., *Peculiar versus Normal Phenomena in A-type and Related Stars, IAU Coll. No. 138*, ASP: San Francisco, 577
Osawa, K.: 1965, *Ann. Tokyo Astron. Obs. (Ser. 2)* **9**, 123
Preston, G.W.: 1974, *ARA&A* **12**, 257
Ramella, M., Gerbaldi, M., Faraggianna, R. & Böhm, C.: 1989, *A&A* **209**, 233
Renson, P., Gerbaldi, M. & Catalano, F.A.: 1991, *A&AS* **89**, 429
Roby, S.W. & Lambert, D.L.: 1990, *ApJS* **73**, 67
Roman, N., Morgan, W.W. & Eggen, O.J.: 1948, *ApJ* **107**, 107
Ryabchikova, T.A.: 1991, 'Magnetic Ap stars: evolutionary status and abundance anomalies' in G. Michaud & A. Tutukov, ed(s)., *Evolution of Stars: The Photospheric Abundance Connection, IAU Symp. No. 145*, Kluwer Academic: Dordrecht, 149
Sadakane, K.: 1991, *PASP* **103**, 355
Sadakane, K. & Okyudo, M.: 1989, *PASP* **41**, 1055
Sadakane, K. & Jugaku, J.: 1981, *PASP* **93**, 60
Sadakane, K., Jugaku, J. & Takada-Hidai, M.: 1985, *ApJ* **297**, 240
Sadakane, K., Jugaku, J. & Takada-Hidai, M.: 1988a, *ApJ* **325**, 776
Sadakane, K., Jugaku, J. & Takada-Hidai, M.: 1988b, *PASP* **100**, 811
Sadakane, K., Takada, M. & Jugaku, J.: 1983, *ApJ* **274**, 261
Sargent, W.L.W.: 1967, in R.C. Cameron, ed(s)., *The Magnetic and Related Stars*, Mono Book Corp.: Baltimore, 329.
Sargent, W.L.W. & Jugaku, J.: 1961, *ApJ* **134**, 777
Savanov, I.S.: 1986, *PhD Thesis*, Crimean Astrophysical Observatory
Savanov, I.S.: 1995, *Astron. Zh.* , in press
Schaller, G., Schaerer, D., Meynet, G. & Maeder, A.: 1992, *A&AS* **96**, 269
Schneider, H.: 1993, 'Statistics of CP stars in a magnitude-limited sample: The Bright Star Catalogue' in M.M. Dworetsky, F. Castelli, R. Faraggiana, ed(s)., *Peculiar versus Normal Phenomena in A-type and Related Stars, IAU Coll. No. 138*, ASP: San

Francisco, 629

Seggewiss, W.: 1993, 'Chemically peculiar stars among spectroscopic binaries revisited' in M.M. Dworetsky, F. Castelli, R. Faraggiana, ed(s)., *Peculiar versus Normal Phenomena in A-type and Related Stars, IAU Coll. No. 138*, ASP: San Francisco, 137

Sharpless, S.: 1952, *ApJ* **116**, 251

Shore, W.: 1978, *PhD Thesis*, University of Toronto

Shore, W.: 1993, 'The magnetospheres of the helium peculiar stars of the upper main sequence' in M.M. Dworetsky, F. Castelli, R. Faraggiana, ed(s)., *Peculiar versus Normal Phenomena in A-type and Related Stars, IAU Coll. No. 138*, ASP: San Francisco, 528

Shore, S.N., Brown, D.N. & Sonneborn, G.: 1988, 'The discovery of a co-rotating magnetosphere in a helium weak star: HD 5737 = alpha Scl' in E.J. Rolfe, ed(s)., *A decade of UV astronomy with the* IUE *satellite, ESA SP-281, vol. 1*, ESA Publ.: Noordwijk, 339

Smith, K.C.: 1992, *PhD Thesis*, University of London

Smith, K.C.: 1993, *A&A* **276**, 393

Smith, K.C.: 1994, *A&A* **291**, 521

Smith, K.C.: 1995a, *A&A* **297**, 237

Smith, K.C.: 1995b, *A&A* , in press

Smith, K.C. & Dworetsky, M.M.: 1990, 'Carbon and nitrogen abundances in HgMn stars' in E.J. Rolfe, ed(s)., *Evolution in Astrophysics:* IUE *Astronomy in the Era of New Space Missions*, ESA Publ.: Noordwijk, 279

Smith, K.C. & Dworetsky, M.M.: 1993a, *A&A* **275**, 335

Smith, K.C. & Dworetsky, M.M.: 1993b, 'Abundance correlations in mercury-manganese stars' in M.M. Dworetsky, F. Castelli, R. Faraggiana, ed(s)., *Peculiar versus Normal Phenomena in A-type and Related Stars, IAU Coll. No. 138*, ASP: San Francisco, 131

Smith, M.A.: 1971, *A&A* **11**, 325

Smith, M.A.: 1973, *ApJS* **25**, 277

Smith, M.A.: 1974, *ApJ* **189**, 101

Stenflo, J.O. & Lindegren, L: 1977, *A&A* **59**, 367

Stępień, K.: 1994, 'Properties and position of CP2 stars on the HR diagram' in J. Zverko & J. Žižňovský, ed(s)., *Chemically Peculiar and Magnetic Stars on and Close to the Upper Main Sequence*, Astronomical Institute of Slovak Academy of Sciences: Tatranska Lomnica, 8

Stibbs, D.W.N.: 1950, *MNRAS* **110**, 395

Stift, M.J.: 1976, *A&A* **50**, 125

Stürenburg, S.: 1993, *A&A* **277**, 139

Takada-Hidai, M., Sadakane, K. & Jugaku, J.: 1986, *ApJ* **304**, 425

Titus, J. & Morgan, W.W.: 1940, *ApJ* **92**, 256

van't Veer, C., Burkhart, C. & Coupry, M.F.: 1988, *A&A* **203**, 123

Vauclair, S.: 1975, *A&A* **45**, 233

Vauclair, S. & Vauclair, G.: 1982, *Ann. Rev. Astr. Astrophys.* **20**, 37

Vauclair, S., Hardorp, J. & Peterson, D. M.: 1979, *ApJ* **227**, 526

Venn, K.A. & Lambert, D.L.: 1990, *ApJ* **363**, 234

Vilhu, O., Tuominen, I.V. & Boyarchuk, A.A.: 1976, 'Abundance studies of peculiar B stars' in W.W. Weiss, H. Jenkner & H.J. Wood, ed(s)., *Physics of Ap Stars, IAU Coll. No. 32*, Universitätssternwarte Wien mit Figl-Observatorium für Astrophysik: Vienna, 563

Wahlgren, G.M., Leckrone, D.S., Johansson, S.G., Rosberg, M. & Brage, T.: 1995, *ApJ* **444**, 438

Wehlau, W. & Rice, J.: 1993, 'Zeeman Doppler Imaging' in M.M. Dworetsky, F. Castelli, R. Faraggiana, ed(s)., *Peculiar versus Normal Phenomena in A-type and Related Stars, IAU Coll. No. 138*, ASP: San Francisco, 247

White, R.E., Vaughan, A.H., Preston, G.W. & Swings, J.P.: 1976, *ApJ* **204**, 131

Willis, A.J.: 1991, 'Abundances in Wolf–Rayet stars, LBVs and OBN stars' in G. Michaud

& A. Tutukov, ed(s)., *Evolution of Stars: The Photospheric Abundance Connection*, *IAU Symp. No. 145*, Kluwer Academic: Dordrecht, 195

Wolff, R.J. & Wolff, S.C.: 1976, *ApJ* **203**, 171

Wolff, S.C.: 1975, *ApJ* **202**, 121

Wolff, S.C.: 1981, *ApJ* **244**, 221

Wolff, S.C.: 1983, *The A-Type Stars: Problems and Perspectives*, NASA SP-463: Washington D.C.

Wolff, S.C. & Preston, G.W.: 1978, *ApJS* **37**, 371

Wolff, S.C. & Wolff, R.J.: 1974, *ApJ* **194**, 65

Zboril, M., Glagolevskij, Yu.V. & North, P.: 1994, 'Properties and position of CP2 stars on the HR diagram' in J. Zverko & J. Žižňovský, ed(s)., *Chemically Peculiar and Magnetic Stars on and Close to the Upper Main Sequence*, Astronomical Institute of Slovak Academy of Sciences: Tatranska Lomnica, 105

LEVITATION

M.J. SEATON

Department of Physics & Astronomy, University College London, Gower Street, London WC1E 6BT, UK

Abstract. Some stars have outer layers which are sufficiently quiescent for diffusion to occur and to modify the relative abundances of the chemical elements. Levitation occurs when the forces due to radiation pressure are larger than those due to gravitation.

The paper describes some recent work on the calculation of the radiative forces using atomic data obtained in the course of the work of the Opacity Project.

Large abundance anomalies are observed for the HgMn stars, which lie on or close to the main sequence and have effective temperatures in the range 11000 to 15000 K. Some results are given for calculated abundances of elements of the iron group in the HgMn stars.

> **levitate:** *To rise or cause to rise and float in the air,*
> *usually attributed to supernatural intervention.*
> (Collins Dictionary of the English Language)

1. Some reminiscences — by way of an introduction

I suppose it was in about 1970, or perhaps a bit earlier, that Bob Wilson came to my room and asked what objects I would like to observe if I had the opportunity to make observations in the ultra-violet. I replied: novae and planetary nebulae. Bob asked for reasons why, then asked me to write it down, saying that would help him in making the case for an international project to put a UV telescope in space. I obliged. And thought no more about it.

It would have been in the spring of 1978 that I received an official notice saying that, as a result of the application which I had made, I had been given a quite substantial allocation of IUE observing time. I had not been aware of having applied, but that is how, for a few years, I became an observational astronomer. The notes which I had given Bob had, presumably, been filed as an application for observing time. Fortunately for me, Julie Lutz was spending the 1977/78 academic year at UCL, on sabbatical leave from Washington State University, and she introduced me to all the skills of finding objects on Palomar, SRC and ESO plates, making finding charts, triple-checking on RA and Dec, and doing all the other things required for preparing observing runs. Julie and I had the very first European Guest Observer shift at Vilafranca and used it for the planetary nebula work. We were ably assisted by David Stickland as Resident Astronomer. It was most exciting to see spectra forming on the monitor screen which had never been seen before. From the start I liked the system of IUE data archives. If one made an observation and did not extract and write up all of the information

Astrophysics and Space Science **237**: 107–123, 1996.
© 1996 *Kluwer Academic Publishers. Printed in Belgium.*

which could be obtained from it, one had the satisfaction of knowing that the data were there, and at some future date some one would make good use of them.

It was more difficult to know what to do with the allocation of telescope time for the nova work. What can you do if you have the allocation but there is no nova in the sky in a really interesting phase? Fortunately, IUE had the answer in the Target of Opportunity Program. Before long an interesting nova did obligingly appear, and in all we had more time than had been allocated. For a while I was the co-ordinator of the European Target of Opportunity Team for the nova work. One of the frustrating aspects of the work was the difficulty in obtaining optical observations at nearly the same time as the UV ones. Most optical observatories have rigid allocation schedules and allow little leeway for observing transient phenomena. A big advance was the introduction by Bob Williams of a program for monitoring novae at Cerro Tololo. Long may that program continue!

Immediately after making my first UV observations I went to Colorado loaded with tapes, and analysed the data using the interactive computer system at LASP (Laboratory for Atmospheric and Space Physics). Returning to London in the autumn, I found that all the work there was being done with batch processing on an ageing IBM 360/65. That seemed incredibly primitive and Bob and I therefore submitted an application to SERC for an interactive system similar to that at LASP, based on a PDP/11. It was one of two or three similar applications made at about that time and was not in itself successful, but it did provide one of the initiatives which led to the eventual setting-up of the UK STARLINK system. The first STARLINK VAX at UCL was in what had been the clean room for developing the IUE cameras, and was unbelievably cramped.

The chance to spend a while as an observer, with a very exciting new instrument, was one which I would certainly not have liked to have missed. On the other hand, before long it became clear to me that we had some very good research students and that they were every bit as good as myself at doing the observational work. Maybe, after all, the craftsman should stick to his last. In 1982 Norman Simon published his paper suggesting that existing tables of stellar envelope opacities might be in error by factors of 3 or so. In the summer of that year I was once again visiting Colorado and Dimitri Mihalas drew to the attention of David Hummer and myself the need for a massive new effort in the opacity work. That led to the birth of The Opacity Project (OP). For many years work in atomic physics had been done both at University College London (UCL) and at Queen's University of Belfast (QUB) with much friendly collaboration and sometimes an edge of competitiveness. It occurred to me that the computations required for resolution of the opacity problem, could be made if UCL and QUB joined forces and pooled all expertise. Everyone concerned, on both sides of the Irish

Sea, agreed enthusiastically. Some 30 people were eventually involved in the work, from 5 different countries. For a period of 10 years* we all met every 6 months to co-ordinate the work. When the main task was finished (see Seaton, Yu Yan, Mihalas and Pradhan, 1994) the team wanted to continue the collaboration, and that led to the birth of the new Iron Project led by David Hummer, which continues to go from strength to strength.

About a year ago it occurred to me that the large amounts of atomic data which we had obtained in the course of the OP work could be used in connection with another problem, the calculation of radiation forces acting on various chemical elements in stars. The classes of chemically peculiar (CP) stars had been known for a long time and there seemed to be little doubt that the basic mechanism was diffusive separation, first discussed in detail by George Michaud in 1970 (The reader wanting to know more about the CP stars need not have far to go: a review by Keith Smith is included in the present volume). Some elements settle under gravitational forces while others levitate to the surface under forces due to radiation pressure. For many CP stars the IUE spectra are particularly rich in spectrum lines which can be used to deduce abundances. Some of those stars had been observed many times and using the archived data it proved possible to use cunning techniques of co-adding the spectra so as to obtain results with remarkably high signal-to-noise ratios. So there was more work to be done in helping to interpret such results, which all goes to explain why, when asked to write this article, I have something new to say.

2. Theory

2.1. OUTLINE OF ESSENTIALS

Let F_ν be the outward-directed flux in a star, radiative energy per unit area, time and frequency. The corresponding flux of momentum is $(1/c)F_\nu$. Let $\sigma_\nu(l)$ be the cross section of an atom of a chemical element l for absorption of radiation. Then the total momentum absorbed per atom per unit time is

$$\mathcal{G}(l) = (1/c) \int \sigma_\nu(l)F_\nu \, d\nu \qquad (1)$$

which is, of course, just the force acting on the atom. The corresponding gravitational force is $M(l)g_{grav}$ where $M(l)$ is the atom mass and

$$g_{grav} = \mathcal{M} \, G/\mathcal{R}^2, \qquad (2)$$

is the gravitational acceleration, $\mathcal{M}$ being the mass of the star, $\mathcal{R}$ its radius and G the gravitational constant. It is usual to put, for the radiative force, $\mathcal{G}(l) = M(l)g_{rad}(l)$, defining the radiative acceleration $g_{rad}(l)$.

---

* The reader would be mistaken to conclude that the OP work took 300 person-years. The 30 participants were by no means all engaged full-time on the OP work

There is not much further simplification which can be made for the case of a stellar atmosphere. One must make a model, calculate the flux F_ν, and evaluate the integral for $\mathcal{G}(l)$. One can make simplifications for the deeper layers in which the equation of radiative transfer can be solved using the diffusion approximation, which gives

$$F_\nu = \frac{\sigma_R}{\sigma_\nu(\text{tot})} f_\nu F \tag{3}$$

where: σ_R is the Rosseland-mean opacity cross-section; $\sigma_\nu(\text{tot})$ is the total monochromatic opacity cross-section;

$$F = \int F_\nu \, d\nu \tag{4}$$

is the total outward flux; and

$$f_\nu(T) = (dB_\nu/dT)/(dB/dT) \tag{5}$$

where $B_\nu(T)$ is the Planck function and $B = \int B_\nu d\nu$. The Rosseland-mean cross section is defined by

$$(1/\sigma_R) = \int \frac{1}{\sigma_\nu(\text{tot})} f_\nu \, d\nu \tag{6}$$

and the Rosseland-mean opacity, per unit length, is $\kappa_R = \sigma_R N$ where N is the number of atoms per unit volume.

From (1) and (3),

$$\mathcal{G}(l) = (1/c)\sigma_R \gamma(l) F \tag{7}$$

where $\gamma(l)$ is the dimensionless quantity defined by

$$\gamma(l) = \int \frac{\sigma_\nu(l)}{\sigma_\nu(\text{tot})} f_\nu \, d\nu. \tag{8}$$

Thus, given that the Rosseland-mean opacity is known, the only other quantity required is $\gamma(l)$. It depends on temperature, density and chemical composition but does not depend on any details of a stellar model.

The total luminosity of the star is $\mathcal{L} = 4\pi\mathcal{R}^2 F$ and the ratio of radiative to gravitational accelerations is therefore

$$\frac{g_{\text{rad}}(l)}{g_{\text{grav}}} = \frac{1}{4\pi cG} \frac{\gamma(l)\sigma_R}{M(l)} \frac{\mathcal{L}}{\mathcal{M}}. \tag{9}$$

Atoms of element l will gravitate if $g_{\text{rad}}(l) < g_{\text{grav}}$ and levitate if $g_{\text{rad}}(l) > g_{\text{grav}}$. It is, of course, here assumed that the layers concerned are sufficiently quiescent for such processes of diffusive separation to occur.

It is sometimes convenient to express g_{rad} in terms of the effective temperature, T_{eff}. The radiative flux is $F = \pi B(T_{\text{eff}}) = C_{\text{SB}} T_{\text{eff}}^4$ where C_{SB} is the Stefan-Boltzmann constant. We therefore obtain

$$g_{\text{rad}}(l) = \frac{\sigma_R \gamma(l)}{cM(l)} C_{\text{SB}} T_{\text{eff}}^4. \tag{10}$$

2.2. SATURATION EFFECTS

The largest contributions to the radiation pressure usually comes from absorption in spectrum lines and, unless the abundance of element l is very small, the flux F_ν has minima in the vicinities of the lines of l. The radiative force acting on l can therefore be very sensitive to the abundance of l.

This can also be seen another way. Let the chemical element k have abundances $A(k)$ by number fraction,

$$\sum_k A(k) = 1. \tag{11}$$

The mono-chromatic opacity cross-section is

$$\sigma_\nu(\text{tot}) = \sum_k \sigma_\nu(k)A(k) + \mathcal{E}\sigma_T \tag{12}$$

where $\mathcal{E}$ is the number of electrons per atom and σ_T the cross section for electron scattering. Using (12), equation (8) can be written

$$\gamma(l) = \int \frac{\sigma_\nu(l)}{\sum \sigma_\nu(k)A(k) + \mathcal{E}\sigma_T} f_\nu \, d\nu. \tag{13}$$

Since $\sigma_\nu(l)$ can be very large in the vicinities of the lines of l, $\gamma(l)$ can be sensitive to the value of $A(l)$.

2.3. VARIATIONS IN ATOM ABUNDANCES

When diffusive separations occur, the abundance of any element l can vary over a wide range of values. We therefore require the quantities σ_R and $\gamma(l)$ for many values of the abundance of element l. Let us start with some standard set of abundances $A(k)$, for example solar-system abundances. We then change the abundance of element l by a factor χ, replacing $A(l)$ by $\chi A(l)$. For all $k \neq l$ we must then replace $A(k)$ by $A(k) \times [1 - \chi A(l)]/[1 - A(l)]$ so as to preserve the normalisation condition (11). We calculate σ_R and $\gamma(l)$ on a mesh of values of χ, usually taken to be $\log(\chi) = -2, -1, 0, 1, 2, 3$ and 4. For each mesh-point for χ we also calculate the derivatives, $\partial\sigma_R/\partial\chi$ and $\partial\gamma(l)/\partial\chi$. Values of σ_R and $\gamma(l)$ for any required value of χ can then be calculated accurately using cubic interpolations.

2.4. MOMENTUM TRANSFER IN PHOTO-IONISATION PROCESSES

In giving an outline of essentials in section 2.1, I have omitted one important point of detail. In a process of photo-ionisation the momentum absorbed is shared between the product ion and the ejected electron (Sommerfeld, 1930; Seaton, 1995a). In calculating the radiative forces acting on atoms of individual chemical elements we must include only the momentum given to the atomic ions. I therefore define an effective cross-section for momentum transfer to the atom, $\sigma_\nu^{\text{mta}}(l)$, which differs from $\sigma_\nu(l)$ in that one subtracts away that fraction of the photo-ionisation cross sections which gives momentum to the ejected electrons. To allow for that correction one must replace $\sigma_\nu(l)$ by $\sigma_\nu^{\mathrm{mta}}(l)$ in equations (1), (8) and (13).

A more detailed account of the theory used for the computation of radiative forces will be given elsewhere (Seaton, 1995b).

3. Atomic data for elements of the iron group

The present contribution gives some results for levitation of iron-group elements.

Elements of the iron group make very important contributions to Rosseland mean opacities in regions of stellar envelopes having temperatures in the vicinity of 2×10^5K (Rogers and Iglesias, 1992; Seaton $et\ al.$ 1994). So long as one is not considering mixtures highly enriched in those elements, they do not make such important contributions in outer layers, with T closer to 10^4K. In the original OP work we therefore included very extensive atomic data for the higher ionisation stages of the iron-group elements but less complete data for the first few stages.

For the outer layers it is essential to include more complete data for those first few ionisation stages if one wishes to calculate the radiative forces for the iron-group elements and to calculate Rosseland means for cases in which the abundancies of those elements are markedly enhanced. I have therefore supplemented the original OP data with data from Kurucz (1988) for the first few ionisation stages in Cr, Mn, Fe and Ni (some 24 million lines in all).

4. Archived atomic data

4.1. ATOMIC DATA

In the course of the OP work we obtained extensive atomic data: energy levels, oscillator strengths, and photo-ionisation cross sections. Since those data can be of value in work on various other problems in physics and astronomy, they have been archived at the Centre de Données Astronomiques de Strasbourg (CDS) in a database system TOPbase (Cunto $et\ al.$, 1993). Selected data and reprints of 31 OP papers have also been published in a book (The Opacity Project Team, 1995).

4.2. THE MONOCHROMATIC OPACITIES

The monochromatic opacities $\sigma_\nu(k)$ for each element k have been calculated assuming local thermodynamic equilibrium (LTE) and have been archived as functions of frequency ν on a mesh of values of (T, N_e) where T is temperature and N_e is electron density. Calculations have been made for 17 elements: H, He, C, N, O, Ne, Na, Mg, Al, Si, S, Ar, Ca, Cr, Mn, Fe and Ni. The amount of data is large, some 650 Mb as binary files.

4.3. ROSSELAND MEAN OPACITIES

In order to calculate Rosseland mean opacities one must specify a chemical composition. It is usual first to define the relative abundances of the metal atoms (all elements other than H and He), then to define the mass-fractions X, Y and Z for hydrogen, helium and metals. Since $X + Y + Z = 1$ one need specify only two numbers, usually taken to be X and Z. Computations have been made for a number of number of metal-mixtures and a range of (X, Z) combinations. These give files of Rosseland mean opacities, in standard OP formats, on the (T, N_e) mesh. A code OPFIT.FOR (Seaton,1993) can be used to read the files in standard OP format and to interpolate Rosseland means to any required values of (T, ρ), where ρ is mass-density. It also gives the derivatives of Rosseland-mean opacities with respect to T and ρ (the derivatives are important for work on stellar pulsations).

4.4. INTERPOLATIONS IN X AND Z

A recent development (Seaton, 1995c) has been to produce files on a mesh of (X, Z) values which give Rosseland means and information required to calculate the derivatives of the means with respect to changes in X and Z. An interpolation routine IXZ.FOR reads these files and provides files in standard OP format for any required values of X and Z, which can be used as input to OPFIT.FOR.

4.5. DATA FOR THE CALCULATION OF RADIATIVE FORCES

Files are created for each initial mixture $\{A(k)\}$ and each chemical element l, giving for each (T, N_e) mesh-point the values of σ_R, $\partial\sigma_R/\partial\chi$, $\gamma(l)$ and $\partial\gamma(l)/\partial\chi$ on each mesh-point for χ. An interpolation code ACCFIT.FOR then enable one to calculate σ_R and $\gamma(l)$ for any required combination of T, ρ and χ.

4.6. ACCESS TO ARCHIVED DATA AT CDS

4.6.1. Atomic data

OP atomic data are archived in TOPbase. To obtain the TOPbase user manual, make an **anonymous ftp** call to CDS, cdsarc.u-strasbg.fr (IP 130.79.128.5), logon as **anonymous**, give e-MAIL address as password, and cd /pub/topbase. Then get userman.tex.

To obtain interactive access to TOPbase, make a **telnet** call to CDS, logon as **topbase** and give password Seaton+.

4.6.2. Opacity data

All OP opacity data are in process of transfer to CDS archives, together with codes for making interpolations. A lot is there at the time of writing, more will be there later. To obtain up-to-date information, make an **anonymous** ftp call to CDS, cd /pub/cats/VI/80, and get the file ReadMe. Then help yourself to whatever data and codes you may require.

5. Diffusion

5.1. THE DIFFUSION EQUATIONS

For the purposes of the present work I take the outward diffusion velocity to be given by

$$v = D \left\{ \frac{M(l)}{kT} (g_{\text{rad}}(\chi) - g_{\text{grav}}) - \frac{\partial \ln(\chi)}{\partial r} \right\} \tag{14}$$

where D is the diffusion coefficient (see Aller and Chapman, 1960)*, $M(l)$ is the mass of atom l and r is the stellar radial co-ordinate. The number-density for element l is $n = \chi(l)A(l)N$ where χ is the abundance-enhancement factor, $A(l)$ the standard abundance, and N the total number-density for all atoms.

The equation for conservation of number of atoms l is then

$$\frac{\partial n}{\partial t} + \frac{\partial(nv)}{\partial r} = 0 \tag{15}$$

where t is the time.

6. The HgMn stars

In a valuable review on the subject of element segregation in stellar outer layers, Vauclair and Vauclair (1982) say that "Mercury Manganese stars probably represent the closest approach to the ideal situation of pure diffusion in stellar atmospheres". These stars lie on or close to the main sequence and have effective temperatures $T_{\text{eff}} = 11000$ to 15000K. Extensive studies of element abundances in HgMn stars have been made at UCL by Mike Dworetsky and Keith Smith: Smith and Dworetsky (1993) gives results for iron-group elements.

6.1. ENVELOPE MODELS

Diffusive separations are of most importance in layers occupying no more than 1 or 2% of a total stellar radius, and one can therefore use plane-

---

* The expression given by Aller and Chapman for D is for diffusion of an ion of charge z. In general we have to consider a range of values of z, z_j for each ionisation stage j. In the present work we calculate g_{rad} using archived monochromatic opacities which have already been summed over contributions from the ionisation stages, and we therefore do not have separate values of g_{rad} for each ionisation stage. We therefore use value of D calculated using mean values for the charge z. In later work it will be desirable to calculate values of g_{rad} separately for each stage. That will take a lot more computer time.

It should also be noted that in the present work we neglect a further term in the expression for v, involving the temperature gradient.

parallel models. I have constructed such models for envelopes of stars with $T_{\text{eff}} = 10000$ to 15000 K in steps of 1000 K and $\log(g_{\text{grav}}) = 3.87$ which is the mean value obtained spectroscopically by Smith and Dworetsky for the HgMn stars (their values of g_{grav} show no systematic dependence on T_{eff}). I take the near-surface conditions to be those for a grey atmosphere in the Eddington approximation, using the Rosseland-mean opacity as the assumed frequency-independent opacity. Comparisons with results from proper stellar atmosphere calculations shows that the models are good for $\tau \geq 1$ where τ is the Rosseland-mean optical depth. In later work it will be desirable to make use of models with more realistic atmospheric structures; meanwhile, the present models permit study of some general trends.

In calculating the models I have assumed that there is no convection. The HgMn stars are observed to have depleted surface abundances of He. An essential feature of diffusion theory for such stars (Michaud, 1986) is to assume that He has settled and that there is no He II convection zone.

One of the main new features of the present work is in the availability of atomic data for the deeper layers. The models go down to depths with temperatures of 10^7K.

6.2. LINE FORMATION

Since the abundances may vary as functions of depth it is necessary to consider the depths of formation of the observed spectral features. The weak lines observed in the visual region are formed comparatively deep in the atmospheres, say at optical depths $\tau \simeq 0.5$. The strong resonance lines observed in the UV are formed higher in the atmospheres. For those lines abundances are obtained using fluxes in line wings, formed at $\tau \leq 0.1$.

6.3. STATIC SOLUTIONS

I have defined an abundance-enhancement factor χ, such that $\chi = 1$ when no enhancements or depletions occur (for that case I use the S92 solar-system abundances $A(k)$ from Seaton $et\ al$ 1994). As χ increases, $g_{\text{rad}}(\chi)$ decreases due to saturation effects and may eventually reach a value of χ giving $g_{\text{rad}} = g_{\text{grav}}$. There are then no forces acting on element l. It must be noted, however, that such a value does not always exist: one can have cases for which $g_{\text{rad}} > g_{\text{grav}}$ even when the lines are fully saturated (pure element l, $\chi(l)A(l) = 1$); and cases for which $g_{\text{rad}} < g_{\text{grav}}$ even when fully desaturated ($\chi \to 0$).

I define χ_{stat} to be the value of χ which gives $v = 0$. It follows from equation (14) that χ_{stat} is then obtained on solving the equation

$$(g_{\text{rad}}(\chi) - g_{\text{grav}}) - (kT/M)\partial \ln(\chi)/\partial r = 0. \tag{16}$$

Solutions are obtained by iteration: a first approximation is obtained by solving $(g_{\text{rad}}(\chi) - g_{\text{grav}}) = 0$ which is then used to obtain a first estimate

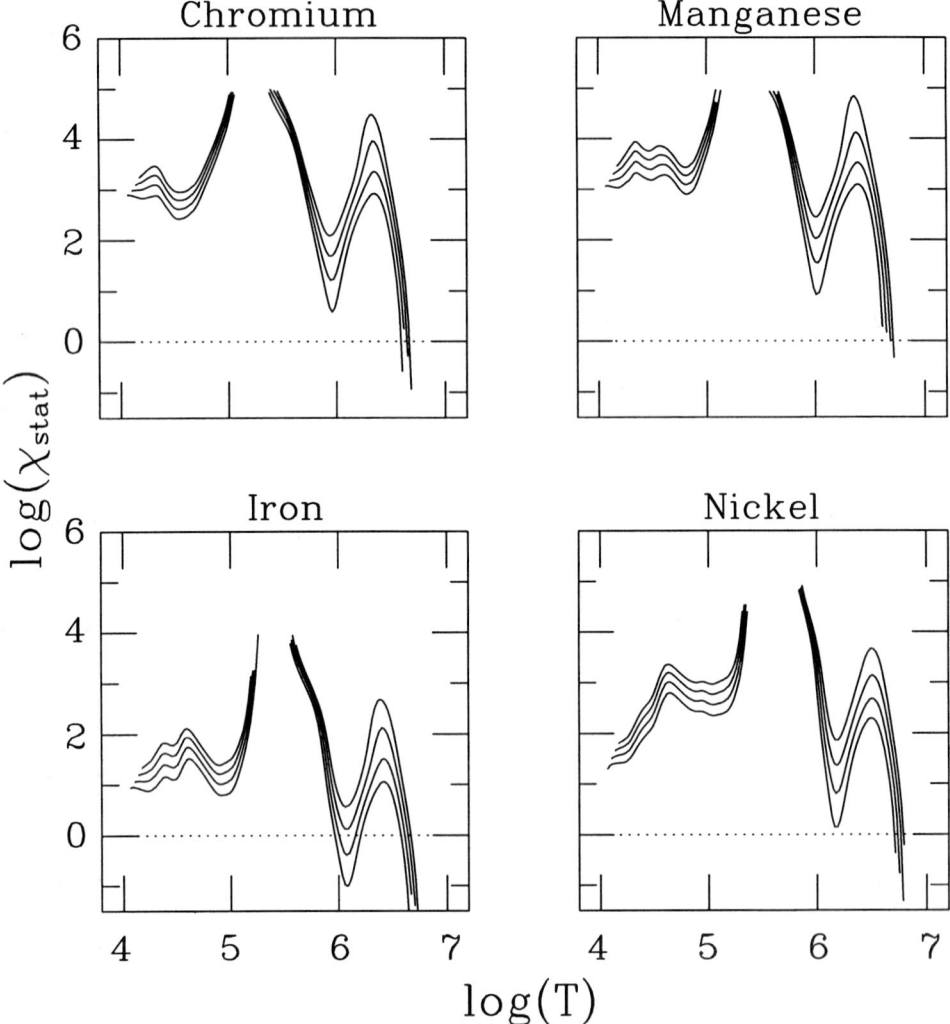

Fig. 1. χ is a factor by which abundances are enhanced or depleted, and χ_{stat} the value of χ which gives zero diffusion velocities v. The plot shows values of $\log(\chi_{stat})$ for Cr, Mn, Fe and Ni for models with $T_{eff} = 11000$, 12000, 13000 and 14000 K, as function of $\log(T)$ where T is the temperature in the stellar envelope. In all cases the lowest curve is for the coolest star, and the highest for the hottest.

for the derivative term in (16). The solutions converge rapidly and the final solutions are never very different from the first approximation.

   Figure 1 shows values of χ_{stat} for Cr, Mn, Fe and Ni in the models with $T_{eff} = 11000$, 12000, 13000 and 14000 K. The variation of χ_{stat} with depth is mainly determined by changes in the ionisation structure of the element concerned. There are deep minima when the dominant ionisation stages are

Ar-like (at $\log(T) \simeq 5.0$ for Fe) and Ne-like (at $\log(T) \simeq 6.1$ for Fe). Between those two minima χ_{stat} becomes very large and there is a region for which no solution exists (solution of (16) would require $\chi(l)A(l) > 1$).

Another point should be noted. The main reason for the differences in the magnitudes of χ_{stat} for the different iron-group elements is due to the differences in the reference abundances $A(l)$. Thus $A(l)$ for Fe is larger than that for Mn by a factor of order 100, and for given values of χ saturation effects for Fe are much larger than those for Mn. If we considered the actual number-fractions giving static solutions, $\chi_{\text{stat}}(l) \times A(l)$, then we would obtain comparable orders of magnitude for the different iron-group elements. Figure 2 shows our values of χ_{stat} for the outer layers, plotted against $\log(\tau)$ where τ is the Rosseland-mean optical depth. For the atmosphere regions, say $\log(\tau) < 0$, our values will not be accurate due to the approximations made in calculating our atmospheric models and to our use of the diffusion approximation to solve the equation of radiative transfer. Nonetheless, the results of Figure 2 may give some indication of possible trends. Much more accurate solutions for Mn in atmospheres of HgMn stars have been made by Alecian and Michaud (1981), who use much better atmosphere models, compute g_{rad} using computed atmosphere fluxes, and make some allowance for NLTE effects in calculating the populations of the Mn energy levels. For each star they give values of g_{rad} for $\chi = 1$ and for one larger value of χ. For $\tau \simeq 1$ our results are in quite good agreement with theirs, but there are substantial differences for smaller values of τ.

6.4. STEADY-STATE SOLUTIONS

Let us define $\mathcal{F} = nv$ to be the particle flux for some element l. A steady-state solution is one which gives $\mathcal{F}(r)$ to be a constant independent of depth r, and hence $\partial n/\partial t = 0$. For such solutions the total number of atoms l may not be conserved.

In considering such solutions I will neglect the pure diffusion term, $\partial \ln(\chi)/\partial r$, in (14) to obtain

$$v = D(M/kT)(g_{\text{rad}} - g_{\text{grav}}). \tag{17}$$

Putting $n = \chi A N$ this gives

$$\mathcal{F} = Hy \tag{18}$$

where

$$H(r) = AND(M/KT) \tag{19}$$

and

$$y(r, \chi) = \chi(g_{\text{rad}}(r, \chi) - g_{\text{grav}}). \tag{20}$$

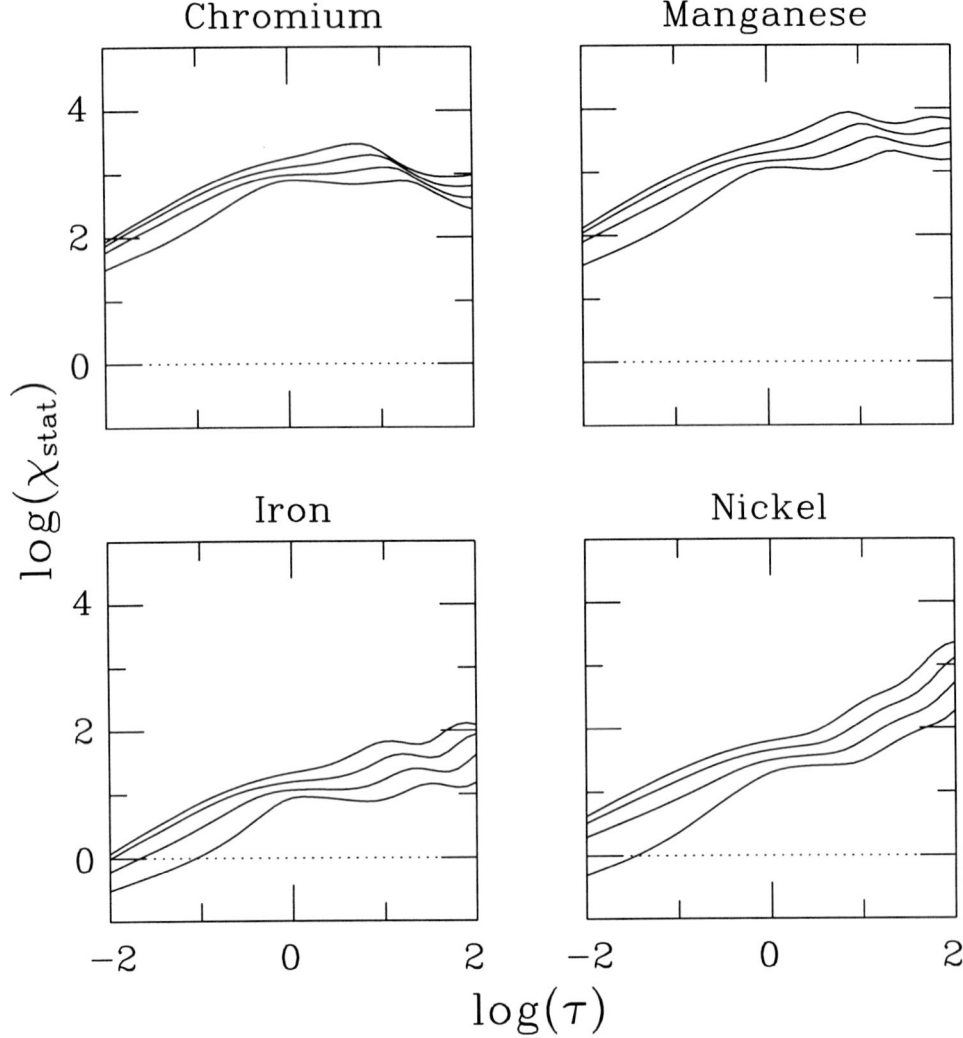

Fig. 2. As for Figure 1, with $\log(\chi_{\text{stat}})$ plotted against $\log(\tau)$ for the outer layers, τ being the Rosseland-mean optical depth.

The procedure is to choose a value $\chi(1)$ of χ at the "surface", $r = r(1)$, and hence to obtain the required value of $\mathcal{F}$, $\mathcal{F} = H(r(1))y(r(1), \chi(1))$. Then for all other depth-points r one has to solve the equation $y(r, \chi) = C$ where $C = \mathcal{F}/H(r)$.

Figure 3 shows the function $y = \chi(g_{\text{rad}} - g_{\text{grav}})$ for Mn in the 11000 K model at a depth-point with temperature $\log(T) = 4.822$. The function rises from zero at $\chi = 0$, passes through a maximum and then decreases as saturation becomes more important, eventually reaching a value of zero

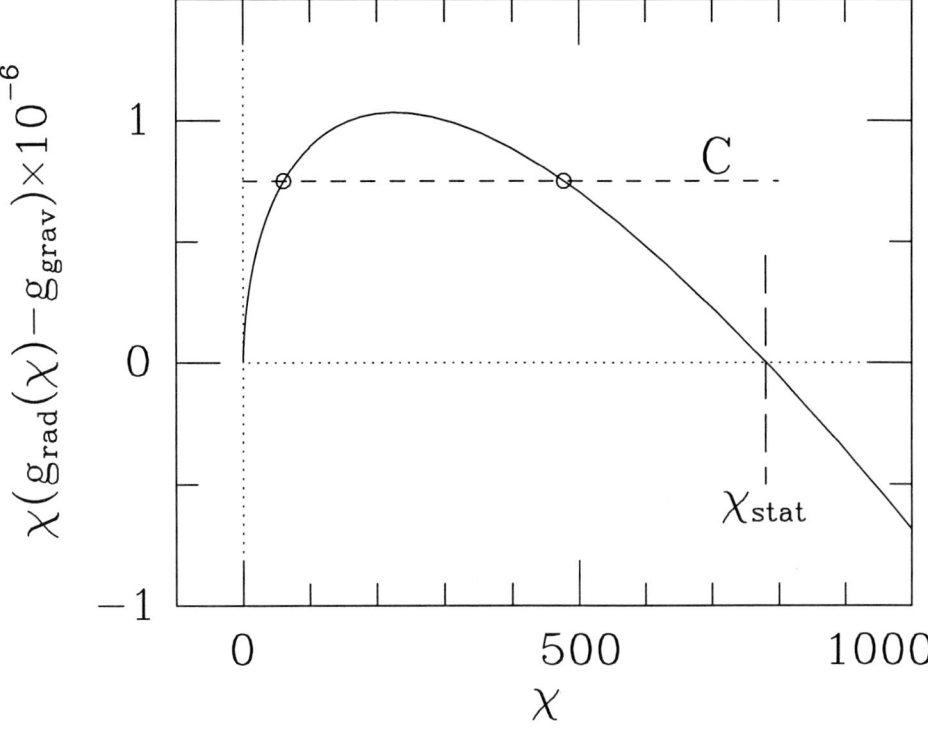

Fig. 3. The function $y(\chi) = \chi(g_{\mathrm{rad}}(\chi) - g_{\mathrm{grav}})$ against χ for Mn in the 11000 K model, at depth giving $\log(T) = 4.822$. For the calculation of steady-state solutions it is required to solve the equation $y(\chi) = C$. Two such solutions are shown by open circles. There are no solutions if C is larger than the maximum value of $y(\chi)$.

at $\chi = \chi_{\mathrm{stat}}$. For a given value of the constant C there are, in general, two solutions of the equation $y = C$ (the open circles on the Figure): a given total flux can be obtained with a small value of n and a large value of v or with a large n and a small v. If, however, C is larger than the maximum value of y there is no solution.

In Figure 4 I consider the same model and plot steady-state values of χ against $\log(T)$ for three values of $\mathcal{F}$ corresponding to $\chi(1) = 20$, 40 and 53.2, where I take the "surface" to be at $\tau(1) = 0.5$ (these three values of $\chi(1)$ are shown by filled circles on Figure 4). For a given flux $\mathcal{F}$ I obtain two solutions, corresponding to the two roots of $y = C$ illustrated in Figure 3. I refer to these as the "upper" and "lower" solutions. Figure 4 also includes, as a thick line, values of χ_{stat}.

We see that χ_{stat} has a minimum in the region of $\log(T) = 4.8$ (the plot of Figure 3 is for that region). For $\log(T) \simeq 4.8$ the lower steady-state solutions have maxima and the upper ones have minima. For the solutions with $\chi(1) = 53.2$ those maxima and minima coincide, corresponding to the

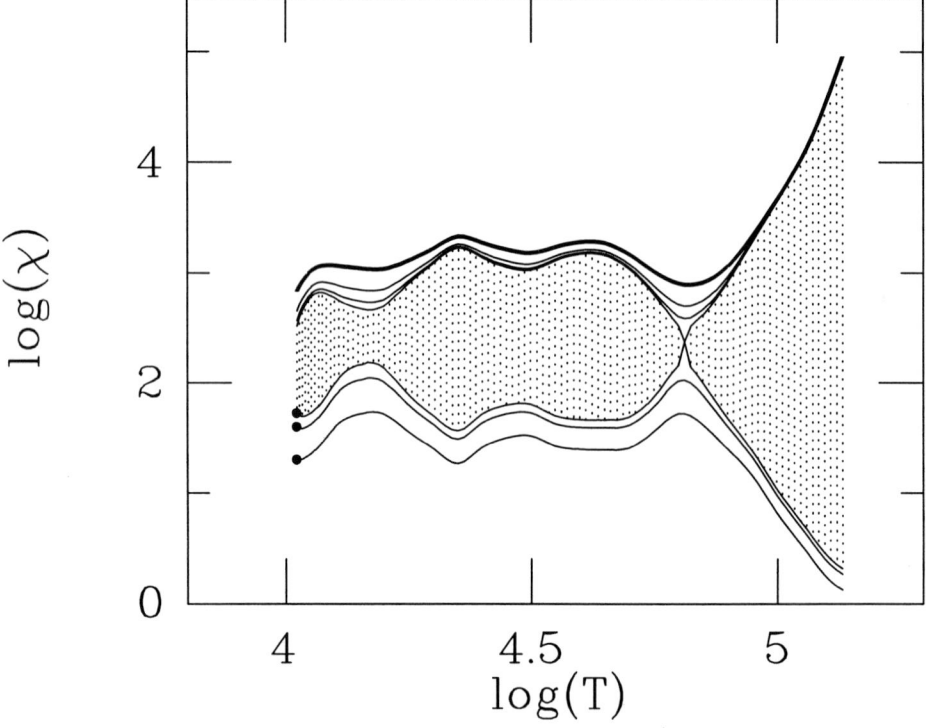

Fig. 4.   Solutions with steady outflow for Mn in the model with $T_{\mathrm{eff}} = 11000K$. There are 3 lower curves and three upper ones. The rates of outflow are determined by the values $\chi(1)$ of χ which occur at depths $\tau(1) = 0.5$. Those values are shown as filled circles, and are $\chi(1) = 20$, 40 and 53.2. The function χ_{stat} is shown by the heavy upper cuve. There are no solutions for steady-state solutions in the shaded area.

maximum value of C for which solutions exist (see Figure 3).

For the case considered there are no steady-state solutions which can be continued to greater depths and which have $\chi(1) > 53.2$. That is to say there are no steady-state solutions in the shaded area of Figure 4. It should be noted that this limitation on the values of $\mathcal{F}$ for which steady-state solutions exist is determined, not by conditions in the atmospheres, but by conditions at considerably greater depths.

Let χ_{ss} be the value of χ corresponding to the maximum value of $\mathcal{F}$ which can be obtained for the lower steady-state solutions. Results for most other cases are similar to those for the case illustrated on Figure 4: the values of χ_{ss} are determined by minima in g_{rad} at depths well below those of the stellar atmospheres. Those minima occur when the dominant ionisation stages are near Ar-like.

In Figure 5 I give, as a histogram, values of χ_{ss} for the four elements Cr, Mn, Fe and Ni and for models with $10^{-4}T = 10$, 11, 12, 13, 14, and 15. I

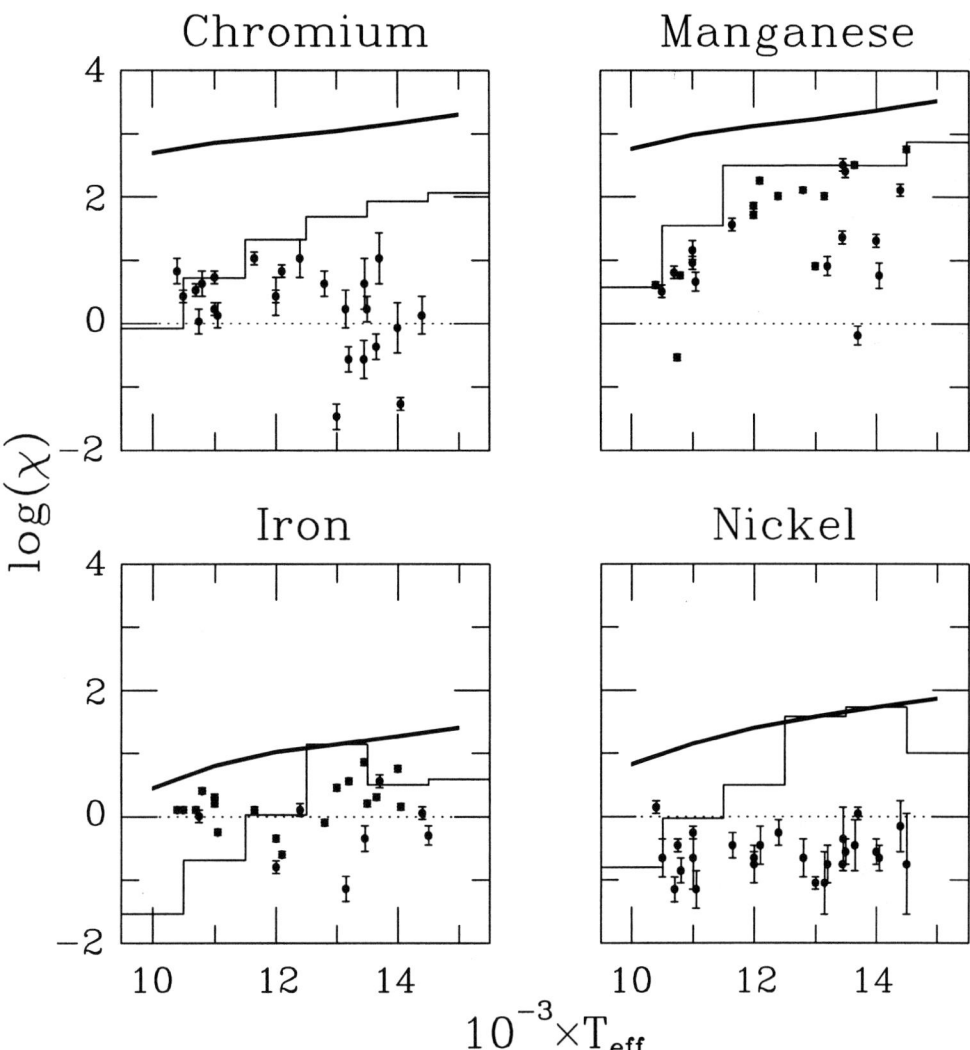

Fig. 5.   Abundances for Cr, Mn, Fe and Ni in HgMn stars with T_{eff} in the range 10000 K to 15000 K. The filled circles with error bars are from the observational results of Smith and Dworetsky (1993). The calculated results are for a layer with Rosseland-mean optical depth $\tau(1) = 0.5$. The upper heavy line shows results from static solutions, χ_{stat}, and the histograms show the maximum abundances, χ_{ss}, which can be obtained from steady-state solutions.

also include the values of χ_{stat} as heavy lines. The results given for χ_{ss} and χ_{stat} are for a depth of $\tau = 0.5$. I also give the values of χ from the analysis of observations made by Smith and Dworetsky, which will be referred to as χ_{obs}. The following comments can be made.

—   In all cases the values of χ_{obs} are smaller than those of χ_{stat}, and in many cases smaller by at least two orders of magnitude. It follows that the observations cannot be explained in terms of near-surface static solutions (the only possible exception would be if the values of χ_{stat} in the line-forming regions were very much smaller than our values of χ_{stat} for $\tau = 0.5$).

- Nearly all values of χ_{ss} are smaller than those of χ_{stat}.
- Nearly all of the observed abundances are either close to χ_{ss} or smaller than χ_{ss}. The only exceptions are some points for Fe with T_{eff} in the vicinity of 11000 K.
- Use of χ_{ss} explains some general trends: for example, that values of χ_{obs} for Mn increase with T_{eff} and reach large values for the hotter stars; and that the values of χ_{obs} for Cr are less than those for Mn, although the values of χ_{stat} are not very different.
- There are a number of cases for which the values of χ_{obs} are a good deal smaller than χ_{ss}. This is particularly marked for Ni in the hotter stars.
- In some cases one obtains very different observed values of χ for two stars having nearly the same value of T_{eff}.

6.5. TIME-DEPENDENT SOLUTIONS

Work on the solution of the time-dependent equation (15) is in progress but not yet complete. Results will eventually be published elsewhere. Let it suffice to give just one comment here: in many cases the time-dependent theory does give solutions with continued outflow and near-surface abundances less than or equal to the values of χ_{stat} discussed in section 6.4.

7. Discussion

The work on radiative levitations depends on the availability of massive amounts of atomic data. In all, many tens of millions of spectrum lines have been included. I am indebted to all participants in the work of the Opacity Project for their help in providing the required data, and also to Bob Kurucz who provided, on CD ROMs, additional data for lines in the first few ionisation stages of the iron-group elements. All of the atomic data and related data for opacities and radiative forces are being made generally available, as described in section 4.

In section 7 I have given some preliminary results from work in progress, concerned with diffusion of iron-group elements in the HgMn stars. If any new progress has been made, it depends largely on the availability of the new atomic data, and particularly on data going down to deeper layers than those which have been considered previously. It appears that the surface abundances may, in many cases, be determined by solutions with surface outflow, and that the rates of outflow and of near-surface abundances may depend on conditions at some distances below the stellar surfaces.

A further point should be noted. As the diffusion proceeds the movements of iron-group elements in the deeper layers are such as to give marked changes in opacities and these can lead to changes in the structures of the models (and possibly introduce additional convective zones). The possible importance of such movements in the layers with temperatures in the range $\log(T)$ between 5 and 6 was first mentioned in a paper by Alecian, Michaud and Tully (1993) using preliminary OP data. There may be other consequences, such as changes in pulsational properties. It will eventually be desirable to re-calculate the structures of the envelope models as the diffusion proceeds.

Acknowledgements

In addition to the acknowledgements already made in section 6, I thank Keith Smith for many discussions about diffusion processes and the properties of CP stars, and Georges Alecian for some much-valued criticism of a first draft of the present paper.

References

Alecian, G. and Michaud, G., 1981. ApJ, **245**, 226.

Alecian, G., Michaud, G., and Tully, J.A., 1993. ApJ, **411**, 882

Aller, L.H. and Chapman, S., 1960. ApJ, **132**, 461

Cunto, W., Mendoza, C., Ochsenbein, F. and Zeippen, C.J., 1993. A&A, 275, L5.

Kurucz, R., 1988. Trans. IAU XXB, p. 168.

Michaud, G., 1970. Ap. J., **160**, 641.

Seaton, M.J., Yu Yan, Mihalas, D. and Pradhan, A.K., 1994. MNRAS, **266**, 805.

Seaton, M.J., 1993. MNRAS, **265**, 25P.

Seaton, M.J., 1995a. J. Phys. B., **28**, 3185.

Seaton, M.J., 1995b. To be submitted

Seaton, M.J., 1995c. In press.

Simon, N.R., 1982. ApJ, **260**, L87.

Smith, K.C. and Dworetsky, M.M., 1993. A&A, **274**, 335.

Sommerfeld, A., 1930. Wave Mechanics, Methuen.

The Opacity Project Team, 1995. The Opacity Project. Institute of Physics, Bristol.

Vauclair, S. and Vauclair, G., 1982. Ann. Rev. Astron. Astrophys., **20**, 37.

WINDS OF LUMINOUS OB STARS

IAN D. HOWARTH AND RAMAN K. PRINJA

Dept. of Physics & Astronomy, University College London

Abstract. We review the historical development of observations of mass loss from hot, luminous stars, and sketch the physical principles underlying the theory of radiatively-driven stellar winds. The theory makes predictions which are shown to be in general accord with the data. We summarize recent results, primarily from the IUE satellite, which demonstrate the importance of time-dependent phenomena in hot-star outflows, and which, in particular, point to effects related to rotational modulation.

1. Introduction

We (that is, the authors of this contribution) owe both a personal and a professional debt to Bob Wilson. The personal debt arises from the privilege of his supervision of our PhD studies, and from the opportunities and encouragement he has given us in our subsequent careers. Professionally, we have benefitted in two rather fundamental ways from Bob's scientific contributions. Most obviously, in common with thousands of astronomers throughout the world, we have enjoyed the unparalleled cornucopia of data which has flowed from IUE. It is not an entirely comfortable feeling to compare how much *we* have worn compared with IUE, since our student years (and yet we now cost more to run, while IUE costs less!).

Perhaps less widely known is Bob's role in the topic of this contribution: his discovery of mass loss from OB-type stars, *via* high-speed winds. (It has been claimed – though we believe apocryphally – that Bob had himself forgotten this result from his early career!) The recognition of mass loss from novae and from Wolf-Rayet stars dates back to the work of McCrea and Beals, respectively, in the late 1920s. However, both these classes of star have enormously strong, broad emission lines; the optical signature of mass loss in OB stars is far more subtle. Its identification required painstakingly careful handling and measurement of photographic data, which Bob carried out as part of his studies while at ROE (achieving, incidentally, signal:noise ratios in excess of $\sim$100 in his mean spectra). He discovered outflow velocities in excess of $\sim$1000 km s^{-1} in several hot stars, such as α Cam (Wilson 1955), and, noting that such velocities exceed the surface escape velocities for the stars he studied, correctly deduced that luminous OB stars have highly supersonic stellar winds. He even identified radiation pressure as the mechanism most likely to be responsible for the winds (Wilson 1958).

Bob moved on to other fields, and the demanding observational requirements of the topic lead to its general neglect for many years (although

Astrophysics and Space Science **237**: 125–143, 1996.

Anne Underhill achieved similar results in her pioneering studies of early-type stars; Underhill 1958). The renaissance came with the first rocket UV spectra of OB supergiants (Morton 1967a). These showed the strong, broad P-Cygni profiles of C IV, N V, and Si IV that have since become so familiar in thousands of IUE spectra of hundreds of OB stars (e.g., Fig. 3). These observations directly stimulated the development of the theory of radiation-driven winds. Today, the importance of stellar winds to the evolution of the most massive stars, and to the physical and chemical processing of the interstellar medium, is an established fact, and their study occupies scientists world-wide.

2. Radiation: the Driving Force

Hot stars possess intense radiation fields. This radiation is, essentially, radial in direction. Its interaction with matter results in absorption or, more typically in OB-star winds, scattering of photons – that is, a photo-excitation followed almost immediately by a radiative decay through the same channel. Because the timescales for radiative decay are usually much shorter than for other processes, most species are found primarily in the ground state; and because of their large cross-sections, most absorptions (or scatterings) are from the ground state to low-lying excited states. Thus the most sensitive spectroscopic signature of hot-star winds is given by resonance lines of abundant ions, and these lines occur in the UV.

Scattering locally degrades the directionality of the radiation field; a scattered photon no longer streams radially with respect to the star. Conservation of momentum therefore requires that the scattering body acquire an outward motion. This is the basic mechanism driving stellar winds in early-type stars. And, just as all observers see all distant points redshifted in an expanding Universe, so a scattering ion in a stellar wind sees all photons from distant sources as red-shifted from their wavelengths at their points of origin (normally the stellar photosphere). Since the scattered photon is re-emitted with its frequency essentially unchanged *in the rest frame of the scatterer* (the 'comoving frame'), the nett effect of the scattering is a red-shift. That is, some of the scattered photon's energy has been transferred to the kinetic energy of the wind.

This effect is directly observable in the P-Cygni profiles which are the standard signature of mass loss. The photons 'missing' from the blue-shifted absorption component of the profile (formed by scattering out of the line of sight from the observer to the photosphere) are compensated for by photons scattered into the line of sight from the 'halo' of wind material around the star. (Since photons are conserved in the scattering process, that compensation should be exact, excepting photons which are scattered back into the photosphere and hence 'lost'.) Those photons are red-shifted from their orig-

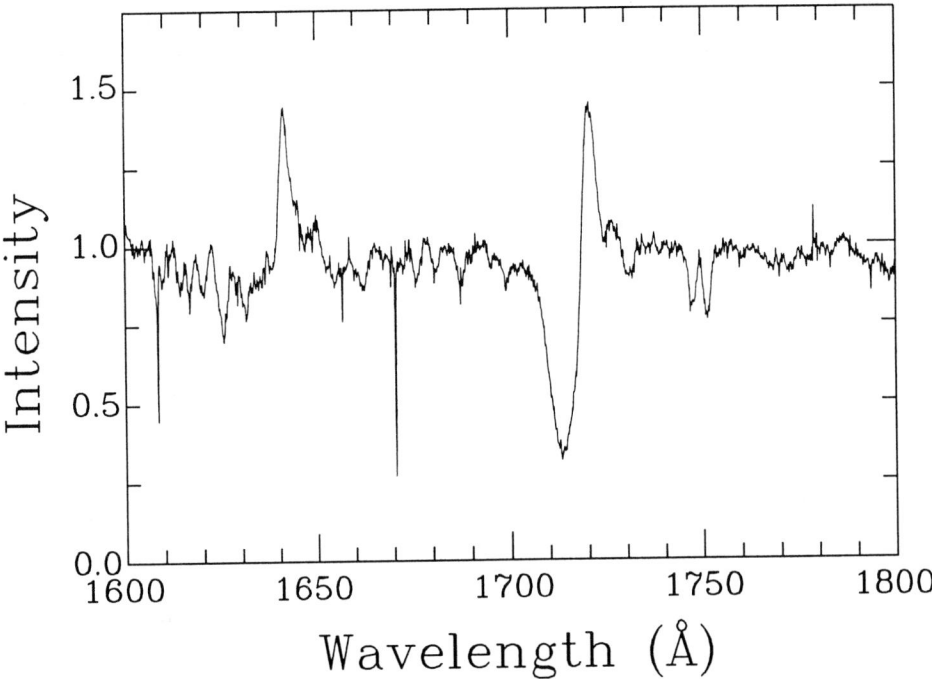

Fig. 1.   Part of the IUE spectrum of the O4 I(n)f star ζ Puppis (averaged over several separate observations). The He II 1640Å line is in emission, and the N IV 1718Å line has a P-Cygni profile, with blueshifted absorption and redshifted emission.

inal wavelengths, and so, in combination with the absorption profile, form the red-shifted emission component of a P-Cygni profile (Fig. 1). The energy 'missing' from the absorption component, less the energy in the emission component, has gone into driving the outflow.

2.1. A SKETCH OF THE THEORY

These qualitative arguments were developed into a quantitative theory by Lucy & Solomon (1970) and by Castor, Abbott & Klein (1975; 'CAK'). The basic theory was subsequently elaborated by Abbott (1978, 1980), Pauldrach, Puls & Kudritzki (1986), Friend & Abbott (1986), and others. Although numerical models of considerable sophistication have now been developed (e.g., Pauldrach et al. 1994; Schaerer & Schmutz 1994), the fundamental characteristics of radiation-driven outflows can be identified from from very simple considerations (cf. Abbott 1978; Kudritzki, Pauldrach & Puls 1986).

The rate at which an isolated strong line, at rest frequency ν_0 and with width $\Delta\nu$, absorbs momentum is

$$\mathrm{d}p/\mathrm{d}t = (L/c)(L_\nu \Delta\nu/L) \tag{1}$$

where the first term on the right-hand side represents the total rate of photon momentum of the star and the second term is the fraction absorbed by the line, and $\Delta\nu = \nu_0 \Delta v/c$. The acceleration is just the rate of momentum absorption per mass element, so we can write the radiative term as

$$g_{\rm rad} = \frac{dp}{dt}/\Delta m \tag{2}$$

where, from mass continuity in steady-state spherical symmetry,

$$\Delta m = 4\pi r^2 \rho(r)\Delta r \tag{3}$$

(that is, the mass-loss rate is $\dot{M} = 4\pi\rho(r)r^2 v(r)$). Combining equations 1–3 we obtain

$$g_{\rm rad} = \frac{L}{c^2}\frac{L_\nu \nu_0}{L}(4\pi r^2 \rho(r))^{-1}\frac{dv}{dr} \tag{4}$$

$$= \frac{NL}{\dot{M}c^2}v(r)\frac{dv}{dr} \tag{5}$$

where we have replaced the final term of equation 1 by a factor N, which represents the fraction of the radiative momentum flux which is intercepted by *all* scattering lines. The equation of motion (ignoring pressure terms) is determined by the outward radiative force and the inward gravitational force:

$$v\frac{dv}{dr} = g_{\rm rad} - g \tag{6}$$

$$= \frac{NL}{\dot{M}c^2}v(r)\frac{dv}{dr} - \frac{GM(1-\Gamma)}{r^2} \tag{7}$$

(where $\Gamma = \kappa L/4\pi GMc$ is the ratio of stellar to Eddington luminosities, and the continuous opacity is $\kappa \simeq \sigma_{\rm e}$, the electron-scattering value). Subsuming 'uninteresting' parameters into a constant we obtain

$$\dot{M} = \frac{L}{c^2}N(1-\epsilon) \tag{8}$$

where

$$\epsilon = \frac{GM(1-\Gamma)}{dv/dr}\frac{4\pi\rho(r)c^2}{NL}. \tag{9}$$

Equation 8 demonstrates the first basic prediction of radiation-driven-wind theory: that there should be a dependence of mass-loss rate on stellar luminosity. (The legerdemain of concealing an additional L term in the ϵ factor is justified by calculations and observations which show that ϵ is of order 10^{-1}, and hence negligible for these illustrative arguments.)

Equations 5 and 8 show that the acceleration $v dv/dr$ is proportional both to g_{rad} and to $g_{rad} - g$. Thus we must also have

$$v(r)\frac{dv}{dr} \propto \frac{GM(1 - \Gamma)}{r^2}. \tag{10}$$

Integrating from the photosphere ($v = 0$) to infinity ($v = v_\infty$) we obtain

$$v_\infty \propto v_{esc} \tag{11}$$

where v_{esc} is the surface escape velocity and v_∞ is the terminal velocity. Equation 11 represents the second basic prediction of radiation-driven wind theory, and survives into more detailed numerical models (albeit with a range of 'constants' of proportionality).

2.2. THEORY MEETS OBSERVATION

2.2.1. Terminal Velocities

Ultraviolet spectroscopy provides an unrivalled method for the measurement of terminal velocities, since only UV resonance lines are sufficiently optically deep to trace the dynamics of the stellar wind to large distances from the photosphere. The first major survey of OB-star winds was carried out by Snow & Morton (1976), using *Copernicus* ultraviolet spectra. They were unable to discern any tight correlation between v_∞ and v_{esc}, but at the time of their study the importance of electron scattering in determining the effective gravity (the '$1 - \Gamma$' term in eqtn. 10) was not widely recognized. Abbott (1978) took this term into account, and found $v_\infty \propto v_{esc}$, as required by theory, with a constant of proportionality of ~ 3.

A much enlarged sample of stars is now available, from IUE observations (Howarth & Prinja 1989; Prinja, Barlow & Howarth 1990; Groenewegen & Lamers 1989). This dataset confirms the correlation between the escape and terminal velocities, and thus supports line radiation pressure as the dominant driving force of hot-star winds. The current generation of stellar-wind models reproduces the observed terminal velocities tolerably well, not only for luminous OB stars, but also for hot subdwarfs and the central stars of planetary nebulae (Pauldrach et al. 1988). Indeed, the success of the models is such that attempts are now being made to use the v_∞/v_{esc} correlation to deduce the escape velocities (and thus the masses) of luminous OB stars.

2.2.2. Mass-loss Rates

The first estimates of mass-loss rates from OB stars were made by Morton (1967b) from strong UV lines, and indicated mass-loss rates of $\sim 10^{-6} M_\odot \, yr^{-1}$ for the Orion-Belt stars. Although the measurement was relatively crude (the analysis of UV spectra continues to present challenges three decades on), it was clearly consistent with radiative driving. If we equate the final mass

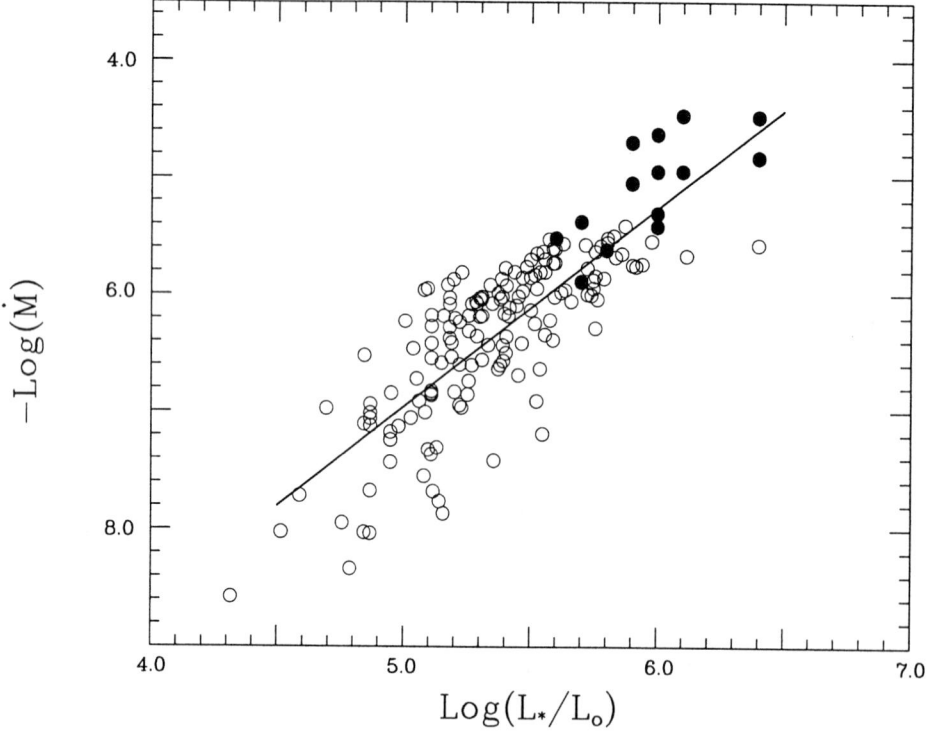

Fig. 2. The relationship between mass-loss rate (in solar masses per year) and luminosity for a sample of O stars, based on data from Howarth & Prinja (1989). Open circles are UV results, and filled circles radio detections. The solid line represents eqtn. 14.

momentum flux of the wind with the momentum transferred by a single scattering of every stellar photon in a doppler-broadened line, then

$$\dot{M}v_\infty = \frac{L_\nu}{c}\nu_0\frac{v_\infty}{c},$$ (12)

so that if ν_0 is near the peak of the stellar energy distribution $\dot{M} \simeq L/c^2$. Covering the spectrum with adjacent, nonoverlapping lines gives a 'single-scattering maximum mass-loss rate' of

$$\dot{M} \simeq L/(v_\infty c)$$ (13)

(cp. eqtn. 8), which provides a reasonable order-of-magnitude estimate of actual mass-loss rates for luminous OB stars. Morton's calculations were consistent with this estimate.

The first systematic survey of mass loss from hot stars was carried out by Barlow & Cohen (1977). By measuring the infrared free-free emission produced in stellar winds, they found $\dot{M} \propto L^{1.1}$ – a striking confirmation of one of the basic predictions of the theory. Subsequent work has tended to revise the constant of proportionality upwards. For example, our analysis of the high-dispersion IUE spectra of more than 200 O stars, externally calibrated in zero-point to radio observations of free-free emission, gives

$$\log(\dot{M}/M_\odot \mathrm{yr}^{-1}) = 1.69 \log(L/L_\odot) - 15.4 \qquad (14)$$

(Howarth & Prinja 1989; Fig. 2). The external calibration is required because lines which are weak enough to be sensitive to column density generally arise in minority ions with small fractional abundances, and stellar-wind models do not, in general, predict these ionization fractions with sufficient accuracy (e.g., Groenewegen & Lamers 1991), although the situation is improving (Pauldrach et al. 1994). Radio observations are well suited to the normalization, in spite of the small fluxes involved, because the radio 'photosphere' – radial optical depth ~unity – forms at many tens or hundreds of stellar radii above the stellar surface (e.g., Bieging et al. 1989). Thus the radio data can be modelled for mass-loss rates independently of the velocity law, unlike the infrared continuum, which for normal OB stars is produced in the accelerating part of the wind (and which is usually strongly contaminated by photospheric emission).

3. The Time-Dependent Behaviour of OB Star Winds

The theory of radiation-driven flows accords so well with observations that there is little doubt that it correctly identifies the dominant physics of hot-star winds. The mismatches between theory and (time-averaged) observation are relatively minor, although they certainly demand attention (Lamers & Leitherer 1993; Schaerer & Schmutz 1994; Puls et al. 1995).

Notwithstanding these successes, the standard theory of radiatively-driven winds is a *steady-state* theory. However, stability analyses of the line-driving mechanism show it to be intrinsically highly unstable (e.g., Carlberg 1980; Owocki & Rybicki 1984, 1985). The instability mechanism is straightforward in the linear regime: a small velocity perturbation will displace a fluid element out of absorption line produced by its neighbours. It therefore 'sees' a stronger driving force, and experiences increased acceleration. If the initial perturbation was to higher velocity, then the displacement increases, and an instability results.

Luminous O and early-B type stars offer observable diagnostics of time-dependent wind phenomena across the electromagnetic spectrum: in X-rays (e.g., Collura et al. 1989; Berghofer et al. 1995), through optical spectroscopy (e.g., Conti & Niemela 1976; Snow, Wegner & Kunasz 1980; Ebbets 1982), and into the radio regime (e.g., Bieging, Abbott & Churchwell 1989). There is no doubt, however, that our present understanding of the nature and importance of stellar-wind variability owes much to UV data provided by the IUE satellite. The long-lived IUE operations have allowed high-resolution time-series data to be collected for P-Cygni profiles in several OB stars.

These observations provide a powerful probe into the time-dependent nature of stellar winds, and have demonstrated both that stellar-wind variability is a widespread phenomenon in hot stars, and that the line-profile variations are systematic, large-scale effects, which challenge any perception of a smooth wind. The structured nature of hot-star winds has important implications for the determination of basic wind parameters, including mass-loss rates, terminal velocities, and the ionization balance, and for theories dealing with the hydrodynamics of line-driven outflows. In the remainder of this review we outline the current observational picture of time-variable phenomena in hot-star winds, emphasizing, where appropriate, results obtained with IUE.

3.1. COHERENT STRUCTURES AND UV DISCRETE ABSORPTION COMPONENTS (DACs)

One of the earliest indications that the stellar winds of OB stars may not be smooth and homogeneous came from the discovery of 'narrow absorption components' in UV P-Cygni profiles. These features – which are 'narrow' only by comparison to full widths of the profiles ($\Delta v/v_\infty \sim 0.1$) – were first observed in rocket spectra and *Copernicus* satellite data (e.g., Underhill 1975; Morton 1976; Snow & Morton 1979). Gathier, Lamers & Snow (1981) and Lamers, Gathier & Snow (1982) carried out the first systematic investigations of the narrow absorption features, using *Copernicus* spectra; these studies were followed by the IUE-based surveys of Prinja & Howarth (1986) and Howarth & Prinja (1989). The features are usually present close to the bluewards edge of the P-Cygni profiles in a variety of wind-formed resonance lines (e.g., C IV, N V, Si IV), with the same central velocity in every line. Some examples are shown in Figure 3.

Although P-Cygni line-profile changes were found in the *Copernicus* spectra of OB stars examined by Snow (1977), York et al. (1977) were among the first to show that the narrow components were variable on timescales of hours (in the O VI $\lambda1031$ line). However, because of the relatively low sensitivity and lack of multiplexing of the *Copernicus* instrument, the nature of these profile changes remained poorly understood until the launch of IUE in 1978. Subsequent studies of individual stars, based on IUE high-resolution time-series data, have established that the variations in the wind lines are not chaotic fluctuations, but are instead very systematic – a discovery which rates as one of the major accomplishments of the IUE satellite.

The most prominent signature of variable wind structure in hot stars is the presence of blueward-migrating 'discrete absorption components' (DACs), which stand out clearly in time-series IUE spectra. An example is given in Figure 4, where a sequence of Si IV $\lambda1400$Å P-Cygni profiles is shown as a

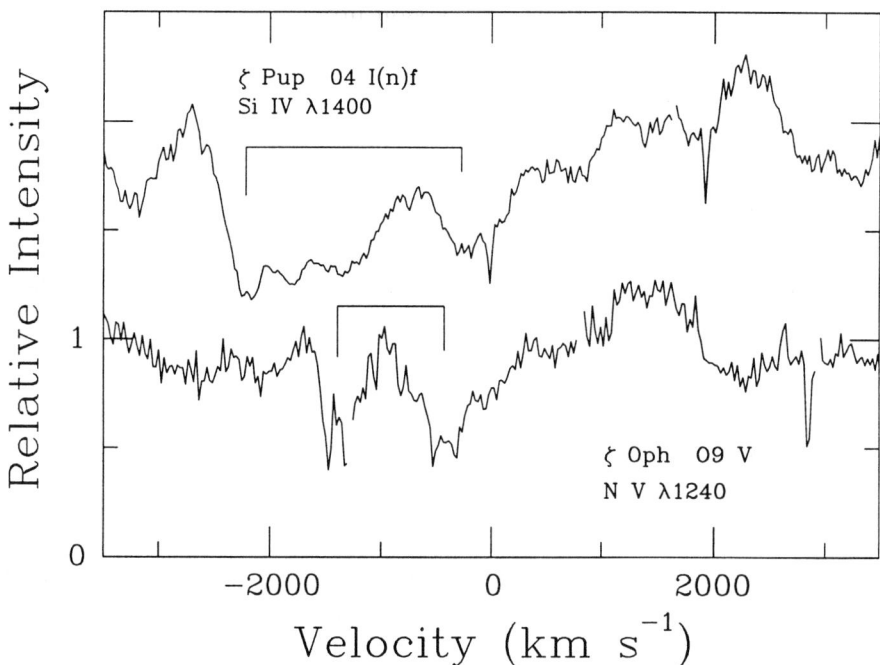

Fig. 3.   High-velocity, narrow absorption features in IUE high-resolution spectra of an early-O supergiant and a late-O main-sequence star. The Si IV and N V lines are both doublets; the zero of the velocity scale is for the bluewards component, and the ticks show the position of the doppler-shifted absorption features.

function of time for the O7 giant ξ Per. The profile variability – corresponding to changes > 20% in Si^{3+} column density – can be interpreted in terms of localized optical-depth enhancements, which are typically first detected at $\sim$ $0.3v_\infty$. The features subsequently accelerate bluewards through the absorption trough, becoming increasingly narrow in velocity as they approach the terminal velocity (see, e.g., Prinja, Howarth & Henrichs, 1987; Prinja & Howarth, 1988; Henrichs 1988; Prinja et al. 1992; Kaper, 1993). The narrow absorption components identified in isolated *Copernicus* and IUE spectra are now seen to be just one aspect of variability across a very broad velocity range in the absorption trough.

The UV observations also provide the best opportunity to track directly the velocity, and hence acceleration, of material associated with hot-star winds. The features evolve over timescales which range from many hours to several days, typically on scales similar to, or longer than, the wind flow-times ($\sim v_\infty/R_*$). In most cases the accelerations are no more than $\sim$50% of the values predicted by steady-state wind models. However, the observed morphology of P-Cygni line profiles (e.g., Groenewegen & Lamers 1989), and

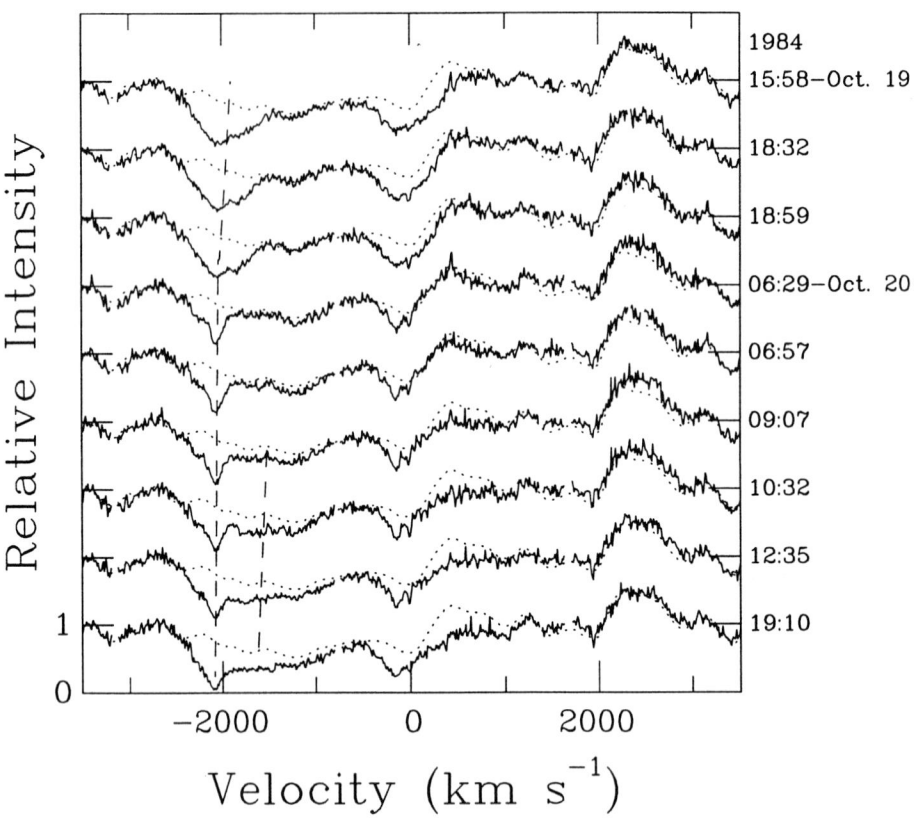

Fig. 4.   Part of a time sequence of Si IV λ1400 P-Cygni profiles in the spectrum of ξ
Per [O7.5 III(n)((f))]. A reference model of the 'underlying' profile is shown in each case
(dotted line) to make clear the variability in the observed profiles. The dashed lines trace
out the development, in the bluewards doublet component, of two discrete absorption
components (DACs). Matching behaviour can be seen in the redwards component, at
~zero velocity (the doublet separation of ~2000 km s^{-1} roughly matches the velocity
displacement of the features in this example).

global wind-density structures derived from Hα emission profiles (Leitherer
1988), support the radial-velocity laws predicted by steady-state models.
This apparent disparity has an important bearing on the physical relation
between the spectroscopic features and the time-averaged wind structure.
The inference is that we are observing the slow evolution of a perturbation
or structure in the wind, through which stellar wind material flows; i.e., we
are not observing a single mass-conserving feature, such as a shell or blob,
when we follow the evolution of a discrete absorption feature.

Migrating DACs have now been reported in the UV wind lines of a wide
range of hot stars, including O and early B-type supergiants, giant and

dwarf O stars, and a WN 7 Wolf-Rayet star. The UV case studies combine with 'snapshot' surveys of profile morphology to suggest that systematically evolving wind structure is widespread in hot stars; its regularity and widespread incidence makes this a new class of stellar variability.

3.2. OPTICAL SPECTROSCOPY AND THE NEAR-STAR WIND STRUCTURE

Swings & Struve (1941) were among the first to note variability in the H and He emission lines of O-type stars. The widespread occurrence of changes in optical wind-lines was established by the surveys of Hutchings & Sanyal (1976), Ebbets (1980, 1982), Snow, Wegner & Kunasz (1980), and Grady, Snow & Timothy (1983). The latter studies, using modern detectors, mostly sampled the winds on daily or longer timescales, and their results pointed to changes in the global mass-flux of less than a factor or 2, a result supported by UV studies (Prinja & Howarth 1986).

The optical waveband provides an important opportunity to test whether the inner wind is effectively stable on ~hour timescales, since many optical wind lines are principally formed by recombination (e.g., He I $\lambda5876$, Hα), and hence preferentially arise in the high-density (near-star) region of the outflows. Another advantage is that these spectral lines can be readily observed with high signal-to-noise and high dispersion. A key issue in time-dependent studies of hot stars is the degree to which the inner stellar wind – i.e., ~1.0–2.0 stellar radii – is affected by rapid changes and systematic structure on timescales comparable to the flowtimes in the region where Hα is formed. Details on the character of the near-star wind may help constrain physical models for the origin of wind variability, including the causal role (if any) played by photospheric variations.

The majority of the UV case studies show systematic wind activity, in terms of localised and evolving optical-depth enhancements, to first be evident at intermediate velocities (i.e., $\sim 0.3 - 0.5 \, v_{\infty}$). * This is true also for Hubble Space Telescope (GHRS) UV spectroscopy (e.g., Heap 1994). In part, this must arise because of the strength of the low-velocity absorption in UV lines. The absorption trough can be intrinsically saturated – and hence insensitive to optical-depth changes – even when the residual intensity appears quite high, because of 'infilling' by forward-scattered photons.

Motivated by these IUE results, several studies have recently been carried out in order to explore possible links between optical and UV wind variability. Henrichs (1991) and Henrichs, Kaper & Nichols (1994) have reported simultaneous UV and optical spectroscopy of mid- and late-O type stars. They found evidence which they interpret as suggesting that the initial appearance of a DAC in IUE data may be accompanied by an enhance-

---

* Henrichs, Kaper & Nichols (1994) and Howarth & Smith (1995) have documented line-profile variability down to velocities of 50–100 km s^{-1}, but this probably only represents the 'tail' of broad features centred on intermediate velocities.

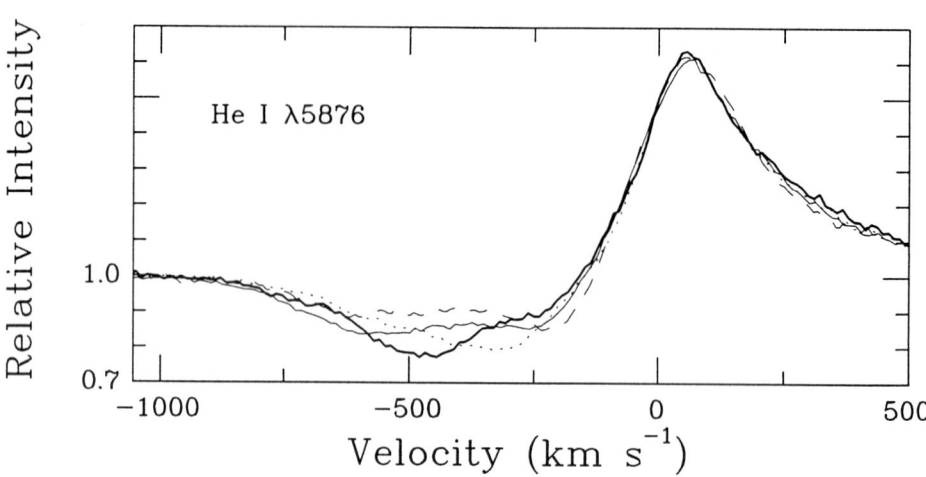

Fig. 5.   Night-to-night variations in the He I λ5876 P Cygni profiles of the O8: Iafpe star HD 152408.

ment in the blueward emission flux in Hα or He II λ4686. Prinja & Fuller-ton (1994) and Prinja, Fullerton & Crowther (1995) have monitored varia-tions in the optical wind-formed lines of the high-mass-loss O8 supergiants HD 152408 and HD 151804, respectively. Time-series monitoring of the He I λ5876 P Cygni profiles revealed systematic variability in both cases, which takes the form of optical-depth enhancements which travel from $\sim 0.15 v_\infty$ to the profile edge (blueward of $\sim 0.8 v_\infty$) over ~ 24 hours (see also Fuller-ton, Gies & Bolton 1992). Relative to hydrogen, the column densities of the optical features are estimated to be an order of magnitude higher than those of DACs seen at low velocities in the UV spectra of OB stars. Sympathetic variations were also evident in the *blueward* Hα emission wings, such that an increase in He I absorption (due to 'optical DACs') is accompanied by a decrease in Hα flux. The results suggest that the two spectral lines react in concert to localized density changes.

It is reasonable to suppose that the migrating spectral features (and relat-ed short-timescale variations) seen in the optical lines represent counterparts of the physical structure diagnosed by the UV DACs. The emission-line for-mation regions for optical helium and Balmer lines are probably within ~ 3.0 stellar radii of the photosphere (e.g., Hillier 1987; Puls et al. 1995). The optical results therefore trace deep-seated variations and show that the near-star winds are unstable on hourly timescales. It seems that departures from steady state are likely to occur, at some level, throughout the winds of OB stars. The overall variations must be due to substantial perturbations in the flow which most likely co-exist with small-scale instabilities which are predicted by time-dependent models of line-driven instabilities (e.g., Owocki, Castor & Rybicki 1988; Owocki 1994). These small-scale structures probably provide the microscopic broadening mechanism needed to give 'soft' blue wings and extended saturated absorption troughs in strong P-Cygni profiles (e.g., Puls, Owocki & Fullerton 1993), and account for the observed stability of X-ray fluxes in OB stars (Berghofer & Schmitt 1994). The UV and optical DACs, however, require structure on much larger scales, and which is spatially coherent across a reasonable fraction of a stellar radius. The origin of this *large-scale* instability is unknown, though, as we outline below, a key constraint is that it may be somehow rooted at the stellar photosphere.

TABLE I

Wind Activity versus Stellar Rotation

Star	Sp. type	$v_e \sin i$ (km s^{-1})	P_{rot} (max) (days)	DAC recurrence time (days)
ζ Oph	O9.5 V	400	1.1	0.8
68 Cyg	O7.5 III:n((f))	315	2.3	0.7
λ Cep	O6 I(n)fp	275	4.0	1.2
HD 64760	B0.5 Ib	238	4.8	1.2
ζ Pup	O4 I(n)f	230	4.8	0.8
ξ Per	O7.5 III(n)((f))	215	2.6	1.3
ζ Ori A	O9.7 Ib	135	13.4	1.6
HD 34656	O7 IIf	105	4.8	0.9
15 Mon	O7 V((f))	100	8.1	>4.5
HD 164402	B0 Ib	90	13.5	>3.0
HD 162978	O7.5 II((f))	80	10.1	>3.0
λ Ori A	O8 III((f))	65	11.5	1.6
19 Cep	O9.5 Ib	40	22.8	>5.
10 Lac	O9 V	30	14.3	>4.

The minimum rotation periods are given by $2\pi R/v_e \sin i$ and are subject to uncertainties of $\sim$20% or more.

3.3. WIND VARIABILITY AND STELLAR ROTATION: THE IUE 'MEGA' CAMPAIGN

A comparison of the observed accelerations and recurrence timescales of DACs in time-series IUE data, as a function of projected rotation velocity, ($v_e \sin(i)$), led Prinja (1988, 1992), Henrichs, Kaper & Zwarthoed (1988), and Kaper (1993) to draw attention to the potential role of stellar rotation in modulating wind activity in OB stars. This connection is summarized for 14 OB stars in Table I, where DAC recurrence times are compared to the projected rotation velocities and estimated maximum rotation periods. There is a trend towards faster-developing, more-frequent DACs with increasing

$v_e \sin(i)$. Typically, however, the individual time-series on which the DAC results are based span only 3–4 days. This means that the behaviour of only 1–2 DACs can normally be followed, that the upper limit for the DAC recurrence time is not constrained for the slower rotators, and that the stellar rotation period is not fully covered.

Recognizing these limitations, the participants at the workshop on 'Instabilities and Variability of Hot-Star Winds' (Moffat et al. 1994) proposed a substantial, collaborative effort to secure intensive IUE time-series data for selected stars, extending over several predicted rotation periods. Even after 18 years of operation, the expectation was that IUE data would provide the new observational constraints to gain greater insight to the physical processes responsible for wind variations. In the resulting IUE 'Mega' campaign (Massa et al. 1995), which took place in 1995 January, the stellar winds of HD 50896 (EZ CMa; WN 5), HD 64760 (B0.5 Ib), and HD 66811 (ζ Pup; O4 If(n)) were monitored almost continuously. The datasets span three or more stellar rotation periods, and amount to ~150 high resolution spectra, for each star.

Initial results from this campaign have been reported by St-Louis et al. (1995; EZ CMa), Prinja, Massa & Fullerton (1995; HD 64760), and Howarth, Prinja & Massa (1995; HD 66811). Two-dimensional, greyscale image representations of the line-profile variability in the Si IV $\lambda\lambda$1393.76, 1402.77 P Cygni profiles of HD 64760 and ζ Pup are shown in Figure 6.

The stellar wind of HD 64760 exhibits two main types of structure, which co-exist but which evolve on different timescales. First, there are relatively numerous, fast-moving features (with observed accelerations of ~1.0–1.8×10^{-2} km s^{-2}), which span most of the absorption trough from an initial appearance redward of ~ –200 km s^{-1}. Secondly, and distinctly, there are two slower-moving features (observed accelerations of ~3–9$\times10^{-3}$ km s^{-2}), which are more reminiscent of the DACs seen in previous case studies of O-type stars. It is the faster evolving, more frequent, optical-depth variations which, when monitored over such an extensive period, reveal a periodic modulation of the wind. Fourier time-series analysis of the N V and Si IV profiles reveals a 1.2-day period which is $\frac{1}{4}$ of the expected stellar rotation period. The line-profile changes are consistent with an interpretation in terms of a set of co-rotating wind features which occult the stellar disk at least 3 times during the observing run (Prinja, Massa & Fullerton 1995).

Time-series analysis of the ζ Pup data set (Figure 6; Howarth, Prinja & Massa 1995) reveals significant power at periods of 19.2 hours and 5.2 days. The former may be identified with the mean recurrence time of the DACs, while the latter is consistent with the photospheric rotation period. It seems that while the DAC recurrence time is *not* an integral fraction of the rotation period for ζ Pup, cyclical patterns of absorption-line variations *are* observed on the rotation period. It is also interesting to note there is

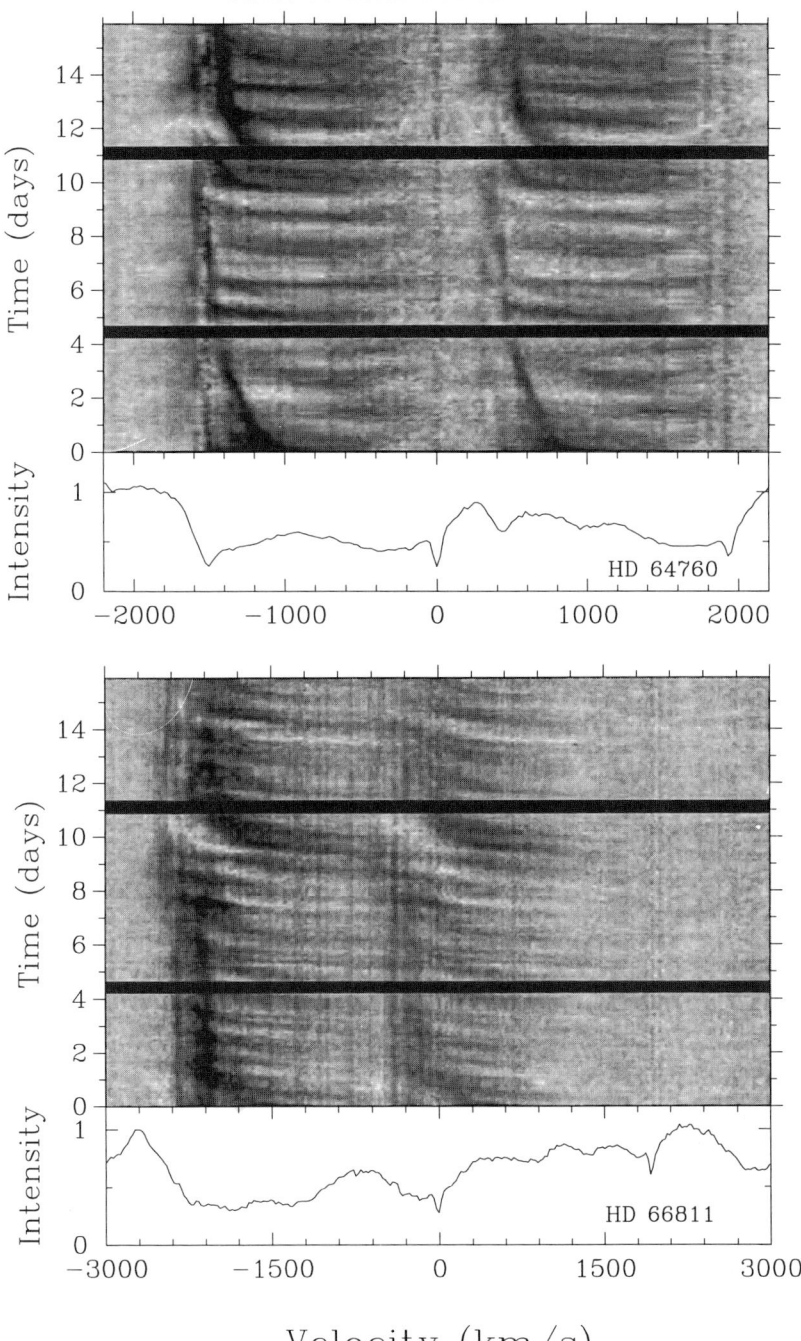

Fig. 6.   Greyscale representation of time variability in the Si IV λλ1400 wind lines of HD 64760 (B0.5 Ib) and HD 66811 (ζ Pup; O4 I(n)f). For each star, every spectrum was normalized to a 'minimum absorption' template constructed from the full dataset, which resemble the mean spectra shown in the lower panels. The dynamic spectra are therefore darkest at times and wavelengths where the line optical depths are greatest.

no significant power in these wind-formed profiles at the 8.5-hour period which has been identified in optical absorption-line variations (Baade 1991, Reid & Howarth 1995), and which may be associated with photospheric velocity fields resulting from non-radial pulsation. However, analysis of Hα observations of ζ Pup shows that both the 8.5-hr and 19.2-hr periods can be recovered in that line (Reid & Howarth 1995) – suggesting that time-dependent wind structure can be traced to 'deep' regions, as for HDs 151408 and 152408.

The IUE 'Mega' campaign has, therefore, demonstrated that substantial structure occurs in the stellar winds of hot stars, and can persist over several rotation cycles. Because this timescale greatly exceeds the flowtimes, it is highly unlikely that a suitable 'clock' can be obtained from a mechanism which is entirely intrinsic to the outflow. The periodic nature of the variations instead points to a mechanism that couples the outer wind layers to modulations at the photosphere. A general model which may address many of the observed features is that of corotating streams, first discussed for early-type stars by analogy with the solar case (Mullan 1984, 1986), and applied specifically to the 'Mega' data by Owocki, Cranmer & Fullerton (1995).

Whatever the proximate cause of the rotational 'clock', it must ultimately be locked to the photosphere. Two possible sources for photospheric modulation in early-type stars are velocity fields resulting from stellar pulsation (e.g., Jerzykiewicz 1994, Gies 1991) and inhomogeneities due to magnetic fields (e.g., Bohlender 1994). In neither case has an explicit, causal connection to the outflowing OB star wind been demonstrated, although the identification of the same 8.5-hr period in the photospheric absorption lines and the wind-formed Hα line of ζ Pup is strongly suggestive (Reid & Howarth 1995), and photospheric pulsations are now generally accepted to occur in some hot stars (e.g., Reid et al. 1993; Fullerton, Gies & Bolton 1995). Observational surveys have not yet revealed any firm detections of magnetic fields at the 300 Gauss level for any normal, single OB star (e.g., Barker 1986; Bohlender 1994), but this does not rule out a possibly significant role for weaker fields.

The question remains as to whether rotational modulation of OB star winds is ubiquitous. Should a causal connection between wind activity and rotation be firmly established, perhaps from further observations of slow, normal, and fast rotators, we will be confronted with a radical and fundamental shift in our understanding of the winds, atmospheres, and photospheres of hot stars. It is remarkable that, even after 18 years of almost uninterrupted, round-the-clock operation, the IUE satellite can still provide such spectacular new insights into the nature of hot-star winds.

References

Abbott, D.C.: 1978, *ApJ* **225**, 893

Abbott, D.C.: 1980, *ApJ* **242**, 1183

Baade, D.: 1991, 'Regular variability of optical lines in ζ Pup' in D. Baade, ed(s)., *ESO Workshop on Rapid Variability of OB-Stars: Nature and Diagnostic Value*, ESO:Munich, 21

Barker, P.K.: 1986, in A. Slettebak and T.P. Snow, (ed(s)., *The Physics of Be Stars* (IAU Coll. 92), CUP: Cambridge, 38

Barlow, M.J. and Cohen, M.: 1977, *ApJ* **213**, 737

Berghofer, T.W., Baade, D., Schmitt, J.H.M.M., Kudritzki, R.-P., Puls, J., Hillier, J., and Pauldrach, A.: 1995, *A&A.* in press.

Berghofer, T.W. and Schmitt, J.H.M.M.: 1994, 'Rosat X-ray light curves of early-type stars' in A.F.J. Moffat, S.P. Owocki, A.W. Fullerton, and N. St-Louis, ed(s)., *Instability and Variability of Hot-Star Winds*, Kluwer: Dordrecht, 309

Bieging, J.H., Abbott, D.C., and Churchwell, E.B.: 1989, *ApJ* **340**, 518

Bohlender, D: 1994, 'Observations of magnetic fields in B stars' in L.A. Balona, H.F. Henrichs, and J.M. Le Contel, ed(s)., *Pulsation, Rotation and Mass Loss in Early-Type Stars, IAU Symp. 162*, Kluwer: Dordrecht, 155

Carlberg, R.G.: 1980, *ApJ* **241**, 1131

Castor, J.I., Abbott, D.C., and Klein, R.: 1975, *ApJ* **195**, 157

Collura, A., Sciortino, A., Serio, S., Vaiana, G.S., Harnden, F.R., and Rosner, R.: 1989, *ApJ* **338**, 296

Conti, P.S. and Niemala, V.: 1976, *ApJL* **234**, L51

Ebbets, D.: 1980, *ApJ* **235**, 97

Ebbets, D.: 1982, *ApJS* **48**, 399

Friend, D.B. and Abbott, D.C.: 1986, *ApJ* **311**, 701

Fullerton, A.W., Gies, D.R., and Bolton, C.T.: 1992, *ApJ* **390**, 650

Fullerton, A.W., Gies, D.R., and Bolton, C.T.: 1996, *ApJ,* in press.

Gathier, R., Lamers, H.J.G.L.M., and Snow, T.P.: 1981, *ApJ* **247**, 173

Gies, D.R.: 1991, 'Nonradial pulsations and line profile variability in early type stars' in D. Baade, ed(s)., *ESO Workshop on Rapid Variability of OB-Stars: Nature and Diagnostic Value*, ESO:Munich, 229

Grady, C.A., Snow, T.P., and Timothy, J.G.: 1983, *ApJ* **271**, 691

Groenewegen, M.A.T. and Lamers, H.J.G.L.M.: 1989, *A&AS* **79**, 359

Groenewegen, M.A.T. and Lamers, H.J.G.L.M.: 1991, *A&A* **243**, 429

Heap, S.R.: 1994, 'Monitoring of winds of hot stars with HST/GHRS' in A.F.J. Moffat, S.P. Owocki, A.W. Fullerton, and N. St-Louis, ed(s)., *Instability and Variability of Hot-Star Winds*, Kluwer: Dordrecht, 87

Henrichs, H.F.: 1988, 'Intrinsic variability in ultraviolet spectra of early-type stars: the discrete absorption components' in P.S. Conti and A.B. Underhill, ed(s)., *O Stars and Wolf-Rayet Stars*, NASA: Washington, 199

Henrichs, H.F.: 1991, 'Why are winds of O stars variable?' in D. Baade, ed(s)., *ESO Workshop on Rapid Variability of OB-Stars: Nature and Diagnostic Value*, ESO:Munich, 199

Henrichs, H.F., Kaper, L., and Nichols, J.S.: 1994, 'Wind variability in O-type stars' in L.A. Balona, H.F. Henrichs, and J.M. Le Contel, ed(s)., *Pulsation, Rotation and Mass Loss in Early-Type Stars*, Kluwer: Dordrecht, 517

Henrichs, H.F., Kaper, L., and Zwarthoed, G.A.A.: 1988, 'Rapid variability in O star winds' in E.J. Rolfe, ed(s)., *A Decade of UV Astronomy with the IUE satellite*, ESA: Noordwijk, 145

Hillier, D.J.: 1987, *ApJS* **63**, 947

Howarth, I.D. and Prinja, R.K.: 1989, *ApJS* **69**, 527

Howarth, I.D. and Smith, K.C.: 1995, *ApJ* **439**, 431

Howarth, I.D., Prinja, R.K., and Massa, D.: 1995, *ApJL* **452**, 65

Hutchings, J.B. and Sanyal, A.: 1976, *PASP* **88**, 279

Jerzykiewicz, M.: 1994, in L.A. Balona, H.F. Henrichs and J.M. Le Contel, ed(s)., *Pulsation, Rotation and Mass Loss in Early-Type Stars* (IAU Symp. 162), Kluwer: Dordrecht, p.3

Kaper, L.: 1993, *Wind Variability in Early-type Stars* (PhD thesis, Univ. of Amsterdam)

Kudritzki, R.P., Pauldrach, A., and Puls, J.: 1986, in E.J. Rolfe, ed., *New Insights in Astrophysics,* ESA: Noordwijk, 247

Lamers, H.J.G.L.M., Gathier, R., and Snow, T.P.: 1982, *ApJ* **258**, 186

Lamers, H.J.G.L.M. and Leitherer, C.: 1993, *ApJ* **412**, 771

Leitherer, C.: 1988, *ApJ* **326**, 356

Lucy, L.B. and Solomon, P.M.: 1970, *ApJ* **159**, 879

Massa, D. et al.: 1995, *ApJL* **452**, 53

Moffat, A.F.J., Owocki, S.P., Fullerton, A.W., and St-Louis, N.: 1994, *Instability and Variability of Hot-Star Winds,* Kluwer: Dordrecht

Morton, D.C.: 1967a, *ApJ* **147**, 1017

Morton, D.C.: 1967b, *ApJ* **150**, 535

Morton, D.C.: 1976, *ApJ* **203**, 386

Mullan, D.J.: 1984, *ApJ* **283**, 303

Mullan, D.J.: 1986, *A&A* **165**, 157

Owocki, S.P.: 1994, 'Line-driven instability and other causes of structure and variability in hot-star winds' in A.F.J. Moffat, S.P. Owocki, A.W. Fullerton and N. St-Louis, ed(s)., *Instability and Variability of Hot-Star Winds,* Kluwer: Dordrecht, 3

Owocki, S.P., Castor, J.I., and Rybicki, G.B.: 1988, *ApJ* **335**, 914

Owocki, S.P., Cranmer, S., and Fullerton, A.W.: 1995, *ApJL* **453**, 37

Owocki, S.P. and Rybicki, G.B.: 1984, *ApJ* **284**, 337

Owocki, S.P. and Rybicki, G.B.: 1985, *ApJ* **299**, 265

Pauldrach, A., Puls, J., and Kudritzki, R.-P.: 1986, *A&A* **164**, 86

Pauldrach, A., Puls, J., Kudritzki, R.P., Méndez, R.H., and Heap, S.R.: 1988, *A&A* **207**, 123

Pauldrach, A.W.A, Kudritzki, R.P., Puls, J., Butler, K., and Hunsinger, J.: 1994, *A&A* **283**, 525

Prinja, R.K.: 1988, *MNRAS* **231**, 21P

Prinja, R.K.: 1992, 'UV P Cygni variability in O stars' in L. Drissen, C. Leitherer, and A. Nota, ed(s)., *Nonisotropic and Variable Outflows from Stars,* ASP Conf Ser: San Francisco, 167

Prinja, R.K., Barlow, M.J., and Howarth, I.D.: 1990, *ApJ* **361**, 607

Prinja, R.K. and Howarth, I.D.: 1986, *ApJS* **61**, 357

Prinja, R.K., Howarth, I.D., and Henrichs, H.F.: 1987, *ApJ* **317**, 389

Prinja, R.K. and Howarth, I.D.: 1988, *MNRAS* **233**, 123

Prinja R.K. et al.: 1992, *ApJ* **390**, 266

Prinja, R.K. and Fullerton, A.W.: 1994, *ApJ* **426**, 345

Prinja, R.K., Fullerton, A.W., and Crowther, P.A.: 1995, *A&A,* in press

Prinja, R.K., Massa, D., and Fullerton, A.W.: 1995, *ApJL* **452**, 61

Puls, J., Owocki, S.P., and Fullerton, A.W.: 1993, *A&A* **279**, 457

Puls, J., et al.: 1995, *A&A,* in press

Reid, A.H.N., et al.,: 1993, *ApJ* **417**, 320

Reid, A.H.N. and Howarth, I.D.: 1996, *A&A,* in press

Schaerer, D. and Schmutz, W.: 1994, *A&A* **288**, 231

Snow, T.P.: 1977, *Ap J* **217**, 760

Snow, T.P. and Morton, D.C.: 1976, *ApJS* **32**, 429

Snow, T.P., Wegner, G.A., and Kunasz, P.B.: 1980, *ApJ* **238**, 643

St-Louis, N.L. et al.: 1995, *ApJL* **452**, 57

Swings, P. and Struve, O.: 1941, *ApJ* **91**, 546

Underhill, A.B.: 1958, *Mem. R. Sci. Liège* **20**, 91

Underhill, A.B.: 1975, *ApJ* **199**, 691

Wilson, R.: 1955, *The Observatory* **75**, 222

Wilson, R.: 1958, *Mem. R. Sci. Liège* **20**, 85
York, D. G. *et al.*: 1977, *ApJ* **213**, L61

WOLF–RAYET STARS

*Physical, Chemical and Mass Loss Properties
and Evolutionary Status*

ALLAN J. WILLIS

*Department of Physics & Astronomy, University College London, Gower Street, London
WC1E 6BT, UK*

Abstract. This paper reviews the current status of knowledge regarding the basic physical
and chemical properties of Wolf–Rayet stars; their overall mass loss and stellar wind characteristics and current ideas about their evolutionary status. WR stars are believed to be
the evolved descendents of massive O–type stars, in which extensive mass loss reveals successive stages of nuclear processed material: WN stars the products of interior CNO-cycle
hydrogen burning, and WC and WO stars the products of interior helium burning. Recent
stellar evolution models, particularly those incorporating internal mixing, predict results
which are in good accord with the different chemical compositions observationally inferred
for WN, WC and WO stars. WR stars exhibit the highest levels of mass loss amongst early-
type stars: mass loss rates, typically, lie in the range $[1–10] \times 10^{-5}$ $M_\odot \mathrm{yr}^{-1}$. Radiation
pressure–driven winds incorporating multi-scattering in high ionisation–stratified winds
may cause these levels, but additional mechanisms may also be needed.

1. INTRODUCTION

I first met Bob Wilson when he became Perren Professor of Astronomy at
UCL in 1972, and soon after he took me on as his first PhD student. At that
time we undertook pioneering ultraviolet spectrophotometric observations of
Wolf–Rayet stars obtained with the S2/68 experiment on the ESRO TD-1
satellite, which was soon to be followed by higher spectral resolution UV
data from the *IUE* satellite – both space projects in which Bob had played
a seminal rôle. The early programmes that Bob and I undertook, led to the
development of a major research group at UCL involved in, *inter alia*, the
investigation of hot, massive stars, their stellar winds, and interaction effects
in binary systems. Extensive programmes in these fields have been ongoing
at UCL since then.

Wolf–Rayet (WR) stars as a stellar class were first identified over 125
years ago (Wolf & Rayet, 1867), through prism spectroscopy of a few stars in
Cygnus which showed strong, broad emission lines superimposed on a rather
weak, underlying blue continuum, in stark contrast to the usual absorption–
line spectrum characterising most stellar types. The identification of further
examples, often linked with OB stars and associations, strongly suggested
a Population I status, but progress in determining their true physical and
chemical nature and evolutionary status was very slow. A major step forward came from the work of Beals (1929) who recognised that the observed
P–Cygni and emission profiles (some flat-topped) could be explained by

Astrophysics and Space Science **237**: 145–168, 1996.

invoking emission line formation in a strong, outwardly accelerating, stellar wind. The observed line widths, interpreted as doppler broadening, indicated outflows of $\sim$ 1000–3000 km s^{-1}, greatly in excess of expected stellar escape velocities – the first strong indication that mass loss is a major issue for WR stars. Beals and Plaskett (1935) further showed that the WR stars could be separated into two well defined spectral sequences: WN stars showing emission lines predominantly from He and N transitions, and WC stars exhibiting emission in He, C and O ions with little evidence for N, developing the first WR classification scheme which (with modifications and updates) we still largely employ to this day. This spectral dichotomy indicated, at least to some, a true chemical abundance separation. The status of WR research, and progress in trying to understand these stars during the next few decades, formed the basis of two milestone conferences: a meeting in Boulder in 1968 (Gebbie & Thomas, 1971) and the first IAU Symposium devoted to WR stars held in Buenos Aires in 1971 (Bappu & Sahade, 1973). Classification schemes, the distribution of WR stars in the Galaxy, estimates of stellar temperatures, luminosities, masses, and wind structure, potential evolutionary scenarios and updates on the (then) latest model atmosphere theories, all figured prominently at these conferences. Since then there has been a veritable explosion of activity in WR research, based on new observational data across the electromagnetic spectrum, coupled with tremendous advances in model atmosphere and stellar evolution theory. Results have been reported regularly at many IAU symposia and other major meetings (IAU 99: de Loore & Willis 1982; IAU 116: de Loore, Willis & Laskarides, 1986; IAU 143: van der Hucht & Hidayat, 1991; IAU 163: van der Hucht & Williams, 1995). Further major reviews of recent WR research can be found in Abbott & Conti (1987), Kudritzki & Hummer (1990), van der Hucht (1992), Crowther & Willis (1994), and Maeder & Conti (1994).

It is now generally agreed that WR stars form an integral, and advanced stage in the evolution of massive stars ($M_{initial} \geq$ 25–40 $M_{\odot}$), representing the evolved descendents of massive O–type stars. Extensive mass loss and/or Roche-lobe overflow in binaries, coupled with the possibility of internal mixing, provides a mechanism whereby the outer atmosphere can be sequentially 'peeled down' to reveal the chemical products of nuclear processing: WN stars exposing CNO–burning products, whilst the more evolved WC stars (and WO stars – see below) show the products of subsequent He–burning. WR stars thus provide a superb test for theories of advanced massive stellar evolution and of nucleosynthesis models. Their heavy mass loss and supersonic winds provide an important source of photon and mechanical energy and chemical enrichment into the ISM. Their number, subtype distribution and properties in different galaxies provides important insight into the initial mass functions and massive star formation rates in different metallicity environments.

In this review I will summarise some of the recent results in WR research that have led to the above conclusions. The sheer volume of research in this field currently being undertaken means that I have had to be highly selective, emphasising progress in the determination of the fundamental physical, chemical and mass loss properties of WR stars stemming from observational programmes and their quantitative analysis.

2. CLASSIFICATION, TEMPERATURES and LUMINOSITIES

2.1. CLASSIFICATION

The basic subtype classification for both WN and WC stars stems from Smith (1968, 1973), who used relative optical lines strength ratios to assign an 'excitation' class, analogous to spectral typing for OB stars. Thus the WN sequence, WN9, WN8, ... WN3 (lower-to-higher excitation) is defined using line ratios of NII/NIII/NIV/NV, whilst for WC stars the sequence WC9, WC8, ... WC4, (again lower-to-higher excitation) reflects the relative strengths of observed CII/CIII/CIV transitions. Subsequent refinements are discussed by van der Hucht *et al* (1981) and Abbott & Conti (1987). The relative appearance of stars in both WN and WC sequences observed in both ultraviolet spectra (eg. Willis et al. 1986), and in infrared spectra (eg. Smith & Hummer 1988, Crowther & Smith 1996a) is generally found to be in accord with their optical classifications. For the WN stars the term WNL ('late') is often assigned to lower excitation subclasses, WN6–WN9 (Vanbeveren & Conti 1980), particularly for stars whose optical spectra show evidence for significant hydrogen emission. Conversely, the higher excitation WN–types, WN3–WN5/6, which usually do not exhibit optical hydrogen emission features, are often designated WNE (early). However, this 'late'/'early'–type designation is *not* intended as a reflection of their evolution timeframes ! The likely connection between O–stars and WN stars (first proposed by Conti 1975) was reinforced by Walborn (1977, 1982) who identified a new spectral subclass, denoted Ofpe/WN9, whose optical spectra contained both typical Of-type emission features in NIII, HeII and low ionisation WN9–like emissions in HeI, NII. Bohannan & Walborn (1989) have identified 10 Ofpe/WN9 stars in the LMC, Willis *et al* (1992) one such object in M33, and Allen *et al* (1990) have suggested that some stars near the Galactic Centre are of this class. Barlow & Hummer (1982) identified a third WR sequence, the WO stars, which show exceptionally strong optical OVI 3811,3834 emission lines as well as unusually strong CIV transitions. Individual WO stars are designated subtypes WO1...WO4, depending on the relative strengths of OIV/OV/OVI.

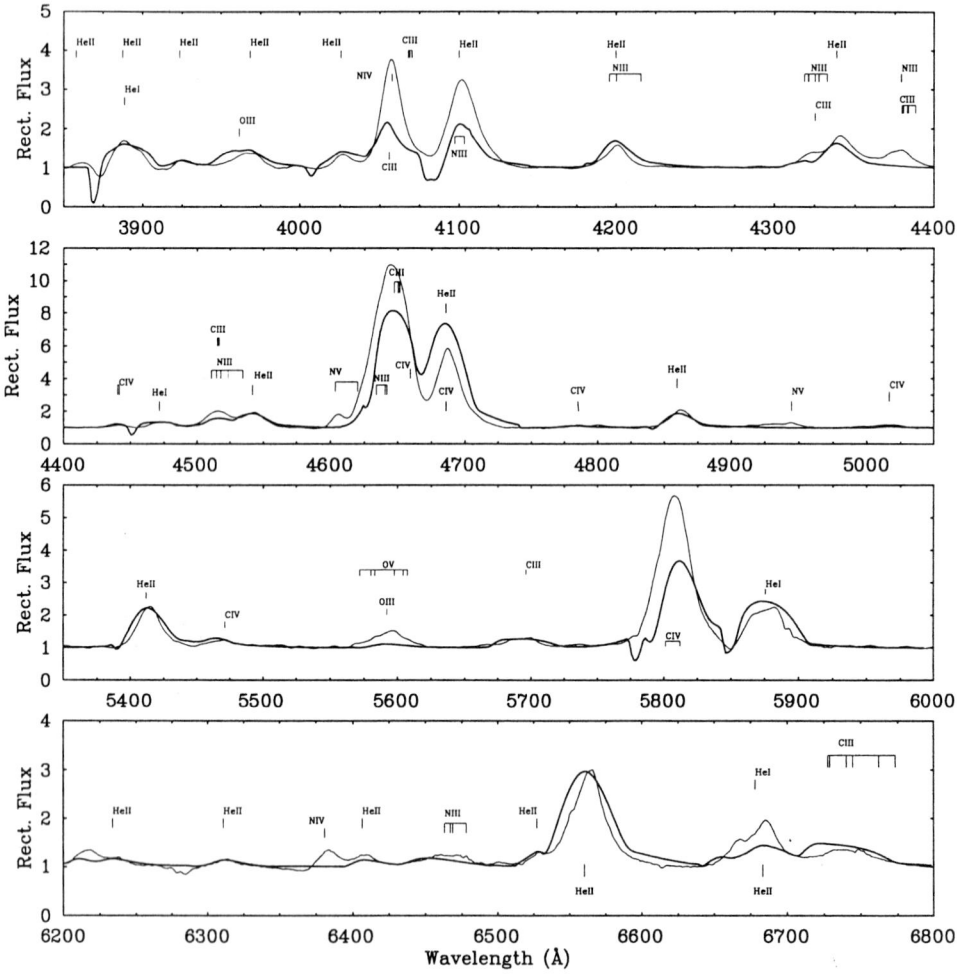

Fig. 1.   Comparison of the observed optical spectrum (thick line) of WR8=HD 62910
(WN6-C4) with the synthetic spectrum (thin line) from the Standard Model (from
Crowther et al. 1995e)

2.2. TEMPERATURES AND LUMINOSITIES

Attempts to determine reliable effective temperatures and radiative luminosities for WR stars based on simple analyses of observed UV–optical continuum energy distributions have been fraught with ambiguity arising through uncertainties in (i) reddening corrections, (ii) the applicability of the simple models used for fitting purposes, and (iii) the fact that the bulk of the emergent radiation is at unobservable wavelengths, below 912 Å. Early results, summarised by Willis & Garmany (1987), indicated values in the range 25000–50000 K, depending on subtype, with large uncertainties for individual stars. Similar values were also found from simple Zanstra analyses of optical and UV HeII emission line strengths (eg. Nussbaumer *et al*, 1982), but again the inherent uncertainties were unacceptably large. Estimated luminosities from these analyses yielded values log $L/L_\odot \sim 5.5$ for WNL stars, and ~ 5.0 for WNE and WC stars. Analysis of the optical/near–IR light curve of the eclipsing binary V444 Cyg (WN5+O5) by Cherepaschuk *et al* (1984) suggested a higher value of $T_{eff} = 900000$ K for the WN5 star, with log $L/L_\odot = 5.69$.

In the past few years, the position has improved substantially, mainly as a result of great improvements in model atmosphere techniques. This involved the development of more sophisticated models involving the simultaneous solutions of the equations of statistical equilibrium, for continuum and line transfer, for dense, spherical geometry, monotonic velocity law stellar winds – the WR *Standard Model* – initially developed independently at JILA (Hillier 1987a,b) and at Kiel (Hamman & Schmutz 1987, Hamann *et al* 1988, 1991) and now also in extensive use at UCL, Pittsburgh, Potsdam and Munich. With the *Standard Model* it is now possible to undertake detailed spectral synthesis modelling of WR continua and line spectra, including the effects of abnormal chemistries in the opacity and radiative transfer computations, with very good fits to the data (see Fig 1), giving considerable confidence in the derived stellar parameters. For WN stars, the atmospheric structure is usually computed using H/He or pure He models (since other elements are essentially trace), whilst for WC stars C and O must be introduced at the outset (Hillier, 1989, Hamann *et al* 1992, Koesterke & Hamann 1995).

Schmutz *et al* (1989) used extensive model grids predicting the equivalent widths of optical HeI and HeII lines, together with the stellar absolute visual magnitude, to interpret measured HeI,II equivalent widths for a sample of some 30 Galactic WN stars. Their results gave values of $T_{eff} \sim 35000$ K and log $L/L_\odot = 5.5$–6.0 for WNL stars, and for WNE stars values of log $L/L_\odot$ = 5.0–5.5 and a wider span of $T_{eff} = 35000$–90000 K. Similar results, using this technique, were derived by Howarth & Schmutz (1992) for galactic WN and WC stars, and by Koesterke *et al* (1991) for WN stars in the LMC.

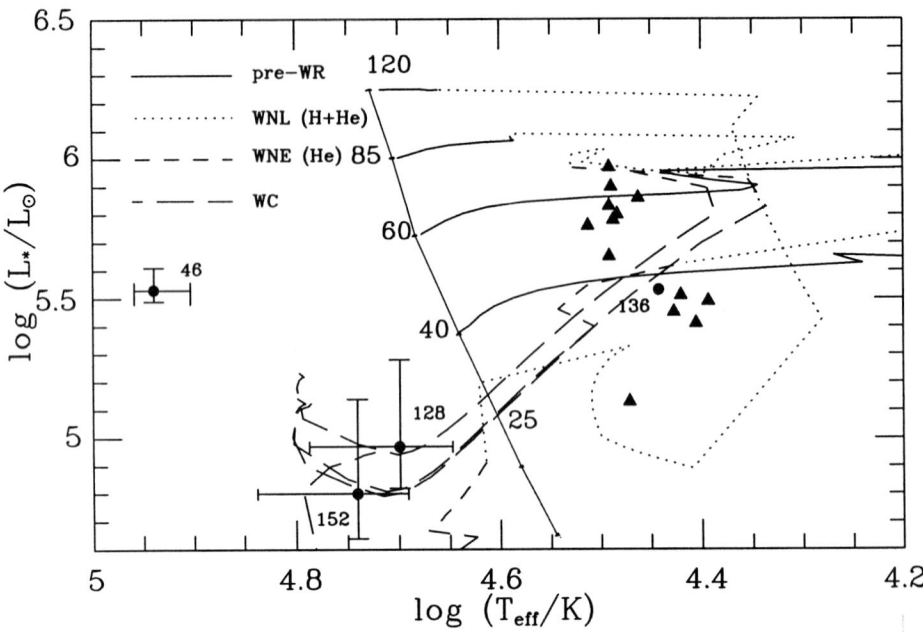

Fig. 2. The location of WNL and WNE stars in the H–R diagram from the results of Standard Model atmosphere analyses (from Crowther et al. 1995d) compared to evolutionary tracks of Meynet et al. (1994)

The employment of the WR *Standard Model* has now been used to perform detailed spectral synthesis analyses (so called 'tailored modelling') of a substantial number of individual WR and related stars, using high quality line profiles in UV (*IUE*), optical and infrared spectra. The first tailored analyses of HD 50896 (WN5, Hillier 1987b, Hamann *et al*, 1988) and HD 192163 (WN6, Hamann *et al*, 1994), have now been superceded by results for over thirty galactic and LMC WN stars in a major series of papers by Crowther and colleagues. Crowther *et al* (1995a) in their analyses of LMC and galactic Ofpe/WN9 stars derive $T_* = 29500$ K and $\log L/L_\odot = 5.65$, in good agreement with the fine analysis of R 84 (Ofpe) by Schmutz *et al* (1991). For the WNL stars (WN7–8), Crowther *et al* (1995b) find values of $T_{eff} \sim 25000$-30000 K for all objects, but a clear separation in luminosity between WNL stars with a low hydrogen content ($X_H \leq 15\%$: $\log L/L_\odot = 5.5$) and those WNL stars with a substantial hydrogen content ($X_H = 50 \%$: $\log L/L_\odot = 5.9$). Similar results were also found by Hamann *et al* (1991). For the WNE stars, which show rather weak optical line emission (often denoted WNE-w), Crowther *et al* (1995d) find a wide spread in temperatures and luminosities: $T_{eff} = 50000$–90000 K, $\log L/L_\odot = 4.8$-5.5. The location on the HR diagram of a sample of WN stars resulting from these analyses is shown in Fig 2., with the implications for evolutionary models addressed in section 5.

For WC stars, the most extensive tailored analyses to date have been carried out by Koesterke & Hamann (1995), following a pioneering study of HD 165763 (WC5) by Hillier (1989). They find that WC stars with relatively weak optical emissions (denoted WC-w) have $T_{eff} \sim 50000$ K, whilst the stronger lined WC stars (denoted WC-s) have higher values, $T_{eff} = 60000$-100000 K. Koesterke & Hamann (1995) find WC luminosities span a substantial range: $\log L/L_{\odot} = 4.7$–5.5.

To date there has been no tailored analyses of WO stars, to yield accurate stellar temperatures and luminosities, although the high spectral ionisation apparent is suggestive of values of T_{eff} greater than for WC stars. Photoionisation modelling of the nebula G 2.4+1.4 surrounding the WO1 star WR 102 ($\equiv$ Sand 4) by Esteban *et al* (1993) yielded $T_{eff} \sim 105000$ K and $\log L/L_{\odot} \sim 5.08$, although the result is somewhat uncertain since a pure helium stellar model was employed.

3. CHEMICAL ABUNDANCES

The general spectral dichotomy of WR stars into the WN and WC sequences (now extended to the additional WO class) has long been qualitatively interpreted as reflecting gross chemical abundance differences. However, progress in determining quantitative abundances has been rather slow because of the underlying difficulty in generating reliable model atmosphere techniques for these stars in which the usual assumptions of plane–parallel geometry, LTE, static atmospheres etc. have to be abandoned. As outlined below, a combination of new, high quality spectroscopy at optical, infrared and ultraviolet wavelengths analysed using highly sophisticated nLTE, moving–atmosphere models has produced rapid progress in this field, and confirmed that the atmospheres of WR stars are chemically evolved. It is found that WN stars show significant hydrogen and carbon depletions, with enhanced helium and nitrogen, indicative of the exposition of the products of interior CNO–cycle burning. For WC (and WO) stars no hydrogen or nitrogen is apparent, with their enhanced helium, carbon and oxygen abundances, indicating He-burning products. Thus WR stars provide a superb astrophysical laboratory for testing the theories of nuclear processing in stars and of advanced stellar evolution.

3.1. WN STARS

3.1.1. Helium abundances
The H/He ratio in WN stars has been determined for a large sample of stars using two basic techniques. The simplest approach (and one with surprising accuracy) is to use the optical He II Pickering Decrement (a plot of measured line intensities observed in the He II (n–2) series, where even–n transitions coincide with Hydrogen Balmer lines) *versus* n, and measure the

odd–even difference (for zero hydrogen all the (n–2) He II lines lie on a single correlation). This technique was applied by Smith (1973), Willis & Wilson (1978) and Smith & Willis (1982, 1983) who concluded that most WN stars were severely depleted in hydrogen. The method was extended by Conti, Leep & Perry (1983, CLP) to 37 WN stars in the Galaxy and 21 WN stars in the LMC, finding little evidence for significant hydrogen in WNE stars, whilst WN7 and WN8 stars generally showed evidence for some hydrogen, with H/He lying in the range 0.4–10 (the bulk at the lower end). Vreux *et al* (1989) extended the approach to near–infrared observations of He II lines and associated Paschen–series H–transitions, confirming the results from CLP.

With the advent of the *Standard Model* these results are now put on a firmer footing, from detailed tailored analyses of optical, IR and ultraviolet spectra. Hamann *et al.* (1990) used the Kiel models to analyse the He/H ratios in 4 WN stars from a fine analysis of optical He II line profiles, confirming the low hydrogen content inferred by CLP. Crowther *et al.* (1995a) derived H/He = 2.5 ± 1 for a sample of LMC Ofpe/WN9 stars (reclassified as WN9–10 stars) from tailored analyses of optical He I and He II line profiles. A value of H/He = 1.5 was derived for the single WN9 star known in the Galaxy (WR108). Crowther *et al.* (1995b) performed a quantitative analysis of optical and near–IR spectroscopy of 9 Galactic WN7-8 stars. They find that stars with strong He I lines show very low hydrogen abundances ($\sim$ 15% by mass) whilst stars classified as WN7+abs (absorption features seen in the higher Balmer series) show a substantial (although depleted) hydrogen content ($\sim$ 48% by mass). For the higher excitation WNE stars tailored analyses have been carried out by Crowther *et al.* (1995c,d). For the weak–lined WN3–4 stars they find ratio H/He $\sim$ 0.8, whilst for the WN6 stars HD 192163 and HD 191765 values of 0.5 and $\leq$ 0.05 are derived respectively.

The basic result is that WN stars are generally severely defficient in hydrogen, with helium substantially enhanced. Recent model atmosphere analyses of some O–type and Of–type stars are also showing that examples of these classes are also chemically peculiar, with moderate He–enhancements. For instance Y = .18 for α Cam (O9.5Ia, Voels *et al* 1989), Y = 0.2 for ζ Pup (O4I(n)f; Bohannan *et al* 1990) – see also results from Herrero *et al* (1992), Kudritzki & Hummer (1990), and Crowther & Bohannan (1996).

3.1.2. CNO Abundances

The first attempt to quantify the C/N ratio in WN stars was carried out by Willis & Wilson (1978) who applied a simplified Sobolev analysis of low resolution UV spectrophotometry of 9 WR stars obtained with the S2/68 experiment in the TD-1 satellite. They estimated a value of C/N $\sim$ 0.01, compared to the solar value of about 3, and concluded that WN stars exhibited abundances reflecting interior CNO-cycle equilibrium products. This work was extended to Sobolev analyses of higher resolution UV spectroscopy obtained with the *IUE* satellite by Smith & Willis (1982, 1983) for both galactic and LMC WN stars. Using important resonance and inter-combination lines of C III]1909, CIII 2296, CIV 1550, NIII 1750, NIV]1486, NIV 1718 and NV 1240, they derived C/N ratios for WN stars accurate to about a factor of two. The results gave: C/N = $[2–6] \times 10^{-2}$ for WNE stars and $[0.6–4] \times 10^{-2}$ for WNL stars. Although more uncertain, estimates of the N/He and C/He ratios clearly pointed to the C/N abnormality being the result of carbon depletion and nitrogen enhancement, as expected for chemical modification in the CNO-cycle hydrogen burning phase of a massive star. Contempory Sobolev analyses of optical spectra of WN stars by Nugis (1982) led to estimates of N/He $\sim [0.2–2] \times 10^{-2}$, comparable to the CNO-cycle equilibrium prediction of 4×10^{-3}.

The above results, from the rather crude Sobolev modelling, have recently been put on a much firmer footing from *Standard Model* analyses of UV, optical and near-infrared spectra. A detailed analysis of HD 50896 (WN5) by Hillier (1987a,b) gave C//N = 0.07, and N/He $\sim [0.4–4] \times 10^{-3}$, whilst Hamann *et al.* (1994) have derived N/He = 0.005 for HD 192163 (WN6). Crowther *et al.* (1995a) have carried out *Standard Model* analyses of several LMC Ofpe/WN9 stars, finding values of C/N ~ 0.1 and N/He ~ 0.003, with some evidence for a somewhat higher N/He value for the galactic counterpart WR 108 (NB. in all cases some hydrogen is also present, typically with H/He ~ 2). For galactic WNL stars, Crowther *et al.* (1995b) find C/N = 0.01 and N/He ~ 0.003–0.008, with evidence for the highest C/N ratios in the coolest WNL stars. For weak line WNE stars, Crowther *et al.* (1995d) find C/N = 0.05 and N/He = 0.004. In addition they derive a rough estimate of C/O = 1, O/N = 0.05 as typical. A somewhat higher O–abundance is estimated for HD 104994 (WN3–pec) with C/O ≤ 0.3 and O/N ~ 1.

3.2. WN–C STARS

Although the great majority of WR stars can be clearly ascribed to either the WN or WC sequences, Conti & Massey (1989) found intermediate classifications for six Galactic and two LMC stars based on the anomalously strong emissions apparent in CIV 5801, 5812 relative to HeII 4686, in optical spectra that otherwise had an WN–like appearance. These are termed WN–C stars. Further examples in the LMC and M33 have been found by Morgan & Good (1987) and Schild *et al.* (1990). The brightest of these stars, HD 62910 (WN6–C4) was studied by Willis & Stickland (1990), who used a simple Sobolev treatment to estimate C/N ~ 0.3 and N/He ~ 0.01 by number. They concluded that HD 62910 is a single star with abundances intermediate between WN and WC phases. Recently, Crowther *et al.* (1995e) have carried out a *Standard Model* tailored analyses of HD 62910, and two further examples of WN-C stars (WR145 in the Galaxy and HDE 269485

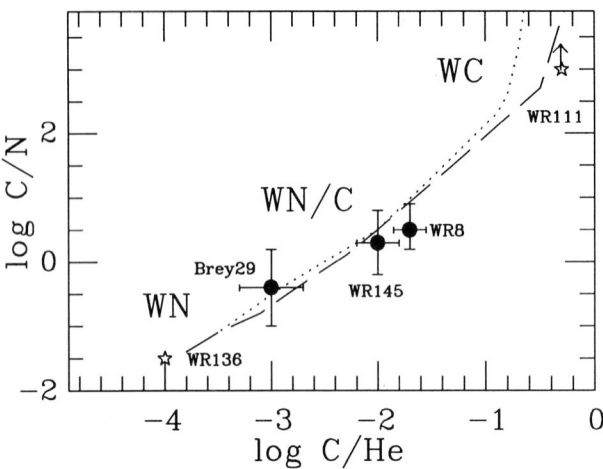

Fig. 3.   The abundance ratios of C/N and C/He derived from *Standard Model* analyses of WN, WN–C and WC stars (from Crowther *et al.* 1995e). The results are compared to theoretical predictions from Langer (private comm.) for stars of $40 M_\odot$ (dotted line) and $50 M_\odot$ (dashed line) during the WR phases.

in the LMC), deriving more accurate atmospheric parameters. Their results give $T_{eff} = 32000$ K, log $L/L_\odot = 5.1$, C/N ~ 3, C/He ~ 0.02, H/He ~ 0 and C/O ~ 4 by number for HD 62910, confirming the intermediate evolutionary status of these stars (see Fig 3).

Most stellar evolution models for massive stars predict a sharp discontinuity in the N–abundance between the WN and WC phases and few, if any expected WN–C objects, whereas the numbers of observed intermediate stars suggests an overall fraction of about 4%. Langer (1991) has shown that the inclusion of semi–convection may provide an additional chemical mixing mechanism and, for a relatively short timescale (about 10^4 yr), significant amounts of N and C can simultaneously be exposed at the stellar surface. The agreement between the predicted abundances and those determined by Crowther *et al* (1995e) for WN–C stars provides strong support for idea that mixing does indeed take place at the boundary of the convective He core in WR stars.

3.3. WC STARS

Despite intensive searches there has been no identification of any hydrogen signatures in the UV, optical or IR spectra of WC stars, and it is generally agreed that a ratio H/He = 0 is appropriate for this class (see Willis 1991b).

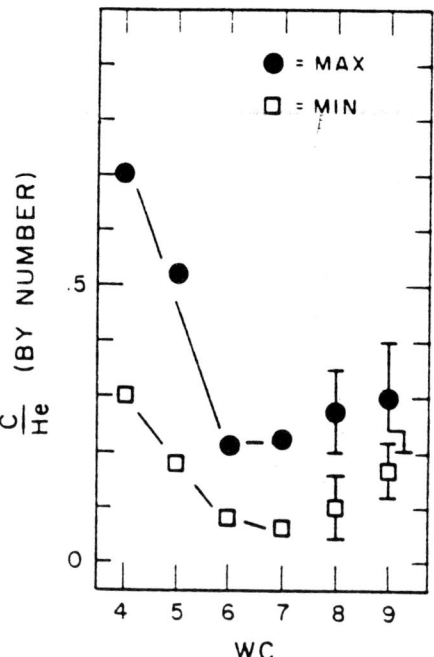

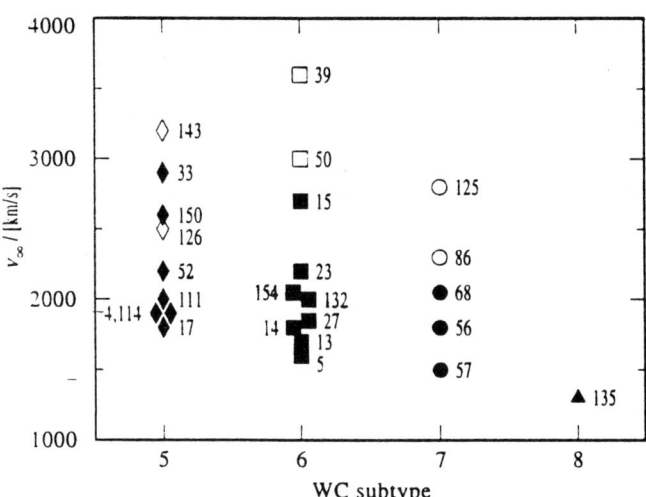

Fig. 4. (a) The C/He ratio (by number) for WC stars as a function of subtype derived from infrared H/K spectroscopy by Smith & Hummer (1988), (b) The carbon mass fraction for WC stars vs. subtype determined from *Standard Model* analyses by Koesterke & Hamann (1995). Not the large scatter at individual subtypes but no apparent correlation in carbon abundance with subtype.

Similarly, no unambiguous identification of any nitrogen lines has been made in WC stars, although very weak features in some optical spectra have been suggested as being due to nitrogen (eg. Bappu, 1973). Smith & Willis (1982, 1983) set an lower limit of C/N $\geq$ 60 from upper limits of $W_\lambda \leq 1$ Å for diagnostics lines of NIV] 1486 and NIV 1718 in *IUE* spectra of WC stars. Stellar evolution models predict effectively zero nitrogen abundances for the WC phases of massive stellar evolution.

The predominance of emission lines due to C, He and O transitions in WC spectra clearly points to enhanced abundances in these species. Evolutionary models, for different stages of exposition of helium burning products, predict values of C/He = $0.1 - 0.7$ (eg. Maeder, 1983). Simplified Sobolev modelling of optical spectra by Nugis (1982) gave estimates of C/He $\sim$ 0.1 for WC stars, whilst a simple recombination treatment of optical WC carbon lines by Torres *et al* (1986) gave values of C/He $\sim$ 0.1–0.8. Prior to the application of the *Standard Model* techniques, the most extensive determination of the C/He ratio was carried out by Smith & Hummer (1988) in their analysis of the near–IR H and K spectra of 17 galactic WC stars. An LTE treatment gave lower limit values of C/He = 0.04–0.3, whilst allowance for nLTE effects gave upper limit values of C/He = 0.1–0.7, with clear evidence for a decrease in this ratio with decreasing excitation subtype (WC4...WC7, see Fig 4a). Similar results were found from IR analyses by Eenens & Williams (1992).

The application of *Standard Model* tailored spectral analyses of WC stars has been more complicated than for WN stars, since the (expected) high C–abundance (and that of oxygen) necessitates the inclusion of these species as opacity sources in the atmospheric structure codes at the outset. Initial results for HD 165763 (WC5) were reported by Hillier (1989) who found C/He $\sim$ 0.5, and O/C $\leq$ 0.2. Hamann (1990) found a similar result for this star. Recently Koesterke & Hamann (1995) have calculated a grid of He–C models for the WC classes and used these to analyse the optical spectra of 25 Galactic WC stars, of types WC8–WC5. They find mass fraction of C varying from 0.2–0.6, (see Fig 4b), but no apparent correlation in the C/He ratio with WC subtype (in contrast to the conclusion of Smith & Hummer, 1988). In general the derived (high) C/He ratios for WC stars are in reasonable agreement with expectations from stellar evolutionary theory for the exposition of He–burning products.

However, at the onset of He–burning, the enhanced nitrogen produced by earlier CNO-cycle processing, is predicted to be very quickly destroyed by the reaction: $^{14}N(\alpha,\gamma)^{18}F(\beta,\nu)^{18}O(\alpha,\gamma)$ $^{22}Ne(\alpha,n)^{25}Mg$ yielding a predicted high ($\times$ 10 solar) abundance of neon in WC stars (eg. Maeder 1983). To date, the neon abundance has only been determined for one WC star – the binary system γ Velorum (WC8+O9I). Barlow, Roche & Aitken (1988) from a detailed study of the fine structure lines of [NeII]12.8 μm and [NeIII]15.5 μm derived a value of Ne/He = 1×10^{-3} – only twice solar, in conflict with the evolution predictions. The binary nature of γ Velorum may be one possible cause for this conflict, or it may be that any enhanced neon has already been processed to ^{25}Mg, or something may be awry with the cross sections for the above nuclear processing chain. To reconcile this conflict it is clearly important to determine neon abundances for a larger sample of (single) WC stars, and this should become possible with spectroscopy in the 2–40 μm region with the *ISO* satellite, launched in November 1995.

3.4. WO STARS

As noted above, the WO stars were first identified as a WR sequence by Barlow & Hummer (1982) as objects which showed exceptionally strong emission line of OVI 3811,3834 in addition to strong emissions of HeII and CIV, OIV, OV transitions. Three Pop I WO stars are known in the Galaxy, one in the LMC and one in the SMC. Kingsburgh *et al.* (1995) have carried out a detailed analysis of the optical and ultraviolet spectra of these five WO stars deriving their physical and chemical properties – the latter from a recombination analysis of selected transitions. The derived abundance ratios (by number) are found to lie in quite restricted ranges. For the Galaxy and LMC stars they find: C/He $\sim$ 0.51, with C/O $\sim$ 4.6–5.2, with the combined (C+O)/He ratio = 0.62. The SMC WO star has C/He = 0.81, C/O = 2.7 and (C+O)/He = 1.10. Kingsburgh *et al.* (1995) conclude that these abundance patterns are consistent with recent stellar evolutionary models for the exposition of He–burning and α–capture products, and moreover that the differences derived between examples in the Galaxy, LMC and SMC are in line with model expectations for different metallicity environments.

4. MASS LOSS PROPERTIES

WR stars exhibit the highest rates of mass loss amongst the early type stars, and it is this characteristic that is fundamental in determining their evolutionary status and overall spectral appearance. Their supersonic stellar winds are so dense that the underlying photosphere is generally not observed directly. Emission lines and P–Cygni profiles produced in the wind dominate the observed spectra at UV, optical and IR wavelengths, and indeed the bulk of the emergent continuum radiation is also produced in the wind. Quantitative determinations of the mass loss rates, wind velocities (especially the wind terminal velocity) and velocity laws for different subtypes in the WN and WC sequences are an essential prerequisite for stellar atmosphere modelling and stellar evolution theory, and substantial progress in these areas has been made in recent years.

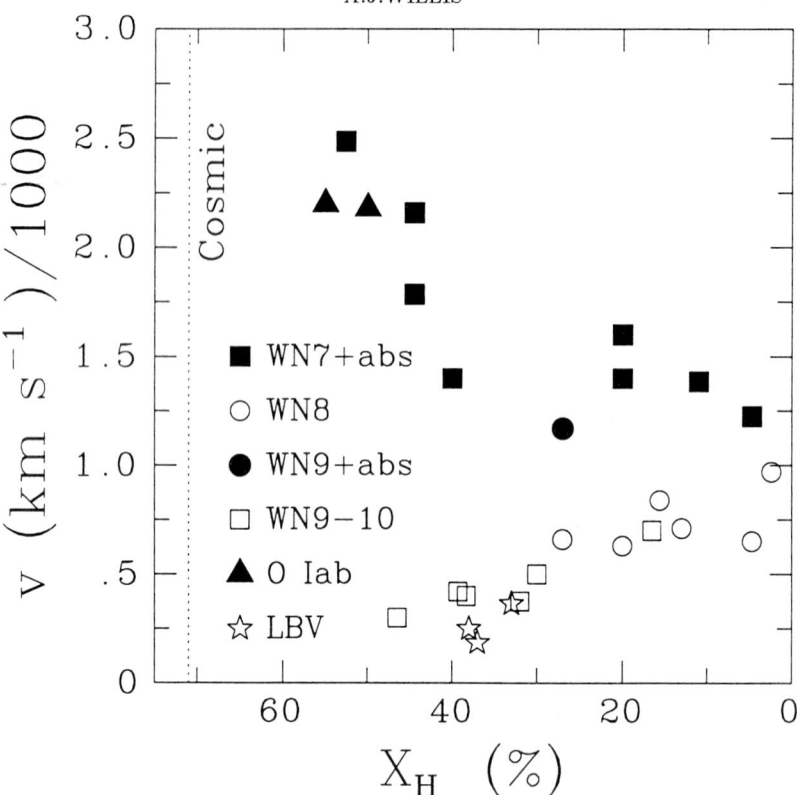

Fig. 5. The terminal velocities, v_∞, of WNL, Of and LBV stars as a function of the H/He ratio (from Crowther *et al* 1995c). The data indicate an evolutionary link between the Of and WN7 and WN9 classes, and between the LBV and WN8 stars.

4.1. WIND VELOCITIES

Radiation pressure wind theory for OB stars (eg. Kudritzki & Hummer, 1990) predicts a velocity law of the form:

$$v(r) = v_\infty \left[1 - r_*/r\right]^\beta$$

with $\beta \sim 1$. This law has been successfully used in the quantitative modelling of OB stellar spectra and has also been adopted in WR *Standard Model* analyses. Although little is known directly about the WR wind velocity laws, Cherepaschuk *et al.* (1984) did find that the above law was consistent with that deduced empirically from modelling the near–UV, optical and near–IR eclipse data for the binary system V444 Cyg (WN5+O6). This result provides evidence that the WR stellar winds may be accelerated through radiation pressure, as for OB stars, although the very high levels of mass loss rates for WR stars may need an additional mechanism (see below).

Determinations of the wind terminal velocities, v_∞, for WR stars have been carried out through a variety of techniques. Ultraviolet measurements of P–Cygni absorption lines in the UV resonance lines in *IUE* high resolution spectra have been used by Prinja *et al.* (1990) to measure the maximum violet extent of the saturated troughs, v_{black}, as a direct measurement of v_∞ for 35 WR stars. These results agreed well with estimates of v_∞ derived by Williams & Eenens (1989) from P–Cygni absorption profiles of the near–IR line of He I 2.058 μm, formed far out in the wind, and also with measurements by Howarth & Schmutz (1992) of the He I 1.0830 μm line in 24 Galactic WR stars. Optical line profile modelling has been used by Hamann *et al* (1993) to estimate v_∞ for 53 Galactic WN stars, and by Koesterke *et al* (1991) for 19 WN stars in the LMC. Their results show *no* significant differences between Galactic and LMC stars of the same spectral types, and further a good correlation of v_∞ with WN subtype, increasing from about 500 km s^{-1} at WN9 to about 3000 km s^{-1} at the highest excitation subclass of WN2. A good correlation of wind velocity with WC subtype has been recognised for many years from both optical and ultraviolet spectra (eg. Willis 1982, Torres *et al* 1986), with values rising from about 800 km s^{-1} at WC9 through to values in excess of 3500 km s^{-1} at WC4. Again no significant differences are found between values of v_∞ for Galactic and LMC WC stars.

Recently, Crowther *et al* (1995c) have found an interesting correlation between v_∞ and the derived H/He abundance ratio for WNL stars, Of and LBV stars (see Fig 5). The data for the WN7 stars shows a clear progression to lower values from that of the Of stars, which are believed to be their progenitors, whilst the data for the WN8 stars indicate a progression to higher values from the LBV stars, which are considered to be their progenitors.

4.2. Mass Loss Rates

Quantitative determinations of the mass loss rates for WR stars have been carried out using a wide variety of techniques, including: (i) analyses of the emergent free-free emission observed at radio and infrared wavelengths, (ii) analyses of eclipse effects observed in some binary systems, (iii) analyses of phase–dependent polarisation data for binary systems, (iv) model atmosphere analyses of observed line spectra, and (v) the period change in the binary system V 444 Cyg.

Observations of the free-free emission at radio wavelengths potentially provides an excellent method of determining mass loss rates, since the emergent flux is produced at large radii in the wind where the velocity will have reached its terminal value. In the assumption of spherical symmetry, the formula developed by Wright & Barlow (1975) may be used to derive $\dot{M}$:

$$\dot{M} = 0.095 v_\infty \left[\frac{S_\nu^{0.75} D^{1.5}}{(g\nu)^{0.5}} \right] \left[\frac{\mu}{Z\gamma^{0.5}} \right]$$

where S_ν is the radio flux in Jy at frequency ν in Hz, v_∞ is in km s^{-1}, D

is the stellar distance in kpc, g is the gaunt factor, and μ, Z and γ are the mean molecular weight, r.m.s. ionic charge and mean number of electrons per ion respectively. At radio wavelengths, assuming the emergent flux is produced by thermal free–free emission, the spectrum is predicted to have a distribution $S_\nu \propto \nu^\alpha$, with a spectral index of $\alpha = 0.6$. The most extensive set of radio observations of WR stars has been compiled by Abbott *et al* (1986) who give 4.9 GHz data for 22 stars measured with the VLA (and upper limits for a further 13 stars), and also 14.9 GHz observations for a subsample of 6 stars. The majority are confirmed to be thermal emitters. Abbott *et al* (1986) deduced mass loss rates lying in a relatively narrow range of $[0.8–8] \times 10^{-5} \ M_\odot \mathrm{yr}^{-1}$, with no significant differences between WN and WC stars, or with subtype in these sequences. Similar values were also derived for single WR stars and those in binary systems. Barlow *et al* (1981) applied this technique to an analysis of infrared 10 μm observations of 21 WR stars, deriving similar results for $\dot{M}$. Some revision to these results has been made by Prinja *et al* (1990) and Willis (1991a) who used updated values of v_∞ and improved estimates of the ionisation balance in the outer stellar winds, but the overall scale of WR mass loss rates has not really altered significantly.

Model atmosphere analyses of optical WR spectra also yields determinations of mass loss rates from tailored studies (see Schmutz *et al* 1989, Koesterke *et al* 1991). Again the results show values lying in the range $\sim 10^{-5}–10^{-4} \ M_\odot \mathrm{yr}^{-1}$. From a spectral analysis Howarth & Schmutz (1992) derive mean mass loss rates of $\sim 4 \times 10^{-5}$ (WNE), $\sim 3 \times 10^{-5}$ (WNL), and $\sim 4 \times 10^{-5}$ (WC), again emphasising the lack of any correlation with spectral type. An analysis of phase–dependent optical linear polarisation variations in 10 WR+O binaries by St Louis *et al* (1988) has also yielded mass loss rates in the range $1–10 \times 10^{-5} \ M_\odot \mathrm{yr}^{-1}$. For the binary system V 444 Cyg, Khaliullin *et al* (1984) have measured the period change in the system (0.22 ± 0.04 sec yr^{-1}), and infer a mass los rate for the WN5 component of $\dot{M} = 1.1 \times 10^{-5} \ M_\odot \mathrm{yr}^{-1}$. An analysis of UV spectral line eclipse data for the WC8+O binary systems γ Velorum (Willis *et al* 1979) and CV Serpentis (Howarth *et al* 1982) has yielded estimates of the WC8 mass loss rates of 9×10^{-5} and $7 \times 10^{-5} \ M_\odot \mathrm{yr}^{-1}$ respectively.

4.3. THE MECHANISM OF WR MASS LOSS

It is well known that there is a tight correlation between mass loss rates and radiative luminosities for OB stars (eg. Howarth & Prinja, 1989), with $\dot{M} \propto L^{1.6}$, in good agreement with the basic predictions of radiation pressure wind theory. Abbott *et al* (1986) suggested that a similar relation also holds for WR stars, but this contention has not been substantiated by later results (see Howarth & Schmutz 1992, and Willis 1991a). Mass loss rate determinations for binary systems (for which WR masses can be estimated) have been used

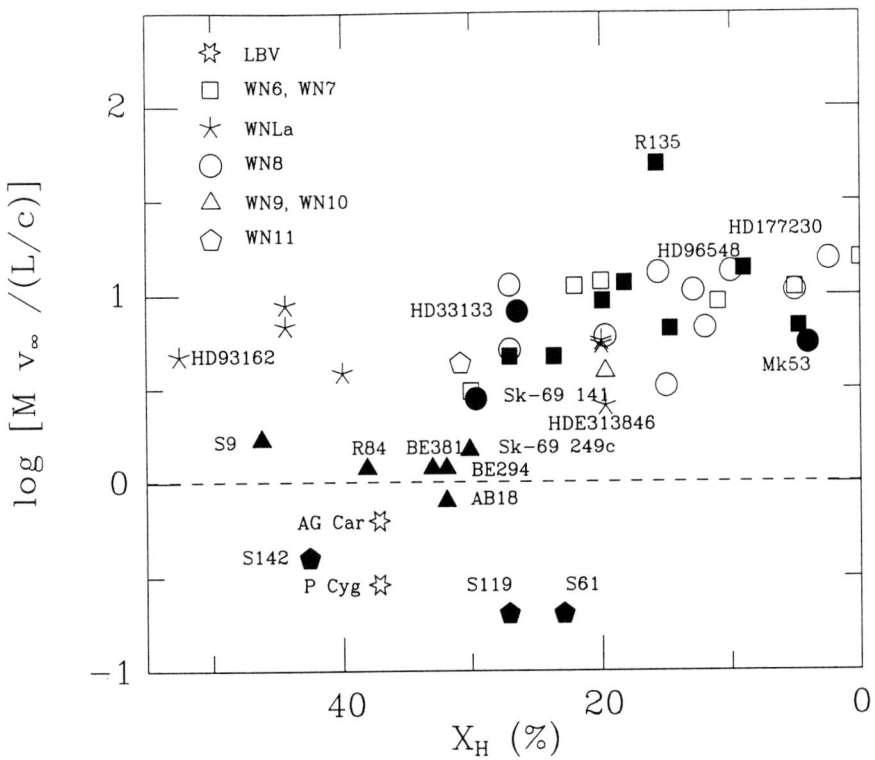

Fig. 6. The relation betwen the wind Γ–parameter and surface hydrogen abundance (by mass) for WNL and LBV stars in the Galaxy (open symbols) and LMC (filled symbols). From Crowther & Smith (1996b)

to test for a relation between mass and $\dot{M}$, but again different studies have yielded conflicting results. Abbott *et al* (1986) suggested a relation $\dot{M} \propto M^{2.3}$, whilst St Louis *et al* (1988) found evidence for a rough scaling of $\dot{M} \propto M^{0.8-1.3}$, the latter being similar to the conclusion of Smith & Maeder (1989) and Howarth & Schmutz (1992). At present there appears to be little consensus on such scaling laws for WR stars. Probably the best available evidence suggests a weak scaling of $\dot{M}$ with luminosity and mass, but with a shallower index than for OB stars, and with a significantly larger spread.

It is very evident that the WR stars show values of wind momentum compared to the 'single–scattering limit', $\Gamma \equiv \dot{M} v_\infty /(L/c)$, which are larger than unity, and substantially greater than for OB stars (for which Γ is $\leq$ 1). Typical values range from $\Gamma \sim 4$ for WNL stars, ~ 15 for weak–lined WNE stars, ~ 30–60 for strong–lined WNE stars, and ~ 60 for WC stars (Willis 1991a, Hamann *et al* 1993, Howarth & Schmutz 1992). Lamers & Leitherer (1993) have shown that their is a good correlation between Γ values determined for O-stars, Of stars and WNL stars, whilst Crowther &

Smith (1996b) find a correlation between Γ and surface hydrogen abundance for LBV stars and WNL stars (see Fig 6). Whilst a clear distinction is found between hydrogen–rich and hydrogen–poor WNL stars, there is *no* apparent differences between the mass loss properties of WN stars in the Galaxy and LMC. This is in contrast to expectations from radiation pressure–driven wind theory which predicts lower terminal vlocities and mass loss rates in lower metallicity enviromnents like the LMC.

The very high values of Γ for WR stars has led to the consideration that other mechanisms than radiation pressure are required to drive their mass loss. Smith & Maeder (1989) suggested that enhanced mass loss rates could be induced by some form of mechanical instability due to the ϵ mechanism, where it is expected that the effect would increase with changing mean molecular weight and mass within the star. Rotational and/or magentic effects have been suggested (eg. Cassinelli 1991), whilst pulsational instabilities have been suggested by Glatzel *et al* (1993), and Langer *et al* (1994). However, radiation pressure may indeed still provide a possible solution, if multi–scattering is taken into account in the highly ionisation–stratified WR winds. For instance, Lucy & Abbott (1993) have shown that values of Γ up to about 10 may be achievable in such circumstances. Hillier (1996) has given a recent review of the various possibilities by which radiation pressure wind theory may be reconciled with WR mass loss rates.

4.4. WIND STRUCTURE AND VARIABILITY

It is now well recognised that the stellar winds of OB stars are intrinsically unstable and permeated by shocked material produced as a natural hydrodynamical phenomenon in the radiation pressure driving process (see Lucy 1982, Owocki *et al* 1988). This provides an explanation for some of the observed phenomena such as their soft X–ray emission, extended saturated troughs in UV resonance line P–Cygni profiles, and the occurence and variability of Discrete Absorption Components (DAC) (see Prinja & Howarth, this volume). It is also evident that WR stellar winds exhibit many similar characteristics and likely have the same type of structure. Prinja & Smith (1992) have observed DACs in the UV spectrum of HD 93131 (WN7+abs), and UV P–Cygni profile variability has been found for a number of other WN stars (eg. HD 50896 (WN5) – Willis *et al* 1989, St Louis *et al* 1993, and HD 192163 (WN6) – St Louis *et al* 1989). Optical line and continuum variability in the emission lines of both WN and WC stars has been observed by Moffat & Robert (1991, 1992), and optical continuum polarisation variations are also seen at the level of $\sigma_p \sim 0.01$–0.15 % (Robert *et al* 1988). Evidence for wind density enhancements or clumps has been found from an analysis of UV eclipse data of WR139 (WN5+O) by Cherepaschuk *et al* 1984).

The first X–ray observations of a small sample of WR stars were obtained with the *Einstein* satellite, and showed luminosities in the range $L_x \sim 10^{30}$–

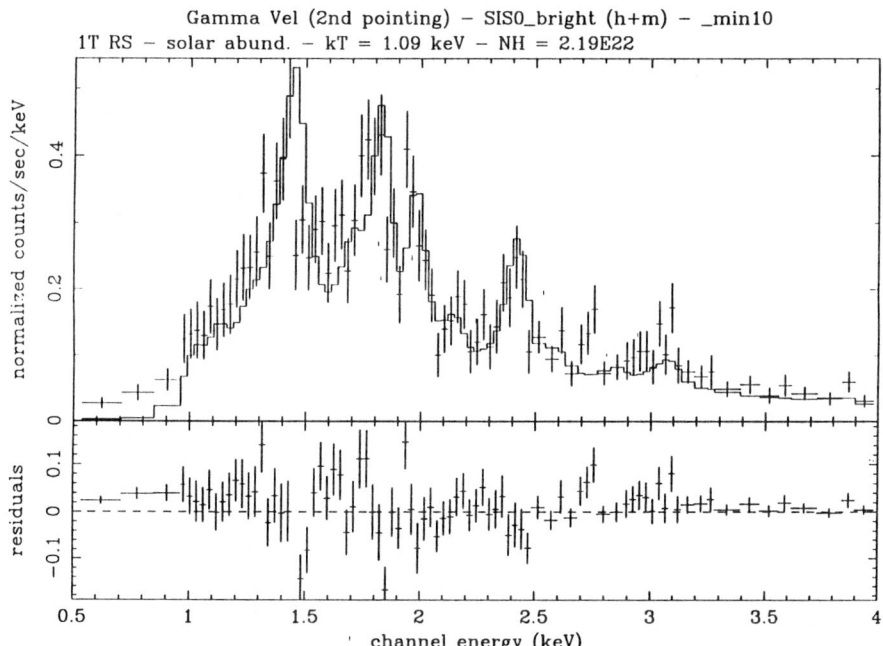

Fig. 7.   The X–ray spectrum of colliding–wind emission in γ Velorum observed with the *ASCA* satellite (from Stevens *et al* 1996)

10^{33} erg s^{-1} in the 0.5–4 keV energy range. As for OB stars, the spectra (poorly defined) appeared quite soft, suggestive of X–ray emission produced by shocked material permeating the winds (see Pollock, 1987). The *ROSAT* satellite is providing more detailed measurements of the X–ray emission from WR stars and confirming the softness of their spectra (Pollock, 1995). A detailed study of the *ROSAT* X–ray spectrum and variability of HD 50896 (WN5) has been reported by Willis & Stevens (1996), who find that $L_x = [2.2–3.8] \times 10^{32}$ erg s^{-1} in the 0.1–2.5 keV energy range, with kT = 0.28 keV. They estimate that the emergent X–ray emission from this source originates in shocked material at very large radii in the wind, $\geq 1000\ R_\odot$. In the case of WR+O binary systems, in which both stars have strong stellar winds, it is theoretically expected that the collision of the two winds can yield a strongly shocked interaction region emitting X–rays (Prilutski & Usov 1976).

Hydrodynamical models of such colliding–wind phenomena have recently been developed by Stevens *et al* (1992). In the case of the WC8+O9I binary γ Velorum the predicted colliding–wind X–ray emission has recently been directly observed from phase–dependent *ROSAT* data (Willis *et al* 1995). In addition to soft X–ray emission from the WC8 component (at kT = 0.19 keV, produced in wind shocks) an additional, harder component (kT $\geq$ 1 keV) is observed at binary phase $\sim$ 0.5, viewed through a cavity in the WR wind, produced by the wind interaction effects. The well–defined X–ray light curve of this emission shows a half–oepening angle of $\sim$ 25°, in good agreement with model predictions. Harder X–ray observations of γ Velorum obtained with the *ASCA* satellite (covering an energy range of 0.5–10 keV) are providing tighter constraints on the luminosity and X–ray *line spectrum* of the colliding–wind emission (Stevens *et al* 1996) which again are found to be in excellent accord with the hydrodynamical models. Indeed the *ASCA* data are providing the first real X–ray spectroscopy of such systems (see Fig 7), and a good 'taste' of the type of new and exciting observational material that will come from future generations of X–ray satellites, such as the ESA *XMM* mission.

5. EVOLUTIONARY STATUS AND SCENARIOS

Conti (1975) first proposed that single WR stars could be formed by mass loss stripping of the most massive O–type stars, with the Of–types forming an intermediate phase. Stellar evolution model calculations, incorporating the effects of ongoing, heavy mass loss have supported this general view (eg Maeder 1983), showing that stars with initial masses $\geq$ 40 $M_\odot$ could be expected to successively reveal nuclear processed material from the CNO–cycle (WN stars) and He–burning and triple–α reactions. Single star evolution models show that the 'Conti scenario' of O $\longrightarrow$ Of $\longrightarrow$ WN can probably only occur for stars of extremely high initial mass ($\geq$ 120 $M_\odot$. Lower mass stars are expected in the models to suffer enhanced mass loss episodes either in a Red Supergiant phase or an LBV phase. The effects of mass loss are anticipated in the evolution models to be strongly dependent on Galaxy metallicity (Maeder & Meynet 1994) in the sense that in low–Z environments evolution to the WR phases only proceeds for the initially most massive stars. This is broadly in line with the observed numbers and distributions of WN, WC and WO stars in different galaxies (see, Maeder & Conti (1994) for a review of massive star populations in galaxies).

Results from the recent extensive *Standard Model* spectral analyses of WR stars have been used by Crowther *et al* (1995c) to propose two distinct evolutionary sequences for the O–star to WR phases, with initial mass of the O–star being the delineating factor:

$$O \longrightarrow Of \longrightarrow WNL \longrightarrow WN6\text{--}7 \longrightarrow WNE \longrightarrow WC \ (M_{int} \geq 60 \ M_{\odot})$$
$$O \longrightarrow LBV \longrightarrow WN9\text{--}11 \longrightarrow WN8 \longrightarrow WNE \longrightarrow WC \ (M_{int} \geq 40 \ M_{\odot})$$

Direct evidence for an evolutionary link between the LBV and WN9–11 phases was discovered by Stahl *et al* (1983) who showed that the spectrum of the LMC star R 127 varied over timescales of years from that typical of these two spectral types. In addition one of the components in the WN binary system HD 5980 has recently been found to have changed its optical spectrum to that of a LBV type (Barba *et al* 1995). Whilst mass loss is clearly the major factor in affecting the evolution of massive stars, it appears that additional considerations need to be taken into account in the stellar structure and evolution models. For instance, it is evident that enhanced He/H ratios are found in the outer atmospheres of many Of and LBV stars, although normal evolution models do not predict this exposition until the WN phase is reached (see Maeder & Meynet 1987). This discrepancy may be accounted for if some form of rotational mixing is included (Fliegner & Langer 1995). Further the inclusion of slow mixing through the process of semi–convection, as proposed by Langer (1991) is probably required to account for the numbers and chemical properties of WN–C stars (Crowther *et al* 1995e).

Clearly there has been substantial progress in recent years in the determination of the basic properties of WR stars, and their evolutionary status. In the near future we can expect further progress on both observational and theoretical fronts. For instance, detailed spectroscopy will become possible in the mid–infrared with the *ISO* satellite, with X–ray spectroscopy at high sensitivity becoming feasible with missions like *XMM*. The *Gemini* telescopes (and other 8–metre class facilities) will provide more detailed studies of WR stars in external galaxies. On the model atmosphere front, we can expect considerable advances with the incorporation of wind–blanketing in the models, and the relaxation of other assumptions, to take account of the realisation that the winds are clumped, and time–variable, and (at least for some stars) their is direct evidence from spectropolarimetry of axisymmetric winds.

References

Abbott, D.C. & Conti, P.S., 1987. *Ann. Rev. Astron. & Astrophys.*, **25**, 113–150.

Abbott, D.C., Bieging, J.H., Churchwell, E. & Torres, A.V, 1986. *Astrophys. J.*, **303**, 239.

Allen, D.A., Hyland, A.R., & Hillier, D.J., 1990. *Mon. Nor. R. Ast. Soc.*, **244**, 706.

Bappu, M.K.V., 1973. *in Wolf-Rayet & High temperature Stars – IAU Symp. 49* (eds. M. K. V. Bappu & J. Sahade), D.Reidel Pub. Co., Holland, p 59.

Bappu. M.K.V., & Sahade, J. (eds.), 1973. *Wolf-Rayet & High Temperature Stars: IAU Symposium 49*, D.Reidel Pub. Co, Holland.

Barba, R.H., Niemela, V.S., Baume, G., & Vazquez, R.A., 1995. *Astrophys. J. Lett.*, **446**, L23.

Barlow, M.J. & Hummer, D.G., 1982. In: *Wolf-Rayet Stars: Observations, Physics, Evo-*

lution, IAU Symposium 99, eds. de Loore, C.W.H. & Willis, A.J., p.387, Reidel, Dordrecht.

Barlow, M.J., Smith, L.J. & Willis, A.J., 1981. *Mon. Not. R. Astr. Soc.*, **196**, 101.

Barlow, M.J., Roche, P.F. & Aitken, D.K., 1988. *Mon. Not. R. Astr. Soc.*, **232**, 821.

Beals, C.S., 1929. *Mon. Nor. R. Astr. Soc.*, **90**, 202

Beals, C.S., & Plaskett, J.S., 1935. *Trans. I.A.U.*, **5**, 184.

Bohannan, B. & Walborn, N.R., 1989. *Publ. Astron. Soc. Pac.*, **101**, 639.

Cassinelli, J.P., 1991. In: *Wolf-Rayet Stars and Interrelations with other Massive Stars in Galaxies, IAU Symposium 143*, eds. van der Hucht, K.A. & Hidayat, B., p. 289, Kluwer, Dordrecht.

Cherepashchuk, A.M., Eaton, J.A. & Khaliullin, , 1984. *Astrophys. J.*, **281**, 774.

Conti. P.S., 1975. *Mèm. Soc. Roy. Sci. Liége*, 6ème série, Tome IX, 193.

Conti, P.S., Leep, E.M. & Perry, D.N., 1983. *Astrophys. J.*, **268**, 228.

Conti, P.S. & Massey, P., 1989. *Astrophys. J.*, **337**, 251.

Crowther, P.A., & Willis, A.J., 1994. *Space Sci. Rev.*, **66**, 85.

Crowther, P.A., Hillier, D.J., & Smith L.J., 1995a. *Astron. Astrophys.* **293**, 172.

Crowther, P.A., Hillier, D.J., & Smith, L.J., 1995b. *Astron. Astrophys.*, **293**, 403.

Crowther, P.A., Smith, L.J., Hillier, D.J., & Schmutz W., 1995c. *Astron. Astrophys.*, **293**, 427.

Crowther, P.A., & Bohannan, B., 1996. *Astron. Astrophys.*, submitted.

Crowther, P.A., Smith, L.J., & Hillier, D.J., 1995d. *Astron. Astrophys.*, **302**, 457.

Crowther, P.A., Smith, L.J., & Willis, A.J., 1995e. *Astron. Astrophys.*, **304**, 269.

Crowther, P.A., & Smith, L.J., 1996a. *Astron. Astrophys.*, **305**, 541.

Crowther, P.A., & Smith, L.J., 1996b. *Astron. Astrophys.*, in press.

de Loore, C.W.H., & Willis, A.J., (eds.), 1982. *Wolf-Rayet Stars: Observations, Physics & Evolution*, D. Reidel Pub. Co., Holland.

de Loore, C.W.H., Willis, A.J. & Laskarides, P. (eds)., 1986. *Luminous Stars and Associations in Galaxies, IAU Symposium 116*, Reidel, Dordrecht.

Eenens, P.R.J. & Williams, P.M., 1992. *Mon. Not. R. Astr. Soc.*, **255**, 227.

Esteban, C., Smith, L.J., Vílchez, J.M. & Clegg, R.E.S., 1993. *Astron. Astrophys.*, **272**, 299.

Fliegner, J., & Langer, N., 1995. *in Wolf-Rayet stars, binaries, colliding winds and evolution* Proc. IAU Symp. No.163 (eds. K A van der Hucht & P M Williams), Kluwer, p 163.

Gebbie, K.B., & Thomas, R.N., (eds.), 1971. *Wolf-Rayet Stars*, NBS Pub. No. 307.

GlatzW., Kiriakidis, M., & Fricke, K.J., 1993. *Mon. Not. R. Astr. Soc.*, **262**, 7.

Hamann, W-R. & Schmutz, W., 1987. *Astron. Astrophys.*, **174**, 173.

Hamann, W-R., Schmutz, W. & Wessolowski, U., 1988. *Astron. Astrophys.*, **194**, 190.

Hamann, W-R., Dünnebeil, G., Koesterke, L., Schmutz, W. & Wessolowski, U., 1991. *Astron. Astrophys.*, **249**, 443.

Hamann, W-R., Leuenhagen, U., Koesterke, L. & Wessolowski, U., 1992. *Astron. Astrophys.*, **255**, 200.

Hamann, W-R., Koesterke, L. & Wessolowski, U., 1993. *Astron. Astrophys.*, **274**, 397.

Hamann, W-R., Wessolowski, U. & Koesterke, L., 1994. *Astron. Astrophys.*. submitted.

Herrero, A., Kudritzki, R.P., Vilchez, J.M., Kunze, D., Butler, K., & Hasser, S., 1992. *Astron. Astrophys.*, **261**, 209.

Hillier, D.J., 1987a. *Astrophys. J. Suppl.*, **63**, 947.

Hillier, D.J., 1987b. *Astrophys. J. Suppl.*, **63**, 965.

Hillier, D.J., 1988. *Astrophys. J.*, **327**, 822.

Hillier, D.J., 1989. *Astrophys. J.*, **347**, 392.

Hillier, D.J., 1991. *Astron. Astrophys.*, **247**, 455.

Hillier, D. J., 1992. In: *Atmospheres of Early-Type stars*, eds. Heber, U. & Jeffery, C.S., p. 105, Springer-Verlag.

Hillier, D.J., 1996. *in Hydrogen Deficient Stars*, ASP Conf. series., in press.

Howarth, I.D. & Prinja, R.K., 1989. *Astrophys. J. Suppl.*, **69**, 527.

Howarth, I.D. & Schmutz, W., 1992. *Astron. Astrophys.*, **261**, 503.

Howarth, I.D., Willis, A.J., & Stickland, D.J., 1982. *ESA Sp-176*, p 331.

Kingsburgh, R.L., Barlow, M.J., & Storey, P.J., 1995. *Astron. Astrophys.*, **295**, 75.

Koesterke, L., Hamann, W-R., Schmutz, W. & Wessolowski, U., 1991. *Astron. Astrophys.*, **248**, 166.

Koesterke, L., & Hamann, W-R., 1995. *Astron. Astrophys.*, **299**, 503.

Kudritzki, R.P. & Hummer, D.G., 1990. *Ann. Rev. Astron. & Astrophys.*, **28**, 303–345.

Lamers, H.J.G.L.M. & Leitherer, C., 1993. *Astrophys. J.*, **412**, 771.

Langer, N., 1991. *Astron. Astrophys.*, **248**, 531.

Langer, N., Hamann, W.R., Lennon. M., *et al*, 1994. *Astron. Astrophys.*, **290**, 819.

Lucy, L.B., 1982. *Astrophys. J.*, **255**, 286.

Lucy, L.B. & Abbott, D.C., 1993. *Astrophys. J.*, **405**, 738.

Maeder, A., 1983. *Astron. Astrophys.*, **120**, 113.

Maeder, A., 1990. *Astron. Astrophys. Suppl.*, **84**, 139.

Maeder, A., & Conti, P.S., 1994. *Ann. Rev. Astron. Astrophys.*, **32**, 227.

Maeder, A. & Meynet, G., 1987. *Astron. Astrophys.*, **182**, 243.

Maeder, A., & Meynet, G., 1994. *Astron. Astrophys.*, **287**, 803.

Meynet, G., Maeder, A., Schaller, G., Schraerer, D., & Charbonnel, C., 1994. *Astron. Astrophys. Suppl.*, **103**, 97.

Moffat, A.F.J. & Robert, C., 1991. In: *Wolf-Rayet Stars and Interrelations with other Massive Stars in Galaxies, IAU Symposium 143*, eds. van der Hucht, K.A. & Hidayat, B., p. 109, Kluwer, Dordrecht.

Moffat, A.F.J. & Robert, C., 1992. In: *Nonisotropic and Variable Outflows from Stars, A.S.P. Conference Series, Vol. 22*, eds. Drissen, L., Leitherer, C. & Nota, A., p. 203.

Moffat, A.F.J., & Robert, C., 1994. *Astrophys. J.,*, **421**, 310.

Moffat, A.F.J, Drissen, L., Lamontagne, R. & Robert, C., 1988. *Astrophys. J.*, **334**, 1038.

Nugis, T., 1982. In: *Wolf-Rayet Stars: Observations, Physics, Evolution, IAU Symposium 99*, eds. de Loore, C.W.H. & Willis, A.J., p. 131, Reidel, Dordrecht.

Nussbaumer, H., Schmutz, W., Smith, L.J., & Willis, A.J., 1982. *Astron. Astrophys. Suppl.*, **47**, 257.

Owocki, S.P., Castor, J.I. & Rybicki, G.B., 1988. *Astrophys. J.*, **335**, 914.

Pollock, A.M.T., 1987. *Astrophys. J.*, **320**, 283.

Pollock, A.M.T., 1995. *in WR Stars – binaries, colliding winds & evolution*, (eds. K. A. van der Hucht & P. M. Williams), Kluwer, Acad. Pub., p 429.

Prilutskii, O., & Usov, V., 1976. Sov. Astron., **20**, 2.

Prinja, R.K. & Howarth, I.D., 1986. *Astrophys. J. Suppl.*, **61**, 357.

Prinja, R.K. & Smith, L.J., 1992. *Astron. Astrophys.*, **266**, 377.

Prinja, R.K., Barlow, M.J. & Howarth, I.D., 1990. *Astrophys. J.*, **361**, 607.

Schaerer, D. & Maeder, A., 1992. *Astron. Astrophys.*, **263**, 129.

Schild, H., Smith, L.J., & Willis, A.J., 1990. *Astron. Astrophys.*, **237**, 169.

Schmutz, W., Hamann, W-R. & Wessolowski, U., 1989. *Astron. Astrophys.*, **210**, 236.

Schmutz, W., Leitherer, C., Hubeny, I. Vogel, M., Hamann, W-R., & Wessolowski, U., 1991. *Astrophys. J.*, **372**, 664.

Smith, L.F., 1968. *Mon. Not. R. Ast. Soc.*, **138**, 109.

Smith, L.F., 1973. *Proc. IAU Symp. No. 49*, (eds. M.K.V. Bappu & J. Sahade), D. Reidel Pub. Co., Holland, p 15.

Smith, L.F. & Hummer, D.G., 1988. *Mon. Not. R. Astr. Soc.*, **230**, 511.

Smith, L.F., & Maeder, A., 1989. *Astron. Astrophys.*, **211**, 71.

Smith, L.J. & Willis, A.J., 1982. *Mon. Not. R. Astr. Soc.*, **201**, 451.

Smith, L.J. & Willis, A.J., 1983. *Astron. Astrophys. Suppl.*, **54**, 229.

Smith, L.J., Crowther, P.A. & Prinja, R.K., 1993. *Astron. Astrophys.*, **281**, 833.

Stahl, O., Wolf, B., & Klare, G., 1983. *Astron. Astrophys.*, **127**, 49.

Stevens, I.R., Blondin, J.M., & Pollock, A.M.T., 1992. *Astrophys. J.*, **386**, 265.

Stevens, I.R., Corcoran, M.F., Willis, A.J., Pollock, A.M.T., Skinner, S.L., Nagase, F., & Koyama, K., 1996. *Astrophys. J.*, in press.

St-Louis, N., Moffat, A.F.J., Drissen, L., Bastien, P. & Robert, C., 1988. *Astrophys. J.,* **330**, 286.

St Louis, N., Smith, L.J¿, Stevens, I.R., Willis, A.J., Garmany, C.D., & Conti, P.S., 1989. *Astron. Astrophys.*, **226**, 249.

St Louis, N., Howarth, I.D., Willis, A.J., Stickland, D.J., Smith, L.J., Conti, P.S., & Garmany, C.D., 1993. *Astron. Astrophys.*, **267**, 447.

Torres, A.V., Conti, P.S. & Massey, P., 1986. *Astrophys. J.*, **300**, 379.

van der Hucht, K.A. 1992. *Astron. Astrophys. Rev.*, 4, 123.

van der Hucht, K.A., Conti, P.S., Lundsytrom, I., & Stenholm, B., 1981. *Space Sci. Rev.*, **28**, 227.

van der Hucht, K.A. & Hidayat, B. (eds)., 1991. *Wolf-Rayet Stars and Interrelations with other Massive Stars in Galaxies, IAU Symposium 143*, Kluwer, Dordrecht.

van der Hucht, K.A., & Williams, P.M., (eds)., 1995. *Wolf–Rayet Stars: Binaries, Colliding Winds & Evolution, IAU Symp. 163*, Kluwer Acad. Pub.

Vanbeveren, D., & Conti, P.S., 1980. *Astron. Astrophys.*, **88**, 230.

Voels, S.A., Bohannan, B., Abbott, D.C., & Humm,er, D.G., 1989. *Astrophys. J.*, **340**, 1073.

Walborn, N.R., 1977. *Astrophys. J.*, **215**, 53.

Walborn, N.R., 1982. *Astrophys. J.*, **256**, 452.

Williams, P.M. & Eenens, P.R.J., 1989. *Mon. Not. R. Astr. Soc.*, **240**, 445.

Willis, A.J., 1982. *Mon. Not. R. Astr. Soc.*, **198**, 897.

Willis, A.J., 1991a. In: *Wolf-Rayet Stars and Interrelations with other Massive Stars in Galaxies, IAU Symposium 143*, eds. van der Hucht, K.A. & Hidayat, B., p. 265, Kluwer, Dordrecht.

Willis, A.J., 1991b. In: *Evolution of Stars: The Photospheric Abundance Connection IAU Symposium 145*, eds. Michaud, G. & Tutukov, A., p. 195, Kluwer, Dordrecht.

Willis, A.J., & Wilson, R., 1978. *Mon. Not. R. Astr. Soc.*, **182**, 559.

Willis, A.J., & Garmany, C.D., 1987. *Exploring the Universe with the IUE Satellite*, (ed. Y. Kondo), D. Reidel Pub. Co., p 157.

Willis, A.J. & Stickland, D.J., 1990. *Astron. Astrophys.*, **232**, 89.

Willis, A.J., & Stevens, I.R., 1996. *Astron. Astrophys.*, in press.

Willis, A.J., Wilson, R., Beeckmans, F., *et al*, 1979. *The First Year of IUE* (ed. A. J. Willis), Univ. Coll. London., p 394.

Willis, A.J., van der Hucht, K.A., Conti, P.S., & Garmany, C.D., 1986. *Astron. Astrophys. Suppl.*, **63**, 417.

Willis, A.J., Howarth, I.D., Smith, L.J., Conti, P.S., & Garmany, C.D., 1989. *Astron. Astrophys. Suppl.*, **77**, 269.

Willis, A.J., Schild, H. & Smith, L.J., 1992. *Astron. Astrophys.*, **261**, 419.

Willis, A.J., Schild, H., & Stevens, I.R., 1995. *Astron. Astrophys.*, **298**, 549.

Wolf, C.J.E., & Rayet, G., 1987. *Comptes Rendues*, **65**, 292

Wright, A.E. & Barlow, M.J, 1975. *Mon. Not. R. Astr. Soc.*, **170**, 41.

ACCRETION AND EVOLUTION IN CLOSE BINARIES

A. R. KING

Astronomy Group, The University of Leicester, Leicester LE1 7RH

Abstract. The discovery of X–ray binary systems in the 1960's opened up stellar evolution theory by revealing further endpoints in addition to white dwarfs. This review summarises recent progress in studies of stellar–evolutionary processes that lead to X–ray binaries themselves, the mass transfer rates that power them, and the accretion processes which convert this into electromagnetic radiation. Particular attention is paid to the topics of mass transfer fluctuations and of the accretion by magnetic compact stars.

1. Introduction

The first X–ray binary, now called Sco X-1, was discovered in 1962 by the first sensitive X–ray detector to be flown above the Earth's atmosphere (Giacconi et al., 1962). This was an entirely unpredicted event; the instrument had been launched in the hope of detecting the effect of solar cosmic ray particles hitting the Moon, ironically a feat only achieved thirty years later with ROSAT. The realisation that most of the bright X–ray sources found with the early experiments were binary systems in our Galaxy did not come instantaneously. This is hardly surprising, as the identification required the recognition both of a new energy source – accretion – and of new, exotic, types of astronomical object, in the forms of neutron stars and black holes. Together with the near–simultaneous discovery of radio pulsars, the advent of X–ray binaries opened up stellar evolution theory by revealing the likely existence of two further endpoints in addition to the white dwarfs. Accretion was first suggested as the power source for quasars by Salpeter (1964), but Shklovsky's (1967) proposal to extend this idea to X–ray binaries allowed the process to be studied in a much more tractable observational environment. We thus arrive at the standard picture of an accreting binary system (Fig. 1), which in essence is the picture already proposed by Kuiper (1941) to explain the non–compact accreting binary system β Lyrae, with the substitution of a neutron star or black hole for the accreting component. It was rapidly realised that the cataclysmic variables, in which this star is a white dwarf, offered an even more tractable (if less spectacular) observational opportunity for studying accretion, and much of the literature on the subject has concentrated on these systems.

The twin aspects I have mentioned, the study of the stellar–evolutionary processes which must have occurred to produce the binaries and the mass transfer rates powering them, and the accretion processes which convert this into electromagnetic radiation, have marked the subject to the present day, and will form the two themes of this chapter. The field is now vast, and there are many specialist treatments covering aspects of it. For accretion theory,

Astrophysics and Space Science **237**: 169–184, 1996.

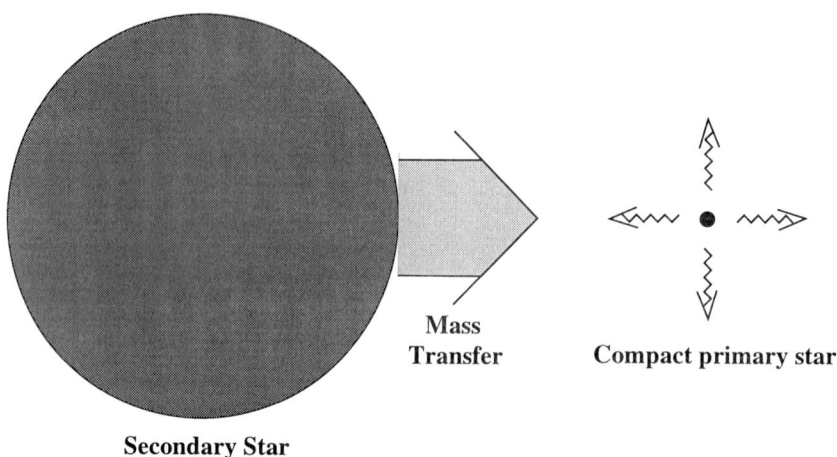

Fig. 1.   An interacting binary system. Mass accreting on to the compact primary star powers the system's luminosity.

the interested reader is referred to Frank et al (1996); for in–depth reviews of many aspects of X–ray binaries, see the collection edited by Lewin, van den Heuvel and van Paradijs (1995). Here, after a general introduction (Sections 2 - 5) I shall confine myself to those aspects which seem currently of greatest interest, with no attempt at completeness.

2. Accretion Power

The discovery of the X–ray binaries posed two immediate problems: astronomers had to explain both the production of X–rays, and the rather high luminosities ($L \sim 10^{38}$ erg s^{-1}) indicated by early distance estimates. The idea of accretion on to a compact object (black hole or neutron star) solves both problems. First, if we naively consider simple radial infall of matter on to a "star" of mass M_1 and radius R_1 (to fix ideas it is easiest to consider a neutron star at this point), we see that the release of gravitational potential energy gives an accretion yield of

$$E_g = \frac{GM_1\dot{M}}{R_1} \simeq 10^{20}\left(\frac{M_1}{M_\odot}\right)\left(\frac{10 \text{ km}}{R_1}\right) \text{ erg g}^{-1}. \tag{1}$$

This is close to the total rest–mass energy $c^2 = 9 \times 10^{20}$ erg g^{-1}, since $R_1 \simeq$ 10 km is only about three times the Schwarzschild radius $R_g = 2GM_1/c^2$ corresponding to the mass M_1. (Strictly speaking we should use general relativity to estimate the binding energy of the accreted matter, but for $R_1 \gtrsim 3R_g$ the deviations from Newtonian theory are small.) We can compare E_g with the maximal yield E_{nuc} from nuclear burning (all the way from hydrogen to iron), which is

$$E_{\text{nuc}} \simeq 0.007c^2 = 6 \times 10^{18} \text{ erg g}^{-1}, \tag{2}$$

and is thus about 20 times lower. Accretion is therefore a much more economical energy source: to supply the typical luminosity $\sim 10^{38}$ erg s^{-1} of a bright X–ray binary requires an accretion rate $\dot{M} \sim 10^{-8} \ M_\odot$ yr^{-1} on to a neutron star or black hole, but would require a mass supply of $\sim 2 \times 10^{-7} \ M_\odot$ yr^{-1} through a nuclear burning process operating at peak efficiency. There is a consequent shortening of the lifetime of the latter type of object by a factor ~ 20 compared with the accreting system.

We can estimate the likely spectral energy E_{ph} of the emission from the accreting star by considering two extreme cases, in which the gravitational energy of each accreting proton is released directly as a single photon, and in which it is thermalized over the surface of the star. These correspond to cases in which the accretion flow is optically thin, with the accretion energy released in some kind of shock, and optically thick flows respectively. In the first case we find

$$E_{\text{ph}} \sim \frac{GM_1 m_p}{R_1} \sim 100 \text{ MeV}, \tag{3}$$

where m_p is the proton mass, and in the second we get

$$E_{\text{ph}} \sim k \left(\frac{L}{4\pi R_1^2 \sigma} \right)^{1/4} \sim 1 \ L_{38}^{1/4} R_6^{-1/2} \text{ keV}, \tag{4}$$

where L_{38} is the luminosity in units of 10^{38} erg s^{-1} and R_6 the radius R_1 in units of 10^6 cm. Note that for a given accretion rate we have $L \propto 1/R_1$, so this estimate of E_{ph} varies as $R_6^{-3/4}$. We thus see that minimum photon energies of 1 keV, i.e. X–rays, are guaranteed provided that $R_6 \lesssim 1$, the radius of a neutron star.

These simple and robust arguments are essentially conclusive in showing that the bright Galactic X–ray sources must involve accretion on to a neutron star or a black hole. If the accreting star is a white dwarf (so that $R_1 \simeq 10^9$ cm) we see from (1) and (2) that nuclear burning is more efficient than accretion by a factor of ~ 30. However, nuclear burning of accreted matter on the surface of a white dwarf is generally explosive, powering a *nova* outburst which lasts $\lesssim 10$ yr. Matter must be slowly accreted over $\sim 10^5$ yr before this can occur again, and for the whole of this time the system is seen via its accretion luminosity alone. These systems are called cataclysmic variables (CVs), and are in many ways ideal for studying both the accretion process and many of the evolutionary phenomena which underlie it. The accretion rates in CVs are more modest, typically $\dot{M} \sim 10^{-10}$ erg s^{-1}, with corresponding luminosities $L \sim L_\odot$. The very large number ($\gtrsim 300$ with known orbital periods) of observed CVs reflects their high space density.

The arguments presented above show that the observational problem of the existence of X–ray binaries and CVs is in essence solved, provided that we can explain the formation of close binaries with compact components and the required accretion rates. It is clear that there is an intimate connection between these two aspects since every accretion rate is a mass transfer rate: the distribution of mass and angular momentum within the binary is continuously changing. Since the changes over observable timescales for an individual system are generally small, and the instantaneous accretion rate can vary widely from the evolutionary mean, we can to some extent decouple the accretion and evolution problems. However we are forced to consider both aspects simultaneously as soon as we consider questions relating to the ensemble of systems, such as the relative proportions in various observed categories.

3. Zoology

The basic picture (Fig. 1) of an accreting binary is realized in a remarkably wide range of forms. The primary star can be a black hole, neutron star or white dwarf, and in the latter two cases can have a strong magnetic field. The secondary star can be main–sequence or in various later stages of evolution; the binary orbit may be eccentric or circular; and finally, mass transfer can be via Roche lobe overflow or capture of part of the secondary's stellar wind. Clearly a large number of combinations are possible, most of which are actually observed. Evidently, arranging all these possible systems into evolutionary sequences which start from a pair of main–sequence stars and reach recognisable endpoints is not a simple task, and certainly not solved. For example, there is no generally agreed picture of the formation of low–mass X–ray binaries (LMXBs) in which a neutron star or black hole accretes from a lower main–sequence companion. The final fate of these binaries is also unclear; they may end as close but detached binaries containing radio pulsars, but this too is controversial.

Even when all the possible ingredients mentioned above appear to be exactly the same, the systems may nevertheless appear in very different observational guises. As an example, in non–magnetic CVs a white dwarf accretes from a lower main–sequence star via Roche lobe overflow, yet this class includes systems whose light is generally steady (UX UMa systems), ones with low states (VY Scl systems), and at least three different types of outbursting systems (dwarf novae of the U Gem, Z Cam and SU UMa types). This huge variety of behaviour seems all the more surprising given the very

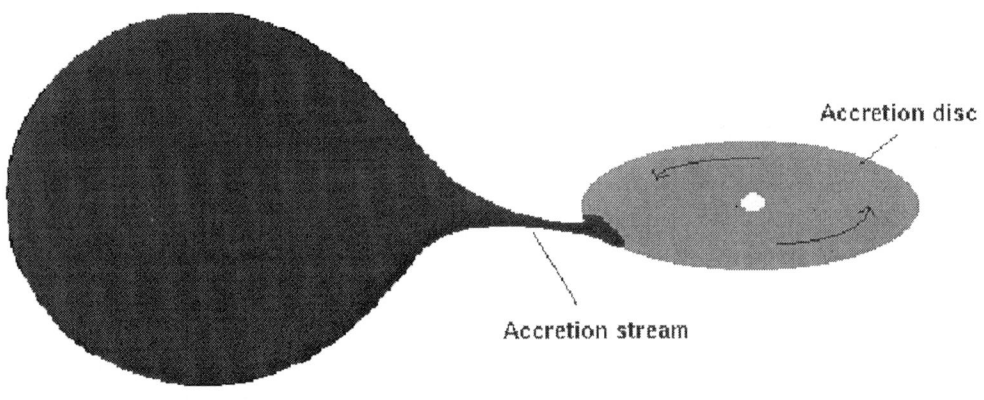

Fig. 2.   A close binary accreting via Roche lobe overflow. Cataclysmic variables (CVs) and low–mass X–ray binaries (LMXBs) are members of this class.

small number of parameters specifying the system (essentially just the two masses and the orbital period). The hidden parameter responsible for the variety may well be time itself: it seems likely that the mass transfer rate in particular can oscillate around the mean value fixed by the evolutionary state, on timescales too long to be directly observable. I shall discuss later some ideas as to how this can happen.

Some nomenclature is vital in classifying interacting binaries. In addition to the LMXBs and CVs we have already encountered, we can distinguish the high–mass X–ray binaries, in which a neutron star (often magnetic) or black hole accretes from a massive early–type companion. These are intrinsically luminous systems, and many of the first X–ray binaries discovered, including the very first X–ray source (Sco X–1) itself, belong to this category. While the orbits of LMXBs and CVs are sufficiently close that tides circularize the system, some of the high–mass systems have wide enough separations (or short enough lifetimes) that their orbits are significantly eccentric. This is especially true of a subgroup in which the companion is a Be star. In further contrast to LMXBs and CVs, accretion is probably from the stellar wind of the companion, rather than Roche lobe overflow as in the former case. Figs. 2 and 3 summarize these differences.

(a)   **(b)**

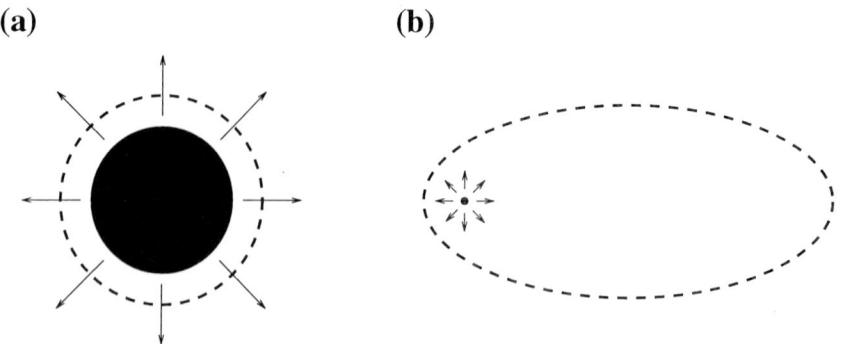

Fig. 3. High–mass X–ray binaries: (a) in those with luminous O, B supergiant companions the neutron star or black hole has a near–circular orbit and accretes from the strong spherical wind of the supergiant. (b) The neutron star has a wide, eccentric orbit in Be–star X–ray binaries and accretes from the disc–like wind around the equator of this star.

4. Evolution

In wind–accreting systems, such as the high–mass X–ray binaries, the evolution through the X–ray phase proceeds as though the compact primary star were absent. The accretion is controlled entirely by the complex properties of the companion's stellar wind. In Roche–lobe overflowing systems such as CVs and LMXBs this cannot be true, and we must seek reasons why the secondary star should fill its critical equipotential surface in this way. The linear scale of the Roche surface is well approximated by the formula

$$R_L = 0.462 \left(\frac{M_2}{M} \right)^{1/3} a, \tag{5}$$

where M_2, M are the secondary mass and the total binary mass $M_1 + M_2$ respectively and a is the separation. This shows that there are two ways to ensure that the secondary star fills its Roche lobe: the star can expand, or the lobe can contract. The first case occurs for example once the companion develops a helium core and begins to expand along the giant sequence. Provided that $M_2 \lesssim M_1$, the effect of transferring mass to the primary in this way will be to widen the separation a (because orbital angular momentum is conserved) sufficiently rapidly that the larger Roche lobe radius R_L can accomodate the expanded star, allowing the process to continue stably. Clearly this causes mass to be transferred on the nuclear evolution timescale $t_{\rm nuc}$ of the secondary, and thus drives typical mass transfer rates $\sim M_2/t_{\rm nuc}$. Evidently the orbital period always increases in this type of evolution.

The second case – shrinking the Roche lobe – requires the loss of angular momentum from the binary. This must occur in CVs for example, as the secondaries are far too low in mass (a few tenths of $M_\odot$) for any significant evolutionary expansion to have taken place. There are various mechanisms carrying off angular momentum from a close binary. In short–period (few hours) systems the most important is probably gravitational radiation. In systems with longer periods, the loss of the secondary's rotational angular momentum via a magnetic stellar wind is can be communicated to the binary orbit via tides. Again, provided that $M_2 \lesssim M_1$ and mass is not lost too rapidly (see below) the star will be able to shrink quickly enough in response to mass loss that it still fits inside the smaller Roche lobe, making the process stable. Given a mechanism carrying off the orbital angular momentum J at a rate $\dot{J}$, we see that mass will be exchanged on the associated timescale $t_J = -J/\dot{J}$, with typical mass transfer rates $\sim M_2/t_J$. The orbital period decreases in this kind of evolution. However, the shrinking of the secondary in response to mass loss occurs at a rate $\sim R_2/t_{KH}$, where $t_{KH} = GM_2^2/R_2 L_2$ is the star's thermal timescale. If mass is lost very rapidly, i.e. $t_J < t_{KH}$, the star will not shrink rapidly and may even expand adiabatically. This increases the mass transfer rate by a factor ~ 2, which is enough to slow the orbital contraction or reverse it, allowing the Roche lobe to accomodate the star. Hence very rapid angular momentum loss can slow or even reverse the evolution to shorter orbital periods.

Applied to CVs, the discussion of the last paragraph is quite success-ful. The proposed angular momentum loss mechanisms do seem to produce roughly the right mass transfer rates for a given orbital period. Moreover, the abrupt termination of the CV distribution below an orbital period of 80 minutes can be attributed to the very long thermal time of the corresponding secondary stars, which have masses $\lesssim 0.1 M_\odot$. Since here $t_J < t_{KH}$, the sys-tems begin to evolve towards longer periods, making 80 minutes a minimum period. Fig. 4 shows a comparison of the theoretically–determined relation between $\dot{M}$ and the orbital period P and observational estimates of these quantities. Despite the reasonable gross agreement, a very large scatter is clearly evident. Although some of this can be attributed to the difficulty of estimating accretion rates from observation, some of the scatter is proba-bly real. This illustrates the remark made above, that "instantaneous" mass transfer rates may differ from the evolutionary mean over timescales too long to be directly observable. This mean itself is enforced on the timescale for the Roche lobe to move about a scaleheight through the secondary, which is $\gtrsim 10^5$ yr, so fluctuations on shorter timescales will not upset the overall evolutionary picture. I shall discuss the possible origin of these fluctuations and their significance below.

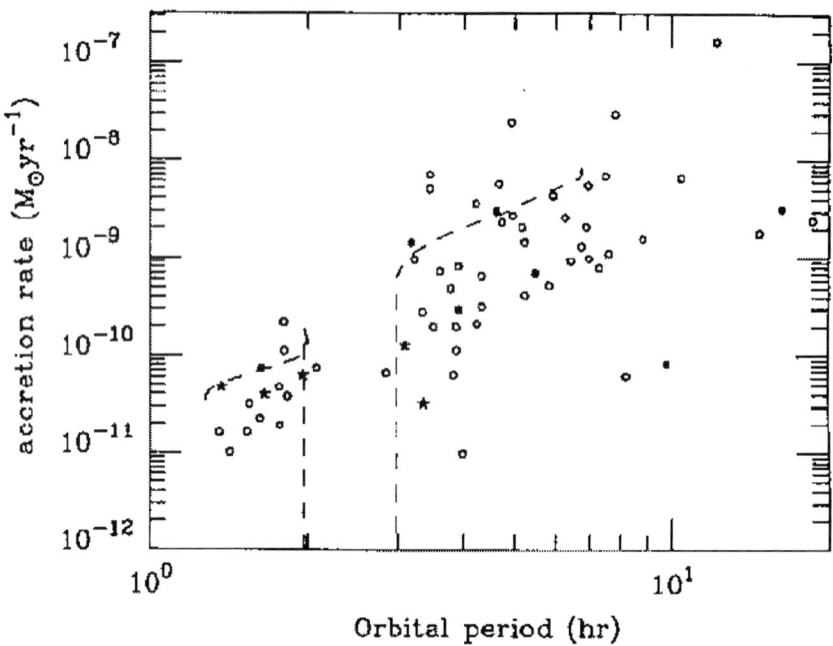

Fig. 4.   Mass transfer rates estimated for individual CV systems by Patterson (1984). The various symbols refer to differnt CV subtypes. The dotted curve is the mass transfer rate predicted by evolutionary calculations (Hameury et al., 1988) as a function of orbital period.

5. Accretion Flows

The discovery of the X–ray binaries stimulated an understanding of accretion flows within a binary system. In particular it was rapidly realized that in many cases the accreting matter had too much specific angular momentum j to hit the accretor directly, i.e.

$$R_1 < \frac{j^2}{GM_1}. \tag{6}$$

In this case the matter must move in orbits about the accreting star. Because the gravitational potential is not exactly $1/r$, these orbits must precess and thus intersect themselves. The intersections must be dissipative (i.e. involve shocks), so the matter will lose orbital energy. On the other hand it is difficult for it to lose angular momentum, so the orbits will be those of minimum energy for given angular momentum, i.e. close to Keplerian circles. The only way that accretion can now occur is for matter on these orbits to interact with matter further out, transferring angular momentum to it. The mechanism by which this occurs is called *viscosity*, and is still far from being understood. However it must clearly be dissipative; provided that the matter can cool efficiently, it will expand rather little out of the orbital plane, and most of the luminosity will be emitted from the two near–plane faces of the matter distribution. Most of the original angular momentum is transported outwards. Ultimately it reaches orbits of sufficiently large radius that the tidal interaction with the companion star becomes appreciable. This removes the transported angular momentum from the accreting matter and deposits it back into the binary orbit, which is where it originally came from.

This type of configuration, called an *accretion disc*, has become a popular ingredient for models of many astrophysical systems. The observational evidence for the existence of discs in some systems, notably non–magnetic CVs, is overwhelming. Unfortunately this is far from true of many other systems: in particular the simple picture presented here is severely modified in several other types of binary system, especially where the accreting object possesses a significant magnetic field.

An obvious consequence of disc accretion is that the central object must spin up, since the material it accretes must have at least the Keplerian specific angular momentum at radii larger than its own, and hence increase the star's angular momentum faster than its mass. The spin rate of the central star is thus a sensitive probe of the mode of accretion. This rate itself is particularly easy to measure if the accreting star is significantly magnetic, which is precisely one of the cases where the standard picture of accretion via a disc, as outlined above, is most suspect.

6. Mass Transfer Fluctuations

As already mentioned above, although binary evolution theory predicts an essentially unique average mass transfer rate for a given orbital period, the instantaneous transfer rate must fluctuate about this value on timescales $\gtrsim 10^5$ yr. A knowledge of the cause and nature of these cycles is vital to a proper understanding of the binary evolution, as they determine the fractions of systems we see in various states; for example, among CVs, the relative numbers of dwarf novae (outbursting) and novalike (steady) systems. Without such knowedge, there is no way of comparing the observed populations with those theoretically predicted, or even of knowing if the observed systems are typical of the entire population.

Until recently the existence of the fluctuations was tacitly assumed, without any real attempt to discover their cause: progress in this direction has occurred only in the last few years. The easiest way of modulating the mass transfer rate $-\dot{M}_2$ from a Roche lobe–filling companion star is to perturb its radius R_2; increases (decreases) of order a few scaleheights H near the

inner Lagrange point (where $H/R_2 = \epsilon \simeq 10^{-4}$) increase (decrease) $-\dot{M}_2$
by orders of magnitude. In most close binaries, irradiation by the accreting
component is by far the largest perturbation that the companion is subject
to, so it is plausible to assume that it can cause radius changes of order
H significantly affecting $-\dot{M}_2$. (In fact I shall indicate below that one can
effectively *prove* that irradiation must be the basic cause of the fluctuations.)
Since the change of $-\dot{M}_2$ in turn affects the irradiating flux, one can see the
potential for some kind of cyclic behaviour here. There are two problems
in carrying out this programme. First, it has to be demonstrated that the
radius is suitably sensitive to the irradiating luminosity, and second, that
$-\dot{M}_2$ is forced to execute a limit cycle, rather than simply settling at a
new equilibrium value under the combined effects of irradiation and angular
momentum loss.

Since irradiation amounts to changing the star's outer boundary condition, and radiative stars are insensitive to this (e.g. Schwarzschild, 1958)
it is clear that we must consider stars with deep outer convection zones
to check the first requirement. This is mildly encouraging, as the companions in CVs and LMXBs are largely convective, at least for orbital periods
$\lesssim 8$ hr. If the star is assumed to be spherically illuminated, it is clear that
there can be a large effect on its radius. In particular, if the irradiation is
intense enough that the radiation temperature at its surface is $\gtrsim 10^4$ K, the
star will lose its outer ionization zone and must adopt a radiative rather
than a convective equilibrium structure. Under such conditions the star's
radius is larger by factors ~ 2 (Podsiadlowski, 1991)than its main–sequence
radius, the increased surface area being required for it to radiate away all
the irradiating luminosity together will its own internally generated nuclear luminosity. The latter is reduced considerably below the main–sequence
value, since the nuclear luminosity L_{nuc} decreases as a high power (~ 7)
of the stellar radius. One can now assume that the change from a convective to a radiative structure removes the orbital angular momentum loss via
magnetic stellar wind braking which was driving the long–term evolution
(see Section 4 above), so that the mass transfer rate declines. This in turn
reduces the irradiating flux and allows the star to collapse back towards
its (convective) main–sequence structure, restarting the cycle (Frank et al,
1992). This scenario offers attractive explanations for the range of mass
transfer rates seen in LMXBs, their orbital period distribution, and their
relative rarity compared with detached systems containing millisecond radio
pulsars, which might have been assumed to descend from them. However,
there are difficulties. First, only the star's intrinsic nuclear luminosity L_{nuc}
participates in bloating it to its new radius, which is achieved on the Kelvin–
Helmholtz timescale corresponding to the nuclear luminosity in the bloated
state. Since a significant radius increase drastically reduces this luminosity
($L_{\mathrm{nuc}} \sim R_2^{-7}$), the latter time can become extremely long, and it is unclear

if the bloated radius can be reached within the evolution time of the binary. Second, the original assumption of spherical illumination appears very questionable. There are several binaries (none of them CVs or LMXBs) in which the companion star is observed to maintain radically different temperatures on its illuminated and unlit faces without apparent difficulty, and estimates of cooling and flow times indeed suggest that very little heated material will circulate over the unlit hemisphere. This unlit portion of the star then offers an easy way for it to rid itself of its nuclear luminosity, which is blocked only over the heated face. Since the unirradiated star is generally close to the Hayashi effective temperature T_H, and the star cannot get cooler than this, one obtains the maximum radius R for the irradiated star by setting its nuclear luminosity (modified by the radius increase) equal to the area of the unlit hemisphere multiplied by σT_H^4, i.e.

$$L_{MS}\left(\frac{R_{MS}}{R}\right)^7 \simeq 2\pi R^2 \sigma T_H^4,$$

where the suffix MS denotes main–sequence values. But since the star is close to the Hayashi temperature when on the main sequence, the right–hand side of this equation is just $(1/2)L_{MS}(R/R_{MS})^2$. Thus

$$R \simeq 2^{1/9} R_{MS} = 1.08 R_{MS}, \tag{7}$$

corresponding to a change in radius far less than that due to the effect of spherical illumination. It is evident that even a considerable reduction of the unlit part of the surface will do little to raise R: the case of spherical illumination thus appears essentially singular. In fact, increases of the same order as that implied by the estimate (7) can already be obtained by the irradiation of one hemisphere of the star with much lower fluxes (Ritter, 1994), which do not disturb its essentially convective character. Further, the stellar radius is actually *more* sensitive to small increases in the irradiating flux when this flux is itself low. In this sense it is (surprisingly) easier to understand mass transfer cycles in CVs than in LMXBs.

It appears then that realistic irradiation can produce radius changes of no more than 10%. Being $\sim 100H$, this is of course more than sufficient to change the mass transfer rates by the required amounts. However we still have to show that these radius changes can follow a genuine limit cycle. The conditions for such cycles have been investigated in considerable detail using a phase–plane analysis (King, 1995; King et al. 1995a,b), and it is shown that such cycles do exist (King et al. 1995a). This analysis can be cast in a very general form, and reveals for example that if the limit cycles are to result from radius changes, the stellar radius must be a function of the mass transfer rate. Since irradiation is about the only physically plausible way of arranging this, this amounts to a proof that irradiation is the basic cause of the cycles. However, from the work of Ritter et al (1995), it appears that radius variations driven by irradiation can only follow a limit cycle when the convective envelope has rather low mass, i.e. for companion masses $0.7 M_\odot \lesssim M_2 \lesssim 1 M_\odot$. This is a disappointing result, as the most pressing need is to understand variations in companion stars of rather lower masses $(0.2 - 0.3 M_\odot)$. The most likely avenue of escape appears to be that the angular momentum losses must also be sensitive to the mass transfer rate: one can imagine this happening for example if variations in the mass loss rate from the companion affect the rate of magnetic stellar wind braking appropriately.

7. Accretion by Magnetic Compact Stars

It has long been recognised that a strong magnetic moment on the accreting star must modify the accretion flow. The most obvious manifestation of this is that accretion occurs only on to restricted regions near the magnetic poles of this star: as it spins, the emission from the surface is modulated at its spin period, making this easy to measure. Magnetic systems thus reveal more about the angular momentum flow within the binary than non-magnetic ones. Interestingly, the nature of these flows is still not fully understood. Until recently the standard approach has been to assume that the magnetic field can be effectively ignored if its pressure $B^2/8\pi$ is less than the ram pressure ρv^2 of the matter, while there is an abrupt transition to flow strictly along field lines if the inequality is reversed. Applied to accretion disc flow, this approach predicts that the flow is disrupted at some "Alfvén radius" R_A where magnetic and material stresses are equal. Once on the field lines, matter feels a centrifugal force $R\omega^2$ as well as gravity GM_1/R^2 per unit mass, and so can only accrete if R_A is inside the "corotation radius" $R_c = (GM_1/\omega^2)^{1/3}$, where matter in Keplerian rotation would exactly keep pace with the rotating field lines. This amounts to an upper limit on the spin rate, which obviously cannot exceed the Kepler value $(GM_1/R_c^3)^{1/2}$ at the corotation radius. Clearly other torques than simple spinup under accretion act on the star: the magnetic field may interact with more slowly-rotating material further out for example. Ultimately the star will reach an equilibrium spin rate at which all these torques balance. It is common to express the result in terms of a dimensionless "fastness parameter" $\omega_s = \omega/(GM_1 R_A^3)^{1/2}$, i.e. the ratio of ω to the Keplerian rate at R_A, with the result

$$P_{eq} = 1.4 \omega_s^{-1} \phi^{3/2} m_1 \mu_{30}^{6/7} R_6^{-3/7} \dot{m}^{-3/7}, \tag{8}$$

where $\phi \sim 1$ is a factor depending on the precise accretion geometry, m_1 is the accreting mass in $M_\odot$, μ_{30} is the magnetic moment in units of 10^{30} G cm^3, R_6 is the radius of the accreting star in units of 10^6 cm ($\simeq 1$ for a neutron star) and $\dot{m}$ the accretion rate in units of 10^{-8} $M_\odot$ yr^{-1} (the

Eddington rate for a neutron star). A theory of the accretion and other torques for disc accretion will predict an equilibrium value ω_s, which must be < 1 to obey the centrifugal limit discussed above. Further, such a theory also gives the spinup rate of the neutron star under the resultant torque when it is not precisely at the value (8).

This kind of approach has had some success in describing the spinup rates of X–ray binaries in which the accretor is a magnetic neutron star (see e.g. Frank et al, 1996; Lewin, van den Heuvel and van Paradijs, 1995), but suffers from several drawbacks. In particular the centrifugal limit on ω_s does not seem to appear in a very natural way in the theory. Perhaps as a result, there have been several revisions of what the equilibrium value of this parameter is likely to be, from the original value $\omega_s \simeq 0.35$ (Ghosh and Lamb, 1979) used to explain the X–ray binaries, to a value much closer to unity used in the application of these ideas to explain quasi–periodic oscillations in LMXBs in terms of a magnetospheric model (Ghosh and Lamb 1991). Recently, enough data have accumulated on another type of system to allow a further test of this type of approach. There are now ~ 16 non–accreting binaries containing millisecond pulsars which are not members of globular clusters and have circular orbits: it is thought that these systems descend from X–ray binaries in which the neutron star was spun up by accretion. The pulsars are effectively frozen at the spin periods they had reached at the end of this phase, as the torques currently acting on them are small. Because they are radio pulsars it is possible to estimate their magnetic fields, and because the orbital evolution is understood one has an idea of $\dot{M}$. One may thus check (8) directly. The result (at least for binaries with long orbital periods) is at first sight quite encouraging, in that the data do appear to fit the slope predicted by (8) in the $\log P - \log B$ plane. However, the corresponding value of ω_s is only $\simeq 0.15$, considerably smaller than any theoretical estimate (Burderi et al, 1995a). The discrepancy leads one to suspect that the $\log P - \log B$ relation may have an entirely different cause. van den Heuvel (1994) noted a correlation between lower magnetic fields and the total mass that the neutron star must have accreted over its binary lifetime. This suggests that accretion may reduce the surface field, at least down to some fixed base value. Incidentally, it is far from clear what happens to all the accreted matter, which can be as much as $1 M_\odot$; certainly the neutron stars do not appear to have masses anywhere close to the $\sim 2.5 M_\odot$ that would be implied if the mass were all retained. Indeed, there is no obvious difference in the masses of the neutron stars inferred to have accreted most (i.e. with the lowest magnetic fields) compared with the remainder. It would appear that the correlation noted by van den Heuvel involves the amount of matter with which the neutron star has *interacted*, rather than necessarily accreted. Whatever the interpretation, if the correlation is real, it implies that the surface field declines at the same time that the neutron star spin rate is trying to reach equilibrium.

If the two timescales become comparable, it follows that the equilibrium can never be truly reached. Evidently this idea requires one to follow the evolution of both the spin and the field simultaneously. It appears (Burderi et al 1995b) that this combined evolution can produce the observed relation between $\log P$ and $\log B$.

A more radical deviation from the picture associated with the standard comparison of magnetic and material stresses appears to be indicated in binaries where the accretor is a magnetic white dwarf. There is considerable observational evidence that the accretion flow is highly inhomogeneous, quite probably in the form of diamagnetic filaments or blobs. Simple estimates (King, 1993) suggest that these blobs may interact with the magnetic field only through a drag resulting from the Lorentz force on the screening currents on their surfaces, and that they can to some extent drift across fieldlines. Specifically, the drag force per unit mass has the form

$$\mathbf{f}_{\text{drag}} = k(\mathbf{v} - \mathbf{v}_B)_{\perp}, \tag{9}$$

where the bracket gives the relative velocity of the blob and field line perpendicular to the latter (the blobs feel no drag moving along field lines), and

$$k = \frac{B^2 l^2}{c_A m}, \tag{10}$$

where B is the local magnetic field value, m, l are the mass and typical length of the blob, and c_A is the Alfvén speed in the interblob plasma (usually $c_A \simeq c$). Since the field lines rotate with the accreting star, this kind of approach allows for exchange of energy and angular momentum between this star and the accreting matter in a more natural way than possible in the standard picture. In particular, blobs either gain or lose orbital angular momentum to the accreting star according to whether their current orbital energy is above or below a critical value of order $-\omega j/2$, where ω is as before the stellar spin rate, and j the blob's specific angular momentum. Blobs with energies above this gain more angular momentum and are ultimately expelled, while those below it lose angular momentum and are ultimately accreted. The drag prescription has the further advantage of being very easy to implement in particle codes simulating the accretion flow, thus allowing one to study the combined effects of magnetic fields and viscosity. For small values of k one finds disc formation in the usual way, while for somewhat larger values the flow is magnetically influenced. Because the accretion or expulsion of matter depends directly on the star's spin ω, the latter adjusts itself until gas can just accrete, i.e. the critical value $\sim \omega j/2$ is equal to the orbital specific energy of gas leaving the companion star, with j also characteristic of this gas. This leads to a direct relation between the spin period P_{spin} and orbital period P_{orb} of the form

$$P_{\text{spin}} \simeq 0.07 P_{\text{orb}}, \tag{11}$$

where the precise coefficient depends on the binary mass ratio. There is some evidence that this relation hold for a class of systems. If k is made still larger, we have flow ultimately along field lines (cf (8)), and the strong magnetic field of the accreting white dwarf is likely to enforce corotation of this star with the binary: these are the AM Herculis systems.

While this behaviour is plausibly similar to what is observed, the most spectacular support for the drag model is provided by the CV system AE Aquarii. This has a rather long orbital period (almost 10 hr) and the shortest observed spin period $P_{\text{spin}} = 33$ s of any magnetic CV. The standard picture would predict that this system has a very weak magnetic field (cf eq 8), while the long orbital period implies large orbital specific angular momentum j. Since the weak magnetic field means that the effective size R_1 of the accreting star is not very large, we would expect from (6) that AE Aqr, of all magnetic CVs, should have a large, well–developed accretion disc. Yet maps of the accretion flow made by Doppler tomography reveal that not only does AE Aqr possess no disc, it actually ejects almost all the matter which attempts to accrete on to its rapidly–spinning magnetosphere. That the expulsion is centrifugally driven is confirmed by the observation that the 33 s spin period is slowly lengthening over a timescale of about 10^6 yr: the corresponding energy loss rate from the white dwarf spin is far greater than the radiative output of the system, and had constituted something of a puzzle.

The drag model gives a natural explanation for all of these features. Unlike the standard procedure of comparing magnetic and material stresses, the criterion for disc formation depends explicitly on the spin rate; if no disc forms, a large amount of expulsion is inevitable as the critical energy $\sim -\omega j/2$ is much more negative than the specific orbital energy of gas leaving the companion. With a suitable choice of drag coefficient k and mass overflow rate $\dot{M} \sim 10^{-10}\ M_\odot$ yr^{-1}, there is full agreement between the predicted and observed spindown power and the detailed Doppler map (Wynn et al, 1994, 1995).

8. Conclusion

This review has concentrated on areas where current progress is most rapid and the author has a personal involvement. However the field of accreting binaries remains one of intense research, and there are several other areas where exciting developments have occurred recently, such as the origin of viscosity in accretion discs, and the possibility of near radial advective accretion in the inner parts of these discs. Limitations of space preclude more than a mention of just these two topics: the interested reader is referred to the books by Frank et al (1996) and Lewin et al (1995) for further details.

Acknowledgements

I thank Graham Wynn for much help in the preparation of this chapter.

References

Burderi, L., King, A.R., & Wynn, G.A., 1995, ApJ, in press

Burderi, L., King, A.R., & Wynn, G.A., 1995, in prep.

Frank, J., King, A.R., & Lasota, J.P., 1992, ApJ 385 L45

Frank, J, King, A.R., & Raine, D.J., 1996, *Accretion Power in Astrophysics*, 3rd Ed, Cambridge University Press

Giacconi, R., Paolini, F.R., Gursky, H., & Rossi, B.B., 1962, Phys. Rev. Lett. 9, 439

Ghosh, P., & Lamb, F.K., 1979, ApJ 234, 296

Ghosh, P., & Lamb, F.K., 1991, in *Neutron Stars: Theory and Observation*, eds. J. Ventura & D. Pines (Dordrecht, Kluwer), 363

Hameury, J.M., King, A.R., & Lasota, J.P., 1988 A&A 195, L12

King, A.R., 1993, MNRAS 261, 144

King, A.R., 1995 to appear in Proceedings of the Abano–Padova Workshop on Cataclysmic Variables, eds. A. Bianchini et al.

King, A.R., Frank, J., Kolb., U, & Ritter, H., 1995, ApJ 444, L37

King, A.R., Frank, J., Kolb, U., & Ritter, H., 1995, in prep.

Kuiper, G.P., 1941, ApJ 93, 133

Lewin, W.H.G., van Paradijs, E.P.J., & van Paradijs, J., eds., 1995 *X–ray Binaries*, Cambridge University Press

Patterson, J., 1984, ApJ Supp 54, 443

Podsiadlowski, Ph., 1991, Nat. 350, 136

Ritter, H., 1994, *Mem. Astr. Soc. It* 65, 173

Ritter, H., Kolb, U., & Zhang, Z., 1995 A&A, in press

Salpeter, E.E., 1964, ApJ 140, 796

Schwarzschild, M., 1958, *Structure and Evolution of the Stars*, Princeton University Press

Shklovsky, I.S., 1967, ApJ 148 L1

van den Heuvel, 1994, to appear in Proceedings of a Workshop on Pulsars, eds. G. Srinivasan et al., Pub. Ind. Acad. Sci.

Wynn, G.A., King, A.R., & Horne, K., 1995a, to appear in Proceedings of the Cape Workshop on Cataclysmic Variables, eds. D. Buckley & B. Warner

Wynn, G.A., King, A.R., & Horne, K., 1995b, in prep.

EXTRAGALACTIC ASTRONOMY

LINE FORMING REGIONS IN ACTIVE GALAXIES AND THEIR NUCLEI

J. E. DYSON and R. J. R. WILLIAMS

Department of Physics and Astronomy, University of Manchester, UK

and

J. J. PERRY

Institute of Astronomy, University of Cambridge, UK

November 22, 1995

Abstract. The spectra of active galaxies and their nuclei are rich in emission and absorption line features. A major aim of present research is the development of self-consistent hydrodynamic models for the production of the line-forming regions. We here review such modelling and stress the central role played by shock phenomena induced by winds and explosions on scales ranging from the circumstellar to the intergalactic.

1. Introduction

Active galactic nuclei (AGN) are a rather poorly defined observational class. The earliest examples of what now constitutes the class were found as bright radio sources, but the class extended to include the bright optical line and continuum sources found in the nuclei of nearby galaxies. They have attracted much theoretical and observational interest, since their properties cannot be explained as conventional stellar systems, and they can be observed from very early cosmic epochs.

AGN are characterized by the breadth of their continuum radiation, which is bright across up to 10 decades in frequency (unlike, for instance, stellar black-body spectra). The spectrum is remarkably uniform in power per decade of frequency from the IR to the X-ray. Superposed on this continuum, there are prominent emission lines, with line widths ranging from hundreds of km s^{-1} up to several hundredths of the speed of light. They provide a rich source of detailed physical information about the very centre of the galaxy from which the lines are emitted. The relative constancy of the line equivalent widths between individual AGNs, and their rapid variability in time, suggest that the lines are produced by gas very close to the central continuum source, and that the line emission must be studied as an intrinsic part of the processes which generate the radiant energy.

The apparent magnitudes of QSOs and AGNs, when corrected for distance and with the assumption of isotropic emission—imply that the integrated luminosities are between 10^{42} and 10^{49} ergs s^{-1} (equivalent to $10^8 - 10^{15} L_\odot$). There are short time scale variations in the continuum emission: in low luminosity objects the X-ray fluxes often vary substantially within hours, and

Astrophysics and Space Science **237**: 187–206, 1996.

sometimes even minutes. *If* the variability is due to a single central object, then that source must, in some objects at least, be smaller than a light-hour, i.e. $\lesssim 10^{14}$ cm. The variability appears to be correlated with luminosity, with the lowest luminosity objects varying the most rapidly, although the relationship has yet to be quantified; the largest inferred sizes of the central continuum source are $\lesssim 10^{17}$ cm. Two central problems in AGN research concern the generation of the extraordinary output of energy from such small volumes in the nuclei of galaxies and its frequency distribution.

Some of the earliest views regarding the nature of AGNs invoked stars, particularly through their evolution to the supernova stage (e.g. Shklovskii 1960; Field 1964; Colgate 1967). However, when severe difficulties arose regarding the necessary efficiency of energy generation and the observed rapid continuum variability, there was a change of opinion in favour of long-lasting gravitational processes (Lynden-Bell 1969). Accretion of matter onto a supermassive black hole (BH) via the intermediate agency of an accretion disc became, and has remained ever since, the favoured model for the power-house driving AGN luminosities. However, there is little strong observational evidence for the existence of supermassive BHs in galactic nuclei. A competing proposal is that the variability arises in compact supernova remnant 'flares' in starbursts (Terlevich et al. 1993, 1995; Filippenko 1992).

The maintenance of energy generation by accretion onto a BH requires a large enough reservoir of accretable material to sustain the source over its lifetime. It appears likely, at least for low luminosity objects, that the interstellar medium (ISM) and/or mass loss from stars in the galaxy that houses the AGN can provide enough mass to fuel the central engine. This view is supported by the rather sparse evidence for extranuclear fuelling (Heckman 1992). In this context, extranuclear implies that the region is not significantly influenced by the central engine—although, as will be discussed later, such regions may be rare. However, the high mass-consumption rates required to explain the energy generation in high-luminosity objects (up to hundreds and even thousands of solar masses of material per year) cannot be readily explained in this manner, and the existence of a dense nuclear star cluster surrounding the BH seems to be necessary (Williams & Perry 1994).

In order to construct physical models of the broad emission line regions (BELRs), which have line widths $\gtrsim 1000$ km s^{-1}, we need to derive densities, temperatures and excitation conditions; emission lines from many ions must therefore be measured simultaneously. Until recently, too few lines were observed in the spectral range available for any single object, and so throughout the history of modelling AGNs it has been necessary to construct 'composite' spectra on the assumption that there is a continuity between objects of different z and absolute luminosity. If, for example, there are systematic differences between high and low luminosity objects, these will be difficult to determine on the basis of the analysis of these composite spectra.

Despite these caveats, a number of very important general features have been established from analysis of the spectra. They are well explained if one assumes that the excitation mechanism is photo-ionization by the broadband 'underlying' continuum; the thermal and ionization structure is determined by the interaction of the continuum with dense ($n_e \gtrsim 10^9$ cm^{-3}) gas. (The continuum here refers both to the central non-thermal radiation and an enhanced flux of UV and soft X-rays usually ascribed to radiation from an accretion disc surface). There are two main physically distinct components to the broad emission line spectra: the low ionization lines (LILs, consisting primarily of the Balmer series and lines from singly ionized species, e.g. CII, SII, FeII) and high ionization lines (HILs, e.g. Ly α and multiply ionized species e.g. CIV, NV, OVI). At least for radio quiet objects (Gaskell 1988, Sulentic et al. 1995), the HILs are systematically blueshifted with respect to the LILs. This can be interpreted as meaning that the HILs are formed in gas having a systematic component of bulk radial motion (probably outflow, possibly inflow), and that there is some material obscuring redshifted emission. The full width at half maximum (FWHM) of the emission lines appears to be correlated with luminosity, which implies that the velocity of the emission line gas is larger in more energetic objects. Curiously, there is also a strong relationship between the widths of the HILs and the LILs (Corbin 1991). The lines vary in response to variations in the ionizing continuum, on time scales ranging from several days in low luminosity objects to more than several years in high luminosity QSOs. This variability can, in fact, be used to probe the structures of the nuclear regions (Peterson 1994).

Rather paradoxically, the sheer success of such photoionization models has been a major hindrance to the development of physical models of the BELR as a whole (Perry 1992). This is because adjustments to the thermal and ionization equilibria take place on time scales which are effectively instantaneous. It is then possible from line ratios, to develop a detailed understanding of the local properties of the clouds (e.g. densities, temperatures, ionizing incident fluxes) without addressing the important issues of the origin and evolution of individual emitting regions. Observational data provide scant direct information on these issues. Yet an understanding of the BELRs is essential since the photoionization models firmly place it in the innermost nuclear regions. Perhaps the central issue is how to account hydrodynamically for gas which occupies only a small fraction of the spatial region where it is found and yet which is moving hypersonically at Mach numbers anywhere from 10 to more than 100 (suggesting that shocks are involved).

Apparent absorption features are seen *within* the profiles of broad lines in the optical and UV in several Seyfert galaxies and at least one QSO (Ulrich 1988; Mathur et al. 1994; cf. however Koratkar et al. 1995).

Broad absorption lines (the BALs) are present in a small percentage (10% or so) of radio quiet high luminosity objects. The velocity widths of the absorption troughs range up to a tenth of the speed of light; their occurrence suggests that violent dynamic effects take place in the nuclear regions. These absorption lines are invariably blue-shifted relative to the corresponding emission lines. If the absorbing gas is in photoionization equilibrium with the continuum radiation field which energises the BELRs, it is probably external to, but contiguous with, the BELRs. The BALs represent one of the most puzzling aspects of all AGN phenomena and it is no exaggeration to say that progress in understanding them has been minimal.

Major developments in AGN astronomy are occurring as a result of ongoing X-ray observations. Technical innovations have resulted in detectors with ever increasing sensitivity and spectral resolution, and revealed the rapid variability of emission in this waveband. The X-ray spectral features observed in AGNs have been reviewed in detail by Mushotzky, Done & Pounds (1993). The underlying continuum in the X-ray waveband is believed to be a smooth power law, the result of processes in a relativistic plasma very close to a central black hole. The narrow range of 2–20keV spectral indices implies such a fundamental explanation. Superposed on this continuum are various features: absorption at low energies by intervening material, excess continuum emission in the 0.1–1keV and >8keV bands, fluorescent lines of low ionization states of Fe at 6.4keV and absorption features due to both low and high ionization material on the line of sight.

While the interpretation of any of these features is subject to change, it seems that the X-ray spectrum may give us much additional information on many constituents of active nuclei. The hard excess and fluorescent Fe lines are generally interpreted as the result of reflection from a large optical depth of cool material – which may well be the accretion disc. Low ionization absorption features will give important insight into the nature of the ISM surrounding the nucleus. Perhaps the most fascinating is the nature of the absorption lines from highly ionized material, the so-called 'warm absorber'. It is possible to identify the absorbing gas with BAL absorbing gas (Reynolds & Fabian 1995), which would then appear to be spatially coincident with the BELRs. In any case, the warm absorber forms an important phase of the ISM in AGNs which has not previously been observable.

The very existence of dynamically active BEL and BAL regions, coupled with a fuelling process for the central source, implies that high speed flows exist in the nuclear regions and that shock phenomena must occur. Moreover, we should not neglect the stellar population within the nucleus. Mass loss in stars occurs throughout the HR diagram. In particular, high speed mass loss occurs in winds from early type stars and in supernova explosions. Clearly, such mass loss in the nuclear regions would produce shocks in any surrounding medium, such as the gas involved in the fuelling and, as we will argue, the shocked gas is central to explanations of the line forming regions of all kinds.

Shocks in active galaxies and their nuclei have several unique features, compared to more conventional interstellar shocks. They have extreme parameters, e.g. velocities up to thousands of km s^{-1}, high incident radiation luminosities, and small filling factors of the shocked gas. Diagnostics of the line emitting gas—line variability and absorption—require that the shocks process gas on observable timescales. Since line emission is usually biased towards the densest gas at the highest pressures it is the properties of this gas which must be modelled successfully in order to account for the observations.

The influence of an AGN is not confined to the nucleus itself, but is felt throughout—and even outside—the host galaxy. Although not discussed further here, the widespread existence of jets on scales ranging from nuclear to intergalactic, have implications for many sorts of shock activity. Direct observational evidence for outflow from the nuclear regions themselves does not exist, and although the BALs are generally assumed to arise in a hypersonic global wind, the distinction between outflow or the effects of outflow is often not made. As we will discuss, the BALs may be indicative of quite complex interactions involving both local and global phenomena.

Strong evidence for the existence of winds generated on circumnuclear or galactic scales comes from optical line emission extended over kpc scales (e.g. Heckman, Lehnert & Armus 1993). Very crudely one might imagine interactions similar in nature to the well studied ones between massive stars and the ISM (e.g. Dyson 1994) where shocks play a major role. The extended optical emission may result from the impact of the wind on interstellar clouds or from the shell driven into the surrounding medium. This latter possibility is supported by the occurrence of characteristic double peaked line profiles (Heckman et al. 1993). In addition to emission line production, the analogue of the shell may contribute to the formation of another type of absorption line, namely the narrow absorption lines (NALs) seen in the spectra of many QSOs (Dyson, Falle and Perry 1979). The NALs have widths of the order of the sound velocity in the absorbing gas (about 20 km s^{-1}) although the velocity of this gas relative to the nucleus of the host galaxy is several thousand km s^{-1}.

Less overly dramatic—but equally important—phenomena are associated with the observed narrow emission lines. Narrow emission lines are observed in most AGN spectra; they have widths less than 1000 km s^{-1}, although their wings may be extensive, ranging up to 2000 km s^{-1}. From the intensity ratios of the forbidden lines, electron densities between 10^3 and 10^4 cm^{-3} are deduced. The narrow line gas must lie in a region (the NLR) up to a thousand times further from the nucleus than the BELR, if they are both photoionized by the same central continuum flux. The narrow emission lines thus 'map' the outer regions of the galactic centres, intermediate between the BELR and the full galaxy. High resolution mapping of the NLRs of several Seyfert galaxies (e.g. Unger et al. 1986) has shown very strong evidence for the association of the optical NLR emission and radio emitting plasma which plausibly may have its origin in the nuclear regions. That, together with the fact that the velocity widths in the NLRs are usually significantly higher than could be expected from a gravitational origin, is suggestive that the NLR is dynamically linked with the escape of plasma from the nuclear regions. In addition to the classic NLRs, further kpc-scale regions of narrow line emission, the extended narrow line regions (ENLRs), have been discovered in narrow band imaging of Seyfert galaxies.

We will not attempt to review all these various phenomena, but will discuss a selection, emphasising the central role of shocks and self-consistent hydrodynamics in their modelling.

2. The Broad Emission Line Regions

Perhaps the most remarkable feature of the BELRs is that, within a fairly wide scatter about the mean, the spectral features of the BELRs of high and low luminosity objects or radio-loud and radio quiet objects show so little systematic variation. There are, however, pronounced differences in line profiles from one object to another (e.g. Robinson 1995). This implies that there is some robust and general formation mechanism for the BELR gas which can work in the context of a range of global structures.

We therefore first explore the role and properties of the BELRs within the context of a model in which the structure of AGNs results from the symbiosis of a black hole and a nuclear starburst cluster. This model was first advanced by Dyson & Perry (1982) and later developed in much greater detail by Perry & Dyson (1985-henceforth PD), Collin-Souffrin et al. (1987) and Williams & Perry (1994).

Very simple considerations show why shocks are inevitable in a system containing a dense stellar cluster and a BH (Perry 1992). If a BH exists at the centre of a galaxy, the Keplerian velocity at distance r_{pc} from a central black hole of mass $M_8 10^8 \, M_\odot$ is $V_k \approx 660(M_8/r_{pc})^{1/2}$ km s^{-1}. The sound speed in the surrounding medium depends on its temperature. Unless some source of mechanical heating dominates, this will almost certainly be the Compton temperature of gas in equilibrium with the non-stellar continuum, $T_c = T_7 10^7$ K (T_7 is typically about unity), so the sound speed is $V_s \approx 380 \, T_7^{1/2}$ km s^{-1}. All stars within a distance $r \approx 3(M_8/T_7)$ pc of the BH therefore move through the local ISM supersonically. It then follows that all mass injection into the ISM must involve shocks, whether through stellar winds, supernovae, or as a result of tidal stripping or stellar collisions. This would be true even in the improbable case that the ISM were completely static.

However, the ISM originates from stellar mass-loss processes (Williams & Perry 1994) and the associated energy and momentum input ensures that typical velocities are roughly three times the sound speed, i.e. the velocities are those which are characteristic of thermal winds.

The main role of a shock is to convert kinetic energy into thermal energy: the post-shock temperature and pressure are increased and in the strong shock limit, the post-shock temperature is $T_s \approx 10^8 \omega^2$ K, where the shock velocity ω is expressed in units of 3000 km s^{-1}. Note that a very large range in shock velocity is expected, since it is a vector sum of the Keplerian velocity, the local ISM velocity and the velocity associated with the injection process.

To appreciate the significance of these shocks for the formation of the BELRs, we first briefly consider the physical state of the emitting gas. The properties of optically thin gas in thermal and ionization equilibrium with a non-thermal continuum radiation source are essentially determined by a single parameter, the ionization parameter, over a broad range of densities characteristic of a BELR. A physically meaningful way of defining this is as the ratio (Ξ) of the local ionizing radiation pressure to the gas pressure. Since the radiation pressure depends on the radiation flux, a direct determination of Ξ from a comparison of observed line spectra with theoretical spectral modelling determines the distance of the irradiated gas from the radiation source. High values of Ξ (i.e. low gas pressures) correspond to gas in which Compton heating balances Compton cooling and the equilibrium temperature is close to the Compton temperature. Low values (i.e. high gas pressures) correspond to gas whose temperature is determined by the balance between photoionization heating and recombination cooling, at a characteristic value of about 2.10^4 K. This latter state is established to be in very good agreement with that deduced from emission line data on the BELR. An immediate inference is that flows of extreme Mach number ($\sim$ 100) are present.

When gas with a high value of Ξ enters a strong shock, the increase in temperature and pressure takes the gas out of equilibrium with the radiation field and it cools, initially by Compton cooling, then by bremsstrahlung, and finally by line emission. Eventually it cools back into equilibrium with the radiation field. If the pressure remains more or less constant as the shocked gas cools, the cooled gas has a much lower value of Ξ than the pre-shock gas. The ionization parameter can be decreased by up to a factor of about $40\omega^2/T_7$—which can, in principle, be very large.

Such radiative shocks are the basis of a mechanism for the formation of the BELR proposed by Dyson & Perry (1982) and PD. Its immediate advantages can be stated simply. Firstly, the gas cooling behind the shock front is thermally unstable once its temperature falls below that at which Compton cooling (which is thermally stable) dominates. The cooled gas will form small clouds whose existence is confirmed by the very small filling factors deduced from emission line fluxes. Secondly, the clouds will retain some component of their pre-shock velocity and thus no acceleration mechanism is needed.

This latter point is coupled with the question of whether small clouds can survive in the ISM of an AGN for any significant time. We know (e.g. Fabian et al. 1986, Mathews & Ferland 1987) that no extensive two-phase equilibrium for the global ISM exists within the active nucleus for Compton temperatures, $T_c \lesssim 10^7$ K. Indeed, limits on the density of the hot ISM phase derived from soft X-ray variability constraints on its optical depth (PD) show that it must have a pressure far lower than that of the line emitting gas. (Although these two considerations rule out the original elegant 'two-phase' BELR model of Krolik, McKee & Tarter (1981), the influence of that paper on subsequent BELR studies cannot be over stressed).

Furthermore, unconfined clouds expand and dissipate in times of the order of their sound crossing time. This is at least an order of magnitude less than the flow time in the broad line region which must be a *minimum* time scale for permanent cloud existence. How then can unconfined clouds last long enough to explain the existence of the observed broad lines? The simplest answer, which escaped notice until pointed out by PD, is that, provided clouds are continuously being created, there is no need for them to be long-lived, i.e. evanescent clouds replace permanent clouds.

The properties of the line emitting clouds depend directly on the structure of the radiative regions behind the shocks. The postshock pressure gradients and flow velocities depend on the size and the energy and momentum input by the obstacles—stars, stellar winds, supernovae—which act as the shock generators in the global flow. The flow times behind small shocks around individual stars themselves are far too short to allow cooling. Larger shocks, where the flow time from the immediate post-shock entry to the sonic point (or line) in the flow is longer than the cooling time can, in fact, maintain the gas at roughly constant pressure while it cools. In the case of QSOs, where the BELRs lie at a characteristic distance of about a parsec from the central source, PD noted that the bow-shocks set up when fast flowing ISM encountered more or less fully expanded supernova remnants in the nuclear stellar cluster, are large enough to provide such constant pressure cooling regions. A comparison of the flow time between the post-shock stagnation zone and the sonic line and the cooling time gave the remarkable result that cool cloud (i.e. BEL) formation would occur everywhere interior to some radius which agreed extremely well with the characteristic BELR radius of a parsec deduced from the ionization parameter Ξ. In fact, cooled gas will be formed also in the earlier stages of the supernova expansion, behind closed shocks in the ISM produced by still-expanding supernova remnants.

A very powerful argument for this general model comes from an examination of the supernova rates required to provide the mass and covering factor observed for the BELR. Williams & Perry (1994) obtained the important result that the rates needed were in harmony with those in a nuclear stellar cluster which was simultaneously providing mass at a rate sufficient to fuel the central engine.

In contrast to high luminosity objects where the cooling shocks are expected to be formed by supernovae, in low luminosity objects (i.e. Seyfert galaxies) supernovae are probably not relevant to the production of the BELR, since a single supernova there can blow away the entire BELR—and may provide an explanation for the change from Seyfert 1 to Seyfert 2 type spectra observed on short time scales in some objects (Seyfert 2s do not have strong BELs). It is much more likely that, in Seyfert nuclei, the cooling takes place essentially as described above in the shocked wind gas from OB or WR stars (Perry, Williams & Dyson, in preparation).

Very detailed modelling of the structure behind a wide range of shocks spanning the range of conditions from the outer to the innermost regions of the BELR has been carried out recently by Innes & Perry (in preparation). They calculated line ratios for the emission from the shocked gas and compared them, where appropriate, with observational data. The agreement of the model results with the data is striking and again provides strong support for shock-generated BELR models.

The kinematic differences between the HILs and LILs (Gaskell 1988; Sulentic et al. 1995) indicate at the very least a two component structure. Collin-Souffrin et al. (1987) suggest that the LILs—which require much higher column densities than the HILs—originate on the surface of the accretion disc which is illuminated by back-scattered X-rays from the central source. The LIL widths must then arise from disc rotation and/or turbulent motions. Provided the HILs are formed in a flow which is dominated by approximately radial outward motion, the obscuration of the receding HIL component by the disc will produce the systematic blue HIL–LIL velocity shift observed in radio quiet QSOs. By definition, these evanescent clouds expand and will heat up to about the Compton temperature. Reynolds & Fabian (1995) have suggested that the warm absorbers may be identified with gas in the intermediate stage between the cool cloud material and the hot ISM.

A variety of other models have been proposed for the BELR. Several authors (e.g. Scoville & Norman 1988; Kazanas 1989) have pointed out that ionized red giant or supergiant winds would have remarkably similar properties to those inferred for the BELR gas. The line widths would result from the stellar velocity dispersion. Although such models are attractive in principle since the stars produce a natural gas reservoir, serious problems arise, not only with the number of such stars required, but also in regard to the actual physical state of the atmospheres in the energetic radiation fields encountered in the BELR (Begelman & Sikora 1992; Hartquist et al. 1995). Roos

(1992) proposed that BEL gas is debris from stars which have been tidally disrupted in the BH gravitational field. Other models for the BELR include magnetic acceleration of clouds off accretion discs (Emmering, Blandford & Shlosman 1992), cloud formation in shocks produced by the interaction of a disc wind with a nuclear wind (Smith & Raine 1985) and the interaction of stars with accretion discs (Zurek, Siemiginowska & Colgate 1994). Our view would be that these processes may make *some* contribution to the BELRs, but they are probably secondary to the shock generation mechanism of PD.

In an extremely stimulating series of papers, Terlevich and his collaborators have revived the idea that AGN activity could be explained by starbursts without the need for a central massive BH. In this scenario, the BELRs form during the latter stage of evolution of the star cluster as a direct consequence of the evolution of supernovae in dense circumstellar material. The BELR clouds form in cooling gas behind the supernova shocks and the cooling gas itself provides the necessary radiation field for photoionization. Heckman (1991) and Filippenko (1992) have judiciously reviewed the evidence for and against the pure starburst model. One serious problem is that this model makes no claim to explain radio loud objects in which the existence of high speed jets strongly suggests the presence of a central massive object. Terlevich et al. (1995) have argued that the observed variability of the BELRs arises naturally as a result of time dependent effects in the cooling region behind the supernovae shocks, and produced remarkable agreement with observed data on a nearby Seyfert galaxy NGC 5548. On the other hand, Filippenko, Ho & Sargent (1993) demonstrated that HST data on a nearby Seyfert galaxy NGC 4395 do *not* support the pure starburst hypothesis. It is important to note, however, that both that model and the starburst-BH model of PD have many interactions in common. Future high spatial resolution data should be a major discriminant of these two scenarios.

A major challenge in AGN studies is to discriminate observationally between the many BELR models. Studies of emission line variability— 'reverberation mapping'—in principle should shed light on the detailed structure of the BELRs. This technique has shown that the BELRs are complex dynamic systems which extend closer to the continuum source and contain gas at higher densities than hitherto thought (e.g. Peterson 1994; Gaskell 1994). Unfortunately, at present, data which are of high enough quality and extensive enough in time coverage for the complex analysis needed are restricted to a few low luminosity objects. The results from these datasets also suggest that the distribution of line-emitting gas is variable on timescales not much longer than the line response time, making it very difficult to extract more than the most sketchy conclusions about the distribution of ionized gas (e.g. Perry, van Groningen & Wanders 1994, Wanders 1995). From the standpoint of a physical understanding of the BELR (and the gas which fuels the luminosity) what is needed is a less intense study of a much wider selection of objects.

3. The Broad Absorption Lines

The BALs are an enigmatic feature in the spectra of roughly 10% of radio quiet QSOs, although they are very occasionally detected towards radio moderate sources. They are often regarded as the most definite evidence for mass loss intrinsic to the QSO. The frequency of their observation can be interpreted as indicating either that the BAL phenomenon occurs in a special class (10% or so) of QSOs in which the absorbing gas has a covering factor of 100%, or that the absorbing gas has a covering factor of about 10% and BALs are present in all radio quiet QSOs. The latter interpretation is currently favoured. BAL profiles have a variety of forms, ranging from almost featureless to highly structured, and either attached or detached from the adjacent emission lines. The ions observed indicate gas temperatures of about 10^4 K and evidence from the lack of associated forbidden line emission suggests electron densities above 10^{6-7} cm^{-3}. This, together with column density estimates, then shows that the absorbing gas must be distributed in thin sheets across the line of sight.

There is rather weak evidence to indicate that some connection can be made between the BELRs and the BALs. Firstly, if the absorbing gas is in photoionization equilibrium with the central continuum, its ionization parameter must be comparable with that in the BELRs. Secondly, estimates of the distance of the BAL gas from the central source put it contiguous but largely external to the BELRs. Finally, correlations exist between BAL line profile properties and those of associated emission lines. However the BALs do have very significantly larger velocity widths than those in the BELRs.

Most theoretical attempts to explain this phenomenon have been in terms of global flow models. For example, Murray et al. (1995) and de Kool & Begelman (1995) have suggested that the absorption lines are formed in a radiatively driven disc wind. Alternatively, the acceleration of clouds by winds (Weymann et al. 1982) or radiation pressure (Arav, Li & Begelman 1994) have been proposed. There are severe hydrodynamic difficulties with either of these last two mechanisms. It is well known that ram pressure acceleration of clouds disrupts them before their velocity has increased much above that of the shock driven into them by the initial wind impact. Radiation driving, whilst efficient in principle, requires a mechanism for cloud confinement which does not simultaneously produce an appreciable drag force on the clouds. Scoville & Norman (1995) have proposed that the BAL clouds originate in dust rich material shed from evolved stars which is accelerated just outside the BELRs by the action of radiation pressure on the dust grains. This latter suggestion does however require the existence of coherent physical structures—tails—behind stars which must extend over long ($\sim$ pc) scales in what is clearly a very dynamically active region.

Perry & Dyson (1991) noted that there was an interesting coincidence between the typical BAL velocity widths and that of the highest velocity ejecta seen in supernovae, and that the frequency with which BAL occur was suggestively similar to the BELR coverage factor in high luminosity objects. They therefore argued that a natural connection between the BELRs and the BALRs arises from the supernovae integral to the BELR model of PD, and that in fact the BALs provide evidence for local, as opposed to global, high velocity flows, although these latter do play a role. A supernova must produce the absorbing material before it slows down appreciably. In this 'Phase I', a fast forward shock is driven into the local ISM: a reverse shock moves into the ejecta. The forward shock moves outwards at the speed of the ejecta whilst the reverse shock accelerates back towards the explosion centre (e.g. Bode & Kahn 1985). Thus the ISM is always shocked to a high temperature. The temperature of newly shocked ejecta is initially low but increases with time as the reverse shock accelerates. In order that this gas produce observable BALs, the cooling time in the shocked gas must be less than the slowing down time and the fast moving cool gas must persist long enough to be observed—in practice for a period greater than a few years.

Simple modelling of Phase I evolution (Perry & Dyson 1992) shows that to satisfy these constraints, supernovae must explode into interstellar material which has a density somewhat lower than that required by the PD models for BEL generation, i.e. somewhat outside the BELRs. The only gas which is cool enough to produce absorption is the 10% or so of ejecta which is shocked only to a low temperature when the two shocks initially move very closely together. This model reproduces well both the column densities of absorbing gas (apart from the very much higher ones associated with the rare Mg II systems) and the timescales of a few years associated with changes in line profile structures. As is well known (e.g. Gull 1975), the ejecta-shocked-ISM interface is Rayleigh-Taylor unstable and the resulting fragmentation and accompanying destruction of fragments may contribute to the velocity structure observed in some line profiles.

This model currently has a number of unresolved problems: the most immediately obvious is that a supernova would produce both blue and red-shifted absorption troughs. Provided that the ISM flow is directed more or less radially outwards in the vicinity of the explosion, the supernova expansion is asymmetric and the shell section moving towards the nuclear regions—that which would produce redshifted absorption—slows down more rapidly and reaches a smaller radius than the shell section moving towards the observer. Moreover, the cooled gas will experience an outward force due to the radiation pressure exerted by the continuum radiation field which will enhance such effects. Although no detailed statistics have been investigated, simple models for asymmetric expansion (Wright, unpublished) show that this is not implausible.

One major problem with this simple idea may lie in the details of the global flows in the nuclear regions. Williams (1993) calculated aspherical radiatively-driven models for these flows (cf. also Williams, Baker & Perry, in preparation). The adopted structure was that of a central massive black hole and an accretion disk. Around this region, a central star cluster injects matter into the ISM. In these models, inflow along the plane of the accretion disk coexists with strong outflows along the axis perpendicular to the disk. The flow therefore has a pronounced bipolar structure. While supernovae occurring in inflow regions will have the wrong behaviour as far as the suppression of red-shifted absorption is concerned, lines of sight through these inflow regions may be obscured by a central molecular torus. Supernovae at least have the merit of not requiring a contentious acceleration mechanism.

The complex structures of these global flows throw up other shock phenomena. Close to the BH, some fraction of an accretion flow along the disk plane can be diverted into an axial wind (Williams 1993). The computed global solutions reflect the interaction between this outflow and the surrounding region where the flow is mass-loaded by stellar mass-loss (cf. Williams, Hartquist & Dyson, 1995). If this central wind is stalled by the mass-loading, a time varying shock forms in the wind. For rather stronger winds, a series of explosive events drives nuclear mass-loss. For the strongest winds, the flow reaches a steady equilibrium state.

In all these cases, strong shocks form in the global flow. In the one case, they are generated by the explosions; in the other, steady shock structures form around the stagnation point as is required, in general, if a strongly aspherical flow is to become fully supersonic at large radii (Williams & Dyson 1994).

4. The Narrow Line Region

It has long been known (Shields & Oke 1975) that in the NLRs, the emitting gas is distributed in an assembly of small clouds, as in the BELRs. The emission spectrum is very well accounted for by photoionization models in which, in a manner analogous to BELR models, the thermal and ionization balance are decoupled from the dynamics and origin of the emitting gas. In recent years, studies of the NLRs in Seyfert galaxies have shown that close links exist between the optical NLR emission and the nuclear non-thermal radio emission. Correlations have been found between the $\lambda21$-cm radio power and both the [OIII] $\lambda5007$ luminosity and line width. The radio emitting plasma appears to be in approximate pressure equilibrium with the thermal gas producing the optical emission. Spatial correlations have also been found between the radio and optical components within the NLRs. Of particular interest are cases in which the optical structure can be resolved and direct spatial association of the optical and radio structures is seen (e.g. NGC 5929—Whittle et al 1986).

These associations led Pedlar, Dyson & Unger (1985) to propose a model for the NLRs in which the optical emission is produced by the interaction of the radio components with the ambient galactic ISM. They proposed that spherically symmetric radio-emitting bubbles of plasma ('plasmons') expanding into the ISM, compressed and accelerated it in a manner entirely analogous to that occurring when supernova ejecta expand into the ISM. The NLRs in this model result from shocked interstellar matter which has cooled back into equilibrium with the nuclear radiation field. In order for optical emission to result, the cooling time in shocked gas has to be less than the expansion time of the plasmon. Again this is analogous to the expansion of supernova remnants, and demands that the plasmon expansion velocity has dropped below a critical value of about 350 km s^{-1} (which is very insensitive to the physical parameters, e.g. ISM density, plasmon energy content etc.). Pedlar et al. (1985) showed that this simple model would reproduce reasonably well the observed filling factors, densities and emission measures of the NLR thermal gas.

The radio components could represent the working surfaces of jets on, say, giant molecular clouds, or it could be that plasmons are emitted directly from galactic nuclei. This latter possibility had already received some theoretical support from Smith et al. (1983) who suggested that plasma ejection from the central regions of low-power AGN takes place in the form of bubbles whereas in high-power AGN it is more likely to take the form of jets. The original work of Pedlar et al. (1985) assumes that the expansion velocity of plasmons is much greater than any linear velocity they might have. However, where the NLRs are resolved, such as in NGC 5929, an NLR has a non-spherical morphology suggesting that they may better be described at the other extreme where the plasmon linear velocity is much greater than its expansion velocity.

The alternative model explored by Taylor et al. (1989) and Taylor, Dyson & Axon (1992–henceforth TDA) therefore is based on the assumption that plasmons are ejected supersonically from the nucleus and that they drive bowshocks into the galactic ISM. Shocked gas which cools to equilibrium with the nuclear UV continuum forms the NLR components, exactly as in the expanding-plasmon model.

Apart from the inclusion of an external continuum radiation field, this model has many similarities to bowshock models of Herbig-Haro (HH) objects (e.g. Hartigan, Raymond & Hartmann 1987) but with one very important difference. A general assumption of the HH models is that the shocked gas cools more or less instantaneously, i.e. the cooling length is much less than the scale size of the shocked zone. However, this is not a valid assumption to make for models of the NLRs since the relatively high velocities into the shocks ($V_s \sim 1000$ km s^{-1}) and the low pre-shock densities (~ 1 cm^{-3}) result in very large cooling lengths. The cooling time of an individual packet of gas depends upon where it enters the bowshock and the models of TDA took this into account.

The finite cooling time combined with the linear motion of the plasmon produces the observed displacement between the radio components and the regions of significant optical emission. Two particularly clear examples are the cases of NGC 5929 and Mkn 78 where the optical sources appear to lie closer to the nucleus than the radio sources, because they have a smaller observed separation than do the radio components. Taylor et al. (1989) successfully reproduced the separation between the radio and optical components of NGC 5929 in terms of this finite cooling length model.

A particularly interesting feature of the model is that the cooling length, defined as the distance from the apex of the shock to where significant optical emission appears, is rather insensitive to the shock velocity (going roughly as only V_s) in contrast to the case of cooling behind a plane shock where the cooling length is roughly proportional to V_s^4 (for Kahn's (1976) $T^{-1/2}$ cooling law). In the TDA model, the concept of a single cooling length must be replaced by that of an effective cooling length, corresponding to the minimum point on the cooling curve; it is a function of where the gas enters the shock and the shape of the shock. The displacement of the optically emitting region from the shock apex is found to be surprisingly insensitive to V_s. For example, for a quadratic shock shape, this effective cooling length is proportional only to $V_s^{16/13}$, and the line profiles calculated depend only on the viewing angle and the value of V_s. This model is very simplified, in that the structure of the shocked region is calculated on the assumptions that the post-shocked gas is well coupled as far as momentum is concerned, but that individual gas elements cool independently of each other, and that the shocked region is optically thin to the central ionizing radiation. Nonetheless, the model predicts, significantly, that the wide diversity of NLR profiles and velocity structures can be understood simply as a two-parameter family, dependent on only viewing angle and plasmon velocity (TDA).

5. Larger Scale Phenomena

There is ambiguous observational evidence for extremely energetic outflows from AGNs. The high velocity blue shifted BALs, as noted above, are usually cited as direct evidence for non-relativistic nuclear outflows. A possibly related phenomenon is that of the galactic superwinds (e.g. Heckman, Armus & Miley 1990) which seem to be associated with the presence of nuclear starbursts. Evidence for these winds comes from large scale ($\geq$ kpc) optical line and X-ray emission perpendicular to galactic discs; line ratios consistent with shock phenomena; double peaked optical emission line profiles extending over kpc which are consistent with expanding bubbles or bipolar flows, with line splitting $\sim$ 200–600 km s^{-1}.

There is a variety of ways in which winds could be driven out of the nuclear regions. Compton heated winds from accretion discs have received considerable attention. These are essentially thermal winds whose terminal velocity will be about three times the sound velocity in gas heated to the Compton temperature, i.e. $\sim$ 2000 km s^{-1}. Starburst winds are driven by the thermalisation of stellar ejecta, i.e. supernovae and stellar winds. Williams (1993) studied radiatively driven winds which have somewhat higher terminal velocities. Terminal velocities of a few 1000 km s^{-1} are produced.

Whatever the wind driving mechanism, the impact of these winds on the galactic ISM has many similarities to the impact of OB star winds on the galactic ISM. A strong shock is driven into ambient gas whilst a reverse shock decelerates the wind. The implications of wind activity are profound. For example, they inject chemically enriched material, energy and momentum into the intergalactic medium (e.g. Heckman, Lehnert & Armus 1993). They have also been suggested as the means of accelerating ISM clouds to typical NLR velocities (Smith 1993), although there are very severe hydrodynamic difficulties with this due to the tendency for the acceleration process to disrupt the accelerated clouds. Here we will concentrate on the suggestion made originally by Dyson et al. (1979) that wind driven shells may be responsible for the class of narrow absorption lines (NALs) which have velocities within a few 1000 km s^{-1} of the associated emission lines.

Statistical studies of the frequency and distribution of narrow metal absorption line systems (e.g. Anderson et al. 1987; Weymann, Carswell & Smith 1981; Perry, Burbidge & Burbidge 1978) provide evidence that absorption with $z_{abs} \approx z_{em}$ shows a statistical excess in radio-loud QSOs and is therefore likely to be intrinsic to the objects. (Most absorbing systems are produced in foreground intervenors, e.g. haloes of galaxies.)

The best studied of these associated systems is the complex in 3C 191 (Williams et al. 1975) which contains at least two absorption components. The main system has a velocity of 0.0027c relative to the emission line red shift (which is, as usual, taken as defining the rest frame). Studies of the absorption spectrum suggest that the absorbing gas is located at a distance of about 10 kpc from the ionizing continuum source, that it has a number density $\sim 10^3$ cm^{-3} and that the gas temperature is $\sim 2.10^4$ K (Williams et al. 1975). Dyson et al. (1979) noted that the gas density is roughly that expected in gas which has cooled to about the estimated temperature behind a shock moving at 0.0027c into the galactic ISM with a number density of unity. They therefore used standard stellar wind driven bubble theory (e.g. Dyson 1981) and showed that the absorbing gas could be the shocked ISM in the wind-driven shell.

The more general problem of the interaction of a nuclear wind and radiation field with galactic ISM was studied by Dyson, Falle & Perry (1980) and Falle, Perry & Dyson (1981—henceforth FPD). The initial stages of the interaction were discussed by Dyson et al. (1979; 1980); FPD showed that the inclusion of the nuclear radiation field had important hydrodynamic consequences for the later evolution of the driven shell, because of radiation pressure exerted on the gas in the shell. Provided that the radiative driving is sufficient, the wind driven shell accelerates. A positive pressure gradient develops in the cooled gas in the shell as the downstream piston pressure (i.e. the shocked wind pressure) decays. The radiative acceleration in photoionized gas is, under the circumstances appropriate here, proportional to the gas density (Röser 1979), and the development of the pressure gradient is thereby amplified. A velocity gradient is simultaneously set up in the shell and, in the shock frame, the gas is moving fastest at the piston. Provided this velocity is subsonic, as in normal stellar wind driven flows, the shock and piston are causally connected. However, when radiation driving is included, the velocity becomes supersonic and the shock and piston disconnect. At large enough times after disconnection, the flow becomes that of an accelerating radiative shock wave which sweeps through the ambient medium. Absorption line formation effectively takes place between the outer shock and the sonic point in the postshock gas, and the velocity dispersion is about equal to the sound speed there, as indicated by the absorption line widths.

There are restrictions on the velocities which can be reached. Firstly, the optical depth to the continuum steadily increases and eventually the radiative acceleration begins to fall. This may not be too serious a restriction since Perry & Dyson (1990) suggested that heating due to turbulent dissipation of energy in the shell might raise the gas temperature high enough ($\sim 2 - 3.10^4$ K) to decrease the neutral hydrogen content and thereby reduce the estimated optical depths. A more serious problem is that for radiative driving to be operative, the cooling time in the post-shock gas must be less than the dynamical timescale. They estimated that it was unlikely that the velocity of the absorbing gas could become greater than about $0.01c$ relative to the rest frame.

As noted earlier, 3C 191 has multiple absorbing systems, and this is also seen in other sources. There are two possibilities for generating these. Firstly, gas subjected to radiative driving which is proportional to the gas density, is subject to radiatively driven instabilities (Mestel, Moore & Perry 1976). Secondly, it is generally agreed that radiative shocks with power-law cooling functions ($L \sim T^\gamma$) in the cooled gas exhibit an oscillatory instability provided $\gamma \approx 1$ and the shock velocity is greater than ~ 140 km s^{-1} (Innes, Giddings & Falle 1987; Gaetz, Edgar & Chevalier 1988). The velocity condition is easily satisfied here and either mechanism may contribute to the appearance of multiple systems.

6. Conclusions

Activity in galactic nuclei appears to be the result of a symbiosis of an accreting black hole at the centre of a dense starburst stellar cluster. The young stars in the cluster lose mass, creating a fast moving, relatively dense ISM which is heated by the radiation produced by accretion of the ISM itself by the black hole. Shocks occur at the interfaces of the stellar mass loss envelopes and the ISM created by the cluster. These shocks are radiative, and the resulting cool, post shock gas is photoionized by the central continuum radiation; the fast moving high pressure photoionized gas radiates the broad emission line spectrum characteristic of an AGN. Similar radiative shocks further out in the galaxy radiate the narrow emission lines, and when the shocks lie on the line of sight to the nucleus they create both broad and narrow absorption features in the spectra.

One of the main challenges to observers is to obtain discriminants of the various scenarios put forward by theorists, and it is clear that multi-wavelength data will be central to meeting this challenge. It can be argued that the great progress in astronomy and astrophysics made over the last two decades stems largely from the major advances in observational techniques and instruments which have come in that period. Bob Wilson played a central role in such advances through his contributions to IUE and UV astronomy as a whole. It is a great pleasure to dedicate this article to him.

References

Anderson, S.F., Weymann, R.J., Foltz, C.B. & Chaffee, F.H.: 1987, *Astron. J.* **94**, 278.

Arav, N., Li, Z.-Y. & Begelman, M.C.: 1994. *Astrophys. J.* **432**, 62.

Begelman, M.C.: 1985, *Astrophys. J.* **277**, 492.

Begelman, M.C. & Sikora, M.: 1992, in S.S. Holt, S.G. Neff & C.M. Urry (eds.), *Testing the AGN Paradigm*, AIP, New York, p.568.

Bode, M.F. & Kahn, F.D.: 1985, *Mon. Not. R. astr. Soc.* **217**, 205.

Colgate, S.A.: 1967, *Astrophys. J.* **150**, 163.

Collin-Souffrin, S., Dyson, J.E., McDowell, J.C. & Perry, J.J.: 1987, *Mon. Not. R. astr. Soc.* **232**, 537.

Corbin, M.R.: 1991, *Astrophys. J.* **371**, L51.

de Kool, M. & Begelman, M.C.: 1995 (preprint).

Dyson, J.E.: 1981, in F.D. Kahn (ed.), *Investigating the Universe*, D. Reidel Publ. Co., Dordrecht, p.125.

Dyson, J.E.: 1994, in T.P. Ray & S.V.M. Beckwith (eds.), *Star Formation and Techniques in Infrared and mm-Wave Astronomy*, Springer-Verlag, Berlin, p.93.

Dyson, J.E. & Perry, J.J.: 1982, in E. Rolfe, A. Heck & B. Batrick (eds.), *Proc. Third European IUE Conference*, Madrid, p.593.

Dyson, J.E., Falle, S.A.E.G. & Perry, J.J.: 1979, *Nature* **277**, 118.

Dyson, J.E., Falle, S.A.E.G. & Perry, J.J.: 1980, *Mon. Not. R. astr. Soc.* **191**, 785.

Emmering, R.T., Blandford, R.D. & Shlosman, I.: 1992, *Astrophys. J.* **385**, 460.

Fabian, A.C., Guilbert, P.W., Arnaud, K., Shafer, R.A., Tennant, A.F. & Ward, M.J.: 1986, *Mon. Not. R. astr. Soc.* **218**, 457.
Falle, S.A.E.G., Perry, J.J. & Dyson, J.E.: 1981, *Mon. Not. R. astr. Soc.* **195**, 397.
Field, G.B.: 1964, *Astrophys. J.* **140**, 143.
Filippenko, A.: 1992, in W.J. Duschl & S.J. Wagner (eds.), *Physics of Active Galaxies*, Spring-Verlag, Berlin, p.345.
Filippenko, A., Ho, L.C. & Sargent, W.L.W.: 1993, *Astrophys. J.* **410**, L75.
Gaetz, T.J., Edgar, R.J. & Chevalier, R.A.: 1988, *Astrophys. J.* **329**, 927.
Gaskell, C.M.: 1988, *Astrophys. J.* **325**, 114.
Gaskell, C.M.: 1994, in P.M. Gondhalekhar, K. Horne & B.M. Peterson (eds.), *Reverberation Mapping of the Broad-Line Region in Active Galactic Nuclei*, ASP Conference Series, Vol.69, San Francisco, p.111.
Gull, S.F.: 1975, *Mon. Not. R. astr. Soc.* **171**, 237.
Hartigan, P., Raymond, J. & Hartmann, L.: 1987, *Astrophys. J.*, **316**, 323.
Hartquist, T.W., Durisen, R.H., Dyson, J.E., Rawlings, J.M.C., Williams, D.A. & Williams, R.J.R.: 1995, *Astrophys. J.* (in press).
Heckman, T.M.: 1991, in C. Leitherer, N.R. Walborn, T.M. Heckman & C.A. Norman (eds.), *Massive Stars in Starbursts*, C.U.P., Cambridge, p.289.
Heckman, T.M.: 1992, in S.S. Holt, S.G. Neff & C.M. Urry (eds.), *Testing the AGN Paradigm*, AIP (New York), p.593.
Heckman, T.M., Armus, L. & Miley, G.K.: 1990, *Astrophys. J. Suppl.* **74**, 883.
Heckman, T.M., Lehnert, M.D. & Armus, L.: 1993, in J.M. Shull & H.A. Thronson (eds.), *The Environment and Evolution of Galaxies*, Kluwer Academic Publ., Dordrecht, p.455.
Innes, D.E., Giddings, J.R. & Falle, S.A.E.G.: 1987, *Mon. Not. R. astr. Soc.* **227**, 1021
Kahn, F.D.: 1976, *Astron. Astrophys.* **50**, 145.
Kazanas, D.: 1989, *Astrophys. J.* **347**, 74.
Koratkar, A., et al.: 1995, preprint.
Krolik, J.H., McKee, C.F. & Tarter, B.: 1981, *Astrophys. J.* **249**, 422.
Lynden-Bell, D.: 1969, *Nature* **223**, 690.
Mathews, W.G. & Ferland, G.J.: 1987, *Astrophys. J.* **323**, 456.
Mathur, S., Wilkes, B., Elvis, M. & Fiore, F.: 1994, *Astrophys. J.* **434**, 493.
Mestel, L., Moore, D.W. & Perry, J.J. 1976, *Astron. Astrophys.* **52**, 203.
Murray, N., Chiang, J., Grossman, S.A. & Voit, G.M.: 1995 (preprint).
Mushotzky, R.F., Done, C. & Pounds, K.A.: 1993, *Ann. Rev. Astron. Astrophys.* **31**, 717.
Pagel, B.F.: 1985, in J.E. Dyson (ed.), *Active Galactic Nuclei*, Manchester Univ. Press, p.373.
Pedlar, A., Dyson, J.E. & Unger, S.W.: 1985, *Mon. Not. R. astr. Soc.* **214**, 463.
Perry, J.J.: 1992, in B. Rocca-Volmerange, B. Guideroni, M. Dennefeld & J. Tran Thanh Van (eds.), *First Light in the Universe: Stars or QSOs?*, Editions Frontières, Gif-sur-Yvette, p.225. Perry, J.J., Burbidge, E.M. & Burbidge, G.R.: 1978, *Publ. Astron. Soc. Pacific* **90**, 337.
Perry, J.J. & Dyson, J.E.: 1985, *Mon. Not. R. astr. Soc.* **213**, 665.
Perry, J.J. & Dyson, J.E.: 1990, Astrophys. J. **213**, 362.
Perry, J.J. & Dyson, J.E.: 1992, in S.S. Holt, S.G. Neff & C.M. Urry (eds.), *Testing the AGN Paradigm*, AIP, (New York), p.353.
Perry, J.J., van Groningen, E. & Wanders, I.: 1994, *Mon. Not. R. astr. Soc.* **271**, 561.
Peterson, B.: 1994, in P.M. Gondhalekhar, K. Horne & B.M. Peterson (eds.), *Reverberation Mapping of the Broad-Line Region in Active Galactic Nuclei*, ASP Conference Series, San Francisco, vol. 69, p.1.
Reynolds, C.S. & Fabian, A.C.: 1995, *Mon. Not. R. astr. Soc.* **273**, 1167.
Robinson, A.: 1995, *Mon. Not. R. astr. Soc.* **272**, 737.
Roos, N.: 1992, *Astrophys. J.* **385**, 108.
Röser, H.-J.: 1979, *Astron. Astrophys.* **80**, 179.
Scoville, N. & Norman, C.A.: 1988, *Astrophys. J.* **332**, 163.

Scoville, N. & Norman, C.A.: 1995 (preprint).

Shields, G.A. & Oke, J.B.: 1975, *Astrophys. J.* **197**, 5.

Shklovskii, I.S.: 1960, *Sov. Astron. A.J.* **4**, 885.

Smith, M.D. & Raine, D.J.: 1985, *Mon. Not. R. astr. Soc.* **212**, 425.

Smith, M.D., Smarr, L., Norman, M.L. & Wilson, R.J.: 1983, *Astrophys. J.* **263**, 432.

Smith, S.J.: 1993, *Astrophys. J.* **411**, 570.

Sulentic, J.W., Marziani, P., Dultzin-Hacyan, D., Calvani, M. & Moles, M.: 1995, *Astrophys. J.* **445**, L85.

Taylor, D., Dyson, J.E., Axon, D.J. & Pedlar, A.: 1989, *Mon. Not. R. astr. Soc.* **240**, 487.

Taylor, D., Dyson, J.E. & Axon, D.J.: 1992, *Mon. Not. R. astr. Soc.* **255**, 351.

Terlevich, R., Tenorio-Tagle, G., Franco, J., Boyle, B., Rozyczka, M. & Melnick, J.: 1993, in B. Rocca-Volmerange, B. Guideroni, M. Dennefeld & J. Tran Thanh Van (eds.), *First Light in the Universe: Stars or QSOs?*, Editions Frontières, Gif-sur-Yvette, p.261.

Terlevich, R., Tenorio-Tagle, G., Rozyczka, M., Franco, J. & Melnick, J.: 1995, *Mon. Not. R. astr. Soc.* **272**, 198.

Ulrich, M.H.: 1988, *Mon. Not. R. astr. Soc.* **230**, 121.

Unger, S.W., Pedlar, A., Booler, R.V. & Hanson, B.A.: 1986, *Mon. Not. R. astr. Soc.* **219**, 387.

Wanders, I.: 1995, *Astron. Astrophys.* **296**, 332.

Weymann, R.J., Carswell, R.F. & Smith, M.G.: 1981, *Ann. Rev. Astron. Astrophys.* **19**, 41.

Weymann, R.J., Schiano, A.V.R., Scott, J.S. & Christiansen, W.A.: 1982, *Astrophys. J.* **262**, 497.

Whittle, M., Haniff, C.A., Ward, M.J., Meurs, E.J.A., Pedlar, A., Unger, S.W., Axon, D.J. & Harrison, B.A.: 1986, *Mon. Not. R. astr. Soc.* **222**, 189.

Williams, R.E., Strittmatter, P.A., Carswell, R.F. & Craine, E.R.: 1975, *Astrophys. J.* **202**, 296.

Williams, R.J.R.: 1993, Ph.D. Thesis, University of Cambridge.

Williams, R.J.R. & Dyson, J.E.: 1994, *Mon. Not. R. astr. Soc.* **270**, L52.

Williams, R.J.R., Hartquist, T.W. & Dyson, J.E.: 1995, *Astrophys. J.* **446**, 759.

Williams, R.J.R. & Perry, J.J.: 1994, *Mon. Not. R. astr. Soc.* **260**, 437.

Zurek, W.H., Siemiginowska, A. & Colgate, S.A.: 1994, *Astrophys. J.* **434**, 46.

THE BROAD EMISSION-LINE REGION IN AGNS AND QUASARS:

THE IMPACT OF VARIABILITY STUDIES

P.M. GONDHALEKAR
Rutherford Appleton Laboratory Chilton, OXON., OX11 0QX.

M.R. GOAD
STScI, 3700 San Martin Drive, Baltimore, MD 21218, USA.

and

P. T. O'BRIEN
Astrophysics, Department of Physics, Keble Road, Oxford, OX1 3RH.

Abstract. Coordinated observations of variability of the continuum and the emission-line luminosities (reverberation mapping) in AGNs and quasars have fundamentally altered our understanding of the broad-line regions in active galaxies. The constraints these observations impose on the models of the BLRs have been demonstrated here by an attempt to model the BLR of NGC5548, the most intensively monitored AGN. Two models of a BLR, with one having a power law radial distribution of density and the other a gaussian radial distribution, are described and the modelled line luminosities and centroid of the line response functions are presented. A self-consistent model is presented for the change in the $C_{IV}/Ly\alpha$ ratio as the continuum luminosity changes. It is shown that BLR gas must be composed of a mixture of optically thin and optically thick gas and the proportion of thick and thin gas alters with the luminosity of the ionizing continuum. The observed centroid or the lag of a line, can be a function of the continuum luminosity.

The variability of the profile of the C_{IV} line in the spectrum of NGC5548 is investigated. This profile is extremely robust and is not significantly affected by changes in the ionizing continuum. Future models of the kinematics of the BLR clouds will have to be based on very stable cloud motions and include anisotropic line emission.

AN HISTORICAL PERSPECTIVE

The launch of the International Ultraviolet Explorer (IUE) on 26 January 1978 heralded a new era in astronomy at ultraviolet (1200Å to 3200Å) wavelengths. IUE was conceived at the then Astrophysics Research Unit (of the then Science and Engineering Research Council) and University College London. The full story of how a small observatory satellite planned in the UK became the International Ultraviolet Explorer, a joint US/ESA/UK project has not yet been told and it is not the purpose of this paper to tell this story. Here we recall the irritating regularity with which some of us, who worked on the IUE detectors, had to reiterate to the astronomical communities (on both sides of the Atlantic) that IUE was going to be capable of first rank extragalactic astronomy. The general perception was that IUE would perhaps be able to observe 3C273 and NGC 4151 and it would be nothing short of a miracle if any other extragalactic object was detected. No amount of

Astrophysics and Space Science **237**: 207–240, 1996.
© 1996 *Kluwer Academic Publishers. Printed in Belgium.*

gymnastics with pre-launch data would convince these sceptics that objects at a redshift of one were entirely within the grasp of IUE. However, IUE confounded these sceptics, and quasars at redshifts of around two have been detected above the 3σ level in the continuum. The observations of the double quasar Q0957+561 A,B (Gondhalekar & Wilson 1980) established IUE as an instrument that was making significant contributions at the forefront of modern extragalactic astronomy.

Soon after the in-flight verification phase, Professor R Wilson (UCL) suggested that the European astronomers with IUE time allocated for extragalactic observations should pool part of their allocations to undertake an intensive study of one or two active galaxies or quasars. This led to the founding of the European Extragalactic Collaboration (EEC) which in its early days was organised and coordinated by P M Gondhalekar (RAL). This collaboration decided to undertake a number of observations of 3C273, NGC 4151 and NGC 1068. The last object had to be dropped as it was not observable with IUE from the European (VILSPA) tracking station. The spectra of 3C273 were combined to produce a high signal-to-noise ratio spectrum (Ulrich et al. 1980). But the observations of NGC 4151 produced the real surprise. This object was found to be violently variable over short time scales (Penston et al. 1981). The large aperture of IUE and the absence of interfering atmosphere left no doubt that the observed variations were intrinsic to this Seyfert galaxy, and the rest, as they say, is history.

1. Introduction

Active galactic nuclei (AGNs) and quasars are the most luminous compact objects in the Universe. The generic model of these nuclei consists of a central super-massive object − probably a black hole (but a super-massive star burst has also been proposed) − surrounded by an accretion disc. This centre is embedded in three principal emission-line regions. The outermost region, the Extended Narrow Line Region (ENLR), has an electron density of $\sim 10^3$ cm^{-3}, the spectral lines formed in this region have Doppler widths of $\sim 10^2$ km s^{-1}, and the radius of this region is ~ 10 kpc. The next region, the Narrow Line Region (NLR), has an electron density of $\sim 10^5$ cm^{-3}, line widths of $\sim 10^3$ km s^{-1}, and a radius of ~ 1 kpc. The inner-most region, the Broad Line Region (BLR), has electron densities $> 10^9$ cm^{-3} and emission-line widths of $\sim 10^4$ km s^{-1} (Netzer 1990, Osterbrock 1993). Both the ENLR and NLR can be spatially resolved in nearby AGNs, but the spatial structure of the BLR cannot be resolved with either existing or proposed astronomical instrumentation.

The BLR, lying closest to the central engine, is the least well understood. Early models of this region were largely concerned with establishing the basic ionization mechanism and an order-of-magnitude chemical composition

(Bahcall and Koszlovsky 1969; Davidson 1972; MacAlpine 1972). These early generations of models established that photoionization was the principle energy source, the clouds producing the lines were narrow filaments and these clouds have (broadly) solar composition (see the review by Davidson and Netzer 1979). For simplicity, and because observations demanded no better, the models were based on the assumption of a single population of clouds, in a spherical geometry, being responsible for all emission-line properties. These photoionization models suggested that the size of the BLR in AGNs is ∼0.1 pc and that in quasars is ∼1.0 pc.

The observations of correlated variability of luminosities of emission lines and the ultraviolet/optical continuum (line reverberation mapping) have brought about a profound re-appraisal of the models of the BLR. Only a few low-luminosity AGNs have been monitored in detail at present (see the reviews by Peterson 1993, 1995; Robinson 1995; and Maoz 1995 for details) and the BLR size, inferred from the time lag between the continuum and line variations, is at least a factor of ten smaller than that deduced from photoionization models. Moreover, these variability studies have demonstrated that the responsivity of the BLR clouds is radially stratified, such that the response of the low-ionization lines (LILs) (*e.g.* Mg II, Fe II, Hβ) originates from larger distances from the central source than the response of the high-ionization lines (HILs)(*e.g.* O VI, Lyα, C IV). The considerably smaller BLR sizes suggested by line reverberation studies implies that the BLR clouds in AGNs and quasars are exposed to a far more intense radiation field than had been previously considered. The implications of this enhanced radiation field have been considered by Ferland and Persson (1989). The high radiation fields would also suggest that a fraction of BLR clouds would be optically thin and the spectrum of an *ad hoc* combination of optically thin and optically thick gas has been investigated by Shields *et al.* (1995).

Despite this increased understanding of the physical state of the gas in the BLR many aspects of emission-line variability are poorly understood. Considerable emphasis has been placed recently on the interpretation of the line response functions (Blandford & McKee 1982, Horne *et al.* 1991, Krolik *et al.* 1991) recovered from variability monitoring data (a full discussion of the response function is given in Section 3). However, implicit assumptions made to recover the response function are poorly understood and do not necessarily reflect the realities of the physical state of the BLR gas. Also, the inversion provides *a* response function of a line, which is not necessarily *the* response function of a line. Moreover, the current quality of the monitoring data (*i.e.* sampling frequency, signal-to-noise, length of the light curves *etc.*) is such that the recovered response function is more a reflection of the inadequacies of the data and not the physics of the BLR. At present it is perhaps more fruitful to model the BLR and challenge these models with *all* data and parameters which result from the monitoring campaign and attempt to constrain the parameter space of the BLR. This approach has the added advantage that the BLR parameters can be directly related to the observed light curves or the measured parameters. In this paper two models of the BLR are proposed and the output from these models is compared with the variability data on NGC5548, at present the most intensively monitored AGN. It is not the purpose of this paper to present a 'complete and final' model of the BLR in NGC5548 because this is not possible at present; instead this paper describes the constraints imposed by the reverberation studies on the proposed and future models of the BLR in this AGN.

An intensive monitoring campaign of NGC5548 was undertaken in 1989: this AGN was observed every four days for eight months with *IUE* and various ground-based telescopes. The bench-mark data, analysis and results obtained during this monitoring campaign have been presented by Clavel *et al.* (1991), Peterson *et al.* (1991), Dietrich *et al.* (1993) Maoz *et al.* (1993). A further short *IUE/HST*/optical campaign was conducted in 1993 and the results of this campaign have been presented by Korista *et al.* 1995. These monitoring campaigns have provided very accurate data on line luminosities and the lag of a line *i.e.* the phase shift between the continuum and the emission-line light curves and also the centroid of the cross-correlation function of the ultraviolet continuum and line light-curves. The mean values of line luminosities and centroids are given in Table 1 where line wavelengths are given in angstroms.

At present only the variability of the integrated line intensities to a change in the ionizing (as measured by the change in the ultraviolet/optical) continuum has been addressed (see reviews by Peterson 1993,1995). However, there is a considerable amount of convolved information locked in the profiles of emission lines, in particular, the kinematics of the radiating clouds can only be inferred from changes in the line profile. The profile variability has not received a great deal of attention up to now. A brief analysis of the variability of the profile of C IV line in NGC5548 is also presented in this paper.

The plan of this paper is as follows: In section 2 a mean continuum energy distribution (CED) of NGC5548 is derived; this CED is based on observations, where available. In subsequent sections this CED is used to model the photoionization of gas in the model BLRs. In Section 3, two BLR models with different spatial structures are described and their model line luminosities and response function centroids are presented. The variation in the C IV/Lyα ratio as a function of continuum luminosity, is also modelled. In Section 4 the C IV line profile (the strongest and the best observed emission line in the spectrum of NGC 5548) and its variability behaviour are investigated. The summary and conclusions are given in Section 5.

TABLE I

Observed and Modelled Line Luminosities and Centroids of NGC5548

	Observed		Model F $n = 10^{11}$ cm^{-3}		Model F $n = 10^{12}$ cm^{-3}		Model G $n = 10^{11}$ cm^{-3}		Model G $n = 10^{12}$ cm^{-3}	
Line Id	R	τ_{cent}^a	R	τ_{cent}	R	τ_{cent}	R	τ_{cent}	R	τ_{cent}
O VI λ1035			1.18	7	1.30	3	1.57	8	0.48	7
Lyα λ1215	1	11.5	1.0	24	1.0	17	1.0	10	1.0	11
O V λ1218			0.16	11	0.23	3	0.25	8	0.13	7
N V λ1240	0.11[b]	6.5	0.21	12	0.48	5	0.28	8	0.31	7
Si IV λ1397			0.07	27	0.16	13	0.05	8	0.16	10
O IV λ1402			0.08	16	0.07	5	0.10	9	0.08	8
C IV λ1549	0.93[b]	12.0	1.06	24	0.92	10	1.18	10	1.32	10
He II λ1640	0.12[b]	5.0	01.7	16	0.29	11	0.27	9	0.24	9
Si III] λ1893			0.02	39	0.02	20	0.01	5	0.02	11
C III] λ1909	0.18[b]	28.5	0.03	20	0.01	9	0.03	7	0.02	9
Mg II	0.18[d]		0.05	49	0.09	39	0.01	9	0.04	10
Hγ	0.05[e]	16.5	0.01	29	0.02	30	0.01	10	0.01	10
He II λ4686	0.04[e]		0.02	15	0.04	12	0.02	9	0.03	9
Hβ	0.11[c]	21.5	0.03	30	0.05	33	0.02	10	0.03	10
He I λ5876	0.03[e]	14.0	0.02	26	0.03	23	0.01	10	0.02	9
Hα λ6563	0.41[e]	20.5	0.08	30	0.10	37	0.05	11	0.05	11
log(Lyα)	42.63		42.42		42.33		42.75		42.75	

τ_{cent}	line centroid (lt-days).
a	Peterson (1995)
b	Korista et al. (1995)
c	Peterson et al. (1991)
d	Clavel et al. (1991)
e	Dietrich et al. (1993)
R	Ratio of the line luminosity relative to the luminosity of the Lyα line.
log(Lyα)	Luminosity of the Lyα line (erg s^{-1}).

2. The Continuum Energy Distribution of NGC5548

A mean 'AGN continuum' has been described by Mathews and Ferland (1987) and has been extensively used to model the photoionization of gas in AGNs. However, observations in the soft X-ray and gamma-ray regions made since 1987 can be used to redefine the 'AGN continuum' in a form which would be more appropriate to model the BLR in NGC5548. Simultaneous *IUE* and *ROSAT/PSPC* observations of NGC 5548 have been presented by

Walter *et al.* (1994). These authors describe the observed CED from about 4 eV (~ 3000Å) to ~ 10 keV by a parametric relation of the form $F_\varepsilon \sim \varepsilon^{(1-\Gamma_{UV})} \times exp(\varepsilon/\varepsilon_{cut}) + \varepsilon^{1-\Gamma_X}$ where Γ_{UV} is the ultraviolet spectral slope, Γ_X is the X-ray spectral slope and ε_{cut} is the cut-off energy in the EUV/soft X-ray. Unfortunately simultaneous hard X-ray (*Ginga*) observations were not available. However, in an extensive study of the hard X-ray (2 keV to 10 keV) emission from Seyfert 1 galaxies, Turner & Pounds (1989) have shown that $\Gamma_X = 1.9$ in most Seyfert 1 galaxies including NGC5548 and this slope of hard X-rays has been used here. The hard X-ray spectrum was normalized to the 2 keV flux detected by *ROSAT*. This hard X-ray spectrum with an ultraviolet slope $\Gamma_{UV} = 2.2$ and a cut-off energy $\varepsilon_{cut} = 415$ eV was assumed to define the continuum energy distribution of NGC 5548 from ~ 3000Å to 10 keV.

Recent observations with the *Compton Observatory* have considerably increased our knowledge of the gamma-ray emission from AGNs and no Seyfert 1 galaxy has been found to have emission above 500 keV. The strongest emission has been observed from the Seyfert galaxy NGC 4151 and this spectrum falls off exponentially with an e-folding energy of 39 keV between 65 keV and 500 keV (Jourdain *et al.* 1992). The shape of the gamma-ray spectrum of NGC 5548 was assumed to be similar to that of the observed spectrum of NGC 4151 and this was normalised to the hard X-ray power law spectrum extrapolated to 65 keV. The *IUE, ROSAT*, hard X-ray and gamma-ray spectrum of NGC 5548 is shown in Fig. 1. The crosses in Fig. 1 are the observed fluxes of NGC 4151. The lower flux of NGC5548 is consistent with the upper limits obtained with the *Compton Observatory*. The location of the source of gamma-rays in an AGN is unknown and it is not at all clear if the BLR in an AGN 'sees' the observed gamma-rays; the gamma-ray flux shown in Fig. 1 should be considered an upper limit of the flux which the BLR clouds would see if the gamma-rays in NGC5548 originate close to the central engine. Compared to the 'AGN continuum', the gamma-ray spectrum of NGC 5548 falls off very fast and the Compton heating due to gamma-rays will be considerably lower than that assumed in the past.

In the optical/near infra-red region (~ 3000Å to 1 μm) a power-law spectrum ($f_\nu \propto \nu^\alpha$) with a power-law index $\alpha = -1.0$ was assumed and this is compatible with observations (Neugebauer *et al.* 1979). The continuum spectrum in the sub-millimetre region of a radio quiet AGN is poorly determined and the frequency at which the infrared spectrum turns over is not well known. A power-law radio spectrum with $\alpha = +2.5$ was assumed here and this is appropriate for self-absorbed synchrotron emission. The energy of the sub-millimeter break in the continuum has a major effect on the line ratios as the free-free heating will alter with this break. Ferland *et al.* (1992) have investigated the change in C IV/Lyα and O VI/Lyα ratios when the

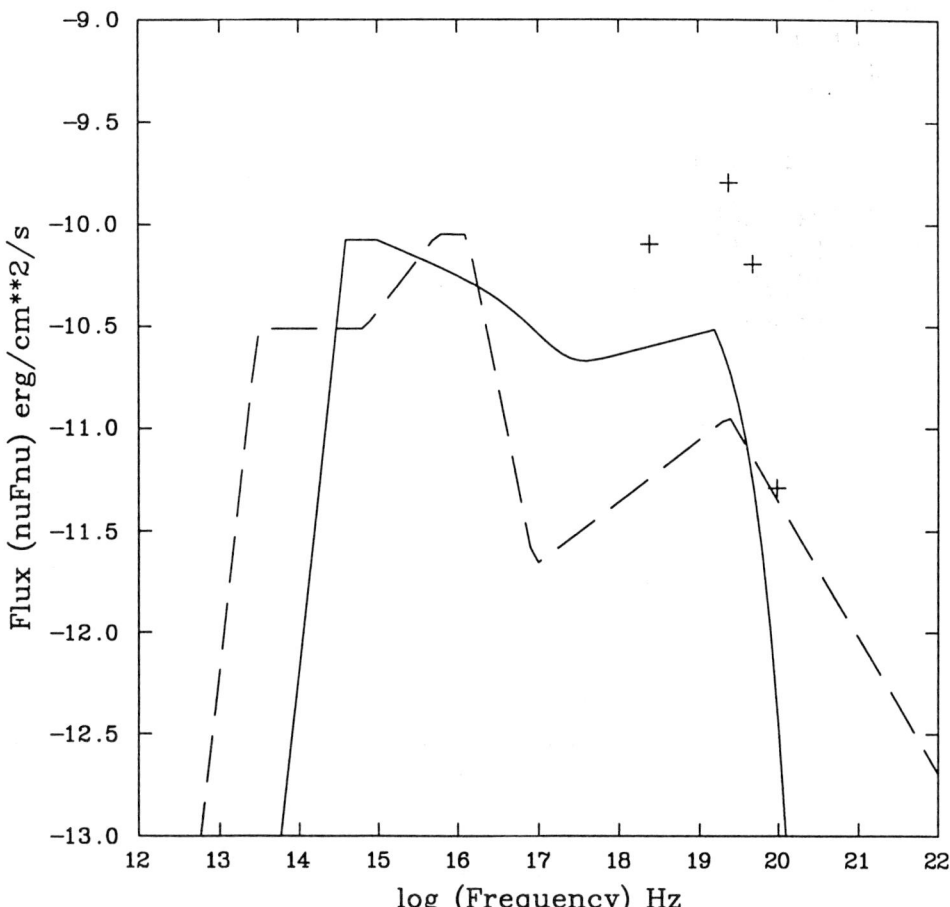

Fig. 1.   Continuum Energy Distribution of NGC5548 (*full line*). The dashed line is the 'AGN Continuum' and the crosses are the gamma-ray observations of NGC4151

radiation field at ionizing energies and the density are held fixed but the frequency of the sub-millimetre break is varied. For the models presented here a sub-millimetre break at 1μm is assumed.

The lack of internal soft X-ray absorption (Walter *et al.* 1994) suggests that the BLR clouds 'see' essentially the observed CED of NGC5548. The flux of ionizing photons, which one obtains by integrating the continuum from 13.6 eV to 2 keV, is 2.62 ph cm^{-2} s^{-1} . The mean intensity of Lyα (Clavel *et al.* 1991) is 0.467 ph cm^{-2} s^{-1}. This suggests a covering factor of the BLR clouds of 0.18, if all the ionizing photons are converted to Lyα photons. The error in this covering factor is at least 15% due to the error in the intensity of the Lyα line. This is a lower limit as the error in the flux of ionising photons has not been included, although this is not expected to be greater than 20%. Moreover, the interpolated EUV spectrum (from $\sim$11 eV to 200 eV) may not be the true EUV spectrum of NGC5548.

3. Models of the Broad Emission-Line Region

Over the last two decades a generic single-slab model of the BLR has been developed and discussed extensively (Davidson & Netzer 1979, Kwan & Krolik 1981, Ferland & Persson 1989). In this model the BLR clouds were assumed to be at a characteristic distance r from a central, single, isotropically emitting source of ionizing radiation. The geometry of the BLR was assumed to be spherical and the line emission from the clouds was also isotropic. The emission from this BLR is mainly a function of a dimensionless ionization parameter

$$U = \frac{Q(H)}{4\pi r^2 nc} \tag{1}$$

where $Q(H)$ is the luminosity of the central source (photons s^{-1}) and n is the electron density in the BLR clouds. In these models a single population of clouds was responsible for all emission properties and large column densities were necessary to produce large optical depths in the Lyman continuum. These models failed to reproduce both the HILs and LILs from the same ensemble of clouds and thus prompted the introduction of two-component models (Collin-Souffrin *et al.* 1986).

In these early models the microphysics of the BLR (atomic physics, radiative transfer *etc.*) was considered in great detail but almost no attention was paid to the macrophysics (*e.g.* geometry, spatial structure *etc.*). Future progress in modelling the BLR is clearly going to be in the development of the understanding of this macrophysics of the BLR. Models of extended and stratified BLRs have been introduced by Rees, Netzer and Ferland (1989, hereafter RNF). These authors assumed a spherical geometry and a simple pressure law to describe the radial distribution of BLR parameters and photoionization of BLR clouds by a single central source. The clouds were assumed to emit isotropically and the line and continuum emissivity from all parts of the BLR was integrated to obtain luminosities. However, these authors did not consider the response of this BLR to a change in the ionizing continuum. A parametric study of an extended BLR has been made by Pérez, Robinson and de la Fuente (1992a,b), who assumed a power-law radial

distribution of line emissivity and investigated the response of an extended
BLR of various radial structures and geometries. Goad, O'Brien and Gondhalekar (1993) (Paper I) and Goad (1995) have undertaken a detailed study
of the response of an extended photoionized BLR. These authors assumed
a pressure law and a radial distribution of BLR parameters, similar to that
assumed by RNF, and proceeded to determine the luminosities and response
functions of prominent emission lines and continua observed in the spectra
of AGNs and quasars. O'Brien, Goad and Gondhalekar (1994) (Paper II)
and Goad (1995) advanced this study to include anisotropic emission from
clouds and O'Brien, Goad and Gondhalekar (1995) (Paper III) and Goad
(1995) extended this study further to include the non-linear response of a
stratified BLR.

The passage of an ionization front through a radially extended BLR can
be described by a time-lag $\tau = (r/c)(1 - \cos\theta)$ where (r,θ) define the position of a cloud in the BLR, r is the radial distance from the source and
θ is the angle between the cloud radius vector and the line of sight to a
distant observer, measured from the side nearest to an observer. The clouds
which respond to a continuum event after a time-delay τ lie on a parabolic
surface of constant delay τ. The emission-line light curve $L(v,t)$, at a line of
sight velocity v produced by an ionizing continuum light curve $C(t)$ passing
through the BLR, can be described by the expression

$$L(v,t) = \int_0^\infty \Psi(v,\tau)C(t-\tau)d\tau\,, \qquad (2)$$

where $\Psi(v,\tau)$ is the BLR 'response function', the response of an emission
line as a function of delay τ, to a δ-function continuum event. The response
function depends not only on the geometry and kinematics of the BLR but
also on the observed line emissivity which is itself a function of the cloud
location within the BLR, and the physical state of the gas at this location.
In practice it is more common to measure the response of the integrated line
flux to a continuum event whereby equation (2) reduces to

$$L(t) = \int_0^\infty \Psi(\tau)C(t-\tau)d\tau\,. \qquad (3)$$

Equations (2) and (3) are based on the assumption that there is a 'linear'
relationship between the continuum and emission line variations (*cf.* paper
III). The aim of reverberation studies of AGNs is to invert equation (2)
(or equation (3)) to determine the response function from the observed
continuum and line light curves (Blandford & McKee 1982, Horne *et al.*
1993). Stable solutions of this inversion are only possible for evenly sampled,
high signal-to-noise ratio data obtained over long periods. With the amount
and quality of data normally available it is more common to cross-correlate
the time series $C(t)$ and $L(t)$ and determine the peak and the centroid of

the cross-correlation function (CCF) (Gaskell & Sparke 1986, Gaskell & Peterson 1987, White & Peterson 1994). The peak delay is a measure of the time-of-flight from the continuum to the BLR and is, therefore, a measure of the scale length ($r = c\tau$) of the BLR (but see Perez, Robinson & de la Fuente 1992a,b). The centroid (τ_{cent}) of the CCF is equal to the centroid of the response function (Penston 1991, White & Peterson 1994). These statements depend very strongly on the assumption that the time-of-flight is the most significant time-scale in the response of the BLR to a continuum event and this has been established only for a few low luminosity AGNs (Peterson 1993,1995).

The centroid of the recovered response functions (*e.g.* Horne *et al.* 1993) is equal to the responsivity and anisotropy weighted radius of the BLR, and is defined as

$$R = \frac{\int_{R_{in}}^{R_{out}} \eta(r) r L_{obs}(r, \theta) dr}{\int_{R_{in}}^{R_{out}} \eta(r) L_{obs}(r, \theta) dr} \, , \tag{4}$$

where $L_{obs}(r, \theta)$ is the observed luminosity and $\eta(r)$ is the responsivity of the BLR. The observed luminosity

$$L_{obs}(r, \theta) = 4\pi \int_{R_{in}}^{R_{out}} \varepsilon_{obs}(r, \theta) A_c(r) n_c(r) r^2 dr \, , \tag{5}$$

where the observed emissivity (Paper II)

$$\varepsilon_{obs}(r, \theta) = \frac{\varepsilon_{totl}(r)}{2} \{1 - (2F(r) - 1) \cos \theta\} \, , \tag{6}$$

where $\varepsilon_{totl}(r) = \varepsilon_{in}(r) + \varepsilon_{out}(r)$, $\varepsilon_{in}(r)$ is the emissivity of the inward cloud-face (facing the source) and $\varepsilon_{out}(r)$ is the emissivity of the outward cloud-face (away from the source). The anisotropy factor

$$F(r) = \frac{\varepsilon_{in}(r)}{\varepsilon_{totl}(r)} \, . \tag{7}$$

$F(r) = 0.5$ for an isotropically emitting cloud, whereas, $F(r) = 1.0$ for a fully anisotropically emitting cloud which emits only towards the source. This is a simple formalism for anisotropic line emission from BLR clouds, the actual form will depend critically on the shape of the clouds.

The responsivity of BLR clouds is defined as the fractional change in the cloud emissivity normalized to a unit fractional change in the ionizing continuum, *for a small change in the ionizing continuum, i.e.*

$$\eta(r) = \frac{\delta\varepsilon(r)/\varepsilon(r)}{\delta U(r)/U(r)} \, , \tag{8}$$

where the change in the ionizing continuum has been represented by a change in the ionization parameter as the ionization parameter is linearly proportional to the ionizing continuum, assuming a fixed spectral shape. The implicit assumption here is that for a small change in the ionizing continuum the BLR clouds will respond linearly. This, of course, is not true for large continuum variations (see section 3.3). Non-linear response of a BLR has been discussed in detail in Paper III.

In this paper the luminosity and the centroid of the response functions of a few prominent lines have been determined for two models of the BLR. The two models differ in their radial distribution of the BLR parameters and are modelled with the code PROSYN (Goad 1995). A spherical geometry is assumed for both models. The cloud elemental abundances were solar (Grevesse & Anders 1989), as is consistent with the available data on Seyfert 1 galaxies (Hamann & Ferland 1993). The emissivities of the photoionized clouds were obtained from the code CLOUDY (version 84.12a, Ferland 1993). The CED of NGC5548 described in section 2 is used for these photoionization calculations.

3.1. A PRESSURE-LAW MODEL

A pressure-law model of an extended BLR has an inner and an outer radius R_{in} and R_{out} respectively and is populated with spherical constant density clouds in pressure balance with the intercloud medium. A strict pressure balance between the BLR clouds and the intercloud medium appears unlikely (Mathews & Ferland 1987), although magnetic pressure may provide the additional support necessary to prevent cloud disruption (Rees 1987). Furthermore, this formalism has the added advantage of reducing the number of free parameters necessary to describe the model. The pressure is assumed to have a power-law radial dependence $P \propto r^{-s}$ and as the gas kinetic temperature is a weak function of U, the gas density $n(r) \propto r^{-s}$ and $U(r) \propto r^{s-2}$. Assuming that the mass of the clouds is conserved, the cloud cross-sectional area $A_c(r) \propto r^{2s/3}$ and the column density $N(r) \propto r^{-2s/3}$. Also the clouds are assumed to move with their virial velocity; consequently the differential covering factor $dC(r) \propto r^{(2s/3)-(3/2)}dr$. A pressure-law model is thus defined by the pressure-law index s, R_{in}, R_{out}, the density and the column density at the inner radius, the total covering factor and the luminosity and spectrum of the ionizing and heating continuum.

A model with pressure-law index $s = 0$ is considered here and this will be referred to as model F, where F refers to the flatness of the distribution. However, we note that the model F discussed in papers I–III has $n = 10^{10}$ cm^{-3}. The inner radius of this BLR was set at 0.3 lt-days; this radius is large enough for the continuum source to be approximated by a point source. Also at this radius the relativistic effects of the central black-hole (if that is the central engine) will be negligible. The outer radius was set at 53 lt-days which con-

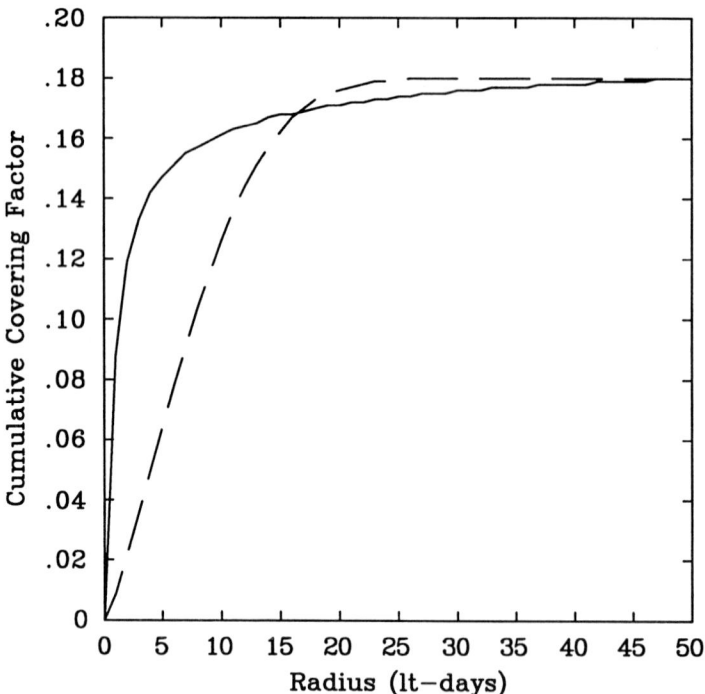

Fig. 2.   Cumulative Covering Factor, for Model F (*full line*) and Model G (*dashed line*).

forms with the criterion defined by RNF, namely [O III] $\lambda4363/H\beta< 0.1$. A column density of 10^{23} cm^{-2} at R_{in} was assumed for this model, and for $s = 0$ the column density, of course, stays constant with radius.

The cumulative covering factor as a function of distance from the source is shown in Fig.2. The covering factor was normalized such that the integrated covering factor at R_{out} was equal to the measured covering factor of 0.18. The radial variation of the ionization parameter (for n=10^{11} cm^{-3} at R_{in}) is shown in Fig.3. At the inner radii the ionization parameter is larger than the values which have been considered for most models. For a column density of 10^{23} cm^{-2} the BLR clouds will be optically thin for an ionization parameter greater than 0.5, and in model F a large fraction of the inner BLR will be radiation bound.

The change with density, of the Lyα line luminosity in model F is shown in Fig.4. The luminosity of this line increases with density and peaks around $n = 10^{11}$ cm^{-3} although this peak luminosity is about 40% lower than the mean luminosity of NGC5548 measured during the 1989 monitoring campaign (Clavel *et al.* 1991). Similarly, the changes with density in the luminosities (relative to Lyα) of six lines are shown in Fig.5. In this model

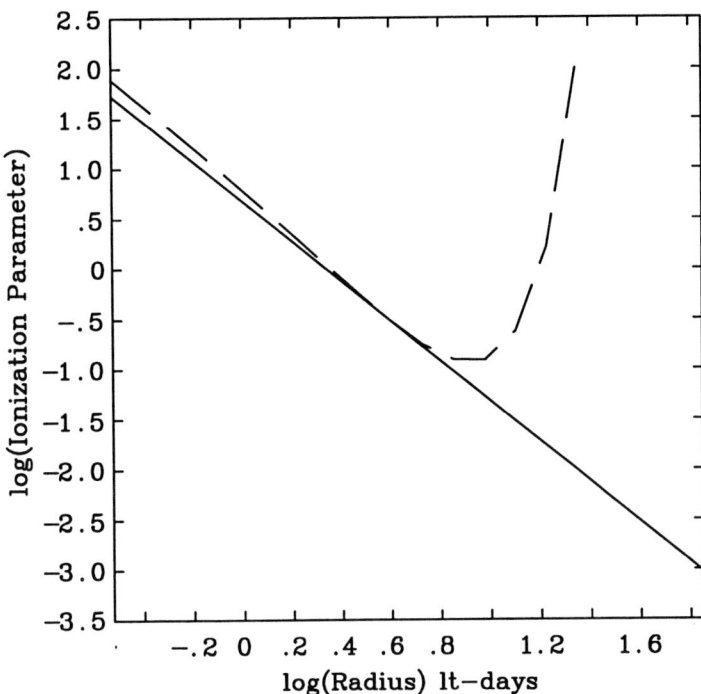

Fig. 3.   Distribution of the ionization parameter as a function of the distance from the central source. Model F (*full line*), and Model G (*dashed line*).

the luminosity of the O VI $\lambda1035$ line is exceptionally high as this line is emitted principally by the highly ionized gas at low radii. From equation (1), for a fixed $Q(H)$, $U \propto n^{-1}$. In model F as the density of the gas increases (*i.e.* U decreases), the emissivity at low radii increases (see Fig.6) and the total luminosity rises.

This can be seen more clearly in the luminosity of the N V $\lambda1240$ line, which rises very rapidly with density because the emissivity of the line at small radii increases with density. The emissivity of the C IV $\lambda1549$ line at small radii also increases with density, but the emissivity at larger radii drops and because the bulk of the gas is at the outer radii (Fig.2) the luminosity of the line falls slightly with increasing density. The luminosity of the He II $\lambda1640$ line also increases with density, but this is not entirely due to the increase in the emissivity at low radii. At higher densities this line becomes more isotropic (Fig.6) and, therefore, the observed luminosity increases. The luminosity of the C III] line drops with density as this line is collisionally de-excited at densities greater than $\sim 10^{10}$ cm^{-3}. The luminosity of the Hβ line is not very sensitive to density.

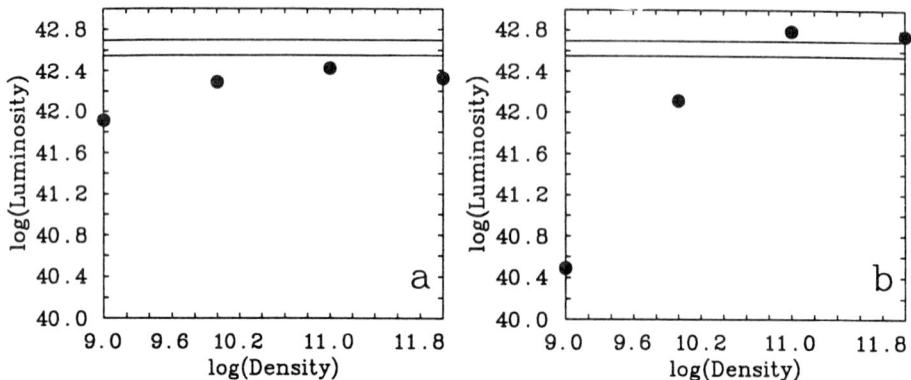

Fig. 4.   The luminosity of Lyα line as a function of density, (a) Model F and (b) Model G. The two parallel lines indicate the upper and lower limit (1 σ) of observations made during the 1989 monitoring campaign of NGC5548.

The change with density, in the centroid of the lines is shown in Fig.7. The centroid of all lines decreases with increasing density. This is due to the increase, at small radii, in the emissivity of a line as the density is increased (Fig.6) and also due to the increase in the responsivity of the lines, at the inner radii, as the density increases. This increase in responsivity will tend to push the centroid of a line to lower values.

The model F luminosity of the Lyα line and the relative (with respect to Lyα) luminosities of fifteen other lines are given in Table 1. The centroids of the response functions of these lines are also given in Table 1. The line luminosities have been given relative to the luminosity of the Lyα line only, but Model F suggests that the O v λ1218 line (and possibly the HeIIλ 1216Å line, whose luminosity has not been calculated here, but see Shields *et al.* (1995)) could be strong and the observed line ratios could be measured relative to the luminosity of Lyα line blended with these two lines and the observed ratios would be lower than the modelled ratios. To estimate the strength of the O v λ1218 line a blend of the modelled Lyα and O v lines was produced; the profiles of these lines were assumed to be similar to the profile of the C IV λ1549 line but the integrated flux of the Lyα line was normalized to unity and that of the O v line was normalized to the modelled line ratio. This blend suggests that for O v/Lyα > 0.25 a shoulder would appear in the red-wing of Lyα and this would be resolved in the *HST/FOS* spectra. This shoulder is not seen in the *HST/FOS* spectra of NGC5548 (Korista *et al.* 1995), suggesting that O v Lyα < 0.25. In practice it should

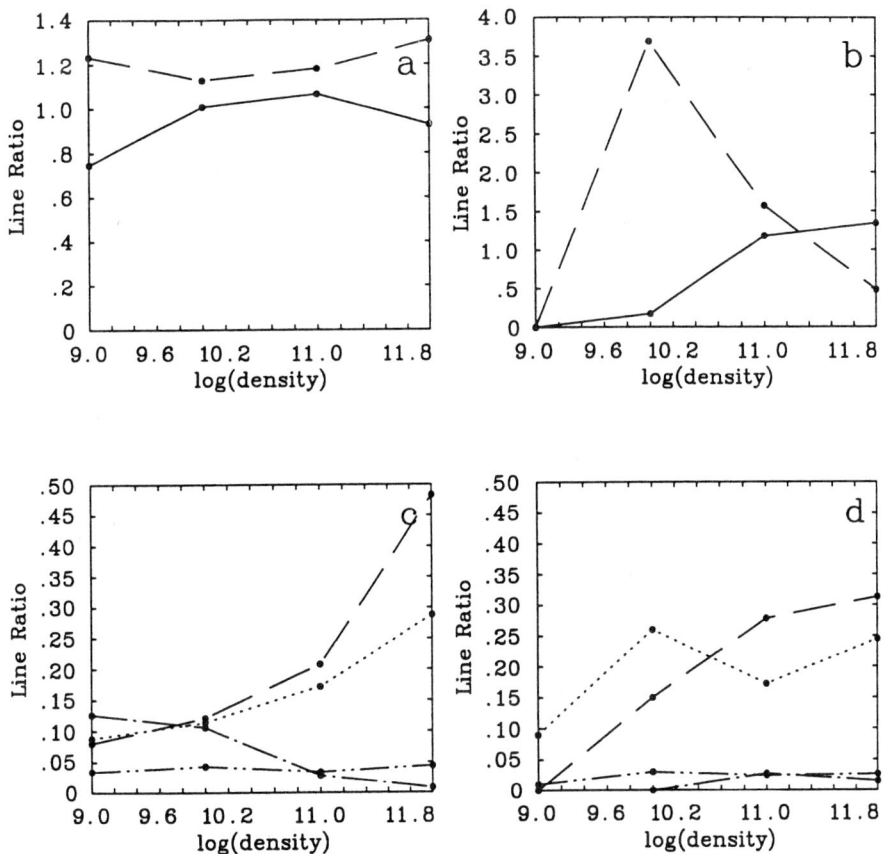

Fig. 5.   The line luminosities (relative to Lyα) as functions of density in Model F (a,c) and Model G (b,d). In (a,b), *full line*-C IV and *dashed line*-O VI. In (c,d), *dashed line*-N V, *dotted line*-He II, *dot-dash line*-C III] and *dot-dot-dash line*-Hβ.

be possible to detect an O V line even weaker than that suggested by this lower limit of the line ratio because the Lyα line is narrower than the C IV line.

The model F line luminosities and centroids have been obtained for densities of 10^{11} cm^{-3} and 10^{12} cm^{-3} respectively. The line emissivity, respon-

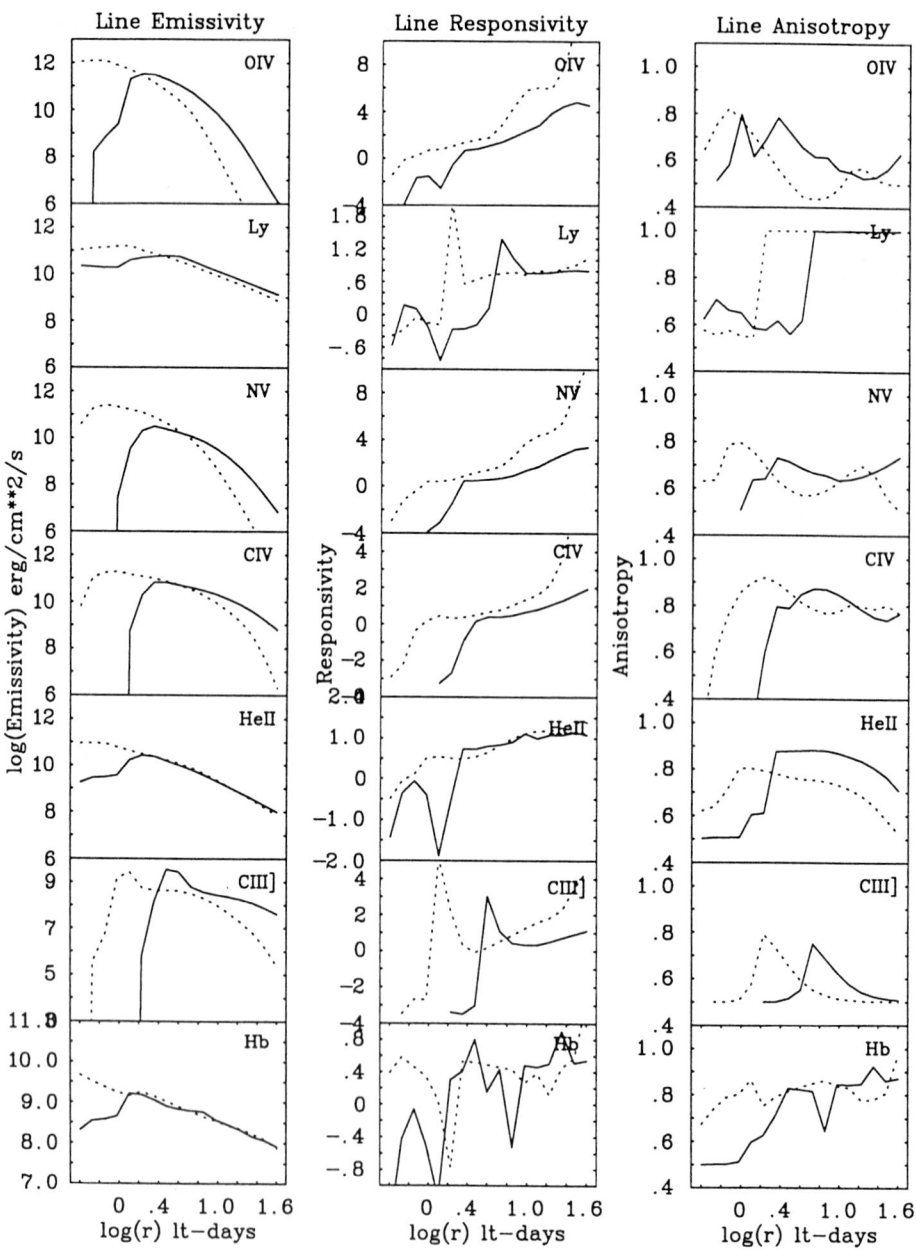

Fig. 6.   The radial distribution of emissivity, responsivity and the anisotropy factor for Model F, at density $n = 10^{11}$ cm^{-3} (*full line*) and $n = 10^{12}$ cm^{-3} (*dotted line*).

sivity and anisotropy factor for each of seven prominent lines are shown, for these two densities, in Fig. 6. As noted above, the luminosity of the O VI line is high in this model. The luminosity of the O VI line in NGC 5548 is unknown but in AGNs in general the O VI/Lyα+N V ratio is between 0.06 and 0.5 (Zheng *et al.* 1995). From the data in Zheng *et al.* the observed $\alpha_{ox} \sim 1$ for NGC5548 which implies O VI/Ly$\alpha \sim 0.4$. The strength of the N V line is also overestimated (by at least a factor of two) but the strengths of all other HILs are compatible with the observations. The strength of all LILs is underestimated by a large fraction; this problem is not new and is certainly not solved in a stratified BLR.

The model F centroids of all lines are higher than the observed centroids by at least a factor of two. This can be understood with reference to Fig. 6. The emissivities of all lines are low at the inner radii and the line responsivities are also low or negative, which will conspire to push the line centroids to higher values. When the density is increased to 10^{12} cm^{-3} the emissivity at inner radii increases and the responsivity at these radii also increases becoming positive for all lines. This will push the line centroids to lower values as can be seen from Table 1. However, at this higher density the luminosity of the N V line (and the O VI line) increases further although the line centroids are now in agreement with observations. The luminosities of other HILs are still compatible with observations but the luminosities of the LILs are still underestimated. In conclusion the model F line luminosities and centroids of HILs are in reasonable agreement with observations for a density of 10^{12} cm^{-3} but the model underestimates the luminosities of LILs and overestimates their centroids.

3.2. A GAUSSIAN MODEL

In this model the density is expressed as

$$n \propto exp(-((r - r_p)/\sigma)^2 \tag{9}$$

Where r_p and σ (=FWHM/2.354) are respectively the radius of the peak and the standard deviation of the density distribution. Note that this is an *ad hoc* assumption and no physical mechanism is offered for this density distribution. In this sub-section the line emission from a spherical distribution of such gas and the response of this gas to a change in the ionizing radiation are investigated.

Assuming constant density spherical clouds of radius R_c and mass conservation we find

$$nR_c^3 = constant \tag{10}$$

The cloud cross-sectional area obeys

$$A_c \propto R_c^2 \propto exp(2/3(((r - r_p)/\sigma)^2)) \tag{11}$$

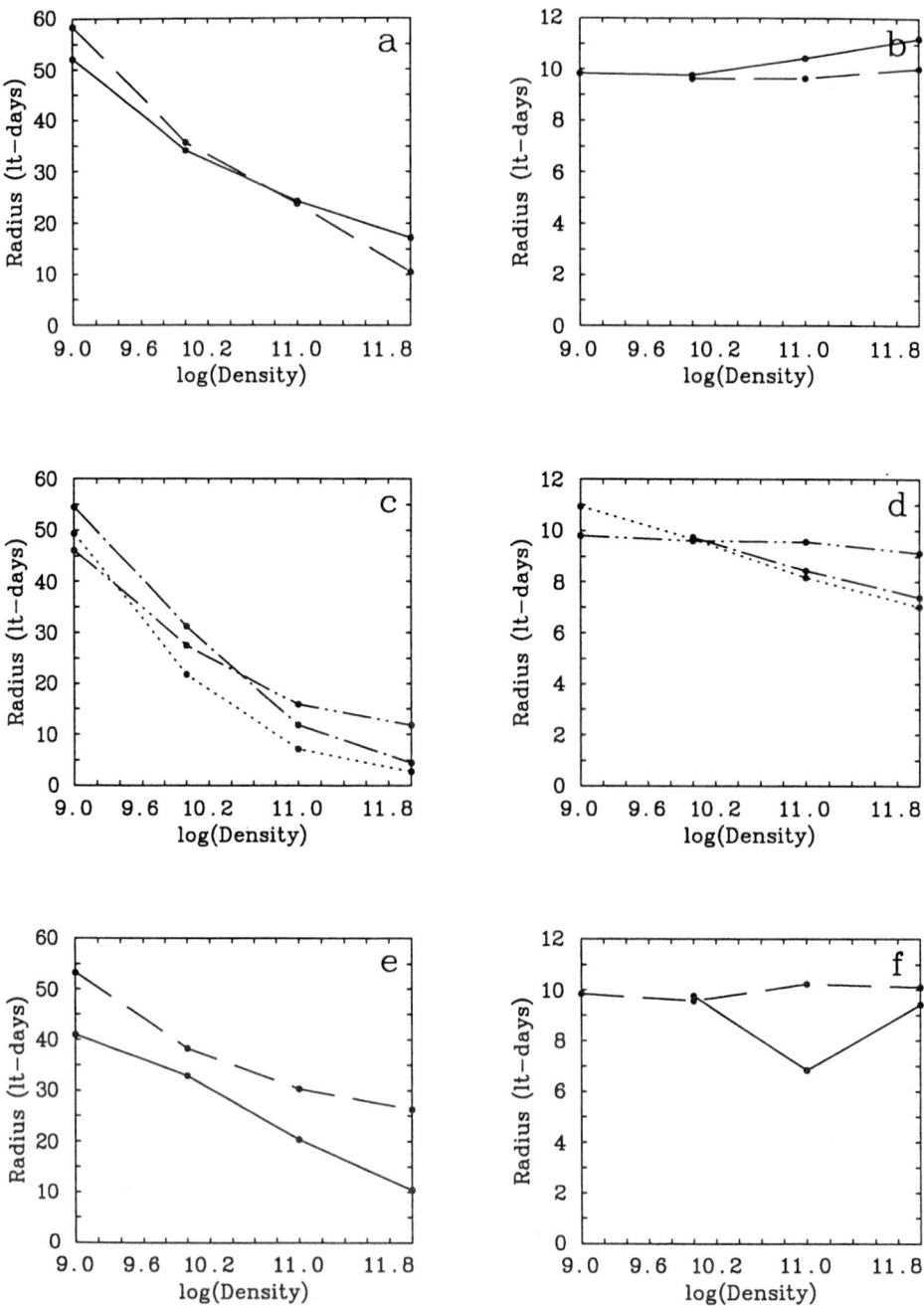

Fig. 7.   Line centroids as a function of density, Model F (a,c,e) and Model G (b,d,f). In (a,b), (*full line*)-Lyα and (*dashed line*)-C IV. In (c,d), (*dotted line*)-O VI, (*dot-dash line*)-N V and (*dot-dot-dash line*)-He II. In (e,f), (*full line*)-C III] and (*dashed line*)-Hβ.

The cloud column density is governed by

$$N = nR_c \propto exp(-2/3(((r - r_p)/\sigma)^2)) \tag{12}$$

The differential covering factor is given by

$$dC(r) \propto A_c(r)n_c(r) \tag{13}$$

Where the cloud number density $n_c(r)$ is also assumed to have the same Gaussian distribution as the density

$$n_c(r) \propto exp(-((r - r_p)/\sigma)^2) \tag{14}$$

and

$$dC(r) \propto exp(2/3(((r - r_p)/\sigma)^2)) \times exp(-(((r - r_p)/\sigma)^2)) \tag{15}$$

The ionization parameter

$$U(r) \propto r^{-2} \times exp(((r - r_p)/\sigma)^2) \tag{16}$$

This model will be referred to as Model G. In this model the 'density' refers to the density at the peak of the density distribution. To obtain integrated line luminosities an inner radius R_{in} of 0.3 lt-days was assumed for reasons given in Section 3.1. The outer radius had to be truncated at 22.5 lt-days as beyond this radius the density drops to a level where meaningful computations cannot be made. This model is thus characterised by the inner and outer radii, R_{in} and R_{out} respectively, the peak (r_p) and the FWHM of the density distribution and the density and the column density at the peak of the distribution. The Model G described here is computed for $r_p = 4$ lt-days, FWHM = 10 lt-days, and a column density of 10^{23} cm^{-2} at r_p.

The cumulative covering factor for Model G is shown in Fig. 2. Compared to Model F there is less gas at small radii in model G. The radial dependence of the ionization parameter, calculated for a density of 10^{11} cm^{-3} is shown in Fig. 3. At small radii the ionization parameter is similar to the ionization parameter for Model F. At larger radii the ionization parameter rises steeply, due to the rapid drop in density, and as a consequence the gas at large radii will be heavily ionized.

The model G luminosity of the Lyα line as a function of density is shown in Fig. 4. The line luminosity rises very steeply with density and peaks around 10^{11} cm^{-3}. The peak luminosity is about 32% higher than the mean luminosity of NGC5548 observed during the 1989 monitoring campaign (Clavel et al. 1991). The relative luminosities of six prominent emission lines, as functions of density, are shown in Fig. 5. The luminosities of the O VI, N V, C IV and He II lines increase with density but that of O VI drops beyond a density of 10^{10} cm^{-3} and the luminosities of all other lines begin to flatten around 10^{11} cm^{-3}. These increases in line luminosities with density are due to an increase, with density, in the line emissivities at small radii (see Fig. 8.).

The line centroids, as functions of density, are shown in Fig. 7. The centroids of both the Lyα and C IV lines increase slightly with density because both lines become more anisotropic as the density increases and this pushes the centroid to higher values. The centroids of the O VI and N V lines, on the other hand, drop with increase in density. This happens because the responsivities of these lines at the inner radii, increase as the density rises and this tends to push the centroids to lower values. The centroids of He II and Hβ are unaffected by an increase in density because the increase with density in the responsivities of these lines at small radii, which would ordinarily decrease the values of the line centroids, are mitigated by the increasing anisotropy of the line emission at small radii. Similarly, the value of the centroid of the C III] line increases at 10^{12} cm^{-3} because the line emission becomes more anisotropic at lower radii as the density increases.

The model G luminosity and centroid of the Lyα line and the relative luminosities of fifteen lines and the centroids of these lines are given in Table 1. These line parameters have also been obtained for densities of 10^{11} cm^{-3} and 10^{12} cm^{-3} respectively. This model also predicts a strong O V $\lambda1218$ line and if this is blended with the Lyα line then the observed line ratios will be smaller than the modelled ratios. However, the *HST/FOS* spectra of NGC5548 suggest that model G also overestimates the strength of the O V line. Similarly, this model overestimates the strength of the O VI and N V lines, and the greater strength of these lines is mainly due to the emission from the highly ionized gas at the inner radii where the ionization parameter is high. The relative strengths of other HILs are compatible with observations, but even in model G the strengths of LILs are underestimated by a large fraction. The centroids of all HILs are in reasonable agreement with observations of these lines. But the centroids of LILs are underestimated by about a factor of two.

At a density of 10^{12} cm^{-3} the luminosities of all the HILs (except O VI) increase, mainly due to the rapid increase in the line emissivities at the inner radii. The centroids of most HILs decrease slightly at this higher density because the responsivities of the lines, at inner radii, increase with density and this tends to push the centroids to lower values. The centroid of the Lyα line, however, increases slightly with density because the line emission becomes anisotropic over a larger radial distance at this higher density. This is true of the C IV line as well but in this case the increase in the responsivity of the line at smaller radii, which would tend to push the line centroid to a lower value, is balanced by the increase in the line anisotropy, over a larger radial region, which pushes the centroid to higher values.

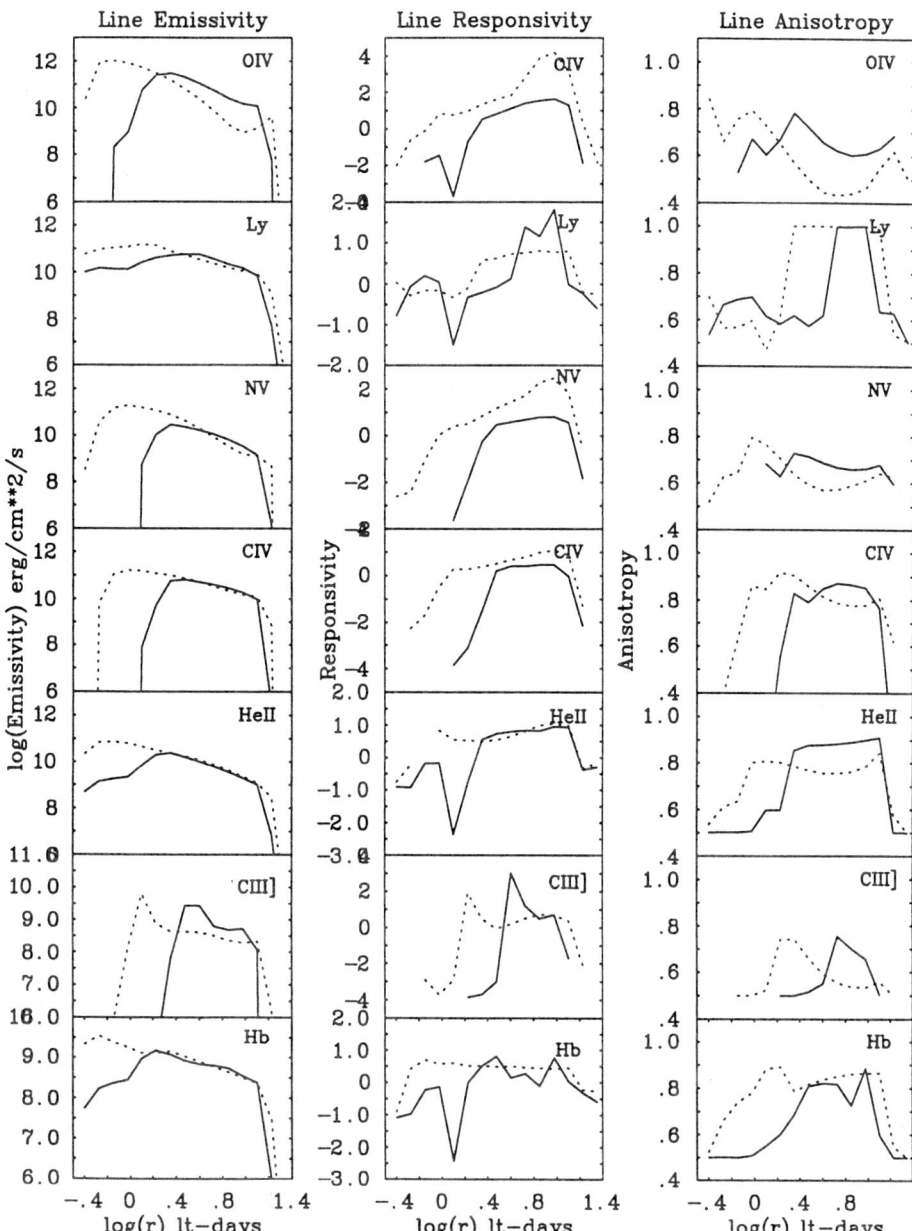

Fig. 8.   The radial dependence of line emissivity, responsivity and the anisotropy factor for Model G, at density $n = 10^{11}$ cm^{-3} (*full line*) and $n = 10^{12}$ cm^{-3} (*dotted line*).

3.3. The CIV/Lyα Ratio in NGC5548

Observationally it has been demonstrated that a negative correlation between
the CIV/Lyα ratio and the continuum luminosity is a common occurence
in variable Seyfert galaxies (Clavel & Santos-Lleó 1990, Kinney, Rivolo &
Koratkar 1990, Pogge & Peterson 1992, Gondhalekar 1992). However, the
single-slab photoionization models of the BLR predict that the CIV/Lyα
ratio should remain constant or increase as the luminosity of the ionizing
continuum increases. This lack of agreement between observations of two
energetically significant lines and model predictions brings into sharp focus
the inadequacy of the single-slab model of the BLRs in AGNs.

The CIV/Lyα ratio is sensitive to the CED of the ionizing continuum
(and the sub-mm breaks in the CEDs of AGNs). The negative correlation
between this ratio and the continuum luminosity can be explained in terms
of a changing shape of the ionizing continuum dominated by a soft excess
(Binette et al. 1989, Clavel & Santos-Lleó 1990, Gondhalekar 1992). However, Shields et al. (1995) have questioned this explanation pointing out that
there is growing evidence that a non-negligible fraction of the BLR cloud
population is optically thin to the Lyman continuum and fully ionized in
hydrogen, and thus a change in the relative proportion of optically thin and
thick gas can explain the negative correlation between the CIV/Lyα ratio
and the continuum luminosity. Shields et al. illustrate this with an ad hoc
combination of optically thin and thick gas.

The negative correlation of the CIV/Lyα ratio with continuum luminosity
is re-examined here within the context of models F and G. In Fig. 9. the
CIV/Lyα ratio in NGC5548 from the 1989 monitoring campaign (Clavel
et al. 1991) is plotted as a function of Lyα luminosity. The ratio has not
been shown as a function of continuum luminosity, as has usually been done
in the literature, because it is now known that there is a lag of about 10
days between a change in the continuum and a corresponding change in
the Lyα and CIV lines. This lag will introduce additional scatter in the
ratio/continuum correlation (Pogge & Peterson 1992). Also Lyα and CIV
lines have been shown to form in the same region of the BLR and the
correlation between the line ratio and the luminosity of the Lyα line refers
to a well defined region of the BLR.

The CIV/Lyα ratio computed for model F, as the continuum luminosity is
increased, is shown in Fig. 9. Note that the shape of the CED was not altered.
The model line ratio and the luminosity of the Lyα line were obtained for a
density of 10^{12} cm^{-3} and the luminosity had to be increased by 0.21 (in the
log) to match the observed luminosity. The lack of agreement between the
observations and the model can be understood by reference to Fig. 10. where
the the luminosities of the Lyα and CIV lines have been shown as functions
of continuum luminosity. As the continuum luminosity is increased in model

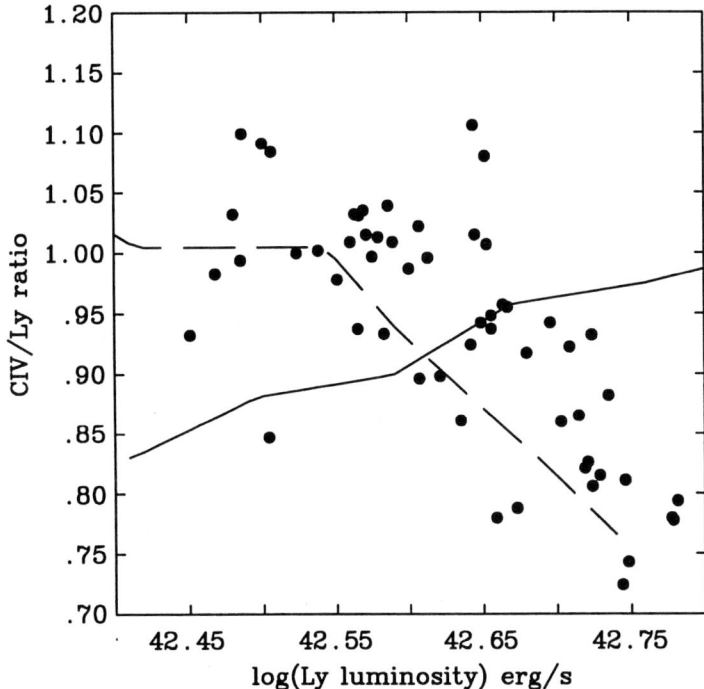

Fig. 9.   The correlation between the C IV/Lyα ratio and the luminosity of the Lyα line in NGC5548. The *dots* are the data obtained during the 1989 campaign to monitor NGC5548. The *full line* represents Model F and *the dashed line* Model G.

F, the rate of increase of the luminosity of the C IV line is higher then that of the Lyα line and this results in an increase in the C IV/Lyα ratio. The line luminosities have been computed for slightly higher continuum luminosity than is required to model the observed C IV/Lyα ratio to demonstrate that in this model, the line luminosities do not saturate, unless of course the luminosity is increased to unrealistic levels.

The C IV/Lyα ratio computed for model G is also shown in Fig. 9. The luminosity of the Lyα line had to be decreased by 0.12 (in the log) and the C IV/Lyα ratio had to be decreased by 0.18 to bring the model into agreement with observations. However, this may just reflect the uncertainty in the cloud covering factor. Thus, although the trend in the modelled C IV/Lyα ratio is in very good agreement with observations, in absolute terms the general agreement is fairly poor. The model G luminosities of the Lyα and C IV lines, as functions of continuum luminosity, have been shown in Fig. 10. Relative to the rate of increase of Lyα luminosity, the rate of increase of C IV luminosity decreases as the continuum luminosity rises and this leads

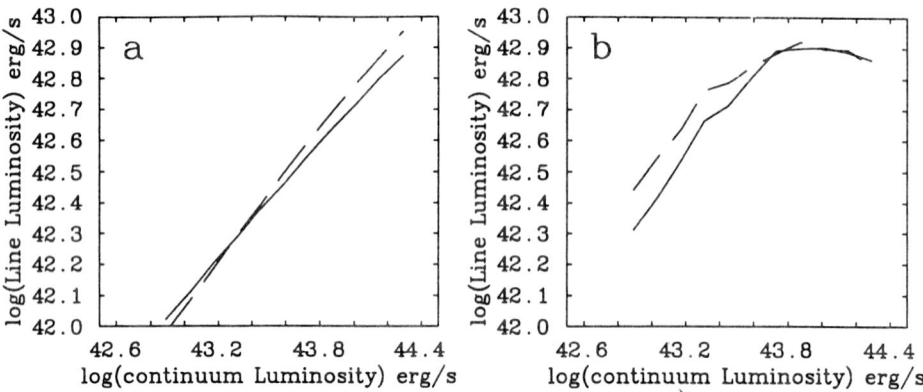

Fig. 10.   The luminosities of the Lyα (*full line*) and C IV (*dashed line*) lines as a function of continuum luminosity, (a) Model F and (b) Model G.

to a decrease in the C IV/Lyα ratio. It is interesting to note that at high continuum luminosities the line luminosities flatten, as has been observed for Fairall 9 (Wamsteker & Colina 1986). This is due to an increasing fraction of optically thin gas in the BLR as the continuum luminosity increases and eventually the BLR becomes radiation bounded, as has been claimed by Wamsteker & Colina (1986). The increase in the amount of optically thin gas with increasing continuum level provides a natural explanation for the observed behaviour of the C IV/Lyα ratio in NGC5548 and other AGNs.

The model F centroids of the Lyα and C IV lines, as a function of continuum luminosity, are shown in Fig. 11. The centroids increase by over a factor of three over the continuum luminosity considered. This increase can be understood from a consideration of the emissivities, responsivities and the anisotropy factors of these two lines shown in Fig. 12. These line parameters are given for both low and high continuum levels. As the continuum luminosity increases the emissivity of Lyα at larger radii increases which leads to an increase in the luminosity of the line. The responsivity of the line at small radii decreases which tends to push the line centroid to higher values, although this is slightly mitigated by the line emission becoming more isotropic as the continuum luminosity rises. The C IV line behaves slightly differently. Although the line emissivity at small radii decreases as the continuum luminosity increases as C^{+3} ions are ionized to a higher state, the emissivity of the clouds at larger radii increases, and as more gas is concentrated at larger radii the line luminosity increases. The line responsivity at smaller radii decreases as the continuum luminosity rises because more

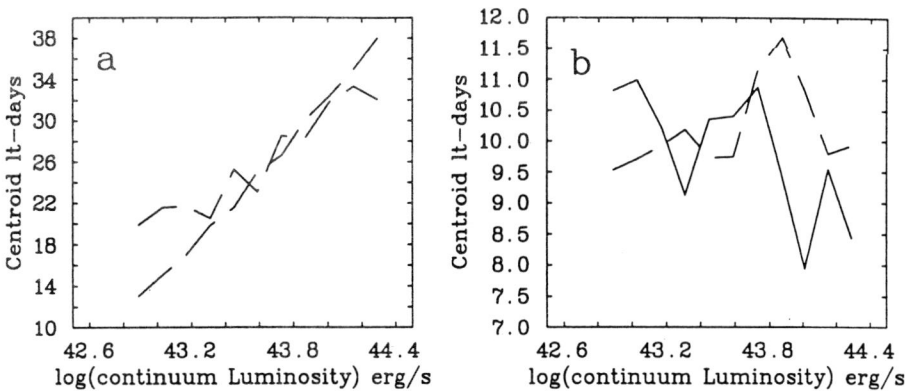

Fig. 11. The centroid of Lyα (*full line*) and C IV (*dashed line*) as functions of continuum luminosity, (a) Model F and (b) Model G.

gas at the inner radii is ionized, and this pushes the line centroid to higher values. For a BLR similar to model F the line centroids, obtained from cross correlation of continuum and line light curves, would be strong functions of the history of continuum variability.

The model G centroids of the Lyα and C IV lines as functions of continuum luminosity, are also shown in Fig. 11. The centroid of Lyα drops from ~11 lt-days to ~8 lt-days as the continuum luminosity increases; however, the centroid of the C IV line is nearly constant (give or take a peak or a drop). This can also be understood by reference to model G line emissivities, responsivities and the anisotropy factors for these two lines shown in Fig. 13. As the ionizing continuum increases the Lyα line gradually becomes optically thin and the line emission becomes isotropic at all radii and this results in a gradual decrease in the value of the line centroid. The C IV emissivity at lower radii drops, because of ionization to a higher state, and the line becomes less anisotropic over a larger radial distance, but the expected drop in the value of the centroid is balanced by the decrease in the line responsivity at the inner radii, which tends to push the centroid to higher values. If the BLR of NGC5548 is similar to model G then the cross correlation of continuum and line light curves would result in values of line centroids which would be close to the centroid of line response functions and would not be a strong function of the continuum light-curve.

In conclusion, model G can reproduce the trend in the observed correlation between the C IV/Lyα ratio and the continuum luminosity although it does slightly over-estimate the line ratio and the line luminosity.

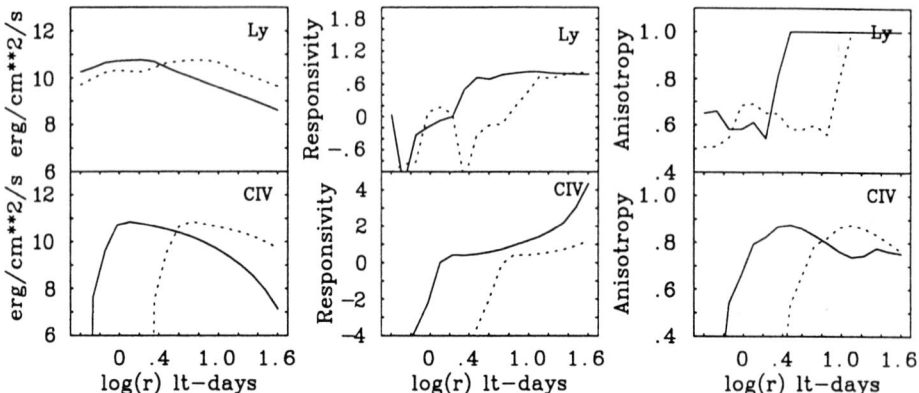

Fig. 12. The radial dependance of line emissivity, responsivity and the anisotropy factor in Model F. The (*full line*) data were obtained for low continuum luminosity and the (*dotted line*) data were obtained for high continuum luminosity.

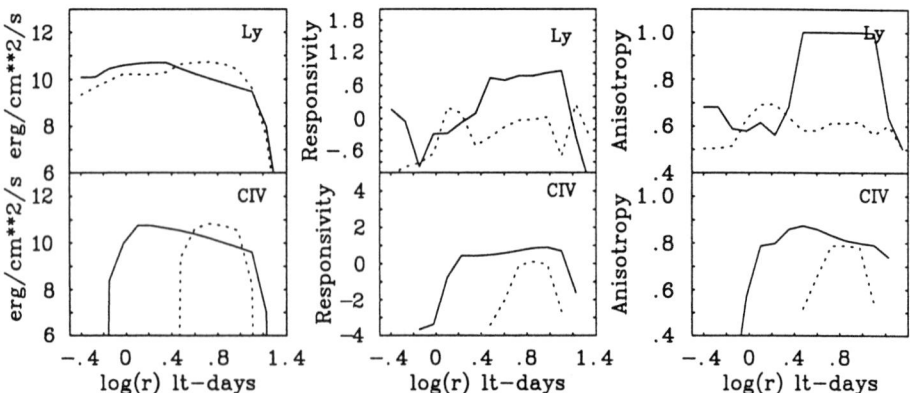

Fig. 13. The radial dependance of line emissivity, responsivity and the anisotropy factor in Model G. The (*full line*) data were obtained for low continuum luminosity and the (*dotted line*) data were obtained for high continuum luminosity.

4. The Profile of the C IV Line

The kinematics of the BLR clouds are poorly understood. Both inflow and outflow have been suggested and random Keplerian orbits are equally likely. The profiles of emission lines are determined by the Doppler motion of the BLR clouds and information on cloud kinematics is convolved in these profiles. Most of the studies of line profiles have concentrated on the Hβ line (eg. Boroson and Green 1989) as this line is accessible to observations at high spectral resolution. However, the reverberation studies suggest that the Hβ line is formed a few tens of lt-days from the source of ionizing radiation and may not be formed in the ensemble of clouds which emit the Lyα and C IV lines. The profiles of these two lines have not been studied in any detail because of their relative inaccessibility to high-resolution observations especially at low redshifts. Wills *et al.* (1993) have analysed the profile of the C IV line in a large sample of high redshift quasars. The profiles of the Lyα and C IV lines in low redshift quasars and AGNs have been investigated by Gondhalekar (1995). This study suggests that the profiles of these lines are very similar and symmetric; however, this study is based on *IUE* data and the low resolution of the *IUE* spectrographs ($\sim$1150 km s^{-1}) is a serious handicap for profile analysis.

A preliminary analysis of the changes in the profile of the C IV line observed in a small sub-sample of data obtained during the 1989 campaign to monitor NGC5548, has been presented by Crenshaw & Blackwell (1990). In this section the profile of the C IV line in NGC 5548, observed during the *IUE/HST* campaign of 1993 (Korista *et al.* 1995), is investigated. The aim here is to identify the temporal variability of the profile; a detailed study of profile variability and the correlation of this variability with line and continuum parameters will be deferred to a later publication. The variability of the profile of the C IV line in the *IUE* and *HST* spectra obtained from Julian date (2440000+)9096.67 to Julian day 9135.12 is considered here (Korista *et al.* 1995). In this analysis the NEWSIPS extraction of the *IUE* spectra has been used and the extraction and calibration of the *HST* data have been described by Korista *et al.* (1995).

In order to investigate the changes in the profile, the line profile was corrected for the underlying continuum and normalised as follows: (1) Both *IUE* and *HST* spectra were resampled on a uniform wavelength scale, wavelength steps of 1.70Å and 0.4Å were used for *IUE* and *HST* spectra respectively. (2) The continuum level was obtained by fitting a straight line between 1375Å and 1800Å. These two wavelength points were selected because in the higher resolution and higher signal-to-noise ratio *HST* spectra the signal level is lowest at these two points. We obtained the mean continuum fluxes at these wavelengths by averaging data in two 25Å wide bands. The interpolated continuum was subtracted from the spectrum between 1375Å and 1800Å.

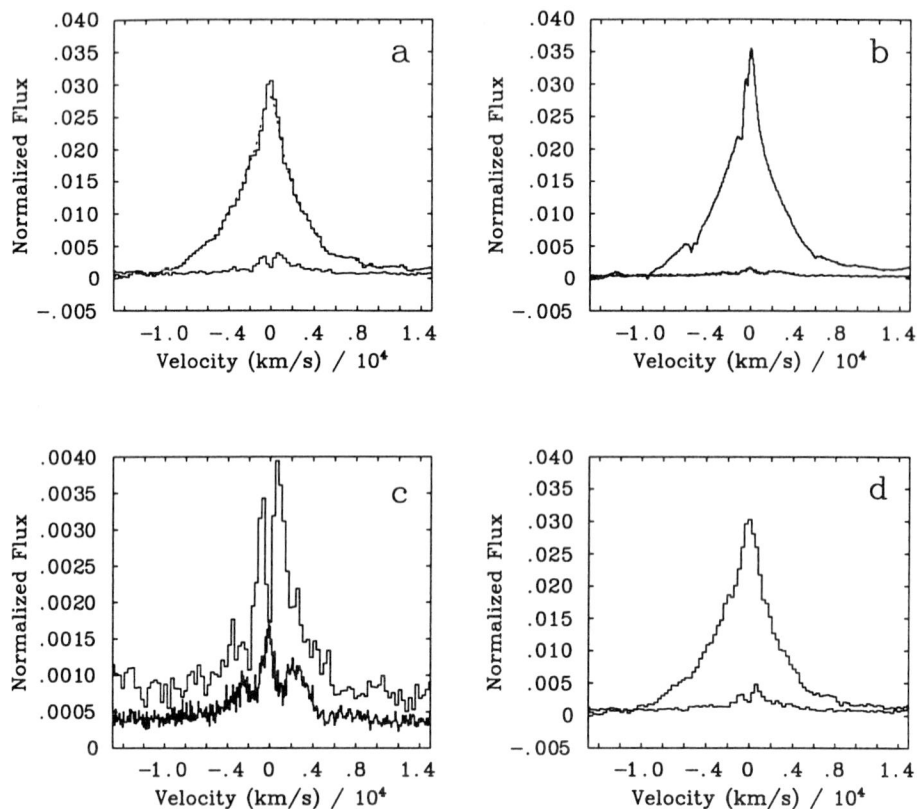

Fig. 14.   The mean profile and the standard deviation profile of the C IV line in NGC5548.
(a) *IUE* data (the degraded *HST* profile is superimposed (*dots*) and is just visible) and
(b) *HST* data. (c) An expanded view of the standard deviation profiles obtained for *IUE*
and *HST* data and (d) The mean profile and the standard deviation for the entire 1993
IUE campaign to monitor NGC5548 (see text for details)

(3) All spectra were then normalised to unit intensity of the C IV line measured between 1500Å and 1630Å. This form of normalisation preserves the
profile of a line independent of the line intensity. (4) The wavelength scale
was converted to a velocity scale with zero at the peak of the C IV line. (5)

We created a mean profile by averaging the normalised flux in each bin, between -15000 km s^{-1} to $+15000$ km s^{-1}, for all spectra. The flux in each bin was given equal weight to avoid a signal-to-noise bias. (6) The standard deviation was obtained for each bin. The 'standard deviation profile' is a measure of the change in the profile of a line. Unfortunately this measure is extremely sensitive to noise as this is added in quadrature. The fixed pattern noise in the *IUE* spectra is an additional 'noise'. Successive *IUE* spectra were obtained over the same part of the *IUE* camera and the fixed pattern noise also adds in quadrature in the standard deviation profile.

The mean *IUE* and *HST* profiles of the C IV line in NGC 5548 are shown in Fig. 14a. and Fig. 14b. respectively. The higher resolution of the *HST* spectra is obvious from the absorption features in the blue wing of the profile. If the resolution of the *HST* profile is degraded with a Gaussian of FWHM = 6Å then the *HST* and *IUE* profiles are identical as can be seen from the superposition of the degraded *HST* profile on the *IUE* profile (Fig. 14a). The standard deviation profiles of the *IUE* and *HST* profiles are shown in Fig. 14c. The higher standard deviation in the *IUE* profile is due to higher noise in these spectra.

TABLE II
Data for Analysis of the Profile of the C IV line

File ID	JD(2440000+)	F_{1375}	$F_{C IV}$
n59097	9096.97	2.38	5.29
n59098	9098.10	2.56	5.14
n59123	9122.94	4.45	6.86
n59124	9123.00	4.28	6.95
n59134	9133.74	2.84	6.83
n59135	9135.12	2.80	6.62

ID	see Korista *et al.* (1995) for details of file ID
F_{1375}	Mean Continuum at 1375Å (10^{-14} erg cm^2 s^{-1} Å^{-1})
$F_{C IV}$	Integrated (from 1500Å to 1630Å) intensity of the C IV line (10^{-12} erg cm^{-2} s^{-1})

However, the general characteristics of these two standard deviation profiles are similar; the changes in the C IV profile are at velocities lower than ~ 4000 km s^{-1} and are not symmetric, and there is little or no change at high velocities. In Fig. 14d. the mean profile and the standard deviation profile of the C IV line observed over the entire *IUE* monitoring campaign of

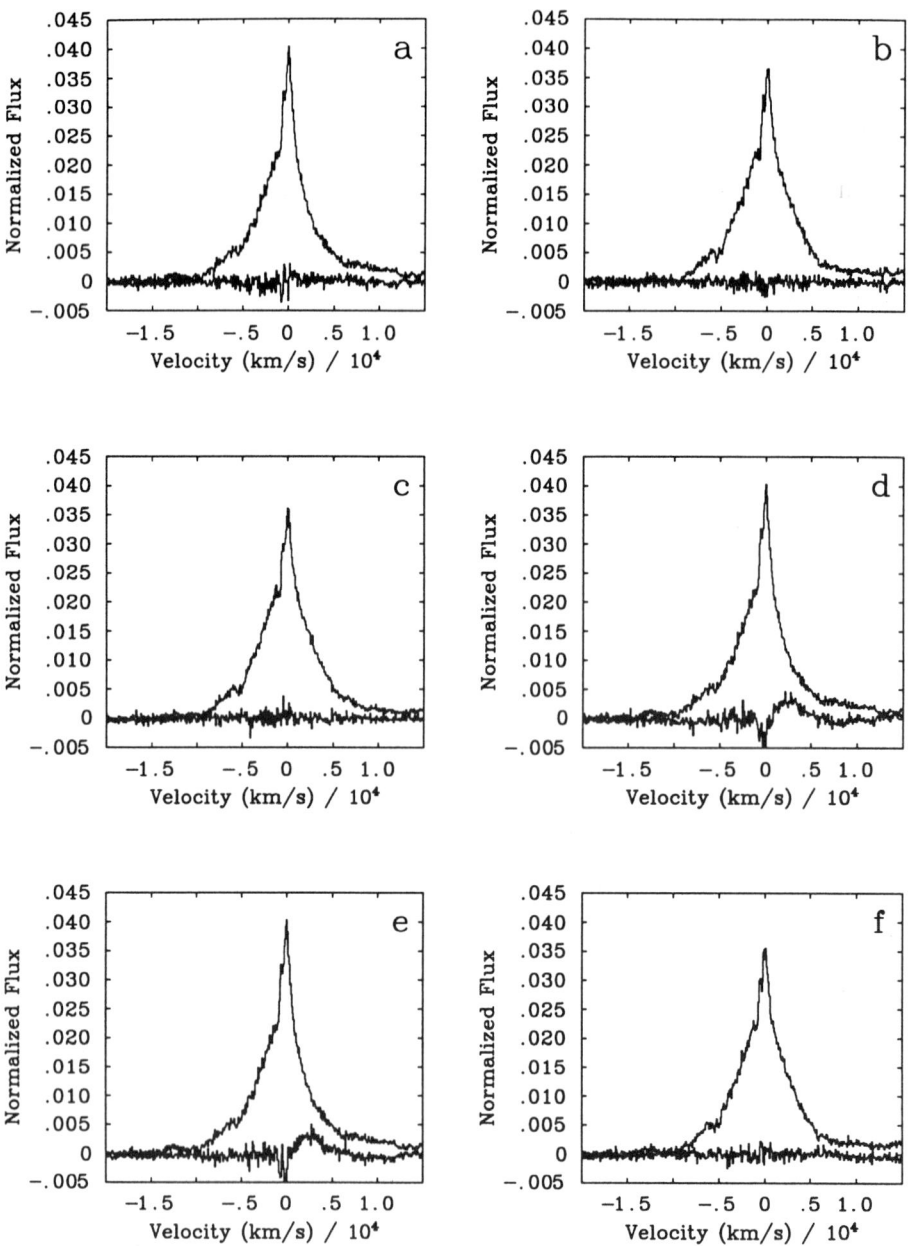

Fig. 15.   The differenced profiles of C IV line in NGC5548. (a) profile in file n59097 and
(n59098−n59097), (b) n59123 and (n59124−n59123), (c) n59134 and (n19135−n59134),
(d) n59097 and (n59124−59097), (e) n59097and (n19135−n59097) and (f) n59123and
(n19135−n59123). See Korista *et al.* (1995) for details of file IDs.

1993 (Korista *et al.* 1995) have been shown, *i.e.* data obtained from Julian day (240000+)9060.64 to 9134.50. This mean profile is almost identical to the mean profile in the latter half of this campaign (Fig. 14a) although the continuum history of this profile is very different. The C IV line in NGC 5548 has an exceptionally robust profile and the kinematics of the BLR clouds do not change significantly even when the flux of ionizing radiation changes by a large fraction, so clearly the motions of the clouds are not influenced by the ionizing radiation.

The standard deviation profile is a measure of the change in the profile but it does not indicate the direction in which the changes have taken place. In order to investigate how the profiles vary, three sets of spectra from the beginning, end and middle of the *HST* monitoring campaign were selected (two spectra in each set). The spectra used, the continuum flux at 1375Å and the intensity of the C IV line in these spectra are given in Table 2. The C IV lines in these spectra were corrected for the background and normalised as described above. The difference in the adjacent profiles is shown in Fig. 15a,b,c. These adjacent spectra were obtained about 24 hours apart and clearly the profile of the C IV line has not changed in this period, nor have the continuum and line luminosities changed over these 24 hr periods. The difference in the profile (n59124−n59097) (see Korista *et al.* 1995, for details of the file nomenclature) is shown in Fig. 15d, and changes can now be seen around 0 km s^{-1} and up to +5000 km s^{-1} in the red-wing. Also subtle small changes can be seen at higher velocities in the red-wing and at low velocities in the blue-wing. Allowing for the delay between the core response relative to the line wings (Korista *et al.*), these changes are consistent with the standard deviation profile in Fig. 14c. The red-wing feature is caused by the higher amplitude responsivity of the red wing as the continuum rises from a local minimum. During this period of about 24 days the continuum luminosity increased by about a factor of two and the line luminosity increased by about 35%. The difference in the profile (n59135−n59097) is shown in Fig. 15e. Differences in the profiles similar to those shown in Fig. 15d can also be seen here. The continuum luminosities in spectrum n59097 and n59135 are similar but the line luminosity is still high by 35%. The difference in the profile (n59135−n59124) is shown in Fig. 15f; these two profiles are very similar. The observed changes in profile shape appear to be confined to the first half of the HST campaign.

5. Summary & Conclusions

In this paper the continuum energy distribution of NGC5548, from radio to gamma-ray energies, is described; this CED is based on observations as much as possible. The crucial ionizing continuum (from ∼10 eV to ∼2 keV) is determined by interpolating between observations at wavelengths longer

than ~ 1200Å and observations at energies higher than ~ 150 eV. The ionizing continuum used in this paper is thus based firmly on observations. We
measured the covering factor of the BLR clouds by equating the integrated
ionizing continuum to the observed intensity of the Lyα line. The value of
the covering factor for NGC5548 is ~ 0.18 and this is very similar to the
covering factor of high redshift and high luminosity quasars (Smith et $al.$
1981;Gondhalekar & Kellett 1995). It would appear that the covering factor
of the BLR clouds in AGNs and quasars is independent of both redshift and
luminosity.

Two models of the BLR in NGC5548 are described; these models differ
in their physical structure but both models are based on the assumption
of spherical geometry. Both models can reproduce the luminosities of the
HILs, except for O VI and N V, which are over-estimated. The centroids of
the response functions of the HILs obtained from these models are in agreement with the centroids of the CCF. However, only model G can reproduce
the observed correlation between the C IV/Lyα ratio and the continuum
luminosity. This suggests that the column density of the region of the BLR
where the Lyα and C IV lines are formed, can not be very high as the gas
in this region must become optically thin for the excursions observed in the
continuum of an AGN. The centroids of the HILs in this model are not a
strong function of continuum luminosity.

Both model F and model G fail to reproduce the luminosities and centroids of the LILs, in both models the LIL luminosities are underestimated
by a factor of two to ten. This disagreement has variously been called the
Lyα/Hβ problem (Baldwin 1978) and the energy balance problem (Netzer
1985). Gondhalekar & Kellett (1995) have shown that the ionizing continuum of an AGN and a quasar (similar to the ionizing continuum of NGC5548
used here) has enough photons to reproduce the observed luminosity of the
Lyα line (for a covering factor of 0.2) but not that of the Hβ line if Case B
conditions are assumed. This is similar to the results of Model F and G, and
clearly suggests an additional source of energy for the BLR. If this energy
input is by absorption of hard X-rays then the region of the BLR where the
LILs are formed must have high column density clouds. The hard X-rays
could be absorbed in an accretion disc as suggested by Collin-Souffrin et $al.$
(1986) (also see Rokaki 1995). It is equally possible that a BLR in which the
density and column density increase with distance from the central source
would meet the conditions required to reproduce the HILs and LILs from
the same ensemble of clouds. Such models will be investigated in future
publications.

The profile of the C IV line in NGC5548 was investigated during a period
when both the continuum luminosity and the luminosity of the C IV line
have increased by a large factor. Only very small changes, around the peak
and in the red-wing of the line, have been observed. This suggests a very

robust profile and the velocity structure of the BLR clouds is not affected by changes in the source luminosity.

In this paper an attempt has been made to model the BLR in NGC5548. It has not been possible to present a 'complete and final' model of the BLR in this AGN, but a number of aspects of the BLR have been identified which are necessary to bring model results into agreement with observations. These aspects will have to be built into future models of the BLR in AGNs and quasars.

Acknowledgements

This paper was produced with the facilities provided by the STARLINK Project, funded by PPARC. G.J. Ferland is thanked for providing a copy of the CLOUDY code and for assistance in using the code. PMG would like to thank B.M. Peterson and K.T. Korista for providing respectively the *IUE* and *HST* data obtained during the 1993 campaign to monitor NGC5548.

References

Baldwin J.A. : 1977, *Astrophysical Journal* **214**, 679
Bahcall J.N. Kozlovsky B.Z. : 1969, *Astrophysical Journal* **155**, 1077
Binette L., Prieto A., Szuszkiewicz E., Zheng W. : 1989, *Astrophysical Journal* **343**, 135
Blandford R.D., McKee C.F. : 1982, *Astrophysical Journal* **255**, 419
Boroson T.A., Green R.T. : 1992, *Astrophysical Journal, Supplement Series* **80**, 109
Cranshaw D.M., Blackwell J.H. : 1990, *Astrophysical Journal* **358**, L37
Clavel J., Santos-Lleó M. : 1990, *A& A* **230**, 3
Clavel J., *et al.* : 1991, *Astrophysical Journal* **366**, 64
Collin-Souffrin S., Dumont S., Joly M., Péquignot D. : 1986, *A& A* **166**, 27
Davidson K. : 1972, *Astrophysical Journal* **218**, 20
Davidson K., Netzer H. : 1979, *Rev. Mod. Phys.* **51**, 715
Dietrich, M., *et al.* : 1993, *Astrophysical Journal* **408**, 416
Ferland G.J., Persson S.E. : 1989, *Astrophysical Journal* **347**, 656
Ferland G.J., Peterson B.M., Horne K., Welsh W.F., Nahar S.N. : 1992, *Astrophysical Journal* **387**, 95
Ferland G.J. : 1993, *HAZY*, University of Kentucky, Department of Physics & Astronomy; Internal Report,
Gaskell C.M., Sparke L.S. : 1986, *Astrophysical Journal* **305**, 175
Gaskell C.M., Peterson B.M. : 1987, *Astrophysical Journal, Supplement Series* **65**, 1
Goad M.R., O'Brien P.T., Gondhalekar P.M.: 1993, *Monthly Notices of the RAS* **263**, 149 (Paper I)
Goad M.R. : 1995, *Ph.D Thesis*, University of London
Gondhalekar P.M., Wilson R. : 1980, *Nature* **285**, 461
Gondhalekar P.M. : 1992, *Monthly Notices of the RAS* **255**, 663
Gondhalekar P.M. : 1995, *Monthly Notices of the RAS* , submitted
Gondhalekar P.M. & Kellett B.J. : 1994, 'The Ionizing Continuum in AGNs and Quasars' in Gondhalekar P.M., Horne K., & Peterson B.M., ed(s)., *Reverberation Mapping of the Broad-Line Region In Active Galactic Nuclei*, Astronomical Society of the Pacific:

San Francisco, 241

Grevesse N., Anders E. : 1989, 'No. 183, Cosmic Abundance of Matter' in Waddington C.J., ed(s)., , AIP: New York,

Hamann F., Ferland G.J. : 1993, *Astrophysical Journal* **418**, 11

Horne K., Welsh W.F., Peterson B.M. : 1991, *Astrophysical Journal* **367**, L5

Jourdain E., *et al.* : 1992, *A & A* **256**, L38

Kinney A.L., Rivolo A.R., Koratkar A.P. : 1990, *Astrophysical Journal* **357**, 338

Korista K.T., *et al.* : 1995, *Astrophysical Journal, Supplement Series* **97**, 285

Kwan J., Krolok J.H. : 1981, *Astrophysical Journal* **250**, 478

MacAlpine G.M. : 1972, *Astrophysical Journal* **175**, 11

Maoz D., *et al.*: 1993, *Astrophysical Journal* **404**, 576

Maoz D. : 1994, 'Echo Mapping of the BLR - A Critical Appraisal' in Gondhalekar P.M., Horne K., & Peterson B.M., ed(s)., *Reverberation Mapping of the Broad-Line Region In Active Galactic Nuclei*, Astronomical Society of the Pacific: San Francisco, 95

Mathews W.G., Ferland G.J. : 1987, *Astrophysical Journal* **323**, 456

Netzer H. : 1985, *Astrophysical Journal* **289**, 415

Netzer H. : 1990, 'AGN Emission Lines' in Courvoisier, T.J.-L., & Mayor, M., ed(s)., *Active Galactic Nuclei*, Springer: Saas-Fee Advanced Course 20,

Neugebauer G., Oke J.B., Becklin E.E., Matthews K. : 1979, *Astrophysical Journal* **230**, 79

O'Brien P.T., Goad M.R., Gondhalekar P.M. : 1994, *Monthly Notices of the RAS* **268**, 845 (Paper II)

O'Brien P.T., Goad M.R., Gondhalekar P.M. : 1995, *Monthly Notices of the RAS* submitted (Paper III)

Osterbrock D. : 1993, *Astrophysical Journal* **404**, 551

Penston M.V., *et al.* : 1981, *Monthly Notices of the RAS* **196**, 857

Penston M.V. : 1991, 'The Trouble with Reverberation' in Miller H.R., & Wiita P.J., ed(s)., *Variability of Active Galactic Nuclei*, Cambridge Univ. Press: Cambridge, 343

Pérez E., Robinson A.R., de la Fuente L. : 1992a, *Monthly Notices of the RAS* **255**, 502

Pérez E., Robinson A.R., de la Fuente L. : 1992b, *Monthly Notices of the RAS* **256**, 103

Peterson B.M., *et al.* : 1991, *Astrophysical Journal* **366**, 119

Peterson B.M. : 1993, *Publications of the ASP* **105**, 247

Peterson B.M. : 1994, 'An Overview of Reverberation Mapping: Progress and Problems' in Gondhalekar P.M., Horne K., & Peterson B.M., ed(s)., *Reverberation Mapping of the Broad-Line Region In Active Galactic Nuclei*, Astronomical Society of the Pacific: San Francisco, 1

Pogge R.W., Peterson B.M. : 1992, **103**, 1084

Rees M.J., Netzer H., Ferland G.J. : 1989, *Astrophysical Journal* **347**, 640

Robinson A. : 1994, 'The LAG Spectroscopic Monitoring Campaign: An Overview' in Gondhalekar P.M., Horne K., & Peterson B.M., ed(s)., *Reverberation Mapping of the Broad-Line Region In Active Galactic Nuclei*, Astronomical Society of the Pacific: San Francisco, 147

Rokaki E. : 1994, 'Reverberation and the Disk Model of AGN' in Gondhalekar P.M., Horne K., & Peterson B.M., ed(s)., *Reverberation Mapping of the Broad-Line Region In Active Galactic Nuclei*, Astronomical Society of the Pacific: San Francisco, 257

Smith M.G., *et al.* : 1981, *Monthly Notices of the RAS* **195**, 437

Shields J.C., Ferland G.J., Peterson B.M. : 1995, *Astrophysical Journal* **441**, 507

Ulrich M.H., *et al.* : 1980, *Monthly Notices of the RAS* **192**, 561

Wamsteker W., Colina L. : 1986, *Astrophysical Journal* **311**, 617

Walter R., *et al.* : 1994, *A & A* **285**, 119

White R.J., Peterson B.M. : 1994, *Publications of the ASP* **106**, 879

Zheng, W., Kriss, G.A., Davidsen, A.F. : 1995, *Astrophysical Journal* **440**, 606

INTERSTELLAR MEDIUM

THE IDENTIFICATION OF DIFFUSE BANDS

D A WILLIAMS

Department of Physics and Astronomy, University College London
Gower Street, London, WC1E 6BT UK

Abstract. Our present knowledge of the diffuse interstellar bands (two of which were first noted by Wilson in 1958) is briefly summarized. Other broad and very broad interstellar features (the 220nm extinction bump, the very broad structure in the extinction curve, the extended red emission in reflection and other nebulae, and the unidentified - sometimes called the "overidentified" - infrared bands) are also briefly described. The origins of all these features remain unknown, in spite of intensive study. Possible relations between these various features and families is briefly discussed. Recent observations are shown to support the hypothesis that the carriers of the diffuse bands are free-flying molecules, whereas the alternative hypothesis of dust grains now seems to be untenable. Candidate types of molecular carriers are mentioned, and the possible source of such molecules in the diffuse interstellar medium is briefly discussed.

1. Introduction

Nearly 40 years ago, Robert Wilson reported some observations of several broad interstellar features (Wilson 1958). He determined the profile of the 443 nm feature (Fig 1), confirmed the suspected 476 nm feature, and detected two new interstellar bands at 489 and 618 nm. The origin of these bands was unknown in 1958. It remains unknown today, in spite of intensive study. The bands observed by Wilson in 1958 are part of a set of unidentified diffuse interstellar bands, (DIBs) which at that time numbered about thirty, but which at present is thought to include some two hundred features. It is surprising, even embarrassing to the science of astrophysics, that a spectrum so rich, so deeply studied over decades should remain unassigned. Some members of the set (at 578 and 579.7 nm) had been observed as early as the end of the nineteenth century as the only dark lines in the spectrum of a Wolf–Rayet star, and it became clear from studies of spectroscopic binary stars (Heger 1922) that the carrier of these bands was non-stellar. By the thirties, Merrill (1934, 1936) had established the interstellar origin of these bands, and of others at 628.4 and 661.4 nm. Therefore, for more than half a century the origin of these bands has been sought, and many theories have been proposed. During the last few years, however, significant progress has been made, and one may reasonably hope that reliable identification of most of these features will soon be made. Robert Wilson might not have realised in 1958 how patient he would have to be to know the carriers of the bands that he observed.

Progress is now being made on several fronts. Firstly, astronomers are benefitting from the use of more reliable and sensitive detectors, and are

Astrophysics and Space Science **237**: 243–266, 1996.

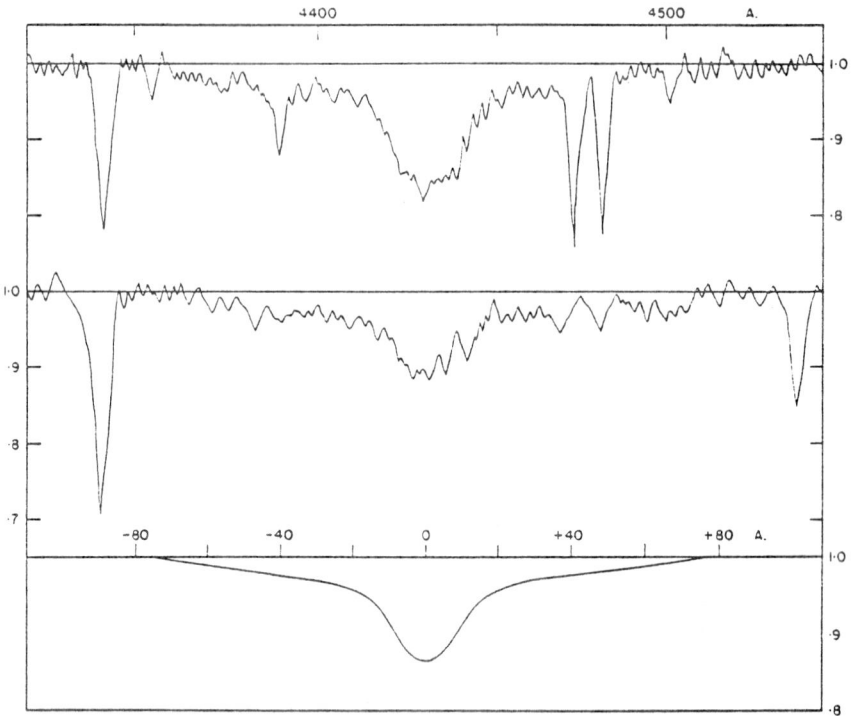

Fig. 1. (from Wilson 1958, Fig. 1). Spectra of HD183143 and HD15570 showing the absorption near 443 nm. The smoothed profile of the feature is also shown. (The wavelengths are in Å).

able to extend the search for DIBs to weaker bands and into the ultraviolet and infrared. The strengths of the bands, used in important correlation studies to indicate band origin, are nowadays much more accurate and the correlations much more reliable. Secondly, laboratory physicists and chemists are addressing this problem, bringing to bear modern laboratory techniques and undertaking studies, particularly of polyatomic molecules, not previously possible. Thirdly, our knowledge of the interstellar medium - though, obviously, incomplete in that we do not know the origin of the DIBs - is nevertheless enormously improved over the description one might have given in 1958. We know the atomic and molecular constituents of the interstellar medium, and can characterize the main components of interstellar dust (see, e.g. Williams 1994). Therefore, we have a better idea as to what may be plausible carriers of the DIBs.

The diffuse bands are so-called because they are generally broader than

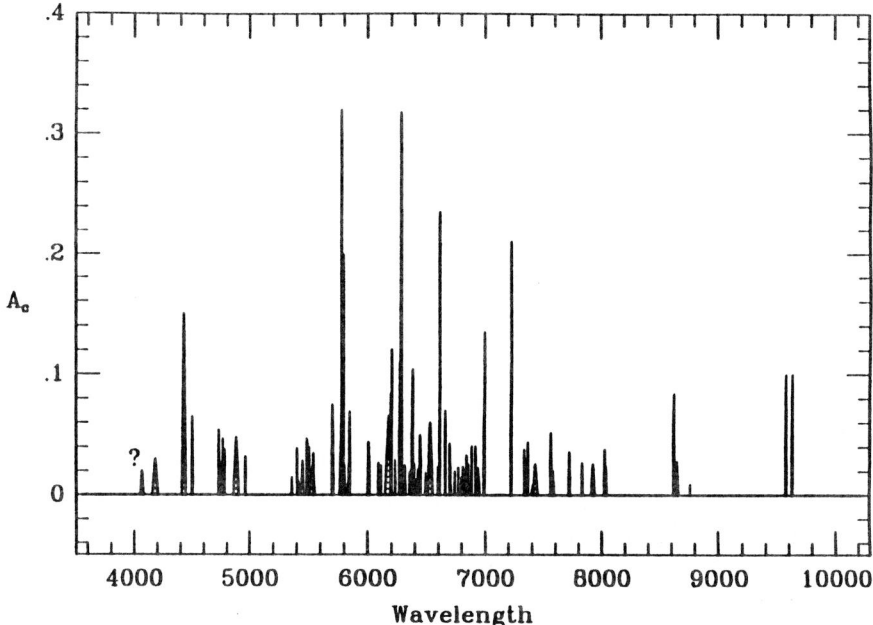

Fig. 2. (from Herbig 1995, Fig. 2). The central depths (values of A_c in magnitudes) of many of the DIBs detected in the spectrum of HD183143, plotted as a function of wavelength (given in Å). Broader bands are shown as shaded triangles.

absorption lines of atoms and molecules in the interstellar gas. FWHMs for DIBs range from about 0.1 nm to a few nm (Herbig 1995). There are, in addition to the DIBs, other spectral features that are also broad, generally much broader. These may or may not be related to the DIBs; they include the interstellar extinction bump at 217.5 nm, the very broad structure (VBS) observed as a reduction of interstellar extinction around 550 nm, the extended red emission (ERE) observed in a variety of reflection nebulae and other sources as a broad emission ($\geq$ 100 nm wide) in the red and infrared, and the unidentified infrared bands (UIBs), emission bands in regions of high excitation at wavelengths ranging from 3.3 μm to about 11.3 μm. In the next section (Section 2) we describe briefly not only the DIBs but also the other broad interstellar features listed above. We shall note the current assignment - if any - of these other broad features. In Section 3 we shall describe the properties of the DIBs in more detail, and consider the evidence for and against an association of other features with the DIBs. In Section 4 we shall discuss some of the models of DIB carriers that have been proposed. In Section 5 we describe problems associated with some of these proposals, and consider how these might be overcome.

2. Diffuse interstellar features

2.1. DIBs

The interstellar spectrum of the B7 supergiant HD183143 has been studied intensively for many years, especially by Herbig. Table 1 of Herbig (1995) lists 127 DIBs present in this spectrum; there are probably a large number of other bands also present, as listed by Jenniskens and Désert (1993). The list of Herbig (1995) is illustrated in Fig 2 (Herbig's 1995, Fig 2) where the central absorption in magnitudes, A_c, is indicated at the wavelength of each DIB. The total equivalent width for this set exceeds 2 nm (Herbig 1967). The set has no members with wavelengths less than 400 nm. The DIB with largest equivalent width (0.34 nm) is at 443 nm; it may also be the shortest wavelength DIB, as the two members at 406.6 and 418 nm are not confirmed as interstellar. Recently, features of wavelengths at around one micron have been detected, extending the range of DIBs well into the infrared. Bands at 957.7 and 963.2 nm (Foing and Ehrenfreund 1994) and 1.1797 and 1.3175 μm (Joblin et al 1990) are present in HD183143.

The range of FWHM for the diffuse bands is remarkable, and it has been suggested that there are two categories based on width, exemplified by the feature at 578 nm, FWHM $\simeq$ 0.2 nm and that at 577.8 nm, FWHM $\simeq$ 2 nm (Wu 1972). An alternative view is that diffuse bands occur in pairs, one broad and shallow with a sharper feature placed on its long wavelength limit. Herbig (1975) identified several such pairs, including the pair comprising the two bands mentioned above. The identification of pairs suggests a possible identification with bandheads and unresolved R branches in gas phase molecules. However, pairs - even if real - are a rare phenomenon among the DIBs and cannot be the complete story.

2.2. THE INTERSTELLAR EXTINCTION BUMP AT 217.5 NM

Stecher (1965) first detected this broad absorption feature in the mid-ultraviolet in a rocket observation, and it has since been extensively studied with astronomical satellites (e.g. Witt et al 1984; Fitzpatrick and Massa 1986). It is found on all galactic lines of sight with E(B-V) > 0.05 mag (though it is not present in the Small Magellanic Cloud; Fitzpatrick 1989). Although the width of the feature can vary widely (see Fig 3), and the far-UV extinction vary in an unrelated way, the central wavelength is nearly constant, within 1 nm of the nominal value (Fitzpatrick and Massa 1986). The strength of the feature is well correlated with reddening, i.e. with the dust causing visual extinction, though some exceptions occur in dense clouds or in regions of high excitation.

The most commonly invoked origin for the bump is small graphite particles (first calculations of the bump were made by Stecher and Donn 1965). From the point of view of elemental abundances, there is ample carbon to

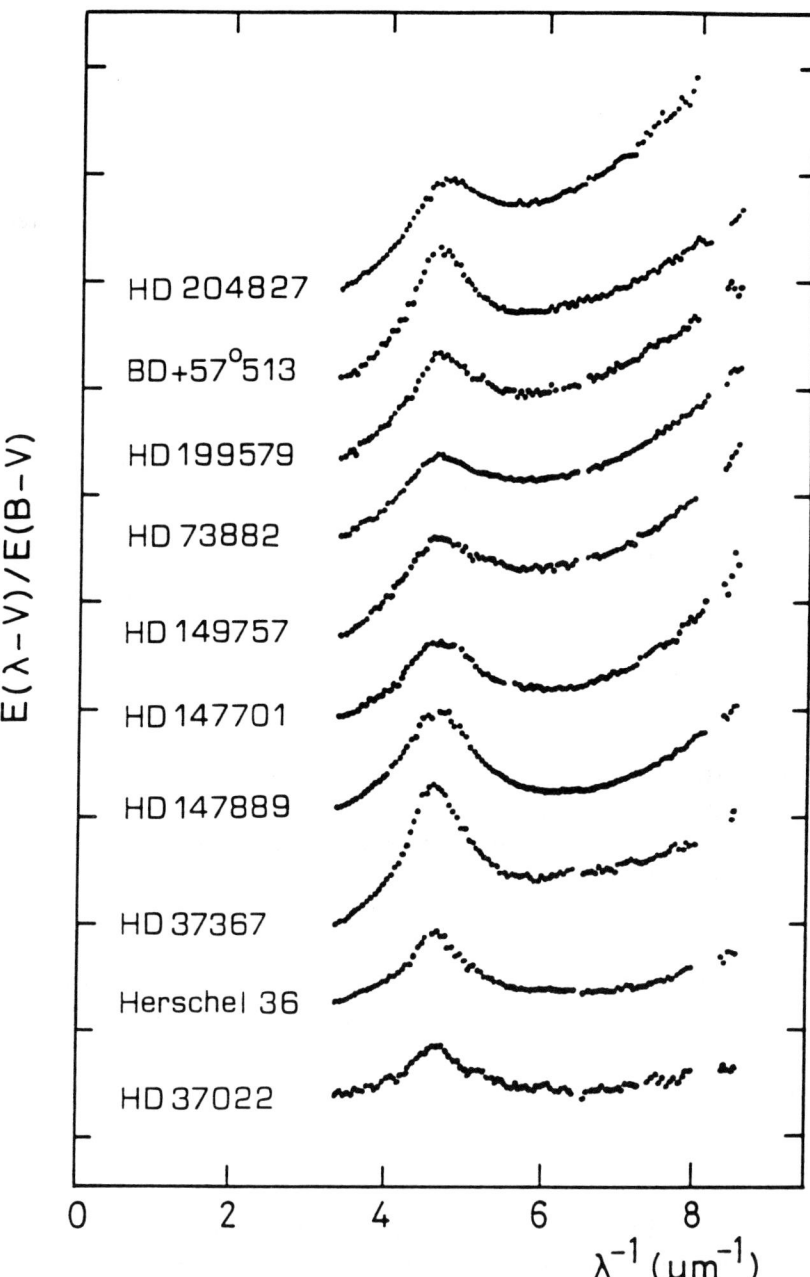

Fig. 3. (reproduced from Whittet 1992, Fig 3.6). A comparison of the ultraviolet extinction curves of 10 stars observed with the International Ultraviolet Explorer (Fitzpatrick and Massa 1986, 1988). The vertical axis is $E_{\lambda-v}/E_{B-V}$ (one division equals four magnitudes), individual curves being displaced vertically for display. The horizontal axis is in inverse wavelengths, where wavelengths are measured in micrometres.

provide the strength of bump. There are, however, some implied restrictions: the graphite particles must be small (radius ≤ 0.01 μm; cf. Draine 1989) though they need not be - and are unlikely to be - spherical. The graphite particles must be fairly pure, though the insertion of hydrogen into the graphitic lattice does broaden the feature (Perrin and Sivan 1990). It may be that amorphous carbon grains (Duley 1984; Jones, Duley, Williams 1990) may be UV annealed in the interstellar medium to graphitic carbon (see also Hecht 1986; Sorrell 1990).

Polycyclic aromatic hydrocarbon molecules (PAHs) may also contribute to the bump (e.g. Salama and Allamandola 1992a; Léger et al 1989), but are expected to provide other features in the UV which are not detected in the interstellar medium. It seems unlikely, however, that the cage molecule buckminsterfullerene, C_{60}, contributes to the bump (Krätschmer 1993; Leach 1993).

2.3. THE VBS AND ERE

It is possible that these two features are manifestations of the same phenomenon. The VBS is detected in plots of the residuals of extinction with respect to a linear plot in the inverse wavelength through the visual region. There is a weakening of the extinction near inverse wavelength 1.75 μm^{-1} and possibly a strengthening near 2.05 μm^{-1}. If the VBS arises from a broad band emission feature, then it may be due to a luminescence band in rehydrogenated amorphous carbon, the band peaking at a wavelength of 0.56 μm (Duley and Whittet 1990). The ERE is also a broad band emission, usually between about 0.6 and 0.8 μm, detected in reflection nebulae (Witt and Schild 1986), planetary nebulae (Furton and Witt 1992), HII regions (Perrin and Sivan 1992) and the galactic cirrus (Guhathakurta and Tyson 1989). The proposed origin of ERE as luminescence from hydrogenated amorphous carbon(Duley 1985; Duley and Williams 1988) is strongly supported by laboratory studies (Furton and Witt 1993). ERE is particularly prominent in the Red Rectangle, a biconical carbon-rich nebula in which the mass loss of an evolved star is illuminated by a B9 star HD44179. Superposed on the broad ERE are a set of atomic, molecular, and unidentified features (Schmidt, Cohen and Margon 1980), see Fig 4.

2.4. THE UIBs

These infrared emission features have principal components at 3.3, 6.2, 7.7, 8.6 and 11.3 μm and widths in the range 0.03-0.5 μm; they were first noted by Gillett, Forest and Merrill (1973); discussions of more recent observations are given by Bregman (1989), and also by Tielens (1993) who explores the PAH hypothesis. There seems little doubt that the features arise in aromatic hydrocarbons. However, no perfect match has yet been obtained between laboratory and observational spectra (see Fig 5), though the comparison is

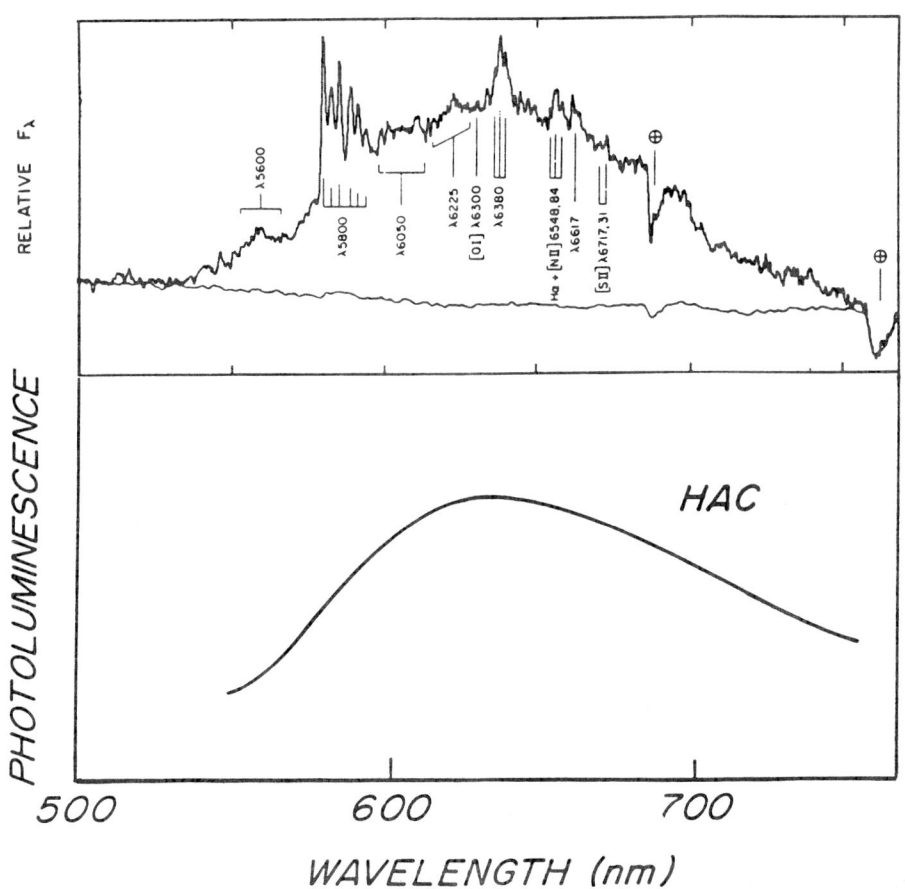

Fig. 4. (from Duley 1985). The upper spectrum is the Extended Red Emission from the Red Rectangle (from Schmidt et al 1980) and the lower spectrum is of luminescence from hydrogenated amorphous carbon at a temperature of 81K (Watanabe et al 1982).

certainly compelling in general. Part of the difficulty is that infrared features shift and change in response to the environment of the active unit in the molecule, and this laboratory phenomenon (e.g. Flickinger, Wdowiak, and Gómez 1991) seems to be evident also in the astronomical spectra (e.g. Tokunaga et al 1991). Nevertheless, some identifications have been claimed. Justtanont et al (1995) report a detection of the PAH chrysene in carbon-rich post - AGB objects.

The PAH hypothesis, although strongly indicated by the observations, cannot be regarded as completely confirmed until a laboratory spectrum of a suitably prepared specification matches in wavelength and intensity the

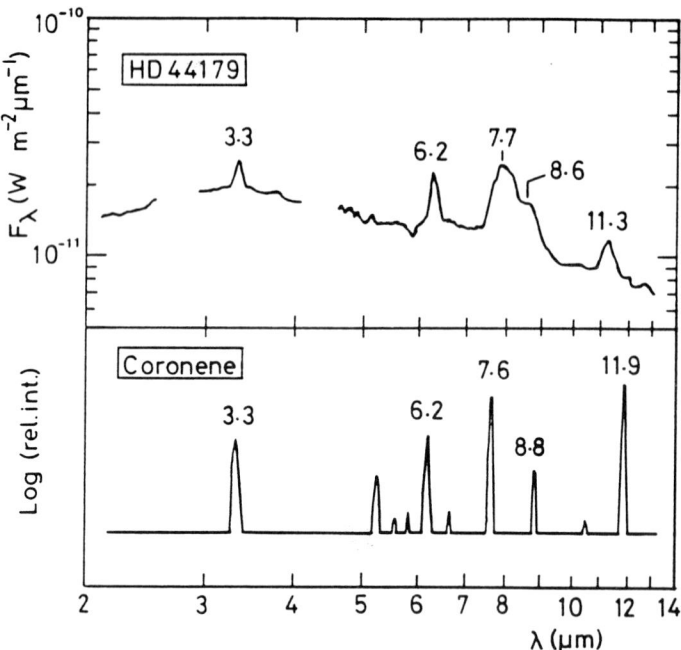

Fig. 5. (from Whittet 1992, Fig. 6.7). The infrared spectrum of HD44179 in the Red Rectangle (Russell et al 1978, upper frame) compared with that for coronene (Leger and Puget 1984). The principal spectral features are labelled by their wavelengths in micrometres.

observed features. The fact that the astronomical 7.7 μm feature is generally stronger than the 11.3 μm feature, while the reverse is true for laboratory measurements (see Fig. 4) has led to the suggestion that interstellar PAHs are photoionized, and some attention has been given to the laboratory spectrum of such species (e.g. Salama and Allamandola 1992; Snow 1992). Other candidates proposed as carriers of the UIBs are partially hydrogenated buckminsterfullerene, $C_{60}H_m$, ($m<60$) (Webster 1992, 1993, a,b,c).

2.5. POSSIBLE ASSIGNMENTS

In summary, the extinction bump at 217.5 nm is plausibly assigned to very small graphite particles. The consequent problem is to understand why the graphitic matter is limited to a small size range. Possibly, the UV annealing of amorphous carbon may be able to promote graphitization up to some limiting size controlled by the photon energy input (cf Duley 1993). The assignment of the ERE (and, possibly, also the VBS) to luminescence in hydrogenated amorphous carbon seems the most secure identification. The

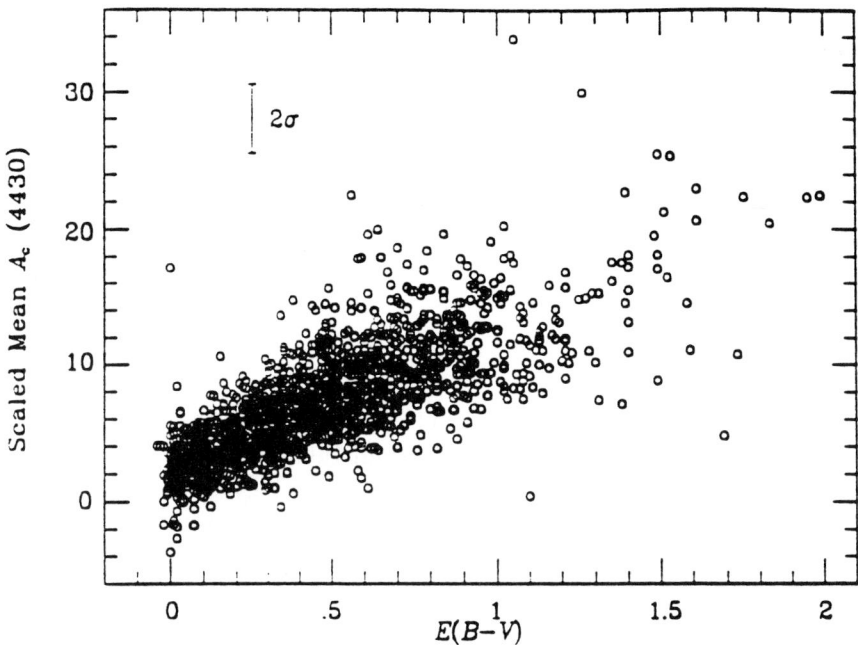

Fig. 6.   (from Somerville 1989). Central depths of the 443 nm feature, A_c, as a function of the corresponding reddening, E(B-V), along lines of sight to 1500 stars.

match between laboratory and astronomical data is good, and there is an immediate interpretation of the variation in peak wavelength and uniformity intensity of emission in this model (Duley and Williams 1995). This interpretation is lent credence by the incorporation of amorphous carbon into most existing grain models (e.g. Mathis 1989; Jones et al 1990). The DIBs are almost certainly associated with PAHs (which themselves probably derive from amorphous carbon). However, it is a concern that a precise and convincing identification of the set of DIBs has yet to be made. Nevertheless, PAHs are very plausible components of the interstellar medium in excited regions - which is, of course, precisely where the DIBs are most readily detected. Thus, good candidates exist for all the very wide features in the interstellar spectrum. By contrast, as we shall discuss in the next Section, the identification of the carriers of the DIBs is much less advanced.

3. Properties of the DIBs

3.1. Correlations

Fig. 6 shows a correlation between the measured central depth, A_c in magnitudes, of the strong band at 443 nm towards 1500 stars with reddening, E(B-V), on the same line of sight. Somerville (1995) notes that the correlation is better than that between simple interstellar molecules, such as H_2 and CO, and reddening, because simple molecules are photodissociated at low redding. Evidently, the carrier of the 443 nm DIB is not photodissociated, but is closely related to the grains that cause the optical extinction. However, this does not imply that the carriers must be in the dust, because the total gas content $(H + H_2)$ is very well correlated with E(B-V), so the carrier may be in the gas. The scatter in Fig. 6 is real, and the DIB strengths generally have a dispersion of about a factor of two about a mean relationship (Herbig 1995).

This result is, apparently, general for all the DIBs that have been extensively studied (Somerville 1995). In addition, there is no evidence for structure in the measured interstellar polarization in the region of DIBs (or of the 217.5 nm bump) according to Martin (1995), Martin et al (1995) and Adamson and Whittet (1995). This implies that the larger grains that cause the polarization cannot be carriers of those DIBs whose polarization curves have been established. Studies in the far UV have shown that there is no correlation of DIB strengths with the amount of slope of the far UV extinction (Wu, York and Snow 1981; Seab and Snow 1984). It seems that DIBs carriers are probably not associated with the population of small grains that cause most of the extinction in the far UV.

These conclusions are confirmed in a detailed study of the DIBs at 578 and 579.7 nm in the spectra of 93 stars by Herbig (1993). The strength of these bands is shown to be linearly related to the column of atomic hydrogen up to a limiting column (Fig. 7, from Herbig 1993) and unrelated to the column of molecular hydrogen. Therefore, the carrier of these bands is not formed in the simple chemistry based on H_2 that gives rise to the known diatomics in diffuse clouds (cf Williams 1994). The strength of these bands does not seem to be influenced by variations in the far UV extinction, nor by the shape of the interstellar extinction curve in the visual and near infrared regions. Therefore, the carrier of these bands is probably neither in large grains nor in small grains, and the good correlation between band strength and reddening, E(B-V), arises simply because reddening and hydrogen column density are closely related. Herbig (1993) also considered whether shocks could release the DIB carriers from dust. He found no systematic variation between band strength and either the depletion of atomic titanium (believed to be released from shocked grains) or the CH^+ abundance, which has been proposed to arise in shocks (though this now seems unlikely, cf. Williams 1994). Finally, from a discussion of the DIB strengths towards several stars in Orion, which

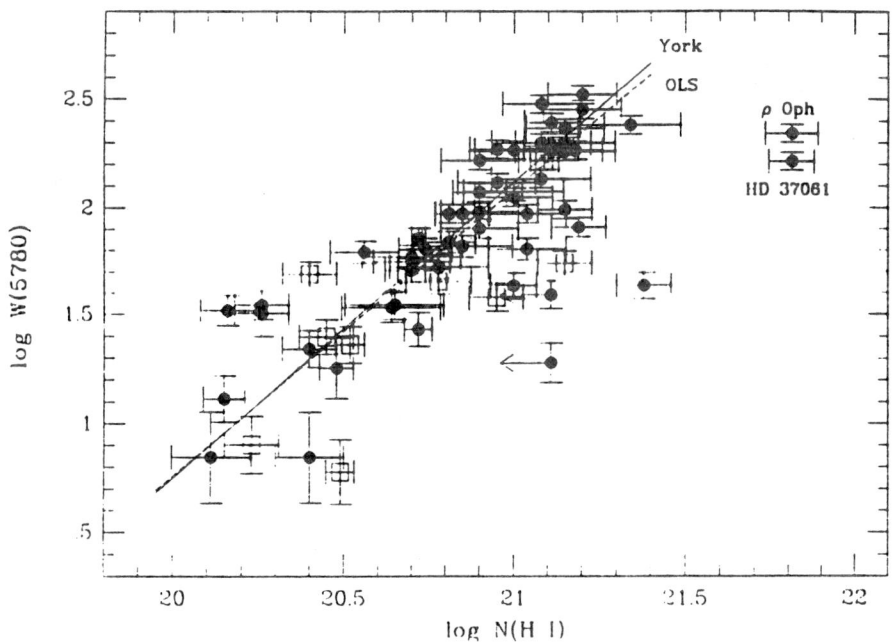

Fig. 7. (from Herbig 1993, Fig. 8). Dependence of the equivalent width of the DIB at 578 nm on the column density of neutral atomic hydrogen, (see Herbig 1993 for further detail).

- like sodium line strengths - appear low relative to atomic hydrogen for these lines of sight, Herbig (1993) deduces that the DIB carriers have a behaviour that is consistent with that of a free, neutral gaseous species responding to the ionization level in the gas as if its ionization/dissociation threshold is somewhat higher than that of atomic sodium, 5.1 eV.

Studies of DIBs in darker clouds tend to show that DIB strengths are weaker than expected from the reddening. Adamson, Whittet and Duley (1991) made a study of several lines of sight through the Taurus Molecular Cloud, and found that the DIB per unit E(B-V) nearly always fell below that expected from diffuse cloud lines of sight. They argue that this indicates that DIB carriers are generally in the surface layers of clouds, and this is where the atomic hydrogen fractions are highest. The failure by Adamson, Kerr, Whittet and Duley (1994) to detect the DIB at 1.3175 μm on lines of sight up to $A_V = 20$ magnitudes in Taurus suggests a significant weakening of that absorber. It seems at least reasonable to conclude that DIB carrier abundances diminish with depth into a cloud. However, higher than expected

THE 6614 DIFFUSE BAND AT R=600,000

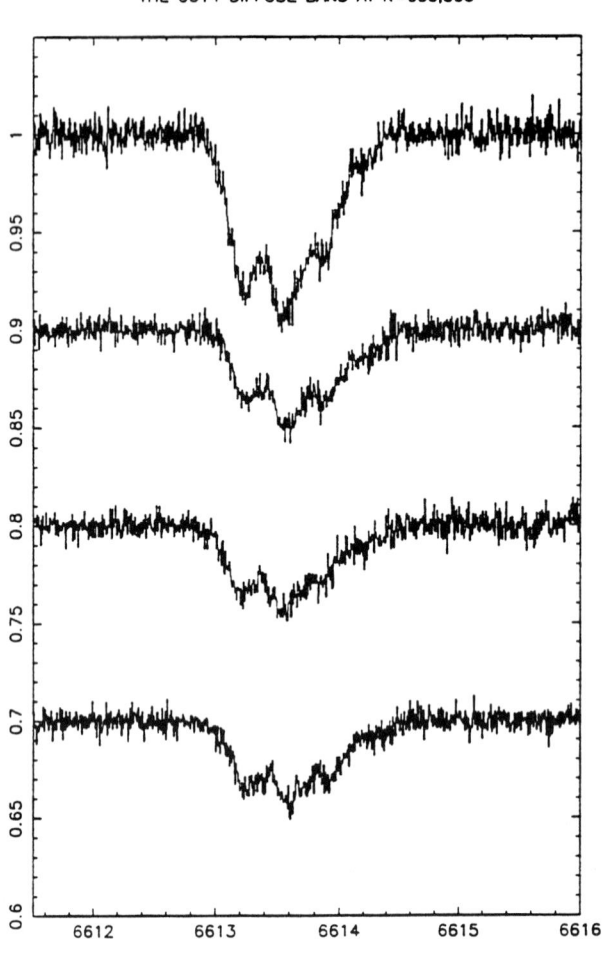

Fig. 8.   (from Hibbins et al 1994). Ultra High Resolution Facility (UHRF) spectra of the
DIB at 661.4 nm for lines of sight towards (top to bottom)μ Sgr, σ Sco, β [1]Sco, and ζ
Oph, shown on a co in Å).

DIB strengths are not accounted for on this picture.

An important and new advance in the study of DIBs has been made with
the introduction of very high resolution observations. Using the Ultra High
Resolution Facility (UHRF) with a wavelength resolution of 600,000 at the
Anglo Australian Telescope (Crawford et al 1994; Diego et al 1995) Hibbins

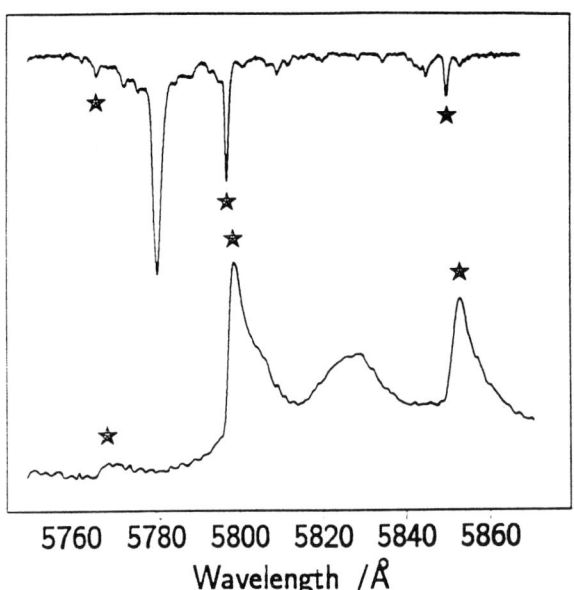

5760 5780 5800 5820 5840 5860
Wavelength /Å

Fig. 9. (from Sarre et al 1995). Comparison of the DIBs towards HD183143 (upper) and the Red Rectangle (lower) in the 580 nm region. See Sarre et al for further details. (Wavelengths are in Å).

et al (1994) reported structure in the 661.4 nm DIB in the interstellar spectra towards μSgr, σSco, β^1Sco and ζOph (see Fig. 8). In each case three components can be clearly seen. Hibbins et al attribute these components to unresolved P, Q, and R rotational branches associated with a molecular transition of a large, almost certainly organic, molecule. Molecular modelling of expected profiles are in progress by Miles and Sarre. It will be exceptionally interesting to see if this important detection of band structure is found for other DIBs and for other lines of sight. If confirmed in this way, it should ultimately be possible to define in some detail the nature of the molecules giving rise to this rotational structure.

Sarre, Miles and Scarrott (1995) have recently made an observational study of *emission* bands in the Red Rectangle. Fig. 4 shows that the ERE from this object has superposed on it a number of emission bands and lines. The UIB features are also strong in this source. Some of these optical features have been identified as atomic and molecular lines; for example, CH^+ lines are found in emission in this object (Balm and Jura 1992), as are lines of CO (Sitko 1983). However, many of the broader features are unidentified, and occur at wavelengths that are offset from those of the DIBs (Warren-Smith,

Scarrott, and Murdin 1981). Scarrott et al (1992) showed, however, that the offsets and the band widths decrease with distance from the star, and that the wavelengths approach those of corresponding DIBs at 579.7, 584.9, 637.6, and 661.3 nm. The study of Sarre et al (1995) explores this spatial variation quantitatively, through high resolution observations in the 510 - 690 nm region. The correspondence of the emission features in the Red Rectangle is compared in Fig. 9 with the spectrum of HD 183143, from Herbig. The DIBs at 579.7 and 585.0 nm lie on the short wavelength edge of the broader Red Rectangle features. Sarre et al measured the peak wavelengths and widths of the prominent Red Rectangle bands as a function of distance from the exciting star, HD 44179, of this nebula. Their results are shown in Fig. 10. The peak wavelengths and widths of the emission bands are seen to decrease with increasing distance from the star and to converge in the most distant (and cooler) location to the corresponding DIB limit. Therefore, the same carriers give rise to both emission and absorption bands, and Sarre et al conclude from the fact that the relative intensities in this group of bands do not very significantly with position in the nebula that a single carrier may be responsible for all the bands in Fig. 9. The band profiles are shown in Fig. 11, the broader profile (top) being obtained for material closer to the star. Sarre et al note that these profiles have the appearance of vibrational band structure with unresolved rotational lines. The steep short- wavelength side is probably a rotational branch head, and part of the red-degraded tail is the extension of the rotational branches to longer wavelengths. Sarre et al point out the similarity of this type of profile to that of a molecule with one of its rotational constants smaller in the excited than in the ground electronic state.

Scarrott et al (1992) had suggested that the 579.9/579.7 Red Rectangle/DIB feature could arise in a large molecule, such as a C_{60} - based species (though not C_{60}). The new observations by Sarre et al (1995) appear to rule out this possibility, as the observed energy intervals in the profile are too small to arise in a C_{60} molecule. A solution in terms of PAH or linear molecules is being sought, and even smaller molecules are not excluded. A recent possible detection of interstellar C_3 may add weight to the suggestion that this molecule be the carrier of DIBs in this spectral region (Schmidt, Cohen, and Margon 1980).

4. The DIB carriers

4.1. DUST GRAINS

Although the weight of evidence given in the previous section certainly supports a free-flying molecular origin for the DIBs, one should note that the observational studies have been concerned with a relatively few lines. It is still conceivable that some DIBs may arise in the dust, and that in some

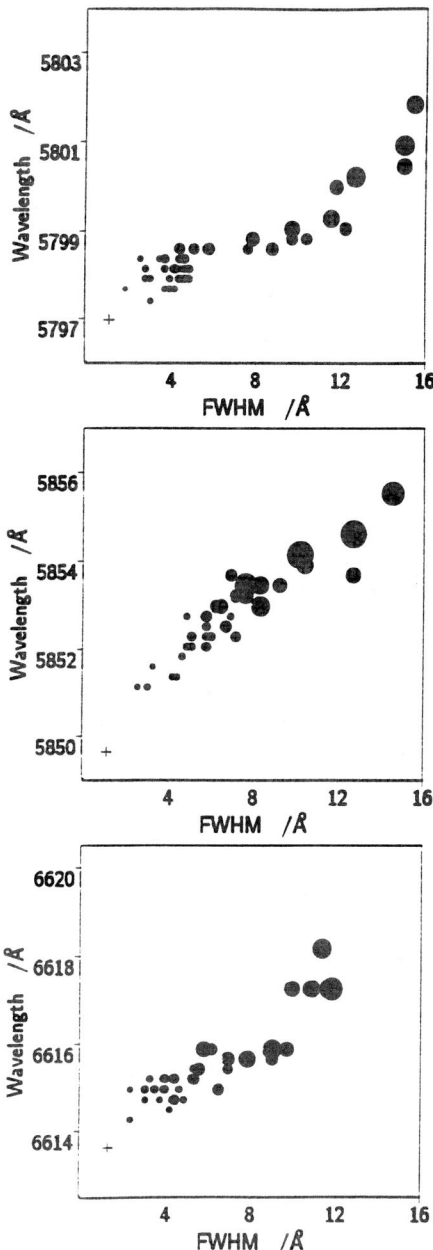

Fig. 10. (from Sarre et al 1995). Peak wavelengths of the three most prominent emission DIBs in the Red Rectangle plotted with FWHM, adopting a velocity shift for nebular species from van Winkel et al (1995). The diameter of the circle is inversely proportional to the distance from the star HD44179, smaller circles therefore representing cooler conditions. The cross marks the corresponding rest wavelength and FWHM for the DIB (absorption) towards HD183143 (Herbig 1995). (Wavelength measurements are in Å).

way the carriers are more abundant in the peripheries of interstellar clouds. Perhaps the presence of a sufficiently strong radiation field is required to promote the dust/carrier system into an appropriate energy state. Most of the early ideas on DIB carriers were concerned with solid state features, and this is reasonable, since the enhanced widths of DIBs relative to free atomic and molecular lines would naturally occur from solid state transitions. Whittet (1992) and Herbig (1995) have reviewed the work on the grain hypothesis, and only a few remarks are made here. The transitions that have been considered include plasma oscillations in metal grains (as involved in graphite as the origin of the 217.5 nm bump) or absorption by embedded or surface impurities. The impurity centres generally absorb in bands that are a few nm wide, but are usually rather weak. In principle, such features could be the origin of some of the broader DIBs, though no specific identifications are accepted. Duley (1979) has noted that interstellar materials may be much more disordered than laboratory crystals, and that as a consequence the f-values of some transitions in iron-group ions might be increased. Hence, it is possible that the general weakness of impurity transitions may be overcome, but the uncertainty of the nature of host grain material makes identification uncertain. To account for the narrower lines, one needs to invoke pure electronic transitions in impurity centres. In such transitions, there is no change in the vibrational site of the host lattice. Detailed studies of assumed impurities in plausible host materials indicate that the band profile is sensitive to grain dimensions (Shapiro and Holcomb 1986 a, b). Symmetric profiles can be produced in the smaller grains (though these may have emission wings), but asymmetries are introduced in larger grains. For sufficiently large grains, the feature becomes an emission line, hence, tight constraints are placed on the size distribution of grains responsible for the asymmetric DIB at 578 nm. One would need to invoke a size distribution for *each* DIB formed in dust to account for the observed profile by the averaging out of unwanted emission wings or asymmetries. Experiments (Duley 1968; Duley and Graham 1969) designed to test the hypothesis that the CaI resonance line at 422.7 nm is shifted in a matrix to 443 nm have shown that a very broad feature arises.

Herbig (1995) lists a number of arguments against the hypothesis that DIBs originate in grains. Firstly, the lack of DIB emission wings seems to require very special choices for particle sizes, and these restrictions must apply on all lines of sight through widely differing physical conditions. This seems implausible. Secondly, the absence of detectable changes in interstellar polarization indicates that the DIB carriers are not present in the grains causing optical polarization. The lack of association between DIB strengths and the far-UV extinction indicates that the carriers are not in the small grains. Finally, the lack of any correlation with indicators of shocks implies either that DIB carrier grains are unaffected by shocks (even though abundances of some elements appear to be modified) or that they are not in the grains.

TABLE I

Tentative assignments of diffuse features

Feature	Assignment	Comment
217.5nm extinction bump	small graphite grains	Significant restrictions on size - how is this achieved?
VBS/ERE	luminescence from hydrogenated amorphous carbon	good match of observations with laboratory data - a plausible model
UIBs	emission from transiently heated PAHs	plausible, but no specific assignments
DIBs	grains	band profiles, carrier abundance problems, restrictions on grain size conclusion: unlikely
	PAHs	plausible, but no specific assignments
	cage molecules - C_{60}^+ - fullerenes?	need confirmation of other bands no lab spectrum
	carbon chains	strong lab support; the best candidate at present. Problem: origin?

4.2. FREE-FLYING MOLECULES: SOME RECENT WORK

The fractional abundance of the carrier of the 443 nm DIB has been shown to be $2.3 \times 10^{-9}/f$, (Whittet 1992), where f is the oscillator strength of the transition. Therefore, for $f \sim 10^{-2}$ (typical of a molecular transition) the fractional abundance is $\sim 10^{-7}$ which is in the range of fractional abundances for simple interstellar diatomic molecules in diffuse clouds. Therefore, there is at least a plausibility argument to consider some small molecules as potential carriers. Herbig (1995) lists a variety of atomic ions and simple molecules whose case has usually been promoted on the basis of some near-

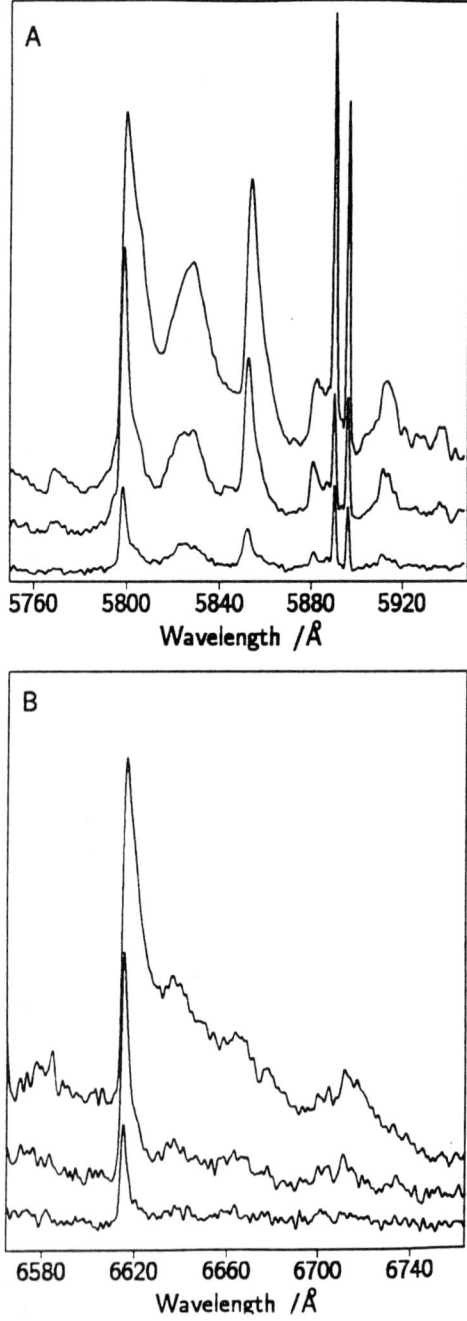

Fig. 11.   (from Sarre et al 1995). Spectra of the Red Rectangle in (A) 580 nm and (B) 660 nm regions at offsets from HD44179 of 7 (top), 12, and 17 (bottom) arc sec. The background ERE has been removed. Each spectrum is offset on a vertical scale and the upper trace is scaled by 0.4. (Wavelengths are in Å).

identity of wavelengths of DIB and proposed feature. The greater width of the atomic or molecular line is seen as arising from the short life time of the upper state. A recent suggestion (Sorokin and Glownia 1995) involving a small molecule is the most abundant interstellar molecule, H_2. These authors claim that two photon absorption

$$^1\textstyle\sum_g^+ v =?, J = 0 - 4 \overset{DIBs}{\leftarrow} C^1\Pi_u v' = 9, J = 1 - 4 \overset{Ly\alpha}{\leftarrow} X^1\textstyle\sum_g^+ v'' = 11, J = 2,3$$

in which the highest electronic state is postulated to exist at the appropriate energy and to have a rotational constant B = 9.0 cm^{-1}. It is a dissociating state. The X $v'' = 11$ is supposed to arise from UV pumping (Stecher and Williams 1967) and absorption of Lyman α radiation induces a transition to the C state which is supposed to be the DIB carrier. While the coincidence of the energy differences with transition energies for the DIBs is an interesting one, and the model should give rise to broad lines since the upper state is short–lived, nevertheless the required abundance of the carrier cannot be met with the proposed mechanism. A laboratory study (Hinnen & Urbachs, 1995) does not find the features predicted by Sorokin & Glownia, in the range 766–788 nm.

Much attention has been given to small carbon molecules as potential DIB carriers, since the original suggestion of Douglas (1977). Douglas noted that a carbon chain of sufficient size would be able to survive photoabsorption by conversion of energy to internal modes, and that would be expected to provide strong absorption in the 400 - 500 nm region. Experiments by Krätschmer, Sorg and Huffman (1985) seemed to be supportive of this point of view.

Recent studies of Fulara, Lessen, Freivogel and Maier (1993) and Freivogel, Fulara, and Maier (1994) provide some very compelling evidence in support of this view. They obtained absorption spectra of mass-selected highly unsaturated hydrocarbons in a neon matrix at a temperature of 5K. The hydrocarbons have been studied under different deposition and radiation conditions. These authors find that for the set of hydrocarbons with between 6 and 16 carbons (with up to two H atoms on each carbon molecule) seventeen bands at wavelengths above 600 nm match the DIBs at within ±36 cm^{-1} (the criterion chosen to accommodate a possible neon matrix shift from the gas phase; it is the measured shift for C_3; Jacox 1988). This is a very striking result, and all the more so since it represents the properties of a *set* of molecules. Absorption spectra for the $C_{14}H_m$ and $C_{16}H_m$ species in a matrix are shown in Fig. 12, together with DIB positions. For the first time, a large group of DIBs seems to have a plausible origin; further, the nature of the carriers - a family of molecules - allows predictions to be made of positions, relative intensities and widths of features. Freivogel et al (1994) note also that the observed bands have large oscillator strengths, possibly

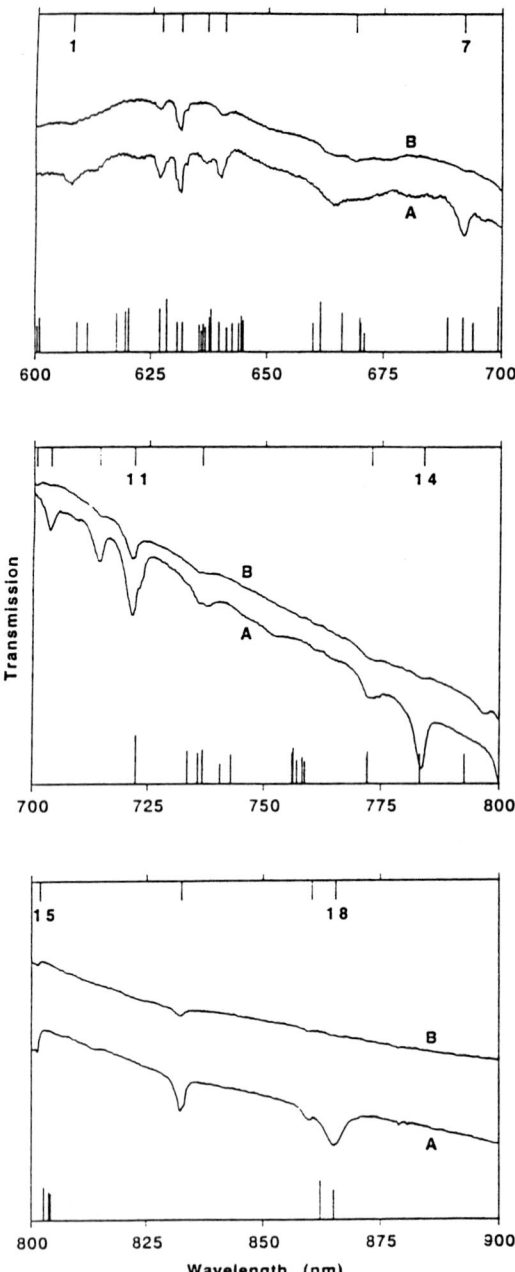

Fig. 12. (from Freivogel et al 1994 Fig. 2). Absorption spectra of (A) C_{14} H_m and (B) C_{16} H_m in 5K neon matrices, in the 600 - marks on the wavelength axis show the DIB positions and intensities on a logarithmic scale of their central depths (Herbig and Leka 1991). The numbered ticks on the upper scale refer to 22 bands identified by Freivogel occurring the laboratory spectra that are within ±0.5 nm of the corresponding DIBs.

in excess of 10^{-2}. Hence, the constraints on abundances of the species in the interstellar medium are not too severe.

Considerable attention has been given to the PAHs as DIB carriers (see Herbig 1995 for a review). These molecules are certainly plausible candidates, since their presence in excited regions is indicated by the UIBs and by the close link between PAHs and amorphous carbon. Large PAHs may survive in the interstellar medium for considerable periods, so their interstellar abundance may not be a problem.

Further, PAHs - both neutral and ionized - have strong transitions in the optical region. However, although some suggestive results have been obtained (e.g. naphthalene; pyrene; Salama and Allamandola, 1992a, b), no convincing matches of more than a few lines have been made. In some cases, other features expected from the carrier were not present in the astronomical spectrum (Snow 1992). Also, even if the pyrene cation is a DIB carrier, then one might expect similar cations also to be carriers but the evidence in support is missing (Ehrenfreund et al 1992). Why should a single specific ion be a carrier? Nevertheless, chlorin has been proposed by Miles and Sarre (1993), and a family of porphyrins in paraffin/pyride grains by Johnson (1994).

The cage molecules, buckminsterfullerene C_{60}, and members of that family have been intensively studied as potential carriers of the DIBs. Attention has recently focused on the ions which are expected to be the dominant form of this material - if present - in diffuse clouds. Laboratory work by Fulara, Jacobi and Maier (1993) involving the ion C_{60}^+ in solid argon detected a doublet at 958.0 and 964.2 nm, and these authors estimated that the gas phase transitions should be in the range 951.0 - 965.0 nm. Subsequently, Foing and Ehrenfreund (1994) have detected two broad features at 957.7 and 963.2 nm, and have assigned these to C_{60}^+. Unfortunately, C_{60}^+ does not have strong transitions in the range 300 - 900 nm and is unlikely to be the carrier of any other DIBs.

5. Conclusion

In Table 1 I attempt to summarize the situation concerning proposed carriers for all the diffuse features; the comments represent my personal view. The table shows that it is possible that the DIBs sample a quite different constituent of the interstellar medium from other diffuse (or narrow) features. This would be a remarkable outcome of this long running investigation. The next year or two will see a continuation of the new and promising lines of study. There will be more information about the high resolution profiles of the bands, and detailed theoretical modelling should indicate the kinds of molecular structures involved. The link between emission and absorption bands will be explored further, and should also reveal much about the molecular structure. The laboratory work on carbon chains should continue to produce more plausible matches with DIBs.

There is a need for modelling the mechanisms by which populations of potential molecular carriers such as carbon chains could be maintained in the diffuse interstellar medium. Hall and Williams (1995) have made a time-dependent study of molecules released from carbonaceous dust at diffuse cloud boundaries and have shown that the concentrations of molecules in otherwise atomic regions required are not implausible as carriers of the DIBs for f-values around 10^{-2}. The Red Rectangle is a region also requiring study from the point of view of dust formation/destruction and chemistry in the carbon-rich outflow. The ideas that have been employed in the study of novae ejecta (Rawlings and Williams 1989) should also be applicable here.

Finally, it is worth noting that if the DIBs are found to originate in carbon chains, then this will re-open the question of the abundance of free-flying PAHs in the interstellar medium. Jones et al (1990) have suggested that they are found maintaining regions of high excitation. Thaddeus (1994) has argued that PAHs are not the dominant large molecule component in the interstellar gas, and poses the question: could a chain molecule/ring molecule population give a satisfactory explanation of the UIB features? Time will tell!

6. Acknowledgment

Two extremely comprehensive and authoritative sources for this review have been Herbig (1995) and Whittet (1992). This article draws heavily on these two references.

References

Adamson, A. J., Whittet, D. C. B. and Duley, W. W., 1991, MNRAS 252, 234

Adamson, A. J. and Whittet, D. C. B., 1995 in "The Diffuse Interstellar Bands" eds A. G. G. M Tielens and T. P. Snow (Kluwer: Dordrecht) in press

Adamson, A. J., Kerr, T. H., Whittet, D. C. B. and Duley, W. W., 1994, MNRAS 268, 705

Balm, S. P. and Jura, M., 1992, AA 261 L25

Bregman, J., 1989 in "Interstellar Dust" eds L. J. Allamandola and A. G. G. M. Tielens (Kluwer: Dordrecht) p109

Crawford, I. A., Barlow, M. J., Diego, F. and Spyromilio, J., 1994, MNRAS 266 903

Diego, F. Fish, A. C., Barlow, M. J., Crawford, I. A., Spyromilio, J., Dryburgh, M., Brooks, D., Howarth, I. D. and Walker, D. D., 1995, MNRAS 272 323

Douglas, A. E., 1977, Nature 269 130

Draine, B. T., 1989 in "Interstellar Dust" eds L. J. Allamandola and A. G. G. M. Tielens (Kluwer: Dordrecht) p313

Duley, W. W., 1968, Nature 218 153

Duley, W. W., 1979, ApJ 227 824

Duley, W. W., 1984, ApJ 287 694

Duley, W. W., 1985, MNRAS 215 259

Duley, W. W., 1993 in "Dust and Chemistry in Astronomy", eds T. J. Millar and D. A. Williams (IOP: Bristol) p71

Duley, W. W. and Graham, W. R. M., 1969, Nature 224 785

Duley, W. W. and Whittet, D. C. B., 1990, MNRAS 242 40P

Duley, W. W. and Williams, D. A. 1988, MNRAS 231 969

Duley, W. W. and Williams, D. A., 1995 in preparation

Duley, W. W. and Williams, D. A., 1988, MNRAS 230 1P

Fitzpatrick, E. L. and Massa, D., 1986, ApJ. 307 286

Fitzpatrick, E. L. and Massa, D., 1988, ApJ. 328 734

Fitzpatrick, E. L., 1989 in "Interstellar Dust", eds L. J. Allamandola and A. G. G. M. Tielens (Kluwer: Dordrecht) p37

Flickinger, G. C., Wdowiak, T. J. and Gmez, P. L., 1991, ApJ 380 L43

Foing, B. H. and Ehrenfreund, P., 1994, Nature 369 296

Freivogel, P., Fulara, J. and Maier, J. P., 1994, ApJ Lett 431 L151

Fulara, J., Jacobi, M. and Maier, J. P., 1993, Chem. Phys. Lett. 211 227

Fulara, J., Lessen, D., Freivogel, P. and Maier, J. P., 1993, Nature 366 439

Furton, D. G. and Witt, A. N., 1992, ApJ. 386 587

Furton, D. G. and Witt, A. N., 1993, ApJ 415 L51

Gillett, F. C., Forest, W. J. and Merrill, K. M., 1973, ApJ 183 87

Guhathakurta, P. and Tyson, J. A., 1989, ApJ. 346 773

Hall, P. and Williams, D. A., 1995, Ap. Sp. Sci. in press

Hecht, J. H., 1986, ApJ. 305 817

Heger, M. L., 1992, Lick Obs. Bull. 10 146

Herbig, G. H., 1967, IAU Symp. 31 85

Herbig, G. H., 1975, ApJ. 196 129

Herbig, G. H., 1993, ApJ. 407 142

Herbig, G. H., 1995, Ann. Rev. Astron. Ap. in press

Herbig, G. H. and Leka, K. D., 1991, ApJ 382 193

Hibbins, R. E., Miles, J. R., Kerr, T. H., Sarre, P. J., Fossey, S. J. and Somerville, w. B., 1994, AAO Newsletter 71 2

Hinnen, P.C., & Urbachs, W., 1995, Chem. Phys. Lett., 240, 351

Jacox, M. E., 1988, J.Phys.Chem. Ref. Data 17 269

Jenniskens, P. and Dsert, F.-X., 1993, A. 274 465

Joblin, C., Maillard, J. P., d'Hendecourt, L. and Lger, A., 1990, Nature 346 729

Johnson, F. M., 1994 in "The Diffuse Interstellar Bands: Contributed Papers" ed A. G. G. M. Tielens, (NASA Conference Publication 10144) p47

Jones, A. P., Duley, W. W. and Williams, D. A., 1990, QJRAS 31 567

Justtanont, K., Barlow, M. J., Skinner, C. J. Roche, P. F., Aitken, D. K. and Smith C. H. 1995 AA in press

Krtschmer, W., Sorg, N. and Huffman, D. R., 1985, Surf. Sci. 156 814

Krtschmer, W., 1993, J. Chem. Soc. Faraday Trans. 89 2285

Leach, S., 1993, J. Chem. Soc. Faraday Trans. 89 2305

Lger, A., Verstraete, L., d'Hendecourt, L., Defourneau, D., Dutuit, O., Schmidt, W. and Lauer, J. C., 1989 in "Interstellar Dust" eds L. J. Allamandola and A. G. G. M. Tielens (Kluwer: Dordrecht) p173

Lger, A. and Puget, J. L., 1984, AA 137 L5

Martin, P. G., 1995, in "The Diffuse Interstellar Bands" eds A. G. G. M Tielens and T. P. Snow (Kluwer: Dordrecht) in press

Martin, P. G., Sommerville, W. B., McNally, D., Whittet, D. C. B., Allen, R. G., Walsh, J. R. and Wolff, M. J., 1994, in "The Diffuse Interstellar Bands" eds A. G. G. M Tielens and T. P. Snow (Kluwer: Dordrecht) in press

Mathis, J., 1989 in "Interstellar Dust" eds L. J. Allamandola and A. G. G. M. Tielens (Kluwer: Dordrecht) p357

Merrill, P. W., 1934, Publ. Astron. Soc. Pacific 46 206,

Merrill, P. W., 1936 ApJ 83 126

Miles, J. R. and Sarre, P. J., 1993, J.Chem. Soc. Faraday Trans. 89 2269

Perrin, J. M. and Sivan J. P., 1992, AA 255 271

Perrin, J. M. and Sivan, J. P., 1990, AA 228 238

Rawlings, J. M. C. and Williams, D. A., 1989, MNRAS 240 729

Russell, R. W., Soifer, B. T. and Willner, S. P., 1978, ApJ 220 568

Salama, F. and Allamandola, L. J., 1992a, ApJ. 395 301

Salama, F. and Allamandola, L. J., 1992b, Nature 358 42

Sarre, P. J., Miles, J. R. and Scarrott, S. M., 1995, Science in press

Scarrott, S. M., Watkin, S., Miles, J. R. and Sarre, P. J., 1992, MNRAS 255, 11P

Schmidt, G. D., Cohen, M. and Margon, 1980, ApJ. Lett. 239 L133

Seab, C. G. and Snow, T. P., 1984, ApJ 277 200

Shapiro, P. R. and Holcomb, K. A., 1986a, ApJ 305 433

Shapiro, P. R. and Holcomb, K. A., 1986a, ApJ 310 872

Sitko, M. L., 1983, ApJ 265 848

Snow, T. P., 1992, ApJ 401 775

Somerville, W. B., 1995 in "The Diffuse Interstellar Bands" eds A. G. G. M Tielens and T. P. Snow (Kluwer: Dordrecht) in press

Sorokin, P. P. and Glownia, J. H., 1995, preprint

Sorrell, W. H., 1990, MNRAS 243 570

Stecher, T. P., 1965, ApJ. 142 1683

Stecher, T. P. and Donn, B., 1965, ApJ. 142 1681

Stecher, T. P. and Williams D. A., 1967, ApJ Lett 149 L29

Thaddeus, P., 1994, in "Molecules and Grains in Space" ed I. Nenner (AIP Press: New York) p711

Tielens, A. G. G. M., 1993 in "Dust and Chemistry in Astronomy", eds T. J. Millar and D. A. Williams (IOP: Bristol) p103

Tokunaga, A. T., Sellgren, K., Smith, R. G., Nagata, T. and Sagata, A., 1991, ApJ 380 452

van Winkel, H., Waelkens, C. and Waters, L B. F. M., 1995, AA 392 L25

Warren-Smith, R. F., Scarrott, S. M. and Murdin, P., 1981, Nature 292 317

Watanabe, I., Hasegawa, S. and Kurata, Y., 1982, Jap. J. Appl. Phys. 21 856

Webster, A., 1992, MNRAS 255 41P

Webster, A., 1993c, MNRAS 264 121

Webster, A., 1993b, MNRAS 263 385

Webster, A., 1993a, MNRAS 262 831

Whittet, D. C. B., 1992, in "Dust in the Galactic Environment" (IOP: Bristol) Williams, D. A., 1994, Contemporary Physics 35 269

Wilson, R., 1958, ApJ. 128 57

Witt, A. N., Bohlin, R. C. and Stecher, T. P., 1984, ApJ. 279 698

Witt, A. N. and Schild, R. E., 1986, ApJ. Suppl. 62 839

Wu C. G., 1972, ApJ. 178 681

Wu, C-C., York, D. G. and Snow, T. P., AJ 86 755

MOLECULAR DIAGNOSTICS OF THE INTERSTELLAR MEDIUM AND STAR FORMING REGIONS

T.W. HARTQUIST
Max-Planck-Institut für extraterristrische Physik
D-85740 Garching, Germany

and

A. DALGARNO
Harvard-Smithsonian Center for Astrophysics
60 Garden Street, Cambridge, Massachusetts 02138 USA

Abstract. Selected examples of the use of observationally inferred molecular level populations and chemical compositions in the diagnosis of interstellar sources and processes important in them (and in other diffuse astrophysical sources) are given. The sources considered include the interclump medium of a giant molecular cloud, dark cores which are the progenitors of star formation, material responding to recent star formation and which may form further stars, and stellar ejecta (including those of supernovae) about to merge with the interstellar medium. The measurement of the microwave background, mixing of material between different nuclear burning zones in evolved stars and turbulent boundary layers (which are present in and influence the structures and evolution of all diffuse astrophysical sources) are treated.

1. Introduction

The development and application of methods for using measured ionic compositions and level populations in laboratory and astrophysical plasmas as diagnostic probes of the physical environment are amongst Sir Robert Wilson's major achievements. We present descriptions of how observationally inferred molecular composition and level populations provide information on the nature of a variety of interstellar sources.

Molecular diagnostic techniques are relevant for the study of many other types of astronomical objects ranging from stellar envelopes to the circumnuclear regions of active galaxies. The abundances and level populations of molecular species play important roles in establishing the physical properties and evolutionary tracks of a wide range of astronomical sources (e.g., Hartquist and Williams, 1995). For instance, the fractional ionization in a collapsing dark molecular cloud is established by chemical processes and, in turn, determines the degree to which magnetic forces affect the collapse; the rate at which radiative losses reduce the thermal pressure in molecular gas shocked by the wind of a young stellar object depends on the composition and the level populations. Because molecular astrophysics is such a rich field we are unable to address here the entire breadth of diagnostic applications that have been made in it or to treat molecular mechanisms that regulate

Astrophysics and Space Science **237**: 267–298, 1996.
© 1996 *Kluwer Academic Publishers. Printed in Belgium.*

the time variability of the physical properties of sources.

Indeed, even though we are restricting consideration to diagnostic functions of molecular composition and level populations in the interstellar medium, we touch on only a limited number of the relevant topics. Our presentation is divided into three main parts. In Sections 2 and 3, we give introductions to the basics of diffuse and dark interstellar cloud chemistry. Sections 4 though 10 are descriptions of selected diagnostic studies of interstellar sources based on determinations of molecular level populations. Sections 11 to 18 comprise the third part which contains examples of the diagnosis of sources from their molecular compositions. Section 19 concludes the paper.

2. The Basic Chemistry of Dark Cool Interstellar Clouds

We define a dark region to be one in which a sufficient fraction of the interstellar background ultraviolet radiation is excluded that it plays little role in the chemistry. We return later to the issue of the necessary conditions for darkness to obtain. Numerous reviews (e.g., Dalgarno, 1987; Millar, 1990) of the basic mechanisms important in dark cloud chemistry exist; at the start of this section we follow the presentation of Hartquist et al. (1993).

We assume dark cool regions to have typically particle number densities in the range of 10^3 cm^{-3} to 10^8 cm^{-3} and temperatures of 10K to 30K. Almost all hydrogen is contained in H_2 and is ionized by cosmic rays at a rate ζ of the order of 10^{-17} s^{-1}; about ninety-seven percent of the ionizations produce H_2^+. The atomic ion He$^+$ is formed by cosmic-ray ionization of He at a rate that is about an order of magnitude smaller. Trace elements are also ionized by direct interactions with cosmic rays, but these direct ionizations are usually chemically insignificant because the elemental abundances of oxygen, carbon, nitrogen, sulfur, and other trace species amount to a fraction of less than 10^{-3} of hydrogen.

The dominant removal mechanism for any species is a reaction with H_2, because of its great abundance, unless the reaction is very slow with a rate coefficient well below 10^{-12} cm^3 s^{-1}. Thus, H_2^+ rapidly forms H_3^+ via the ion-molecule reaction

$$H_2^+ + H_2 \rightarrow H_3^+ + H. \tag{1}$$

Proton transfer from H_3^+ initiates the formation of other species. H_2O and OH are produced by proton transfer to form OH$^+$,

$$O + H_3^+ \rightarrow OH^+ + H_2, \tag{2}$$

followed by the abstraction sequence

$$OH^+ + H_2 \rightarrow H_2O^+ + H \tag{3}$$

$$H_2O^+ + H_2 \rightarrow H_3O^+ + H \tag{4}$$

Reactions of the types represented above and many other ion-molecule reactions often proceed with rate coefficients at or near the Langevin value of order 10^{-9} cm^{-3} s^{-1}. The saturated structure of H_3O^+ prevents it from reacting with H_2. It is removed by dissociative recombination with electrons

$$H_3O^+ + e \rightarrow H_2O + H \tag{5}$$

$$H_3O^+ + e \rightarrow OH + H_2 \tag{6}$$

Most dissociative recombination rate coefficients are of the order of 10^{-7} to 10^{-6} cm^3 s^{-1} in cold clouds.

The chemistry of carbon, like that of oxygen, is initiated by a proton transfer reaction

$$C + H_3^+ \rightarrow CH^+ + H_2 \tag{7}$$

Hydrogen abstraction reactions form CH_3^+ which reacts very slowly with H_2. CH and C are produced by the dissociative recombination of CH_3^+ with electrons.

The products of these simple proton transfer, hydrogen abstraction and dissociative recombination sequences react amongst themselves and generate species containing more than one atom heavier than He. For instance, HCO^+ is formed by the reaction

$$CH_3^+ + O \rightarrow HCO^+ + H_2 \tag{8}$$

¿From its dissociative recombination CO is produced,

$$HCO^+ + e \rightarrow CO + H. \tag{9}$$

After H_2, CO is the most abundant molecular in dark clouds because many reactions in which it is a potential reactant are endothermic. CO is destroyed in reactions with He^+,

$$CO + He^+ \rightarrow C^+ + O + He. \tag{10}$$

It is dissociated by photons emitted following the excitation of H_2 by collisions with cosmic rays (Prasad and Tarafdar, 1983; Gredel et al., 1989)

$$CO + h\nu \rightarrow C + O \tag{11}$$

CO is reformed in a number of ways including the reaction

$$C^+ + H_2O \rightarrow HCO^+ + H \tag{12}$$

followed by (9).

C^+ also reacts slowly with H_2 in a radiative association reaction

$$C^+ + H_2 \rightarrow CH_2^+ + h\nu \tag{13}$$

which has a rate coefficient of about 10^{-15} cm^3 s^{-1}. CH_2^+ reacts with H_2 to form CH_3^+. The radiative association of CH_3^+ with H_2, H_2O and other neutral species is important for the generation of more complex species but we do not consider them here.

The chemistry of some elements deviates from the proton transfer, hydrogen abstraction, and dissociative recombination initiated scheme. The proton transfer reaction

$$S + H_3^+ \rightarrow SH^+ + H_2 \tag{14}$$

proceeds, but the hydrogen abstraction reactions in the sequence that would lead to SH_3^+ are endothermic and SH^+ dissociatively recombines with electrons to re-form S. In low temperature dark interstellar regions the production of sulfur bearing molecules is dominated by neutral-neutral reactions (Oppenheimer and Dalgarno, 1974; Millar and Herbst, 1990). O_2 is abundant in dark cold regions and is produced by the neutral-neutral reaction

$$O + OH \rightarrow O_2 + H \tag{15}$$

Rapid neutral-neutral reactions proceed with rate coefficients of the order of 10^{-10} cm^3 s^{-1}. While the reaction of S with CO is slow, its reaction with O_2 has a rate coefficient of the order of 10^{-12} cm^3 s^{-1} (Baulch et al., 1982).

Through reactions initiated by

$$S + O_2 \rightarrow SO + O \tag{16}$$

much of the sulfur is contained in SO and SO_2 with a significant component in CS formed by the reaction

$$C + SO \rightarrow CS + O \tag{17}$$

The proton transfer reaction

$$H_3^+ + CS \rightarrow HCS^+ + H_2 \tag{18}$$

creates HCS^+ which dissociatively recombines with electrons to give CS again.

The chemistry of nitrogen also differs from that of oxygen and carbon. The proton transfer reaction

$$N + H_3^+ \rightarrow NH^+ + H_2 \tag{19}$$

is slow at low temperatures. The nitrogen chemistry is consequently initiated by a neutral-neutral reaction (e.g. Hartquist and Dalgarno, 1980).

$$N + CH \rightarrow CN + H \tag{20}$$

CN reacts with a number of species to produce nitrogen-bearing molecules. The reactions

$$N + CN \rightarrow N_2 + C \tag{21}$$

and

$$H_3^+ + CN \rightarrow HCN^+ + H_2 \tag{22}$$

$$HCN^+ + H_2 \rightarrow H_2CN^+ + H \tag{23}$$

$$H_2CN^+ + e \rightarrow HCN + H \tag{24}$$

$$H_2CN^+ + e \rightarrow CN + H_2 \tag{25}$$

are important for determining the abundances of CN, N_2, and HCN. Though interstellar N_2 is undetected, the proton transfer reaction

$$N_2 + H_3^+ \rightarrow N_2H^+ + H_2 \tag{26}$$

produces N_2H^+ and N_2H^+ has been observed in interstellar clouds. NH_3 is produced following the destruction of N_2 by He^+

$$N_2 + He^+ \rightarrow N^+ + N + He \tag{27}$$

The hydrogen abstraction reaction

$$N^+ + H_2 \rightarrow NH^+ + H \tag{28}$$

is endothermic by 20 meV but the energy of N^+ upon formation by reaction (27) is sufficient to overcome the endothermicity (Adams, Smith and Millar, 1984). A sequence of hydrogen abstraction reactions ending with

$$NH_3^+ + H_2 \rightarrow NH_4^+ + H \tag{29}$$

leads to the formation of NH_4^+ which dissociatively recombines to form NH and NH_2 and the observationally important species NH_3,

$$NH_4^+ + e \rightarrow NH_3 + H. \tag{30}$$

It is apparent from the study of the above schemes that reactions of H_3^+ with the stable molecules CO, N_2, CS, H_2O, and NH_3 are major sources of the interstellar molecular ions HCO^+, N_2H^+, HCS^+, CH_3^+, and NH_4^+. Often the dissociative recombination reactions of HCO^+, N_2H^+, and HCS^+ simply reproduce the neutral species that reacted with H_3^+. The dissociative recombination of each of H_3O^+, NH_4^+, and CH_3^+ ions with electrons leads to several product species. Thus the branching ratios for dissociative recombination into the different channels are important parameters in interstellar

chemistry. Some limited experimental information is available on the branching ratios (Adams et al., 1991) and new experiments with storage rings are in
progress (Zajfman et al., 1995). Additional mechanisms removing the least
reactive neutral species include reactions with He^+ such as reaction (10) and
dissociation by the absorption of ultraviolet photons emitted due to cosmic
ray interactions with H_2 (e.g. reaction (11)).

Neutral species are removed from the gas phase in collisons with dust
grains on a timescale of the order of 10^{13} s$(n_H / 10^4$ cm$^{-3})^{-1}$, where n_H
is the number density of hydrogen nuclei, if the sticking coefficient with a
dust grain is unity. Chemistry on the surface of dust grains in a hydrogen-
rich environment tends to produce CH_4 from C, CH, CH_2, and CH_3; similar hydrogen saturation occurs following the collision of other unsaturated
species with grains. Various mechanisms, including cosmic ray induced heating of grains (Lèger, Jura and Omont, 1985; Hartquist and Williams, 1990),
photodesorption by infrared photons (Williams, Hartquist and Whittet,
1992) and spot heating by the formation of H_2 on grain surfaces (Duley and
Williams, 1993) have been considered. The gas phase chemical consequences
of depletion onto dust grains without subsequent desorption have been investigated (e.g. Rawlings et al., 1992) as have the effects on the gas phase chemistry of depletion onto dust grains, surface chemical processing and desorption back to the gas phase (Willacy and Williams, 1993; Nejad, Hartquist
and Williams, 1994; Shalabiea and Greenberg, 1994; Bergin, Langer and
Goldsmith, 1995).

Though close to a hundred molecular species in the interstellar medium have been detected (e.g. Irvine, Goldsmith and Hjalmarson, 1987) the
abundances of a much more limited number of particularly simple species
whose formation and removal are best understood constitute the most secure
compositional diagnostics. Table 1 lists the measured fractional abundances
(Irvine et al., 1987) relative to H_2, toward the dark source TMC-1 of some
of the species possessing diagnostic utility. The uncertainty in each of these
fractional abundances is roughly an order of magnitude. Chemical models
(e.g. Nejad et al., 1994; Shalabiea and Greenberg, 1994; Bergin et al., 1995;
and references in those works) compatible with these measured abundances
have been constructed. Source to source variations of the molecular composition occur and require different regions to possess different features. For
instance, the high gas phase H_2O and CH_3OH abundances in some sources
(e.g. the Orion hot core in which the grain temperatures are close to 100K)
have been attributed to the melting of grain mantles following the onset of
dust heating by infrared radiation in the chemically recent past (e.g. Charnley et al., 1992).

TABLE I

Measured TMC-1 Fractional Abundances

Species	Measured
CO	8×10^{-5} [a]
CH	2×10^{-8}
OH	3×10^{-7}
C_2H	$5\text{-}10 \times 10^{-8}$
H_2CO	2×10^{-8}
HCN	2×10^{-8}
HNC	2×10^{-8}
NH_3	2×10^{-8}
CN	3×10^{-8}
HCO^+	8×10^{-9}
N_2H^+	5×10^{-10}
CS	1×10^{-8}
HCS^+	6×10^{-10}
SO	5×10^{-9}

[a] The value of the CO fractional abundance is assumed to be comparable to those inferred for it towards other sources. The value of each of the other fractional abundances was obtained through the multiplication of the assumed CO fractional abundance by an observationally derived value of the ratio of the abundance of the appropriate species to the abundance of CO.

3. The Basic Chemistry of Diffuse Interstellar Clouds

A diffuse interstellar cloud is one in which photoabsorption of the background interstellar ultraviolet radiation field is an important dissociating and ionizing process. Typical cold diffuse molecular regions have number densities of 10^2 to 10^3 cm^{-3} and temperatures of 20K to 100K. Even in diffuse clouds cosmic ray induced ionization is chemically significant. van Dishoeck (1990) has reviewed low temperature diffuse cloud chemistry. In a diffuse cloud with a visual extinction due to dust of less than one magnitude, corresponding to the attenuation of the optical intensity by a factor of $10^{0.4}$, the photodissociation of CO prevents it from containing most of the gas phase carbon. The neutral atomic carbon produced by the photodissociation of CO is photoionized to produce C^+

$$C + h\nu \rightarrow C^+ + e \qquad (31)$$

and reaction (31) is a significant source of gas phase ions.

Usually, whenever C^+ is the dominant carbon-bearing species, photodissociation of CO, OH, H_2O, and O_2 result in O being the most abundant oxygen-bearing species. The ionization potential of atomic oxygen is about

2 meV greater than that of atomic hydrogen and the absorption of electromagnetic radiation in HII regions excludes the photoionization of O as a source of ions in clouds where the photoionization of H is a negligible generator of H^+.

H^+ is formed in diffuse molecular clouds by the cosmic ray induced ionization of H (important where the ratio of the abundance of H to the abundance of H_2 is not much less than unity). H^+ and H_2^+ are formed by the cosmic ray induced ionization of H_2, as in dark clouds. The photodissociation of H_2^+ is a source of H^+. Reaction (1) is a source of H_3^+. Because H_3^+ is removed in diffuse clouds primarily by rapid dissociative recombination and H^+ recombines radiatively with electrons with a rate coefficient of the order of 10^{-12} cm^3 s^{-1}, the H^+ abundance ratio is sufficiently high that the role of H^+ in initiating chemical sequences, as well as that of H_3^+, must be considered.

In diffuse clouds the most important removal mechanism for H^+ is the charge transfer reaction

$$H^+ + O \rightarrow O^+ + H. \tag{32}$$

It is followed by

$$O^+ + H_2 \rightarrow OH^+ + H \tag{33}$$

and then by reactions (3) through (6) and by

$$H_2O + h\nu \rightarrow OH + H \tag{34}$$

to produce OH. The sequence (2) through (6) and (34) is also significant for OH production. In diffuse clouds, OH is removed by

$$OH + h\nu \rightarrow OH + H \tag{35}$$

and

$$C^+ + OH \rightarrow CO^+ + H \tag{36}$$

Reaction (36) is followed by

$$CO^+ + H_2 \rightarrow HCO^+ + H \tag{37}$$

and reaction (9) to produce CO. CO is removed primarily by dissociation following the absorption of interstellar background ultraviolet radiation.

CH in diffuse interstellar clouds is formed as a consequence of reaction (13) which is followed by a hydrogen abstraction reaction to form CH_3^+ which then dissociatively recombines with electrons to produce C, CH, and CH_2,

$$CH_3^+ + e \rightarrow C + H_2 + H \tag{38}$$

$$CH_3^+ + e \rightarrow CH + H_2 \tag{39}$$

$$CH_3^+ + e \rightarrow CH_2 + H \tag{40}$$

In diffuse clouds, any CH_2 formed is photodissociated to form CH, which is in turn photodissociated to produce neutral carbon and photoionized to produce CH^+. However, no chemical scheme has been proposed that can explain the observed abundance of CH^+ by reactions taking place in a quiescent cold molecular cloud.

Reactions of C^+ with CH and CH_2 initiate a sequence which gives rise to the production of C_2 and C_2H

$$C^+ + CH \rightarrow C_2^+ + H \tag{41}$$

$$C^+ + CH_2 \rightarrow C_2H^+ + H \tag{42}$$

$$C_2^+ + H_2 \rightarrow C_2H^+ + H \tag{43}$$

$$C_2H^+ + H_2 \rightarrow C_2H_2^+ + H \tag{44}$$

$$C_2H_2^+ + e \rightarrow C_2 + H_2 \tag{45}$$

$$C_2H_2^+ + e \rightarrow C_2H + H \tag{46}$$

$$C_2H + h\nu \rightarrow C_2 + H. \tag{47}$$

C_2 is removed primarily by photodissociation.

Table 2 gives measured column densities of some of the diagnostically useful diatomic molecular species in a diffuse cloud along the line of the sight to the star ζ Ophiuchi (e.g. van Dishoeck, 1990). The total interstellar hydrogen nucleon column density along the line of sight to the star is 1.4 x 10^{21} cm^{-2}. (We consider the formation of HD, one of the molecules included in Table 2 in section 12.) Order of magnitude agreement between model and observational results for the species other than CH^+ listed in Table 2 have been achieved (e.g. van Dishoeck, 1990; Wagenblast, 1992).

We have divided clouds into two classes distinguished by whether the interstellar background ultraviolet radiation plays a significant role in the ionization and dissociation of chemical species. A typical rate for the photoionization or photodissociation of a species exposed to the unattenuated average background lies in the range of 10^{-11} s^{-1} to 10^{-10} s^{-1} (e.g. Roberge et al. 1991). A typical rate decreases by roughly an order of magnitude with each magnitude of extinction. The cosmic ray induced ionization rate per nucleon heavier than helium is on the order of 10^{-14} s^{-1}; this is the greatest value that the characteristic removal rate of any species containing an element heavier than helium is likely to be. It follows that a region is dark if at least about 4 magnitudes of extinction occurs on all lines of sight from that region to the boundary of the cloud with the more tenuous interstellar medium.

TABLE II

Measured Column Densities towards
ζ Oph

Species	Column Density (cm^{-2})
CH	$2.5 \pm 0.3 \times 10^{13}$
C$_2$	$2.5 \pm 0.2 \times 10^{13}$
CH$^+$	$2.9 \pm 0.3 \times 10^{13}$
CH	$4.8 \pm 0.5 \times 10^{13}$
HD	$2.1 \pm 1.0 \times 10^{14}$
CO	$2.0 \pm 0.3 \times 10^{15}$
CN	$2.9 \pm 0.3 \times 10^{12}$

4. H$_2$ Level Populations on Diagnostics

After having provided a basic introduction to the chemistries that form the simplest molecules in dark and diffuse interstellar regions, we give examples of the utility of level populations of interstellar molecules as diagnostics. We start with the H$_2$ level populations, which may be strongly affected by the processes of formation and removal of H$_2$.

The column densities of the v=0, J=0 through J=6 levels of the lowest electronic state of H$_2$ have been measured in ultraviolet absorption towards a number of bright stars and have been used to probe the physical conditions in the diffuse clouds in these directions (e.g. van Dishoeck and Black, 1986; Wagenblast, 1992). All of these levels may be populated following the formation of H$_2$ on grain surfaces at a rate per unit volume of roughly 3×10^{-17} cm^{-3}s^{-1} $(n_H/1$ cm$^{-3}) (n(H)/1$ cm$^{-3})$, where n_H is the hydrogen nuclear number density and $n(X)$ the number density of the atomic, ionic or molecular species X. The details of the H$_2$ formation on grain surfaces are uncertain, but presumably some fraction of the H$_2$ leaves the grain surfaces in each of these rotational levels. The more highly excited rovibrational levels may be depopulated by radiative cascade routes passing through those levels whose populations have been measured. The radiative decay of each of the J=2 through 6 levels affects its population as well as that of the lower J levels to which the decay takes place. The temperature structures of diffuse clouds can be probed because inelastic collisions involving H$_2$ dominate the J=0 through 2 populations and affect the J=3 population. The absorption of an ultraviolet photon from the interstellar background reduces population in the initial level and results in the population of a rovibrational level in an excited electronic state; about ten percent of the ultraviolet radiative decays from the excited electronic intermediate states are to the rovibrational continuum of the ground electronic state, leading to the dissociation of H_2 at a rate of about 5×10^{-11} s^{-1} in the unshielded standard interstellar background ultraviolet radiation field. The dissociation rate decreases with increasing H_2 column density since the dissociation process involves absorption in lines resulting in H_2 screening itself against dissociating photons. In the other ninety percent of the ultraviolet radiative decays, the excited electronic state connects to a discrete rovibrational level in the ground electronic state; radiative cascade routes through the rovibrational levels then populate v=0, J=0 through 6 levels.

Steady state equilibrium models in which the ultraviolet pumping-subsequent radiative cascade process plays an important role in governing the J=0 through 6 level populations have been developed for diffuse molecular clouds (e.g. van Dishoeck and Black, 1986). The timescale for H_2 formation is of the order of 10^7 years $(n_H/100 \text{ cm}^{-3})^{-1}$ and it is likely that many diffuse molecular clouds do not have steady state H_2 abundances. Wagenblast and Hartquist (1989) have constructed a series of nonequilibrium models for clouds of different ages and densities to show that the H_2 level population data provide insufficient information for the inference of the exact age and number density for a cloud. Wagenblast (1992) has argued that observed J > 4 level population ratios deviate sufficiently from those expected if the ultraviolet pumping-subsequent radiative cascade process dominates their population that the levels are more likely populated by radiative cascade following the formation on grain surfaces of H_2 in particular excited rovibrational levels. Despite the existing uncertainties, the use of population of the J=0 through 6 to diagnose clouds gives estimates of diffuse cloud density and incident ultraviolet radiation field that are each probably reliable to roughly a factor of three or possibly better.

Recently the detection by ultraviolet absorption of vibrationally excited H_2 in the directon of ζ Ophiuchi has been achieved with the Hubble Space Telescope (Federman et al., 1995). The measured population indicates a radiation field intensity of a factor between 1 and 2 times that of the average interstellar radiation field, lower than some earlier estimates (Black and Dalgarno, 1976; van Dishoeck and Black, 1986) but higher than that of Wagenblast (1992).

Shocks have been proposed to be prevalent in diffuse interstellar clouds in order to account for the formation of CH$^+$ (Elitzur and Watson, 1978) by the reaction

$$C^+ + H_2 \rightarrow CH^+ + H \tag{48}$$

The reaction is endothermic by about 0.4 eV and is slow in low temperature cloud material but rapid enough in nondissociative shocks with speeds greater than about 7 km s^{-1} to be an important source of CH$^+$. The temperatures of more than 700 K in shocks that would lead to sufficient CH$^+$

production result in the collisional excitation of the $3 \leq J \leq 6$ H_2 levels. The populations of these levels provide strong constraints on the models for shock production of CH^+ (e.g. Hartquist, Flower and Pineau des Forêts, 1990). Duley et al. (1992) and Hartquist, Dyson and Williams (1992) have explored the alternative possibility that CH^+ forms in turbulent heated boundary layers at the interfaces between molecular clouds and flowing intercloud gas. Pressure differences exist along the surface of a cloud due to the flow around it and result in the evaporation of a cloud into the region of lowest pressure bringing up H_2 to the cloud surface more rapidly than it is photodissociated. H_2 at the neutral cloud surface reacts with C^+ to form CH^+. Collisional excitation of H_2 occurs in the warm region. The H_2 $J=3$ through $J=6$ populations constrain this model but the constraints are not as severe in the interface models as in the shock models, in part because the warm interface regions are at a pressure comparable to the molecular cloud, whereas the shocked gas is at a higher density and much higher temperature than the ambient molecular cloud. Because of the lower pressure in the interface, radiative decay is more effective in maintaining low $J \geq 3$ populations than is the case in the denser gas of comparable temperature in a shock.

Measured H_2 level populations are used to diagnose regions near (≤ 1 pc) bright young stars which are often at much higher densities ($n_H \geq 10^4$ cm^3) than the diffuse clouds. The level populations are determined from observations of infrared emission features at wavelengths between about 2 and 4 microns (e.g. Geballe, 1990). The emission features are associated with transitions between rovibrational levels with vibrational quantum numbers differing by one and between very high ($J > 15$) rotational sublevels in the $v=0$ level. The levels on which the observed transitions originated can be populated by all of the mechanisms mentioned above in the discussion of the $v=0$, $J=0-6$ levels. Radiative cascades following the formation of H_2 in high rovibrational levels and the absorption of ultraviolet radiation and collisions in hot gas all populate the levels involved in the infrared transitions.

Detailed models have been constructed of regions in steady state equilibrium and in which the infrared emitting levels of H_2 are populated due to ultraviolet pumping followed by radiative cascades (Black and Dalgarno, 1976; Tielens and Hollenbach, 1985; Black and van Dishoeck, 1987; Sternberg and Dalgarno, 1989). They are important for understanding the way in which the formation of a star causes a physical response (whch may either induce or inhibit further star formation) in the nearby cloud material.

Much attention has been given to the collisionally excited H_2 line emission from the Orion Becklin-Neugebauer, Kleinmann-Low star forming region (e.g. Brand,1995). Its ultimate cause is a wind from a young stellar object which is impacting a region in which the birth of massive stars is occurring. Various shock models have been proposed to account for the H_2 line emission, but the explanation of the observed broad line profiles, which have widths corresponding to speeds above 100 km s^{-1} much higher than that which would be associated with a nondissociating shock propagating into the ambient material, the line strength ratios, and the positional and velocity invariance of those line ratios in terms of shocks have proven challenging. Malone, Hartquist and Dyson (1994) and Dyson et al. (1995) have suggested that the broad widths of the lines may arise due to the gradual acceleration and entrainment of molecular material by the wind at wind-clump boundary layers. Recently available multiline and high velocity, high spatial resolution H$_2$ infrared data for an individual clump (Tedds and Brand, 1995) in the region should be of considerable utility in diagnosing the response of the region to star formation.

5. CN Level Populations and the Microwave Background

Along with CH and CH$^+$, CN was one of the first molecules detected in the interstellar medium. Optical absorption spectra obtained towards nearby bright stars yielded the ratio of the populations of different rotational levels in the lowest vibrational level of the ground electronic state. From the ratio an excitation temperature of about 3° K was inferred and attributed to the levels being in radiative equilibrium with a background microwave radiation field of originally unknown origin. The use of the CN rotational level populations to probe the cosmic microwave background has continued (Roth, Meyer and Hawkins, 1993).

6. NH$_3$ Level Populations and "Dark Cores"

NH$_3$ is a symmetric top molecule. Two rotational quantum numbers J and K are defined such that $J(J+1)h^2/4\pi^2$ and $Kh/2\pi$ are the square of the total angular momentum and the projection of the total angular momentum onto the molecule's symmetry axis. The probability distribution for the location of the nitrogen atom peaks at positions on both sides of the plane containing the hydrogen atoms; the wavefunction must be even or odd with respect to inversions through that plane, and the wavefunction associated with a state has either two maxima of the same magnitude and located on the opposite sides of the plane, or a maximum and a minimum of the same magnitude located on opposite sides of the planes. The energies of the even and odd levels with the same rotational quantum numbers differ and a radiative transition between them gives rise to the emission or absorption of a photon with a wavelength of about 1.3 cm.

Observations of inversion transitions in the (1,1), (2,2) and (3,3) rotational levels provide considerable diagnostic scope (Cheung et al., 1969a). The fact that the transitions are observed at all implies that $n_H \gg 10^3$

cm^{-3} in the regions containing the detected NH_3 since at lower densities the inversion level populations within a J,K rotational level would come into equilibrium with the microwave background. The metastable nature of the J=K rotational levels means that the (2,2) to (1,1) excitation temperature is close to the kinetic temperature when the kinetic temperature is sufficiently low ($\lesssim 20°$ K) that collisionally induced (2,2) $\rightarrow$ (2,1) transitions are unimportant. (The (2,1) level will radiatively decay rapidly to the (1,1) level so that the (2,2) to (1,1) excitation temperature is below the kinetic temperature.)

Though we do not describe the diagnostic analysis of NH_3 inversion radiation in detail (e.g. Ho and Townes, 1983), we stress the importance that such analysis has played in the discovery of and the elucidation of the properties of the so-called dark cores (e.g. Myers, 1987), the progenitors of stars. The coldest of these cores have temperatures, number densities and masses of roughly 10K, 10^4 cm^{-3} - 10^5 cm^{-3} and 1 - 10 $M_\odot$ respectively, and are most likely the sites of low mass ($\leq 4M_\odot$) star formation. Most of the warmer ones have temperatures of about 30K, number densities between 10^5 and 10^7 cm^{-3} and masses up to several hundred $M_\odot$; some (e.g. the Orion Becklin-Neugebauer, Kleinmann-Low region hot core) are heated by integrated stellar radiation to about 100K. A major mystery is the absence of NH_3 line wings indicative of collapse (Myers and Benson, 1983); it is addressed by Rawlings (1995) in this volume.

7. HC_3N, CS and HCO^+ Rotational Level Populations and the Densities of Star Forming Cores, Disks and Outflows

Though NH_3 has often been used in the initial studies of dark cores, radio and millimeter wave observations of other species, including but not limited to HC_3N, CS and HCO^+, and the population distributions in their rotational levels have been employed to study the structures of dark cores in more detail.

TMC-1 is the most observed dark core in a region of low mass star formation. It lies at a distance of about 100pc, has angular dimensions of roughly 5' x 15' and possesses several fragments. The emissivity peaks of different species are in some cases displaced with respect to one another (e.g. Hirahara et al., 1992). Schloerb, Snell and Young (1983) mapped the J=5 $\rightarrow$ 4, J=9 $\rightarrow$ 8 and J=12 $\rightarrow$ 11 emissions of HC_3N in TMC-1 and used the data to infer that the number density in TMC-1 is 5-10 x 10^4 cm^{-3} and varies little along the object. In contrast, Hirahara et al. (1992) found evidence for a density gradient directed towards the northwest in an analysis of their $C^{34}S$(J=1 $\rightarrow$ 0) and $C^{34}S$(J=2 $\rightarrow$ 0) data; the number densities of 4-40 x 10^4 cm^{-3} that they derived also ranged higher than that obtained by Schloerb et al. (1983). The differences in the results of Schloerb et al. (1983)

and Hirahara et al. (1992) may arise because the CS and HC_3N fractional abundances peak at different densities in TMC-1.

B335 is another well observed dark core, but whereas TMC-1 has no embedded star, B335 contains a far-IR source. Hasegawa, Rodgers and Hayashi (1991) have mapped the HCO^+ (J=3 → 2), HCO^+ (J=4 → 3) and $H^{13}CO^+$(J=3 → 2) emissons of B335. The J=4→3 emission region is more centrally peaked on the IR source than the J=3→2 emission region; near the J=4→3 peak the line emission strength ratios indicate a number density of greater than 10^6 cm^{-3}, and though a significant part of J=4→3 emission may arise in a disk, Hasegawa et al. (1991) concluded that the dense gas is not responsible for the collimation of the outflow whose presence is apparent by the existence of wings in CO emisson characteristic outflows. The HCO^+ (J=3 → 2) emission possesses such wings as well, but the HCO^+ (J=4 → 3) emission does not. Even so, the existence of the wings in the HCO^+ (J=3 → 2) features implies that the outflowing number density exceeds 10^4 cm^{-3}.

8. Submillimeter Emissions of CS, HCN, HCO^+ and CO and Warm, Dense Gas in Regions of Star Formation

The detection of radiation from progressively higher rotational levels requires that observations be conducted at correspondingly shorter wavelengths. Because the excitation energies of the levels on which the transitions originate are larger, the emissions from higher rotational levels sample warmer gas. In addition, because the radiative lifetimes of higher levels are shorter than those of lower levels, greater densities are required to populate them and the radiation from the higher rotational levels samples denser gas. Submillimeter observations of radiation from highly excited rotational levels permits the diagnosis of the density and temperature of warm dense gas in regions of star formation.

Observations of radiation from lower rotational levels of CS and of the emission from the CS(J=7 → 6) transition have revealed the number density and temperature distributions up to roughly 3x10^6 cm^{-3} and in 30-40 K molecular cloud cores possibly containing several young high mass stars and continuing as sites of active star formation (e.g. Yamashita et al., 1989; Moriarty-Schieven, Snell and Hughes, 1991). For instance, in the NGC 2071 IRS region the density is roughly inversely proportional to the first to second power of the distance from the center of the core.

The HCN(J=9 → 8) and the HCO^+ (J=9 → 8) emissions arise in even denser, warmer regions. The detection of HCN(J=9 → 8) emission from the Orion hot dense core, mentioned in the last paragraph of section 2, led to the conclusion that some gas in it has a number density of order 10^8 cm^{-3} or more. The detection of HCO^+ (J=9 → 8) resulted in the realization that in that source HCO^+ has a much higher fractional abundance at number densities in excess of 10^7 cm^{-3} than expected (Jaffe et al., 1992). The high HCO$^+$ abundance remains unexplained and is of considerable interest because it is closely linked to the ionization balance in the dark region and, thus, to the rate at which ambipolar diffusion allows collapse of the neutral component of a magnetized clump in response to its own gravity.

Observations of CO(J=7 → 6), CO(J=6 → 5) and ^{13}CO(J=6 → 5) have been made to study the distribution of gas with T $\gtrsim$ 100 K and $n_H \gtrsim$ 10^5 cm^{-3} in regions of high mass star formation (Graf et al., 1990; 1993). The emission is generally distributed over a zone of about 1 pc. size and has widths of less than 5 km s^{-1}, implying that shocks are unlikely to be responsible for heating the gas. The measured dust temperatures are usually around 50K and as they are significantly below the gas temperatures, gas-dust collisions can not be responsible for the heating of the gas. Regions heated to 100K by the absorption of stellar ultraviolet radiation are predicted to be much less extended than observed at the high densities associated with the CO(J=6 → 5) emission if the gas is distributed uniformly; however, in a clumpy medium ultraviolet heating could operate in the edges of clumps distant from the central stars (Graf et al., 1990; 1993).

Another heating mechanism operating in a clumpy medium is the dissipation, within a boundary layer, of turbulence driven by shear at the interface of a clump and hotter flowing gas powered by the mechanical energy input of the stars. Because of their high radiation fields, massive stars influence their environments through their radiative luminosities, and regions of high mass star formation are not good laboratories for the study of the response of a medium to the mechanical luminosity of a star. Regions of low mass star formation, because the stars have much weaker ultraviolet radiation fields, are better suited to the diagnosis of the mechanisms governing the response of a medium to mechanical luminosity. TMC-1 lies within roughly half a parsec of a young low mass star. The wind of that star most likely has a substantial influence on the outer boundary of parts of TMC-1. If a substantial turbulent boundary layer exists around the densest part of TMC-1, it may be detectable in CO(J=6 → 5) emission. Given the intractability of theoretical description of boundary layers based on first principles and the huge roles that mass, momentum and energy transfer in them may play in controlling the evolutionary development of diffuse astronomical sources of many types (Hartquist and Dyson, 1993) the detection of submillimeter CO emission in the direction of TMC-1 would be of considerable importance.

9. Radio, Millimeter and Submillimeter Studies of OH and H$_2$O Masers

Maser emissions from a number of molecules including OH and H$_2$O originate in the vicinities ($\leq$ 0.01 parsec) of many young massive stars still surrounded by the remnants of the cores in which they formed. The analysis of radio and submillimeter emissions of the OH and H_2O masers yields information about the number densities near the young stars and the strength of the magnetic fields which dominate the pressures in the cores and, thus, greatly influence core collapse to form stars.

Each rotational level of the ground electronic state of OH is split into two sets of two sublevels. The projection of the total angular momentum onto the internuclear axis differs in sign between the two sets of sublevels. Further degeneracy is removed by hyperfine splitting. Several emission lines at wavelengths of about 18 cm are observable from the transitions between sublevels in the lowest rotational level. Superthermal population distributions in those sublevels are achieved through the collisionally or radiatively induced excitation of higher rotational levels and subsequent cascades to repopulate the sublevels of the lowest rotational level. The superthermal population distributions lead to masing. Zeeman splitting is measured in observations of the masing transitions and yields magnetic field strengths of about 3-10 milligauss (García-Barreto et al., 1988). Higher rotational levels are also split into sublevels, and radio observations of transitions between the sublevels in each of several higher rotational levels of OH provide information on the populations of those rotational levels, thus allowing the diagnosis of the number density, OH fractional abundance, and temperature in an OH maser region. Nonmasing features are formed in more widely distributed gas than the masing features, but in the analysis the properties of the different line forming regions are assumed to be the same. Number densities, temperatures and OH fractional abundances of 1-3×10^7 cm^{-3}, 100-200K and of order 10^{-6} respectively are inferred (Guilloteau, Baudry and Walmsley, 1985; Cesaroni and Walmsley, 1991; Gray, Field and Doel,1992).

H_2O maser emission at a wavelength of about 1.35 cm and arising in the $6_{16} \rightarrow 5_{13}$ (where the notation J_{K^+,K^-}, with J being the quantum number associated with the total angular momentum and K^+ and K^- being associated with its projections onto two molecular axes, is used) transition was detected in 1968 (Cheung et al., 1969b). Pumping due to the collisionally induced excitation of higher rotational levels and subsequent cascades give rise to inversion of the $6_{16} \rightarrow 5_{23}$ level populations (e.g. Neufeld and Melnick, 1991). The use of a collisional pump model and the data for the $6_{16} \rightarrow 5_{23}$ and other observed maser features arising in the $3_{13} \rightarrow 2_{20}$, $10_{29} \rightarrow 9_{36}$ and $5_{15} \rightarrow 4_{22}$ transitions of H_2O at wavelengths of about 1.6 mm, 0.92 mm and 0.93 mm permit diagnosis of the water maser regions (Neufeld and Melnick, 1991). Water masers have number densities and temperatures respectively of 10^8 to 10^{10} cm^{-3} and 400 to 1000 K.

10. Thermal Submillimeter and Infrared Line Emission of H$_2$O

Submillimeter and infrared line emission of H$_2$O is likely to be the dominant cooling mechanism in many high density ($n_H > 10^6$ cm^{-3} regions with temperatures of the order of 10^3 K (e.g. Draine, Roberge and Dalgarno, 1983). Observations of the H$_2$O submillimeter and infrared emission will serve as important diagnostics of dynamics in star forming regions. Currently, such observations have not been made, but orbiting submillimeter and infrared observations (including SWAS and ISO), with which such data will be obtained, are scheduled to become operational in the next several years.

11. Cosmic Rays and Deuterium in Diffuse Clouds

While Sections 4 through 10 were concerned with the diagnosis of regions through the analysis of level population distributions of molecules, sections 11 through 18 contain examples of the diagnosis of regions from their molecular compositions.

Waves in the solar wind interact with cosmic ray protons with energies below about 1 GeV sufficiently to prevent many of them from reaching the Earth (e.g. Jokipii and Davila, 1981). These low energy cosmic ray protons play major roles in driving chemistry (even in diffuse clouds) and in establishing the ionization structure (and, consequently, influencing the effects of magnetic fields on star formation) in dark regions. Thus, more information about them than is obtained in near-Earth in situ measurements is desirable. Furthermore, the determination of the characteristics of their spectra in different physical regions, in principle, gives additional insight into the ways in which they propagate and are accelerated.

Consideration of the discussion around reactions (32) through (35) and in the paragraph preceding them shows that the OH fractional abundance provides a measure of the cosmic ray induced ionization rate (Black and Dalgarno, 1973) if the cloud physical conditions have been inferred from the H$_2$ level populations.

HD is too abundant in diffuse clouds to be formed only at a rate per unit volume equal to the product of the formation rate per unit volume with the abundance ratio of atomic deuterium to atomic hydrogen. Hence, in addition to its formation on grain surfaces like H$_2$, it is formed in diffuse clouds by the sequence

$$H^+ + D \leftrightarrow D^+ + H \tag{49}$$

$$D^+ + H_2 \rightarrow HD + H^+ \tag{50}$$

(Dalgarno, Black and Weisheit, 1973). The H$^+$ formation in molecular clouds is due to cosmic ray induced ionization (see the paragraph preceding that containing reaction (32)). Thus, once OH is used to obtain the cosmic ray induced ionization rate, the observed HD abundance can be used to derive the ratio of the deuterium nuclear number density to the hydrogen nuclear number density, a quantity of well-known cosmological significance. Values of the ratio inferred from the construction of cloud models and consideration of the OH and HD chemistries are compatible with those obtained by other methods for other types of sources (e.g. van Dishoeck and Black, 1986).

Alternatively, one can assume a uniform deuterium to hydrogen ratio, and use HD as well as OH as measures of the cosmic ray induced ionization rate. When this is done the ionization rates in models of various diffuse clouds are generally found to within a factor of a few of 3×10^{-17} s^{-1}.

Hartquist and Morfill (1983) considered the CO, OH and HD data for several lines of sight through the shell of an old supernova remnant in the Per OB2 region. We do not repeat their arguments here, but they concluded that the significantly higher cosmic ray ionization rate inferred for shell material on this side of the star o Per than in material behind it or in shell material in the directions of two other stars in that region is compatible with a reasonable magnetic field geometry and the bulk of the ionizing cosmic rays having energies of only a few MeV. Hartquist and Morfill (1994) showed that the energy to which cosmic rays are accelerated by the second-order Fermi process is in the MeV range if the interstellar turbulence spectrum obeys a Kraichnan power law from the wavelengths of waves resonant with 1 GeV protons down to the wavelengths of waves resonant with MeV cosmic ray protons.

12. Deuterium, the Fractional Ionization and Depletions in Dark Regions

Many deuterated molecules, including DCO$^+$, DCN, N$_2$D$^+$, HDCO, C$_3$HD, C$_4$D, NH$_2$D, CH$_2$DCN and DC$_3$N, and the doubly-deuterated species D$_2$CO (Turner, 1990) have been detected in dense clouds by their emission lines at radio and millimeter wavelengths. Their abundances depend on the galactic [D]/[H] abundance ratio, the ionizing flux, the element depletion and the fractional ionization. The abundances of the deuterated species are enhanced at low temperatures by chemical fractionization processes and the ratio of the deuterated species relative to their hydrogenic counterparts may be considerably larger than [D]/[H]. The enhancements depend on the efficiencies of dissociative recombination and chemical reactions.

The chemistry of interstellar deuterated molecules is complicated, involving a considerable array of gas phase and grain surface reactions (Brown and Rice, 1986; Millar, Bennett and Herbst, 1989; Brown and Millar, 1989 a,b; Turner, 1990). Of the many deuterated species, the chemistry of DCO$^+$ may be the least complex and the ratio of DCO$^+$ to HCO$^+$ may be the most useful as a diagnostic probe.

The hydrogenic counterpart HCO$^+$ is formed mostly by

$$H_3^+ + CO \rightarrow HCO^+ + H_2 \tag{51}$$

with contributions from reactions (8) and (12) and

$$CH_5^+ + CO \rightarrow HCO^+ + CH_4 \tag{52}$$

$$C_2H_2^+ + O \rightarrow HCO^+ + CH \tag{53}$$

$$CH + O \rightarrow HCO^+ + e, \tag{54}$$

depending upon the physical conditions in the interstellar cloud. By substituting D for any one of the hydrogen atoms in these reactions, DCO$^+$ can be formed.

The major source of DCO$^+$ is usually taken to be the reaction

$$H_2D^+ + CO \rightarrow DCO^+ + H_2 \tag{55}$$

because

$$HCO^+ + HD \rightarrow DCO^+ + H_2 \tag{56}$$

is very slow. We ignore other sources of DCO$^+$ and assume that DCO$^+$ is removed by dissociative recombination

$$DCO^+ + e \rightarrow D + CO \tag{57}$$

at the same rate as HCO$^+$ (c.f. reaction(9)). We also assume that both reactions in each pair of reactions such as

$$HCO^+ + H_2O \rightarrow H_3O^+ + CO \tag{58}$$

$$DCO^+ + H_2O \rightarrow H_2DO^+ + CO \tag{59}$$

that remove HCO$^+$ and DCO$^+$ have the same rate coefficient. Then we obtain for the ratio of number densities

$$\frac{n(DCO^+)}{n(HCO^+)} = \frac{1}{3} \frac{n(H_2D^+)}{n(H_3^+)}. \tag{60}$$

Watson (1974, 1976, 1977) pointed out that the exchange reaction

$$H_3^+ + HD \rightarrow H_2D^+ + H_2 \tag{61}$$

is exothermic and could lead to an enhancement in H$_2$D$^+$ at low temperatures that would be reflected in the DCO$^+$/HCO$^+$ ratio. The loss mechanisms for H$_2$D$^+$ are the reverse of reaction (61)

$$H_2D^+ + H_2 \rightarrow H_3^+ + HD, \tag{62}$$

dissociative recombination,

$$H_2D^+ + e \ \rightarrow \ HD + H \tag{63}$$
$$\rightarrow \ H_2 + D \tag{64}$$
$$\rightarrow \ H + H + D \tag{65}$$

and ion-molecule reactions such as (55) and

$$H_2D^+ + O \ \rightarrow \ OD^+ + H_2 \tag{66}$$
$$\rightarrow \ OH^+ + HD. \tag{67}$$

If we equate the rates of formation and destruction of H_2D^+, we obtain the steady state abundance ratio

$$\frac{n(H_2D^+)}{n(H_3^+)} = f \frac{n(HD)}{n(H_2)} \tag{68}$$

where f is the enhancement factor (Watson, 1976)

$$f = \frac{k_{61}}{k_{62} + \bar{k} + \alpha n_e / n(H_2)}. \tag{69}$$

In (69), k_{61} and k_{62} are the rate coefficients of reaction (61) and (62) respectively, $\bar{k}$ is a mean rate coefficient for reactions of the kind (55), (59), and (67), defined by

$$\bar{k}n(H_2) = \sum_x k_x n(x), \tag{70}$$

and α is the sum of the recombination coefficients of (63), (64) and (65).

The ratio of the rate coefficients k_{61} and k_{62} is a known function of temperature (Smith, Adams and Alge, 1982; Herbst, 1982; Sidhu, Miller and Tennyson, 1992). It varies at low temperatures approximately as $\exp(140/T)$ so that in the absence of the mediating dissociative recombination and ion-molecule reactions, extreme fractionation would occur in cold clouds.

The effective ion-molecule rate coefficient $\bar{k}$ can be determined from the individual rate coefficients k_x and it is directly proportional to the oxygen and carbon depletion factors.

In calculations of deuterated molecule abundance, it has been generally assumed that the rate coefficient for the dissociative recombination of H_2D^+ is equal to that for H_3^+ which has itself had a chequered history.

The earlier high values of $\alpha(H_3^+)$ were later attributed to the presence of vibrationally excited ions in the experimental afterglows (Adams, Smith and Alge, 1984, Adams and Smith, 1987, 1988) and they were replaced by values an order of magnitude smaller. Recently the high values have been reinstated by storage ring experiments (Larsson et al., 1993) and by further afterglow measurements (Amano, 1988, 1990; Canosa et al., 1992), though debate persists (Canosa et al., 1992; Smith and Spanel, 1993; Gougosi, Johnsen and Golde, 1995). The storage ring cross sections yield a rate coefficient for H_3^+ of $1.15\text{x}10^{-7}(300/T)^{1/2}$ cm^{3}s^{-1} (Larsson et al., 1993, Sundström et al., 1994; Datz et al., 1995a). For H_2D^+, storage ring measurements lead to a rate coefficient of $6\text{x}10^{-8}(300/T)^{1/2}$ cm^{3}s^{-1} (Larsson et al., 1995; Datz et al., 1995b), a factor of two smaller than $\alpha(H_3^+)$.

It was pointed out by Dalgarno and Lepp (1984) that the Watson fractionation scheme must be extended to incorporate exchange reactions of deuterium atoms such as

$$H_3^+ + D \leftrightarrow H_2D^+ + H \tag{71}$$

$$HCO^+ + D \leftrightarrow DCO^+ + H, \tag{72}$$

because the supply of deuterium atoms is significantly enhanced by the chemical fractionation of DCO$^+$.

Dissociative recombination of H_2D^+, the branching ratios of which have been measured (Larsson et al., 1995), also contributes to the supply of deuterium atoms, the degree depending on the heavy element depletion.

Simple formulas can be constructed that include the deuterium atom contributions (Dalgarno and Lepp, 1984; Opendak, 1993) but more precise conclusions can be derived using comprehensive models of DCO$^+$/HCO$^+$. Using reliable values of the rate coefficients and branching ratios of the dissociative recombination coefficients, Larsson et al. (1995) have reproduced the observational data on the cloud L1529 (Wootten, Loren and Snell, 1982) to infer a value of $1.65\text{x}10^{-5}$ for the cosmic [D]/[H] ratio, a fractional ionization of $8\text{x}10^{-8}$ and a cosmic ray ionization rate ζ obeying $(\zeta/n(H_2)) = 6\text{x}10^{-21}$ cm^{3}s^{-1}, which corresponds to $\zeta = 3\text{x}10^{-17}$ s^{-1} for $n(H_2) = 5\text{x}10^3$ cm^{-3}. The derived [D]/[H] ratio equals that obtained for regions near the Sun from the atomic D/H ratio (Linsky et al., 1993, 1995). Of particular interest is the large ratio $2.3\text{x}10^{-3}$ for the relative abundances of atomic deuterium and atomic hydrogen D/H, which is 150 times [D]/[H].

13. Nonequilibrium Chemistry and the Ages of Sources

Several timescales are associated with the evolution of the chemistry in a dark region where the gas is initially primarily atomic. If the hydrogen is initially atomic, the timescale required to be converted to H_2 is comparable to the timescale for a H atom to strike a grain surface. From a knowledge of the extinction law (see Spitzer, 1978) this timescale is estimated to be $t_H \approx 10^{12}$ s $(n_H/10^5$ cm$^{-3})^{-1}$ in 10K gas.

The chemistry of species containing heavy elements cannot reach equilibrium until sufficient cosmic ray induced ionization has occurred to form at least one H_3^+ ion for each atom heavier than helium. For a cosmic ray induced ionization rate of ζ and a number density of nuclei more massive than helium of $n(z)$, the timescale required for this is roughly $t_i \approx 10^{13}$ s $(\zeta/10^{-17} \text{ s}^{-1})(n(z)/10^{-4}n(H_2))$.

The timescale for heavier neutral species to collide with grains is several times t_H. If, upon collision, the species are not returned to the gas phase, no equilibrium will be reached until the gas phase contains only hydrogen and helium. If species heavier than helium are desorbed from grains, an equilibrium with species heavier than helium can be attained.

In many studies of dark region nonequilibrium chemistry, the hydrogen has been assumed to be initially primarily in molecular form. As time passes, the chemistry goes through several stages. We first consider a model in which collisions of gas phase species with grains are neglected (Millar, 1990). At an evolutionary time comparable to t_i, the chemistry has developed but has not evolved to equilibrium, which takes several times longer. In equilibrium, most gas phase carbon is in CO, but at earlier times much of the gas phase carbon remains in atomic form, facilitating the formation (following the initiation of a reaction sequence by reaction (7)) of much more CH than exists in equilibrium. Similarily much higher abundances of species such as CN, HCN and HC$_3$N exist at some "early times" than in equilibrium.

Chemical variation is significant in a systemic way along TMC-1 (Hirahara et al., 1992), an object which we mentioned in Section 7. For instance, the peaks in NH$_3$ and in HC$_3$N emission are distinct. The NH$_3$ peak may be located in material that collapsed from a diffuse state longer ago than that at which the HC$_3$N did. Modelling of the TMC-1 chemistry on the assumption that collapse proceeded sequentially along its length would be instructive.

Studies have been made of the evolution of dark region gas phase chemistry as depletion onto grains continues. Hartquist and Williams (1989) pointed out that CH is one of the species that actually increases in abundance as depletion of elements heavier than helium continue to the time that these species are depleted by roughly a factor of 100 relative to their solar abundances. Brown and Charnley (1990), Rawlings et al. (1992) and Nejad et al. (1994) have followed chemistry as depletion occurs unchecked. The relevance of these calculations for diagnosis are considered by Rawlings (1996) in this volume.

14. The Interclump Medium of the Rosette Molecular Cloud

Molecules may be present in low ($n_H \approx 10$ cm^{-3}) density material as well as the dark cores which have been addressed in many of the sections. Blitz and Stark (1986) have conducted large scale mapping of the ^{12}CO (J=1 $\rightarrow$ 0) and ^{13}CO(J=1 $\rightarrow$ 0) emissions of the Rosette Molecular Cloud, a so-called giant molecular cloud of 10^5 $M_\odot$ containing a region of high mass star formation. Their survey revealed the presence of about 100 clumps in which $n_H \approx 10$ cm^{-3}, ranging in mass from about 10 $M_\odot$ to several thousand $M_\odot$. It is likely that many such clumps are bound by the pressure of a surrounding medium (Blitz and Stark, 1986; Bertoldi and McKee, 1993), and Blitz and Stark (1986) detected emissions which they attributed to an interclump medium. If the medium is uniform, its number density can be estimated to be about 10 cm^{-3} (Taylor et al., 1996) from the ^{12}CO (J=1 $\rightarrow$ 0) and ^{13}CO (J=1 $\rightarrow$ 0) emissions, and its thermal pressure is probably at least comparable to the thermal pressure of a clump (the magnetic pressure of a clump greatly exceeds its thermal pressure), implying that the interclump medium is at a temperature of the order of 10^3 K. The emissions may originate entirely in denser, inhomogeneous, cooler gas (Schneider et al, 1996) and the firm establishment of the existence of low density molecuar gas at 10^3K may require considerations of its chemistry (Taylor et al., 1996).

At such a temperature reaction (48) is rapid and triggers a sequence that results in the formation of CH. Also at such temperatures the neutral-neutral reactions

$$O + H_2 \rightarrow OH + H \tag{73}$$

$$OH + H_2 \rightarrow H_2O + H \tag{74}$$

become rapid as they are in diffuse cloud shocks (see the next section.) Photodissociation is probably nonnegligible because the interclump medium does not have a sufficiently high extinction to shield itself from interstellar ultraviolet photons. As the precise properties of the interclump medium are unknown, an accurate prediction of the CH and OH column densities in it cannot be made, but it is possible that they approach 10^{16} cm^{-2} which would mean that observations of CH and OH would lead to their detection and allow the diagnosis of the interclump medium in a giant molecular cloud.

The properties of that interclump medium are probably of considerable importance for the propagation of waves of star formation. The birth of stars in one part of a giant molecular cloud results in energy being injected into the interclump medium. A blast wave develops, and clumps engulfed by the blast wave are compressed. The extent of the compression (and whether the compression can trigger star formation by making the clump sufficiently dark that its fractional ionization falls enough to allow ambipolar diffusion to reduce the support of the clump by the magnetic field or to damp turbulence supporting the clump along the field lines) depends on the properties of the blast wave which in turn depend on the nature of the interclump medium. The properties of the interclump medium influence whether star formation in one clump in a giant molecular cloud will trigger star formation only in the nearest-neighbor clump, over the whole cloud, or not at all.

15. Compositional Diagnostics of Shocks in Molecular Gas

In the previous section we mentioned OH and CH as possible diagnostics of ≈ 1000 K, ≈ 10 cm^{-3} gas and in section 4 the suggestion that CH$^+$ observed along lines of sight through diffuse clouds is formed in shocked gas. Shock models of CH$^+$ formation must meet the constraints provided by the observed CH and OH in the directions in which CH$^+$ is detected (e.g. Hartquist et al., 1990; Crane, Lambert and Sheffer, 1995).

Compositional differences between 10K and 10^3 K dark regions have been considered. For example, the sequence

$$S + H_2 \rightarrow SH + H \tag{75}$$

$$SH + H_2 \rightarrow H_2S + H \tag{76}$$

is slow at low temperatures and rapid at high temperatures and is responsible for a number of major differences between the sulfur chemistries of low and high temperature dark regions (Hartquist, Oppenheimer and Dalgarno, 1980). The presence of H$_2$S emission in the Orion Kleinmann-Low, Becklin-Neugebauer region with wings several tens of km s^{-1} wide was interpreted as evidence for the presence of a shock. However, as suggested in section 4, gas may be heated to high temperatures by the dissipation of shear driven turbulence in boundary layers between cold clumps and hotter flowing gas. The detection of the H$_2$S emission indicated the presence of hot gas, but that gas is not necessarily heated by a shock.

Chernoff, Hollenbach and McKee (1982) have suggested that H$_2$ emission line wings in that source (see section 4) originate in an H$_2$ formation region in dusty stellar wind gas that has passed through a starwardly facing termination shock. Dalgarno (1993) has pointed out that the detection of cospatial SO and SO$^+$ emissions would confirm the presence of such a region behind an ionizing shock or a shock in material that is already ionized far upstream.

16. Compositional Diagnostics of Turbulent Boundary Layers

We have mentioned turbulent boundary layers as sites of heating and regions in which the chemistry is similar to that taking place in shocks. For a given initial cold molecular state, hot gas is less dense in a turbulent boundary layer than in gas at the same temperature in a shock; as mentioned in section 4, a turbulent boundary layer is at about the same pressure as the cloud gas, whereas postshock gas is at a very elevated pressure relative to the preshock medium. The chemistry in hot gas is sensitive to the pressure.

For instance, in hot diffuse cloud regions the CH to CH$^+$ abundance ratio varies rapidly with pressure (Duley et al., 1992). As a consequence, the detection of CH^+ absorption features for which no corresponding CH features are seen (Crane et al., 1995) is more easily understood in the context of a turbulent boundary layer model for the origin of CH^+.

Obviously, because CH^+ observations are of absorption features against bright background stars, they contain almost no information about the spatial location of the CH^+. Consequently, the observation of CH^+ will never provide much insight into the natures of boundary layers. Rather, chemical diagnosis of the boundary layer in TMC-1 (c.f. the last paragraph of section 8) potentially could give information about the dissipation rate and whether mixing of ionized material with neutral material occurs as a consequence of ambipolar diffusion or turbulent driven reconnection in a magnetic turbulent boundary layer. Mixing of He^+ from an ionized shocked wind into the neutral gas would lead to the destruction of CO by reaction (10) making possible a richer carbon chemistry (Charnley et al., 1990; Nejad and Hartquist, 1994) than obtains in equilibrium (c.f. section 13). Possibly the high HC_3N abundance at the corresponding emission peak in TMC-1 is due to mixing of ions and neutrals (Chièze, Pineau des Forêts and Herbst, 1991; Williams and Hartquist, 1991) in a boundary layer (Williams and Hartquist, 1991). Heating to temperatures approaching 10^3 K and the presence of the ions, which remove saturated molecules, result in the formation of high abundances of CH and OH (Charnley et al., 1990; Nejad and Hartquist, 1994), due to sequences triggered by reaction (48), the endothermic reaction

$$C + H_2 \rightarrow CH + H \tag{77}$$

and reactions (73) and (74). Measurements with Arecibo of the CH and OH spatial distributions in TMC-1 may yield information about boundary layers.

17. The Stellar-Interstellar Transition

Planetary nebulae, which are ejecta of low mass stars, are traditionally considered to be part of the interstellar medium. Dyson et al. (1989) pointed out that globules, observed as dark prints in emission line pictures of the Helix nebula (NGC 7293) a nearby planetary nebula, must be molecular in order for them to have survived. (Their existence requires that the sound crossing time in each is comparable to or greater than the age of the nebula. Otherwise, the pressure differences over one's surface due to its motion relative to the planetary nebulae's more diffuse gas would result in its evaporation.) Dyson et al. (1989) estimated that the globules' central temperatures and number densities must be several tens of degrees and close to 10^6 cm^{-3}, respectively. A determination of whether the globules are oxygen-rich or carbon-rich is desirable in order to gain insight into the natures of their formation in the planetary nebula's progenitor and of the mixing of nucleosynthesis products formed in different regions of the star. Howe, Hartquist and Williams (1994) have modelled the chemistry in the globules. If they are oxygen-rich, they are probably undetectable in any molecular emission other than that of CO which has already been observed (Cox et al., 1992). However, if a globule has a higher carbon elemental abundance than oxygen elemental abundance, carbon that is not bound in CO is available to form other molecules containing carbon, and carbon-rich globules are most likely detectable in CN and C_2H emissions.

Stellar ejecta mix with the interstellar medium and affect its elemental abundances. The interstellar medium of the nucleus of NGC 1068, a Seyfert galaxy, is the source of HCN emission that appears to be far stronger relative to the CO emission than is typical in the Milky Way molecular clouds. Possibly, the observed region in NGC 1068 is carbon-rich (Sternberg, Genzel and Tacconi, 1994). As in a carbon-rich planetary nebula globule, the availability of carbon not bound in CO results in the enhancement of the abundances of many carbon-bearing species relative to those that would obtain in oxygen-rich environments having similar physical conditions. Chemistry in stellar ejecta may play a role in creating the carbon-richness of the Seyfert molecular regions. The radiation of an active galactic nucleus creates carbon ionization zones that extend inwardly beyond the radii at which dust condensation would otherwise occur in giants and supergiants nearer to the AGN than its associated narrow line emission regions; the ionization of carbon probably suppresses the formation of dust resulting in a substantial drop in the stellar mass loss rates and longer times spent in the phases of evolution during which carbon is dredged up from the stellar interiors and mixed into the envelopes that are ejected (Hartquist et al., 1995).

18. Molecules in Supernova Ejecta

Infrared emission from vibrationally excited carbon monoxide and silicon monoxide was observed in the ejecta of supernova 1987A some 100 days after the explosion, CO by Oliva, Moorwood and Danziger 1987, Spyromilio et al. (1988), Meikle et al. (1989) and Bouchet and Danziger (1993) and SiO by Aitken and Roche (1988), Roche et al. (1989), Roche, Aitken and Smith (1993) and Wooden et al. (1993). The possible contribution of CO and SiO to the emission from type 1a supernovae has been explored by Höflich, Khokhlov and Wheeler (1995).

The presence of the molecules alters the thermal structure of the oxygen core and creates inhomogeneities in the temperature distribution (Liu and Dalgarno, 1995).

In the ejecta, the gas is hydrogen-poor and in SN 1987A no dust was present at the time CO appeared. Above about 3000K, thermal dissociation limits molecular abundances but at lower temperatures molecules can survive. The ionization level is kept high by the action of gamma rays produced by the radioactive decay of nuclei produced by explosive nucleosynthesis.

Several gas phase mechanisms contribute to the formation of CO and SiO. It appears that direct radiative association of C and O (Dalgarno, Du and You, 1990) and of Si and O (Liu and Dalgarno, 1994; Andreazza, Singh and Sansovo, 1995) is the most efficient in SN 1987a. In a different environment, negative ion reactions (Lepp, Dalgarno and McCray, 1990; Liu, Dalgarno and Lepp, 1992)

$$O + e \rightarrow O^- + h\nu \tag{78}$$

$$C + e \rightarrow C^- + h\nu \tag{79}$$

$$O^- + C \rightarrow CO + e \tag{80}$$

$$C^- + O \rightarrow CO + e, \tag{81}$$

would be significant. Although the initiating reactions (78) and (79) are faster than radiative association,

$$C + O \rightarrow CO + h\nu \tag{82}$$

mutual neutralization reactions such as

$$He^+ + O^- \rightarrow He + O \tag{83}$$

$$O^+ + O^- \rightarrow O + O \tag{84}$$

$$C^+ + O^- \rightarrow C + O \tag{85}$$

and photodetachment, such as

$$O^- + h\nu \rightarrow O + e \tag{86}$$

limit the effectiveness of associative detachment as a source of molecules.

Depending on the mixing of the heavy and light elements, the sequences

$$He^+ + H \rightarrow HeH^+ + h\nu \tag{87}$$

$$HeH^+ + H \rightarrow H_2^+ + He \tag{88}$$

$$H_2^+ + H \rightarrow H_2 + H^+ \tag{89}$$

followed by reaction (73) to form OH and by

$$OH + C \rightarrow CO + H \tag{90}$$

$$OH + Si \rightarrow SiO + H \tag{91}$$

also lead to CO and SiO.

If helium is mixed in with the C and O, reaction (10) dominates the destruction of CO. Indeed, it is so effective that the observed abundance of CO can be used to exclude any significant microscopic mixing of the helium into the C/O layer (Liu, Dalgarno and Lepp, 1992; Liu and Dalgarno, 1994).

In the absence of helium, CO is destroyed by the reaction with O^+, by photodissociation by ultraviolet photons and by fast electron impact. The predicted and measured abundances of CO agree closely (Liu and Dalgarno 1995), confirming the absence of microscopic mixing.

19. Conclusion

By necessity, we have had to omit any mention of a large number of important molecular line diagnostics and applications of compositional diagnostics of the interstellar gas. No doubt, others would have presented different selections of topics while showing convincingly the major role in astrophysics that molecular diagnosis of interstellar sources and the processes governing their properties has to play.

References

Adams, N.G., Herd, C.R., Geoghegan, M., Smith, D., Canosa, A., Gomet, J.C., Rowe, B.R., Queffelec, J.L. and Morlais, M.: 1991, *J. Chem. Phys.* **94**, 4852.

Adams, N.G. and Smith, D.: 1987, in M.S. Vardya and S.P. Tarafdar (eds.) *Astrochemistry IAU Symp. 130*, Reidel, Dordrecht, p. 1.

Adams, N.G. and Smith, D.: 1988, in T. J. Millar and D.A. Williams (eds.) *Rate Coefficients in Astrochemistry*, Kluwer, Dordrecht, p. 173.

Adams, N.G., Smith, D. and Alge, E.: 1984, *J. Chem. Phys.* **81**, 1778.

Adams, N.G., Smith, D. and Millar, T.J.: 1984, *Mon. Not. R. astr. Soc.* **211**, 857.

Aitken, D.K., Smith, C.H., James, S.D., Roche, P.F., Hyland, A.R. and McGregor, P.J.: 1988, *Mon. Not. R. astr. Soc.* **231**, 7P.

Amano, T.: 1988, *Astrophys. J.* **329**, L121.

Amano, T.: 1990, *J.Chem. Phys.* **91**, 6492.

Andreazza, C.M., Singh, P.D. and Sansovo, G.C.: 1995, *Astrophys. J.* **451**, 889.

Baulch, D.L.,Cox, R.A., Hampson, R.F., Jr., Kerr, J.A., Troe, J. and Watson, R.T.: 1982, *J. Phys. Chem. Ref. Data* **11**, 327.

Bergin, E.A., Langer, W.D. and Goldsmith, P.F.: 1995, *Astrophys. J.* **441**, 222.

Bertoldi, F. and McKee, C.F.: 1992, *Astrophys. J.* **395**, 140.

Black, J.H. and Dalgarno, A.; 1973, *Astrophys. J.* **184**, L101.

Black, J.H. and Dalgarno, A.; 1976, *Astrophys. J.* **203**, 132.

Black, J.H. and van Dishoeck, E.F.: 1987, *Astrophys. J.* **322**, 412.

Blitz, L. and Stark, A.A.: 1986, *Astrophys. J.* **300**, L89.

Bouchet, P. and Danziger, I.J.: 1993, *Astron. Astrophys.* **273**, 431.

Brand, P.W.J.L.: 1995, Astrophys. Spac. Sci., in press.

Brown, P.D. and Charnley, S.B.: 1990, *Mon. Nat. R. astr. Soc.* **244**, 432.

Brown, P.D. and Millar, T.J.,: 1989a, *Mon. Not. R. astr. Soc.* **237**, 661.

Brown, P.D. and Millar, T.J.: 1989b, *Mon. Not. R. astr. Soc.* **240**, 25P.

Brown, R.D. and Rice, E.H.N.: 1986, *Mon. Not. R. astr. Soc.* **233**, 429.

Canosa, A., Gomet, J.C., Rowe, B.R., Mitchell, B.A. and Queffellee, J.L.: 1993 *J. Chem. Phys.* **97**, 1028.

Cesaroni, R. and Walmsley, C.M.: 1991, *Astron. Astrophys.* **241**, 547.

Charnley, S.B., Dyson, J.E., Hartquist, T.W. and Williams, D.A.: 1990, *Mon. Not. R. astr. Soc.* **243**, 405.

Charnley, S.B., Tielens, A.G.G.M. and Millar, T.J.: 1992, *Astrophys. J.* **399**, L71.

Chernoff, D.F., Hollenbach, D.J. and McKee, C.F.: 1982, *Astrophys. J.* **259**, l97

Cheung, A.C., Rank, D.M., Townes, C.H., Knowles, S.H. and Sullivan, W.T.: 1969a, *Astrophys. J.* **157**, L13.

Cheung, A.C., Rank. D.M. Townes, C.H., Thornton, D.D. and Welch, W.J.: 1969b, *Nature* **221**, 626.

Chièze, J.P., Pineau des Forêts, G. and Herbst, E.: 1991, *Astrophys. J.* **373**, 110.

Cox, P., Omont, A., Huggins, P.J., Bachiller, R., Forveille, T.: 1992, *Astrophys. J.* **266**, 420.

Crane, P., Lambert, D.L. and Sheffer, Y.: 1995, *Astrophys. J. Suppl.* **99**, 107.

Dalgarno, A.: 1987, in G. E. Morfill and M. Scholer (eds.) *Physical Processes in Interstellar Clouds*, D. Reidel, Dordrecht

Dalgarno, A: 1993, *J. R. Soc. Chem: Faraday Transactions* **89**, 2111.

Dalgarno, A.: 1994 *Adv. Atom. Mol. Opt. Phys.* **32**, 57.

Dalgarno, A., Black, J.H. and Weisheit, J.C.: 1973, Astrophys. Lett. **14**, 77.

Dalgarno, A., Du, M.L. and You, J.H.: 1990, *Astrophys. J.* **349**, 675.

Dalgarno, A. and Lepp, S.: 1984 *Astrophys. J.* **287**, L47.

Datz, S., Sundström, G., Biedermann, C.L., Broström, L., Danared, H., Mannervik, S., Mowat, J.R. and Larsson, M.: 1995a, *Phys. Rev. Lett.* **74**, 896.

Datz, S., Larsson, M., Stromholm, C., Sundström, G., Zengin, V., Danared, H., Källberg, A. and af Ugglas, M.: 1995b, *Phys. Rev. A.* **52**, 2901.

Draine, B.T., Roberge, W.G. and Dalgarno, A.: 1983, *Astrophys. J.* **264**, 245.

Duley, W.W., Hartquist, T.W., Sternberg, A., Wagenblast, R. and Williams, D.A.: 1992, *Mon.Not. R. astr. Soc.* **255**, 463.

Duley, W.W. and Williams, D.A.: 1993, *Mon. Not. R. astr. Soc.* **260**, 37.

Dyson, J.E., Hartquist, T.W., Malone, M.T. and Taylor, S.C.: 1995, *Rev. Mex. A.A. (Serie de Conferencias)* **1**, 119.

Dyson, J.E., Hartquist, T.W., Pettini, M. and Smith, L.J.: 1989, *Mon. Not. R. astr. Soc.* **241**, 625.

Elitzur, M. and Watson, W.D.: 1978, *Astrophys. J.* **222** , L141.

Federman, S.R., Cardelli, J.A., van Dishoeck, E.F., Lambert, D.L. and Black, J.H.: 1995 *Astrophys. J.* **445**, 325.

Garcia-Barreto, J.A., Bruke, B.F., Reid, M.J., Moran, J.M., Haschick, A.D. and Schilizzi, R.T.: 1988, *Astrophys. J.* **326**, 954.

Geballe, T.R.: 1990, in T.W. Hartquist (ed.) *Molecular Astrophysics*, Cambridge University Press, Cambridge.

Gougousi, T., Johnsen, R. and Golde, M.F.: 1995, *Phys. Rev. A* in press.

Graf, U.U., Eckart, A., Genzel, R., Harris, A.I. Pulitsch, A., Russell, A.P.G. and Stutzki, J.: 1993, *Astrophys. J.* **405**, 249.

Graf, U.U. Genzel, R., Harris, A.I., Hills, R.E., Russell, A.P.G. and Stutzki, J: 1990, *Astrophys. J.* **358**, L49.

Gray, M.D., Field, D. and Doel, R.C.: 1992, *Astron. Astrophys.* **262**, 555.

Gredel, R., Lepp, S., Dalgarno, A. and Herbst, E.: 1989, *Astrophys. J.* **347**, 289.

Guilloteau, S., Baudry, A. and Walmsley, C.M.: 1985, *Astron. Astrophys.* **153**, 179.

Hartquist, T.W. and Dalgarno, A.: 1980, in: P.M. Solomon and M.G. Edmunds (eds.), *Giant Molecular Clouds in the Galaxy*, Pergamon Press, Oxford.

Hartquist, T.W., Durisen, R.H., Dyson, J.E., Rawlings, J.M.C., Williams D.A. and Williams, R.J.R.: 1995, *Astrophys. J.* **453**, 77.

Hartquist, T.W. and Dyson, J.E.: 1993, *Quart. J.R. astr. Soc.,* **34**, 57.

Hartquist, T.W., Dyson, J.E. and Williams, D.A.: 1992, *Mon. Not. R. astr. Soc.* **257**, 419.

Hartquist, T.W., Flower, D.R. and Pineau des Forêts, G.: 1990, in: T.W. Hartquist (ed.),

Molecular Astrophysics, Cambridge University Press, Cambridge.

Hartquist, T.W. and Morfill, G.E.: 1983, *Astrophys. J.* **266**, 271.

Hartquist, T.W. and Morfill, G.E.: 1994, *Astrophys. Spac. Sci.* **216**, 223.

Hartquist, T.W., Oppenheimer, M. and Dalgarno, A.: 1980, *Astrophys. J.* **236**, 182.

Hartquist, T.W., Rawlings, J.M.C., Williams, D.A. and Dalgarno, A.: 1993, *Quart. J.R. astr. Soc.* **34**, 213.

Hartquist, T.W. and Williams, D.A.: 1989, *Mon. Not. R. astr. Soc.* **241**, 417.

Hartquist, T.W. and Williams, D.A.: 1990, *Mon. Not. R. astr. Soc.* **247**, 323.

Hartquist, T.W. and Williams, D.A.: 1995, *The Chemically Controlled Cosmos*, Cambridge University Press, Cambridge.

Hasegawa, T.I., Rodgers, C. and Hayashi, S.S.: 1991, *Astrophys. J.* **374**, 177.

Herbst, E.: 1982, *Astron. Astrophys.* **11**, 76.

Hirahawa,Y., Suzuki, H., Yamamoto, S., Kawaguchi, K., Kaifu, N., Ohishi, M., Takano, S., Ishikawa, S.-I. and Masuda, A.: 1992 *Astrophys. J.* **394**, 539.

Ho, P.T.P. and Townes, C.H.: 1983, *Ann. Rev. Astron. Astrophys.* **21**, 239.

Höflich, P., Khokhlov, A.M. and Wheeler, J.C.: 1995, *Astrophys. J.* in press.

Howe, D.A., Hartquist, T.W. and Williams, D.A.: 1994, *Mon. Not. R. astr. Soc.* **271**, 811.

Irvine, W.M., Goldsmith, P.F. and Hjalmarson, Å.: 1987, in D.J. Hollenbach and H.A. Thronson, Jr. (eds.) *Interstellar Processes*, D. Reidel Publishing Co., Dordrecht.

Jaffe, D.T., Graf, U.U., Harris, A.I., Stutzki, J. and Lepp, S.H.: 1992, *Astrophys. J.* **385**, 240.

Jokipii, J.R. and Davila, J.M.; 1981, *Astrophys. J.* **248**, 1156.

Lèger, A., Jura, M. and Omont, A.: 1985, *Astron. Astrophys.* **144**, 147.

Larsson, M., Danared, H., Mowat, J.R., Sigray, P., Sundström, G., Broström, L., Filevich, A., Källberg, A., Mannervik, S., Rensfelt, K.G. and Datz, S.: 1993, *Phys. Rev. Lett.* **70**, 430.

Lepp, S., Dalgarno, A. and McCray, R.: 1990, *Astrophys. J.* **358**, 262.

Linsky, J.L., Brown, A., Gayley, K., Diplas, A., Savage, B.D., Ayres, T., Landsman, W., Shore, S.N. and Heap, S.R.: 1993, *Astrophys. J.* **402**, 694.

Linksy, J.L., Diplas, A., Wood, B.E., Brown, A. and Savage, B.D.: 1995, *Astrophys. J.* **451** 352.

Liu, W. and Dalgarno, A.: 1994, *Astrophys. J.* **428**, 769.

Liu, W. and Dalgarno, A.: 1995, *Astrophys. J.* **454**, 472.

Liu, W., Dalgarno, A. and Lepp, S.: 1992, *Astrophys. J.* **396**, 679.

Malone, M.T., Dyson, J.E. and Hartquist, T.W.: 1994, *Astrophys. Spac. Sci.* **216**, 143.

Meikle, W.P.S., Allen, D.A., Spyromilio, J. and Varani, G.-F.: 1989, *Mon. Not. R. astr. Soc.* **238**, 193.

Millar, T.J.: 1990, in: T.W. Hartquist (ed.) *Molecular Astrophysics*, Cambridge University Press, Cambridge.

Millar, T.J., Bennett, A. and Herbst, E.: 1989, *Astrophys. J.* **340**, 906.

Millar, T.J. and Herbst, E.: 1990, *Astron. Astrophys.* **231**, 466.

Millar, T.J., Bennett, A. and Herbst, E.: 1989, *Astrophys. J.* **340**, 906.

Moriarty-Schieven, G.H., Snell, R.L. and Hughes, V.A.: 1991, *Astrophys. J.* **374**, 169.

Myers, P.C.: 1987, in: M. Peimbert and J. Jugaku (eds.) *Star Forming Regions-Proceedings of IAU Symposium 115*, D. Reidel Publishng Company, Dordrecht.

Myers, P.C. and Benson, P.J.: 1983, *Astrophys. J.* **266**, 309.

Nejad, L.A.M. and Hartquist, T.W.: 1994, *Astrophys. Spac. Sci.* **220**, 253.

Nejad, L.A.M., Hartquist, T.W. and Williams, D.A.: 1994, *Astrophys. Spac. Sci.* **220**, 261.

Neufeld, D.A. and Melnick, G.J.: 1991, *Astrophys. J.* **368**, 215.

Oliva, E., Moorwood, A.F.M. and Danziger, I.J.: 1987, *Messenger* **50**, 18.

Opendak, M.: 1993, *Astrophys. J.* **406**, 548.

Oppenheimer, M. and Dalgano, A.: 1974, *Astrophys. J.* **187**, 231.

Pagani, L., Salez, M. and Wannier, P.G.: 1992, *Astron. Astrophys.* **258**, 479.

Prasad, S.S. and Tarafdar, S.P.: 1983, *Astrophys. J.* **267**, 603.

Rawlings, J.M.C.: 1996, this volume.

Rawlings, J.M.C., Hartquist, T.W. Menten, K.M. and Williams, D.A.: 1992, *Mon. Not. R. astr. Soc.* **255**, 471.

Roberge, W.G., Jones, D., Lepp, S. and Dalgarno, A.: 1991, *Astrophys. J. Suppl.* **77**, 287.

Roche, P.F., Aitken, D.K. and Smith, C.H.: 1993, *Mon. Not. R. astr. Soc.* **261**, 522.

Roche, P.F., Aitken, D.K., Smith, C.H. and James, S.D.: 1989, *Nature* **337**, 533.

Roth, K.C., Meyer, D.M. and Hawkins, I.: 1993, *Astrophys. J.* **413**, L67.

Schloerb, F.P., Snell, R.L. and Young, J.S.: 1983, *Astrophys. J.* **267**, 163.

Shalabeia, O.M. and Greenberg, J.M.: 1994, *Astron. Astrophys.* **290**, 266.

Sidhu, K.S., Miller, S. and Tennyson, J.: 1992, *Astron. Astrophys.* **254**, 453.

Smith, D. and Spanel, P.: 1993, *Int. J. Mass. Spectr. Ion Proc.* **129**, 163.

Smith, D., Adams, N.G. and Alge, E.: 1982, *Astrophys. J.* **263**, 123.

Schneider, N., Stutzki, J., Winnewisser, G. and Blitz, L.: 1996, *Astron. Astrophys.* submitted.

Spitzer, L.: 1978, *Physical Processes in the Interstellar Medium*, John Wiley & Sons, New York.

Spyromilio, J., Meikle, W.P.S., Learner, R.C.M. and Allen, D.A.: 1988, *Nature* **334**, 327.

Sternberg, A. and Dalgarno, A.: 1989, *Astrophys. J.* **338**, 197.

Sternberg, A., Genzel, R. and Tacconi, L.: 1994, *Astrophys. J.* **436**, L131.

Sundström, G., Mowat, J.R., Danared, H., Datz, S., Broström, L., Filevich, A., Kälberg, A., Mannervik, S., Rensfelt, K., Sigray, P., Ugglas, M., and Larsson, M.: 1994, *Science* **263**, 785.

Taylor, S.D., Hartquist, T.W., Blitz, L., Dyson, J.E. and Williams, D.A.: 1996 *Astrophys. J.* submitted.

Tedds, J.A. and Brand, P.W.J.L.: 1995, *Astrophys. Space Sci.* in press.

Tielens, A.G.G.M. and Hollenbach, D.: 1985, *Astrophys. J.* **291**, 747.

Turner, B.E.: 1990, *Astrophys. J.* **362**, L29.

van Dishoeck, E.F. and Black, J.H.: 1986, *Astrophys. J. Suppl.* **62**, 109.

van Dishoeck, E.F.: 1990, in T.W. Hartquist (ed.) *Molecular Astrophysics*, Cambridge University Press, Cambridge.

Wagenblast, R.: 1992, *Mon.Not. R.astr. Soc.* **259**, 155.

Wagenblast, R. and Hartquist, T.W.: 1989, *Mon.Not. R.astr. Soc.* **237**, 1019.

Watson, W.D.: 1974, *Astrophys. J.* **188**, 33.

Watson, W.D.: 1976, *Rev. Mod. Phys.* **48**, 513.

Watson, W.D.: 1977, in J. Audouze (ed.) *CNO Isotopes in Astrophysics*, Reidel, Dordrecht, p. 105.

Willacy, K. and Williams, D.A.: 1993, *Mon. Not. R. astr. Soc.* **260**, 635.

Williams, D.A. and Hartquist, T.W.: 1991, *Mon. Not. R. astr. Soc.* **251**, 351.

Williams, D.A., Hartquist, T.W. and Whittet, D.C.B.: 1992, *Mon. Not. R. astr. Soc.* **258**, 599.

Wooden, D.H. Rank, D.M., Bregman, J.D., Witteborn, F.C., Tielens, A.G.G.M., Cohen, M., Pinto, P.A. and Axelrod, T.S.: 1993 *Astrophys. J. Suppl.* **88**, 477.

Wootten, A., Loren, R.B. and Snell, R.L.: 1982 *Astrophys. J.* **255**, 160.

Yamashita, T., Hayashi, S.S., Kaifu, N., Kameya, O., Ukita, N. and Hasegawa, T.: 1989, *Astrophys. J.* **347**, L85.

Zajfman, D., Heber, O., Andersen, L.H., Vejby-Christensen, L., Peterson, H.B., Schmidt, H.T. and Kella, D.: 1995, private communication.

SPECTROSCOPIC EVIDENCE FOR PROTOSTELLAR INFALL

J M C RAWLINGS

*Department of Physics and Astronomy, University College London, Gower Street,
London WC1E 6BT*

Abstract. The observational evidence for infall associated with star formation is discussed. Whilst spectral energy distributions of young protostellar objects are consistent with infall, the best direct evidence comes from millimetre and sub-millimetre spectral line observations. Considerations of the formation of the line profiles and the chemical effects of gas-grain interactions suggest that there is only a very short 'window' in the evolutionary track of a protostellar object during which infall is directly observable. This may explain why so few infall candidates have been detected. It is argued that self-consistent models of the dynamical and chemical evolution of collapsing cores, coupled to multiple high resolution line observations, will provide definitive evidence for the presence of infall in these objects.

1. Introduction

The search for strong direct evidence for inflow associated with the formation of a star remains one of the major quests of modern astronomy. For at least fifteen years Sir Robert Wilson has directed the attention of UCL astronomers to this question, and we are now in a position to promise him a quantitative answer within the next couple of years. Whilst there are a handful of well known protostellar inflow "candidates", as yet no protostellar object has been observed to show unequivocally the signatures associated with inflow and gravitational collapse.

It is generally accepted that stars form following the collapse of cold molecular cloud material. Stars can either form in isolation, or else in complex dynamic structures within large dark cloud complexes. Within these large (5-50pc) complexes (such as Orion, Taurus-Auriga-Perseus) there exist clumps. Some clumps are relatively smooth and starless, whilst others (sometimes referred to as 'core complexes') consist of fragments or 'cores'. B5 and HCL2 are examples of these fragmented clumps. These core complexes may contain several infrared (IRAS) sources as well as gas condensations and evidence of dynamical activity. On the smallest scale there are individual cores of 0.1–0.4pc size and mass 0.3–10M$_\odot$. Typical examples include TMC1 and B1. The cores may be cold and quasi-static, showing no evidence for collapse – or else they may contain infrared sources indicating that collapse to form a star has occurred or is in process. B335 and NGC1333-IRAS2 are examples of cores with embedded sources. Bok globules, which appear to be cold and quiescent are often associated with star formation and about 23% of them possess IRAS point sources. In the discussion that follows, the term

Astrophysics and Space Science **237**: 299–318, 1996.

'core' signifies the gravitationally bound clumps of gas that may eventually collapse (in part) to from a star. The term 'protostellar core' refers to the centrally condensed quasi-static object that will become the star itself.

In the search for evidence of infall, the observational problem is immediately apparent. Cores are cold and nearly isothermal (with the dust continuum and molecular line emissions, such as from ^{12}CO, radiating away much of the infall energy) so that the low lying rotational levels of molecular tracers are optically thick. In addition, the velocity gradients through the core and intercore gas surrounding it are not large (usually less than a few kms^{-1}) and there is little velocity dispersion, implying that the emission lines have limited doppler shifts and broadening. Taken together, these facts imply that the different regions of a collapsing core and its surroundings are radiatively coupled and the interpretation of a line profile can be very uncertain.

Additional complications arise from the presence of bipolar outflows, with the result that the spectral signatures of collapse can be confused with the kinematics of outflow and rotation. Whilst outflows are always present, they are far less of a problem for the youngest low mass protostellar clouds than they are for the more evolved and more massive protostars, where the spectral indicators of infall may be completely swamped by emissions from the protostar and its outflow.

In most of the discussion that follows, we therefore restrict our arguments to low mass star formation. There is a large body of observational data which covers the evolution of low-mass stars from pre-protostellar clouds through to naked T-Tauri stars. For much of the pre-protostellar period, the star forming cores remain cold (<20K) and isothermal. For these reasons, and others, the collapse dynamics of low mass star-forming regions are easier to understand. In addition, low mass stars are more likely to form in isolation than their high mass counterparts which are often associated with external influences in 'star-burst' phenomena.

2. The dynamics of collapse

To place the observationally detectable signatures of collapse in context it is important to establish a consistent theoretical working model of the early stages of protostellar evolution. There are many uncertainties in our understanding of the dynamics of collapse; it is not clear, for example, why so many cores appear to be stable against collapse, or why only a relatively small fraction of a collapsing clump ends up in the protostellar core. These are problems which we do not intend to address here, but we note that their existence limits the validity of the models described below.

It is possible that many of the extended dusty clouds surrounding young stellar objects (YSOs) are envelopes in the state of collapse. However, the distinction between the collapsing cloud and the protostellar accretion disk is not at all clear. Some 50% of low-mass pre-main sequence stars have excess near-infrared and millimetre emissions (Strom et al. 1989; Beckwith et al. 1990) which may be indicative of compact accretion disks. But, if these disks exist, they are small and can only be resolved with high resolution interferometric observations (a typical object at 150pc has an angular size of less than an arcsecond).

In practice, the dynamics of even the most simple case of a perfectly spherical isothermal gravitational collapse are poorly understood. These dynamics are essentially determined by the balance of gravity against pressure. That pressure can be thermal as well as non-thermal. The non-thermal contributions may be provided by magnetic support, turbulence (probably Alfvénic) and, at later stages, by rotation. These contributions will also tend to break the spherical symmetry of the collapse. A variety of models have been developed; mostly describing the simplest case of thermally supported clouds. The problems and differences between these models mainly arise from the assumptions involved and the initial conditions that are adopted. The earlier models of Larson (1969,1972) and Penston (1969) have now largely been superceded by the "inside-out" isothermal collapse models of Shu (1977) et seq. These models have the appeal of numerical simplicity and apparent compatibility with a number of observations. As a result the 'inside out' collapse model has has become widely accepted. Prior to collapse, the core is approximated by a singular isothermal sphere in precarious hydrostatic equilibrium. Such an equilibrium is only possible if the mass of the core is less than some critical value (Bonnor 1956, Ebert 1955). In the limit of infinite central density the configuration approaches the singular form with the density proportional to r^{-2}. The sphere is truncated at a finite radius determined by the external pressure. This configuration is believed to result from sub-sonic quasi-static contraction during which the cores are cold and essentially quiescent. After about 10^6 years some, unspecified, mechanism then initiates collapse at the centre of the core. As material is drawn in to the flow, the outer regions are undermined and a 'collapse expansion wave' moves outwards through the cloud at the local sound speed. The outer regions remain static with $n \propto r^{-2}$ whilst the inner parts approach free-fall with $n \propto r^{-3/2}$. The effect that this has on the density structure of the collapse core is shown in Fig 3a. During this epoch the cloud remains isothermal and the extinction of the central regions rises to ≥ 1000. Eventually, a protostellar core is formed, isothermality and spherical symmetry break down in the core regions, and infalling material encounters a centrifugal barrier as a result of rotation. A disk-like geometry is then established on solar-system size length scales, and a highly collimated (neutral) molecular outflow develops (traced by ^{12}CO emission). Both inflow and outflow are present at this time. By this stage, a further 10^5 years will have elapsed since the end of the quasi-static phase. Finally, the outflow "opens up", inflow is halted, and the newly formed star is seen in the visible and the UV.

3. Spectral energy distributions and protostellar classification

Before the advent of high resolution millimetre and sub-millimetre spectroscopy, the main clues as to the presence of protostellar collapse was almost entirely derived from mid- to far-infrared continuum studies. However, the spectral energy distribution (SED) of a source may have several sub-components originating from regions of very different physical characteristics. Thus whilst a SED may be *consistent* with a given physical model of protostellar collapse, in general it cannot be used to establish the presence of infall. Nevertheless, there has been considerable success in matching the observations with the standard model described above. In particular, Adams, Lada and Shu (1987) have used a radiative transport model to predict the SED from a protostellar cloud undergoing inside-out collapse. In this model the radiation field is dominated by the reprocessing of the accretion luminosity by the surrounding infalling dusty envelope.

Adams, Lada and Shu modelled six sources (including the well-known protostellar candidate L1551–IRS5) and consequently defined a protostellar classification scheme based on the infrared spectral index;

$$n = \frac{d(log\lambda L_\lambda)}{d(log\lambda)}$$

evaluated for λ=2–20μm. The youngest objects (with spectra that rise at far-infrared wavelengths, i.e. $n \geq 0$) were designated as 'Class I' and are generally seen to have a close association with molecular gas. Class I objects were identified as the earliest protostars and are very faint at optical wavelengths having spectra that peak in the mid- to far-infrared. The models are generally successful in explaining the mid- and far-infrared observations but fail to predict the strong near-infrared emission (see Figure 1). It is, however, clear that Class I objects derive most of their luminosity from accretion. The extra near-infrared emission probably originates from light from the accretion disk that is scattered out of a hole in the envelope. These regions of relative infrared transparency are possibly due to jets punching holes through the infalling envelope (e.g. Shu, Adams, & Lizano, 1987; Whitney & Hartmann 1993), or alternatively result from asymmetric collapse (Galli & Shu 1993a,b). The progressively less embedded sources were designated Class II (-1.5<n<0) for which the SED can be modelled by a simple disk model and Class III (-1.5>n) which possess optically thin circumstellar disks. Class II objects can be identified with classical T-Tauri stars and Class III objects with naked, or weak line, T-Tauri stars.

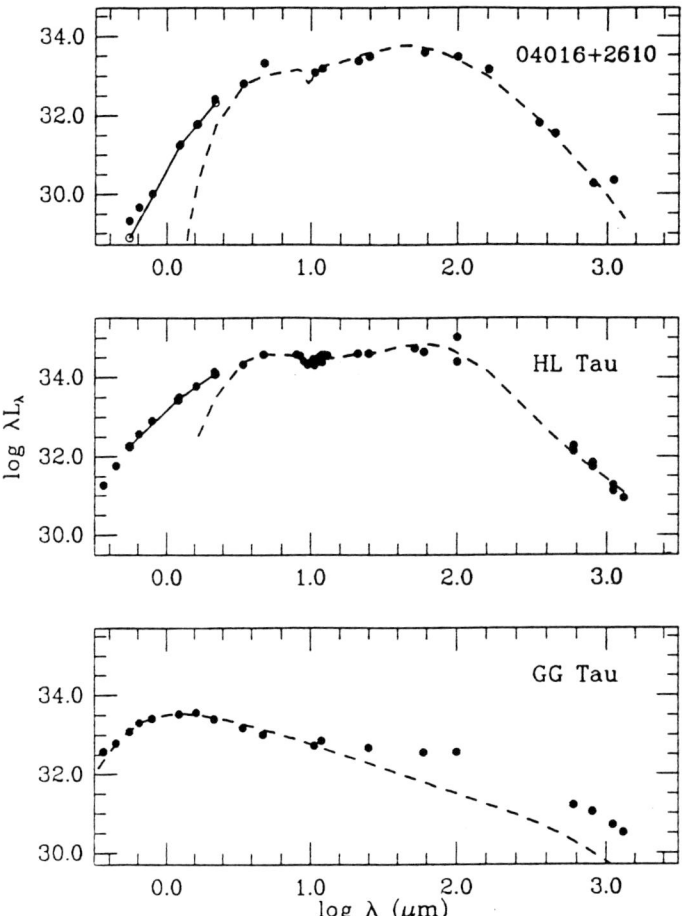

Fig. 1.   Spectral energy distributions (SEDs) of sources in Taurus (from Calvet et al. 1994): The source 04016+2610 (upper panel) is a Class I protostar and is well fitted by a model of radiative equilibrium emission from a dusty envelope (dashed line) with a contribution from extra light scattered out of a hole in the envelope (solid line). HL Tau (middle panel) is a "flat spectrum" T-Tauri star, and GG Tau is a Class II, classical T-Tauri star (the dashed line shows the fit for a simple disk model). HL Tau is apparently very similar to a Class I source, but requires more scattered light to explain the high near infrared fluxes

HL Tau is a Class II object that has attracted much attention. Calvet et al. (1994) found the SED to be flatter than expected for a Class II object and actually quite close to that of a Class I source (see Figure 1) and is thus more consistent with the expected spectrum of an infalling dusty envelope than a disk. In addition, millimetre interferometric mapping of ^{13}CO emission by Hayashi et al. (1993) found that the largest velocity gradient in the flattened distribution (of about 2000au extent) is in the direction of the minor axis and not the major axis as would be expected if rotation alone were the cause of the velocity shift. This implies that radial motion dominates over rotation and that the gaseous disk is in fact contracting. An infall rate of $\sim 4 \times 10^{-6} M_\odot yr^{-1}$ is suggested.

Recent millimetre interferometry reveals that the disk masses of Class I objects are small (Terebey, Chandler & André 1993) (with $M_{CS} \sim 0.1$–0.3 $M_\odot$). Class I objects thus appear to have only remnant envelopes, are not isothermal, and are thus probably fairly well evolved in the context of the standard model. Calvet (1995) has shown that, in embedded and flat spectra clouds, large rotation rates and a large zone of departure from spherical symmetry in the central regions are required in order to explain the SED. In general SEDs (particularly in the mid-infrared region) are very dependent on the physical characteristics of the source, such as the assumed dust opacity law, the source geometry and the size of the core (relative to the telescope beam). In some young very optically thick cores (eg. B335) the SEDs are consistent with a very flat density profile, and indeed there is direct evidence (see section 6 below) that the density profiles in these objects may deviate significantly from that predicted by the standard model. This both implies that the standard model may not accurately describe the evolution of the youngest protostellar objects and that the classification system is not yet fully justified.

4. The youngest protostellar objects

Thermal dust emission is optically thin at millimetre wavelengths for H_2 column densities up to $10^{26} cm^{-2}$. Recent developments in millimetre (typically probing material at $\sim 10K$ and $n(H_2) \sim 10^4 cm^{-3}$) and sub-mm ($T > 20K$, $n(H_2) > 10^5 cm^{-3}$) astronomy, with sensitive, low noise receivers attached to large single dish and array telescopes allow us to look deeper into cold cores with high spectral and spatial resolution (eg. the JCMT 15m and CSO 10.4m telescopes, which offer beam sizes of about 15–20″, corresponding to about 3000au for the nearest clouds). There are two approaches to searching for collapsing cores. One is to look for luminous, deeply self-embedded YSOs. The other is to identify cold, dense cores that are Jeans unstable. It is this latter approach that is best employed in a systematic search for cores in their earliest stages of evolution.

If we define a protostar as an object that is in the process of assembling the bulk of its matter before turning on to the main sequence, then in the absence of a central heat source, the bolometric luminosity is given by;

$$L_{bol} = \frac{GM_\star(t)\dot{M}}{R_\star}$$

where $M_\star$ and $R_\star$ are the mass and radius of the protostellar object, and $\dot{M}$ is the infall rate. Thus, the youngest objects will have the lowest L_{bol}.

However, the youngest objects have large circumstellar masses (M_{cs}). The sub-millimetre luminosity (L_{submm}) is proportional to M_{cs} (due to thermal dust emission) and the ratio L_{bol}/L_{submm} increases as the object evolves.

A more appropriate definition of a protostar may therefore be one where the mass of the centrally condensed protostellar core is less than that of the collapsing envelope. Such objects have been labelled 'Class 0' by André, Ward-Thompson and Barsony (1993) who deduce that $L_{bol}/L_{submm} \leq 200$, where L_{submm} is the luminosity at $\lambda > 350 \mu$m. Surveys of Class 0 objects show them to have centrally condensed cores, with sizes of typically 0.01–0.03pc, masses of between 0.5 and 20 $M_\odot$ and densities in the range 10^7–10^8 cm^{-3}. These objects also appear to be colder ($T_{BB} < 30$K) and more isothermal (the SED being well matched by single temperature gray-body) than the Class I cores and are characterised by extremely high extinctions. Class 0 sources are only seen at sub-mm or longer wavelengths. As they are undetected at wavelengths less than 10μm they are outside the classification scheme of Adams, Lada and Shu. Nevertheless they do have centrally peaking bolometric luminosities and invariably have bipolar molecular (i.e. neutral) outflows which imply the presence of dynamical activity. In the scheme of the standard model it would seem likely that Class 0 cores are those undergoing initial isothermal collapse. A list of the most well known Class 0 cores is given in Table 1. The prototype of the class, VLA1623, was identified by André et al. (1993) in sub-mm continuum mapping observations (350–1300μm) of ρOph A. The circumstellar envelope has dimensions of about 1000au, a mass of 0.6 $M_\odot$, a bolometric luminosity of 1$L_\odot$, a density of 10^8cm^{-3} and drives a highly collimated bipolar outflow. An infall rate of $4 \times 10^{-6} M_\odot < \dot{M}_{infall} < 4 \times 10^{-6} M_\odot$ is implied. In general, Class 0 sources are highly obscured ($A_v > 1000$), have massive envelopes ($M_{env} \geq 0.6 M_\odot$), are roughly spherical and show no evidence for the presence of a compact protostellar disk.

Whilst these are the youngest objects *that are undergoing collapse*, they too have to evolve from a 'pre-protostellar' state. A pre-protostellar core (sometimes referred to as 'class -I' !) would thus be a centrally condensed, gravitationally bound object that is either contracting quasi-statically, perhaps with the aid of non-thermal support, or has not yet started to collapse – if it ever does. Objects belonging to this latter class are relatively common and are usually identified with the 'Myers cores' (Following the survey of NH$_3$ cores by Myers and Benson 1983). This, and similar surveys (e.g. Ungerechts, Walmsley & Winnewisser, 1982) are usually based on NH$_3$ (1,1) and (2,2) observations. They identify pre-protostellar cores as having $r \sim 0.05$–0.3pc, $n \sim 10^4$–10^5 cm^{-3} and $T \sim 10$K. The cores show no evidence for local heating by a protostellar source, and all have sufficient thermal energy for them to be stable against gravitational collapse.

A Class 0 source can be distinguished from a Class I object by its nar-

TABLE I

Some physical parameters of possible Class 0 protostars (from André, 1995 and references therein)

Source	$L_{bol}(L_\odot)$	$M_{envelope}(M_\odot)$	$T_{dust}(K)$
B335	3	0.8	25
L1527	2	0.4	–
IRAS16293	23	2.3	39
VLA 1623	1	0.6	20
NGC1333-IRAS4A	14	7	37
NGC1333-IRAS4B	14	2.7	27
HH24MMS	5	4	20
RNO43-MM	5	0.6	33
L1448-N	10	2.3	–
L1448-C	9	1.4	–
NGC2024-FIR5	$\gtrsim 10$	15	20
NGC2024-FIR6	$\gtrsim 15$	6	20
Serpens-SMM4	2–30	2	15–20
L723-MM	3	0.6	–
S106-SMM	$\gtrsim 24$	$\lesssim 10$	$\gtrsim 20$

row, single temperature SED and its lack of emission at near- or mid-infrared wavelengths. The distinction between Class 0 objects and static, pre-protostellar clouds is not so clear but can be established by the effects of the presence of a central protostellar core in the former. Circumstellar gas falling onto this core is arrested in an accretion shock and is heated to X-ray emitting temperatures (Winkler & Newman 1980). The X-rays photoionize the innermost regions and free-free radio emission is generated. This emission has been detected in many of the Class 0 sources using VLA resolution at centimetre wavelengths. There are bound to be objects that are close to the borderline of the classification system. VLA1623 is such an object, as is the more recently observed cold fragment HH24MMS in the HH24–26 region of the L1630 Orion molecular complex (Bontemps, André & Ward-Thompson 1995). Both these Class 0 objects are centrally condensed dusty clumps within cloud cores which are so cold (T<20K) and/or heavily obscured that they are undetected at $\lambda < 350\mu$m. Class 0 objects are also relatively rare – there are about 12 times as many Class I objects as there are Class 0. The standard model suggests that the lifetime of Class I cores should be $\lesssim 10^5$ years implying that Class 0 objects are very transient, with lifetimes that are typically $< 10^4$ years.

5. Direct spectral evidence for infall

Of the Class 0 objects which are, to date, the best candidate objects for the detection of infall, only three show the characteristic broad self-reversed line profiles that are indicative of infalling envelopes. These are B335 (Zhou et al. 1993), IRAS-16293 (Walker et al. 1994) and L1527 (Zhou et al. 1994). There are many reasons why this should be the case – in particular, confusion with outflowing material (which is the dominant kinematic feature in Class I and more evolved objects) and the possibility that chemical depletion effects may be important (see section 7).

In searching for a possible infall candidate then the (assumed) spherical symmetry of the collapse process results in a very unclear distinction between infall and outflow for optically thin lines. Both dynamical processes will give rise to red- and blue- shifted wings in the line profiles, so that by looking at just one velocity broadened line it is not possible to distinguish between infall and outflow. (Although, by looking at the profiles of several molecular species, chemical modelling can be used to distinguish between the two processes).

Under certain circumstances, optically thick lines can demonstrate very noticeable asymmetries which can be used to distinguish between inflowing gas and outflowing protostellar winds. We must first of all make some very basic assumptions about collapsing clouds;

— There is a rising temperature gradient, so that the core regions are warmer than the outer envelope

— The emission line is optically thick in the infall region

— There is a region of near static material surrounding the collapsing core with the relevant molecular species sufficiently abundant and well populated in the lower level of the transition.

— The inner regions of the cloud are collapsing with a velocity gradient increasing towards the cloud centre

With this set of assumptions, and with the further assumption that the molecular fractional abundance is constant throughout the core, the line profile will have a very characteristic shape. The inflowing material will result in a broadened line profile with red- and blue- shifted wings corresponding to emission from the front and rear hemispheres of the collapsing cloud (as seen by an observer) and a fairly sharp self-absorption feature at the systemic velocity of the object, caused by the cooler material in the surrounding, static envelope. The temperature structure and the optical depth assumptions made above will result in the profile being asymmetric with the blue peak being stronger than the red peak. This is very clearly noticeable in the protostellar candidate B335 (see Figure 2, from Choi et al. 1995) and at first seems counter-intuitive, as the dominant effect is for the general envelope of the emission to be blue-shifted rather than red-shifted. In simple terms, this behaviour can be explained by the effects of the large optical depths; assuming that the infall velocity is sufficiently large that the blue and red-shifted hemispheres are not strongly radiatively coupled, we can consider each separately. Because the line is optically thick, the emission from each hemisphere will be dominated by the region that is closest to the observer. For the blue-shifted wing, this is the region nearest to the core centre and hence the emission is strong (coming from a region of high temperature and density) and samples high infall velocities. By contrast, the red wing is dominated by the outer parts of the collapsing core, where the emission is weak and not strongly velocity shifted. Whilst this description is simplistic and, for instance, does not account for the radiative coupling in the core, it explains why the blue peak is stronger than the red peak. The more opaque the transition, the more asymmetric the line profile. If either of the assumptions that there is a positive temperature gradient or that the wings of the line profile are optically thin in any part of the infalling region is relaxed, then the profile loses its asymmetry. Such symmetric, double peaked profiles could also be formed in cores that have optically thin outflows, or even that are static with a turbulent velocity component that increases towards the cloud centre. Outflows and spherical expansion give rise to asymmetric line profiles, but with the red wing stronger than the blue. The asymmetric profiles that have blue wings that are more prominent than the red are hard to describe by processes other than infall. They could be mimicked, in a contrived fashion, by a static cloud with density fluctuations, but by observing several lines of different molecular species it is possible to rule out such alternatives.

For the lines that are optically thin, or where there is not a significant population in the lower level of the transition in the cooler, static part of the cloud, the profile will not be self-reversed and will simply show the effects of systemic velocity broadening. Taken on their own, such lines cannot be used to detect the presence of inflow. Indeed, they do not necessarily imply the presence of a systemic velocity gradient at all. Similar broadening effects could conceivably be caused by thermal structure, turbulence or cloud rotation. However, when observed in conjunction with other optically thick lines of the same molecular species (or other isotopic variants), they give vital information about the dynamics of collapsing cores. An example of this is very well illustrated in the case of the CS lines in B335 (Figure 2). The CS J=5→4 transition is very optically thick and shows the characteristic asymmetric self-reversal described above. On the other hand the CS J=7→6 transition is optically thin and shows no sign of self-reversal. However, the greater temperatures and critical densities required to populate the higher rotational levels imply that hotter, more rapidly infalling parts of the flow are sampled by this line which is therefore significantly broader than the CS J=5→4 transition. The peak of the CS J=7→6 line naturally occurs at or near the systemic velocity of B335 and thus coincides with the trough in the CS J=5→4 line. As well as being a good indicator as to the presence of infall, the observation of several transitions in any one molecular species can be used to probe the density, velocity and temperature structures of the innermost regions of the infalling cloud. Until recently, the analysis of these lines was almost exclusively carried out using Large Velocity Gradient (LVG) or microturbulent codes. To be valid, the LVG approximation requires that there is no radiative coupling between different parts of the collapsing cloud and that the level populations at a given point are determined by the physical conditions at that location. Thus the thermal line width must be very much less than $R(dv/dr)$ where (dv/dr) is the velocity gradient and R is the linear size of the object. The method is relatively easy to apply but it is obvious that the LVG approximation is not strictly valid in young protostellar cores where the microturbulent width is greater than the systemic velocity gradient and up to 90% (by volume) of the infall region is radiatively coupled to the envelope.

5.1. B335

Infall was identified in IRAS 16293-2422 on the basis of the line shapes and the relative widths for the CS and $C^{34}S$ J=2→1 and J=5→4 lines (Walker et al. 1986). However, the presence of rapid rotation in this object makes it hard to interpret in terms of the standard model. B335 has simpler dynamics, a well defined morphology, and of the three sources with direct spectral evidence for infall, apparently shows the clearest kinematic evidence of collapse. The core is apparently evolving towards the formation of a solar-type star with $L\sim3L_\odot$ (Zhou et al. 1993,1994). As with most Class 0 objects, the dominant kinematic signature is outflow (Lada 1985), but it is possible to disentangle the inflow from the outflow if spherical symmetry breaks down in the central regions with mass inflow directed onto an accretion disk and outflow occurring along jets that are spatially well defined. This is found to be the case in B335, which has outflowing jets that are almost in the plane of the sky, and with an opening angle of approximately 45° (Cabrit et al. 1988). High spatial and spectral resolution studies of the H_2CO J=$2_{12}-1_{11}$,$3_{12}-2_{11}$ and CS J=2→1, J=3→2, and J=5→4 lines were obtained by Zhou et al. (1993) and clearly demonstrate the various features that are indicative of collapse as described above. Previous studies of B335 were either not at sufficiently high resolution, or else they concentrated on commonly used tracers, such as CO which doesn't probe high densities, or else NH_3 which is almost certainly depleted at high densities as a result of freezing out on to the surface of cold dust grains (see section 7).

As B335 has only quite subsonic turbulence and is reasonably spherical, it is relatively easy to model the line profiles. Adopting density and velocity profiles from the self-similar isothermal collapse model of Shu (1977), Zhou et al (1993) modelled the CS and H_2CO line profiles with the infall radius (the boundary of the collapse expansion wave) and the molecular abundances as the only free parameters. The abundances were assumed to be constant and are mainly constrained by the line strengths and the infall radius/collapse age is strongly constrained by the profile shapes. The infall radius is also constrained by 6cm VLA H_2CO observations. The mean temperature was taken to be 12K. A simple radiative transfer model, based on models of the dust continuum SED (from Zhou et al. 1990) was used to establish $T(r)$ (assuming the gas and the dust to be well thermally coupled). Other parameters, such as the sound speed and the mean density were drawn from previous studies. A uniform turbulent speed of 0.12 km s^{-1} was used.

The results from this approach gave good fits to some of the lines (such as the H_2CO J=$2_{12} \rightarrow 1_{11}$), whilst others (such as the H_2CO J=$3_{12} \rightarrow 2_{11}$) were observed to be stronger than predicted by the model. Many of the discrepancies were found to be due to the authors employing the LVG approximation. Choi et al. (1995) have employed a Monte-Carlo modelling technique, with the infall radius and the molecular abundances as free parameter as before. They obtain extremely good fits to the CS and H_2CO profiles (see Figure 2) and deduce that R_{infall}=0.03pc, X(CS)=5.2×10^{-9} and X(H_2CO)=4.6×10^{-9}, where X(A) is the fractional abundance of species A, by number, relative to hydrogen nuclei. These models are grounded on some basic assumptions about the general applicability of the Shu solution; in particular, the model is applied to an object that is not isothermal and has a large truncation radius, but the combined observations and modelling provide the most persuasive evidence for the presence of infall to date and give support to the standard model of protostellar collapse. However, some caution is required; results are *consistent* with the model but do not prove its uniqueness. Some degree of confirmation of the validity of the adopted velocity and density profiles is provided by the successful modelling of several molecular species, but, at best, it can only be said that the line profiles are consistent with an assumed velocity and density structure. They can not give information, on their own, about the evolution of the structure of a collapsing cloud. Conclusive evidence for the validity of any model can only be obtained by self-consistent radiative transfer and chemical modelling (see section 7.1).

6. Alternatives to the 'standard' model

So far we have assumed the validity of the standard model for protostellar collapse, but a number of problems are apparent from surveys of Class 0 and pre-protostellar objects. Firstly, it seems that all Class 0 sources have highly efficient outflows. The equivalent outflow luminosity can be up to 50% of the total bolometric luminosity and is actually more efficient than

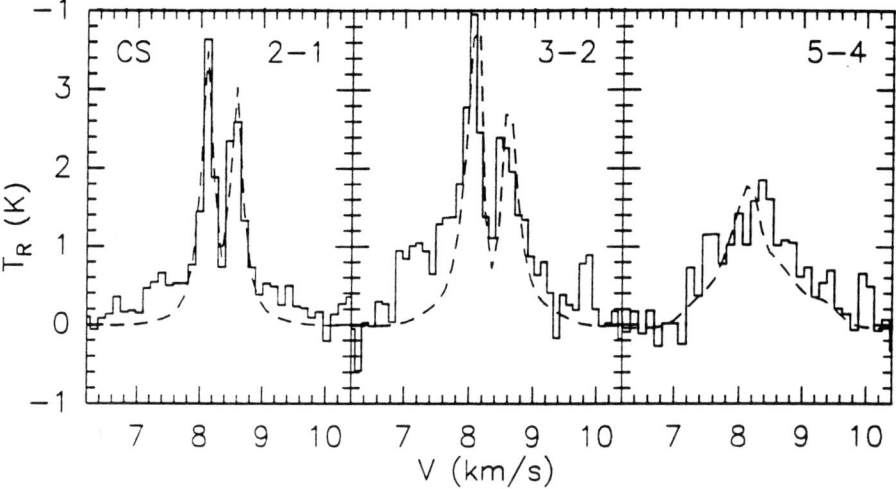

Fig. 2.   CS line profiles. Solid lines are the observations of Zhou et al. (1993). Dashed lines are the best fit model (from Choi et al., 1995) with a CS fractional abundance of 5.5×10^{-9} and $R_{infall} = 0.030$pc.

that for Class I sources. The dynamical age of the outflows suggests that they originated at around the same time that collapse started. They are also highly collimated, typically with opening angles of less than 30°. This suggests that mass ejection is initiated before a protostellar core has formed and is hard to explain in the standard model. Secondly, far infrared observations imply that Class 0 sources may be more flattened than Class I sources. A morphological survey of 16 dense cores by Myers et al. (1991) found the cores to have an average axial ratio of between 0.5 and 0.6, independent of the presence of a YSO. For the cores that are parts of filamentary structures (such as L1498, B35, L1251, L1535) the axes of elongation are reasonably well aligned (typically within 20°), perhaps indicative of magnetic support.

A major problem with the standard model is that the isothermal sphere is both singular and critically balanced against collapse, which from an observational point of view requires that the observed linewidths should be exactly thermal (a key assumption of the model being that thermal pressure alone balances gravity). Internal motions and density fluctuations in the cloud will upset the assumptions made with regards to the formation, configuration and stability of the pre-collapse quasi-static cloud core. The discrepancies appear to be largest for high mass cores where the relationship between linewidth and cloud radius suggests that they may be governed by very different physical conditions, with steeper density profiles, infall rates and pressure structures (Caselli & Myers 1995). In particular, the observed linewidths are dominated by non-thermal supersonic motions (eg. Ho et al.

1977), to the extent that the thermal motions only become comparable to the non-thermal motions at $r \lesssim 7 \times 10^{-3}$pc. A recent high resolution 800μm survey of low mass pre-protostellar cores by Ward-Thompson et al. (1994) found that none of the cores could be fitted by a single power law. Assuming isothermality, the observations imply $n(r) \propto r^{-2}$ and $r^{-1.25}$ in the outer and inner parts ($\theta < 20''$ in the case of L1689A) of the cores respectively. The objects are all dim, with $L \leq 2L_\odot$ so, if a protostellar core were present, then the objects would be very young (10^3–10^4 years). This would seem to be ruled out on statistical grounds. It thus appears that the cores in the sample are quiescent, pre-protostellar and that non-thermal support is playing an important role in their structure.

A possible explanation for these observations is that magnetic field support may play an important role in the early stages of protostellar collapse. In a sample of clouds, cores and OH maser sources, Mouschovias & Psaltis (1995) have shown that there is a strong correlation between the non-thermal linewidths, the magnetic field strengths and cloud radii;

$$(\Delta v)_{wave} \sim 1.4 \left[\frac{B}{30\mu G}\right]^{1/2} \left[\frac{R}{1pc}\right]^{1/2} kms^{-1}$$

where B is the mean magnetic field and R is the cloud radius. The non-thermal motions are interpreted as large amplitude, long wavelength Alfvén waves in self-gravitating, magnetically supported clouds. This is consistent with the collapse models of Ciolek & Mouschovias (1994). In these models a magnetic field with a mean strength of about 30μG supports clouds as large as $5 \times 10^5 M_\odot$ against collapse. Collapse on smaller length and mass-scales is allowed by the redistribution of the magnetic flux within the cloud as a result of ambipolar diffusion and the consequent damping of the magnetohydrodynamic (MHD) waves. The evolution of the density profile is depicted in Figure 3(b) and is very different to that of the inside-out model (shown in Figure 3a). In particular, the density profile is very much flatter in the inner regions and the density at a given point **rises** as time progresses rather than falling as it does in the Shu model. Initially, ambipolar diffusion increases the mass–flux ratio in the central regions. The quasi-static contraction of the magnetically subcritical, but thermally supercritical core, occurs on a timescale that is some 10–20$\times$ as long as the free-fall timescale. Eventually the core regions become both thermally and magnetically supercritical and the collapse becomes more hydrodynamic in the core regions, although the envelopes remain well supported by the magnetic field. If this model is correct then the implications for the observational detection of infall are not encouraging. Because so much of the contraction is quasi-static rather than dynamic, low infall velocities are predicted in all but the most dense parts ($n \gtrsim 10^{10}$cm^{-3}) of the flow. In contrast, most (non-magnetic) models suggest that high velocities should be present once a stellar core is formed. At

(a)   (b)

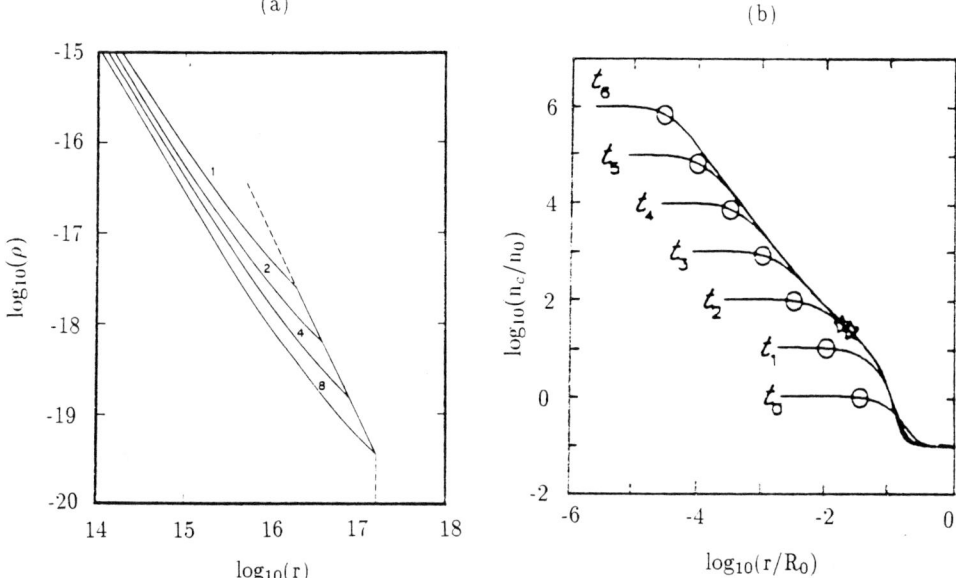

Fig. 3.   Density profiles as functions of position and time: (a) An 'inside-out' collapse model (from Shu, 1977). The labels on the curves correspond to different times (in units of 10^{12} s) after collapse initiates. (b): An ambipolar-diffusion controlled collapse model (from Ciolek & Mouschovias, 1994). R_0 and n_0 are the initial cloud radius and central density respectively. The times t_{0-6} correspond to 0, 1.42×10^7, 1.768×10^7, 1.813×10^7, 1.819×10^7, 1.8205×10^7 and 1.8208×10^7 years.

present we only have a limited quantity of relevant observational data but, if the presence of high velocity inflowing material continues to elude detection then we may be left with the possibility that in most protostellar objects, collapse is controlled by magnetic support. An alternative possibility is that some chemical process (such as freeze-out) depletes molecular material from the denser, faster moving parts of the inflow.

7. Chemical effects

The interpretation and modelling of the line profiles has been based on the assumption that the chemical fractional abundances are constant throughout the envelope; only the populations of the various rotationally and vibrationally excited levels are assumed to vary with position. It is often helpful to talk of molecular transitions as having a critical density (n_{crit}), below which the upper level will not be sufficiently populated by collisional excitation for significant emission to occur. Thus, $C^{18}O$ J=1→0 ($n_{crit} \sim 1.9 \times 10^3 cm^{-3}$) is seen to trace low density material, whilst low lying transitions of NH_3, CS, HCN and H_2CO have higher values of n_{crit}. We should, however, be wary of simply using these molecular transitions as density tracers. In a morphological survey of 16 dense cores in dark clouds Myers et al. (1991) found the mean extent of the NH_3, CS and $C^{18}O$ emissions to be 0.15pc, 0.27pc and 0.36pc respectively. This may seem curious since (n_{crit}(CS J=2→1)=5.8×10^5cm^{-3}, somewhat larger than for the NH_3(1,1) transition. The relative sizes of the NH_3 and CS distributions can be explained by morphological arguments (eg. the presence of small-scale clumping), but the discrepancy is more likely to be caused by the different radiative transfer characteristics of the transitions or by some chemical differentiation effect. The situation is therefore likely to be more complex than is suggested by the morphologies and again implies that a self-consistent dynamical/chemical approach is required in the interpretation of the cores. In the gas phase alone, chemical processes are continually at work, so that molecular abundances vary with time, density and temperature. Cloud cores are dense ($n \gtrsim 10^4$cm^{-3}) and very cold so that the freeze-out of gaseous species onto icy mantles on the surface of grains is likely to be the dominant chemical effect. The timescale for the depletion of gas-phase molecules onto grain surfaces is comparable to the dynamical collapse time (e.g., Iglesias 1977) and is inversely proportional to the density. As collapse proceeds (from 10^4 to 10^8 cm^{-3} within a few 100au) the freeze-out process accelerates. Early observational evidence for the depletion of complex molecules onto icy mantles (e.g. Wootten, Snell & Evans 1980) has now become quite secure (e.g., Mauersberger et al., 1992). High resolution VLA studies are now revealing direct evidence for substantial NH_3 depletions (Wootten 1995): In the coldest cores (such as IRAS 16293A) very cold NH_3 exists but is not well correlated with the cold dust structures. If it is assumed that the sub-mm emission traces the dust and that the dust column density can be used to infer the H_2 column density, then it is found that the fractional abundance of NH_3 is typically a few×10^{-11} or less in these objects; more than two orders of magnitude lower than in the envelopes of the cores. Strong evidence for freeze-out is also seen in other young sources, including NGC2024-FIR5, IRAS 05338-0624 & Serpens FIRS1. An additional complication caused by depletion is that the interpreted density structures of the inner regions of collapsing clouds is very uncertain. The assumption that CO traces the gas density is not necessarily valid and there is clear evidence that the CO:H_2 ratio is a function of A_v in dark cores (Bachiller and Cernicharo, 1986). Minchin, Ward-Thompson & White (1995) also suggest (from 800μm studies of S140/L1204) that the dust-to-gas ratio rises in the regions of highest extinction (A_v=10–100).

At later stages of evolution, as the protostellar core starts to evolve and to become more luminous, isothermality breaks down and the inner parts of the collapsing cloud are eventually heated up so that sublimation of the ice mantles occurs. For polar ices, such as HCN, CH_3OH & H_2O, this temperature is typically in the region of 95–150K, but for non-polar ices, such as N_2, CO & CH_4, it may be as low as 20–30K. The resulting chemically enrichment is short-lived ($t \leq$ few$\times 10^3$years) but dramatic, and is seen in the so-called "hot cores" that exist in the inner regions ($n \sim 10^6 - 10^8 cm^{-3}$) of some high mass O-B star-forming regions (eg. Orion/KL; Blake et al., 1987). There is also strong evidence that similar processes occur in low mass star-forming regions. Blake et al. (1994a) has identified separate 'core' and 'wing' components of emission lines observed towards NGC1333-IRAS4. The former show the abundance of most molecular species (including CO) to be depleted by $\sim$ 20–50$\times$ (as compared to the values seen in typical molecular clouds), whilst the wings show marked enhancements of SiO, HCN & CH$_3$OH, perhaps indicative of desorption processes. In IRAS 16293-2422 three different components are identifiable (Blake et al. 1994b), suggestive of a molecular rich hot core, an accreting envelope and a photochemically dominated outer envelope. In the models of line formation, it is assumed that the molecular fractional abundances are independent of position. Even if gas-phase chemistry does not alter this, then the epoch during which the assumptions of the model may be valid is extremely short-lived: Young Class 0 objects are too isothermal for asymmetric line profiles to be possible, whilst late Class 0 and Class I objects deviate sufficiently from isothermality for 'hot core' conditions to pertain in the inner regions of the core.

At a more complex level, chemical processes may take an active, as opposed to purely diagnostic, role in the collapse process. The pressure that opposes gravitational collapse consists of thermal and non-thermal components. Both of these are to varying degrees dependent on the chemistry. The thermal pressure is determined by the cooling efficiency. This is in part dependent on molecular line emissions. The non-thermal pressure may be dominated by magnetic support and/or by Alfvénic turbulence. Magnetic support can be relieved by redistributing the flux by ambipolar diffusion. The efficiency of this process and also the stability, against damping, of long wavelength MHD waves is determined by the ionization level, which is itself sensitive to the abundances of molecular ions (such as HCO$^+$) which have large recombination rate-coefficients. These effects have yet to be taken into consideration in the models, but suggest that chemical processes may control the very nature of the collapse (Hartquist et al., 1993).

7.1. SELF-CONSISTENT MODELLING – THE WAY FORWARD

Molecules which appear to be good tracers of high density condensations are themselves heavily depleted in the innermost regions of the cores, where the dynamical activity is greatest. At first sight it might seem that nature has conspired to make the detection of infall almost impossible. This is not necessarily the case: By modelling spectra at on- and off-centre positions Choi et al. (1995) have deduced that the CS abundance ($\sim 9 \times 10^{-9}$) in the envelope regions is twice as large as in the infall region. But, by observing several transitions in different species, Zhou et al. (1993) have presented a convincing case for collapse in B335.

Prompted by ammonia observations towards L1498 (Myers and Benson 1983), a possible collapse candidate, chemical models of infall regions were developed by Rawlings et al. (1992) in an attempt to explain the absence of high velocity wings in the NH_3 (1,1) and (2,2) line profiles and to predict which species might have broad wings. Ongoing depletion of species heavier than helium was included and the model parameters were adjusted so that depletion accounted for the observed absence of broad NH_3 wings. An important finding of Rawlings et al. was that some molecular species (most notably HCO^+, N_2H^+, CH, OH, HCO, H_2S and HS) may be enhanced (albeit briefly) as a result of depletion. To explain how this might occur, consider what is the main loss channel for HCO^+ if $X(H_2O)/X(e^-) \gtrsim 100$:

$$HCO^+ + H_2O \longrightarrow H_3O^+ + CO$$

As freeze-out ensues, the H_2O is rapidly depleted, the HCO^+ loss route is inhibited and its abundance rises. The abundance peak occurs at a characteristic radius corresponding to a certain (non-zero) infall velocity. It is thus possible to identify molecular species with broad or double-peaked velocity distributions and thus to directly diagnose the infall characteristics. The problem with this approach is that, in addition to the instantaneous density, velocity and temperature profiles, the chemical abundances are sensitively dependent on the evolutionary history of those profiles. There are therefore very many free parameters in these models that can only be constrained by the observation of a large number of chemical species. In order to be descriptive, models have to consider the global chemical evolution of a star-forming region. There are several distinct phases in the early chemical evolution of a protostellar cloud. These include: (a) The dynamical origin and chemical evolution of the ambient interclump medium, from which protostellar cores eventually form, (b) Quasi-static evolution to a bound, 'pre-protostellar', state, (c) Before dynamic collapse initiates, the core may remain in a quasi-static state for a period of time, (d) The isothermal collapse phase, during which the density rises from 10^4 to $>10^8 cm^{-3}$ and freeze out is the dominant chemical process, and (e) Non-isothermal collapse. Chemical evolution continues throughout all of these phases and for most of the early (isothermal) evolution of the core is dominated by low temperature ion-molecule reactions and freeze-out. In modelling L1498, Rawlings et al. assumed the pre-collapse phase to be governed by a simple, retarded free-fall to a truncated isothermal sphere, while the Shu model was employed for the dynamic collapse. The initial chemical conditions were established using a dynamic cycling model (Nejad, Williams & Charnley 1990), with the chemical constraints primarily determined by the observed NH_3 abundance and profile shape.

In finding excellent agreement with the observations, it was also determined that L1498 is chemically young and that dynamic infall must have started shortly after the formation of the isothermal core. A similar conclusion was reached in models of B335 (Rawlings, Evans & Zhou, 1994) which were constrained by the abundances and line profiles of NH_3, H_2CO & CS. Even though such models are preliminary, they are already shedding light on the validity of the dynamical models.

In conclusion, it cannot be over-emphasised that a fully self-consistent approach is essential in identifying the presence and characteristics of protostellar infall. Individual observations, whether they be of morphologies, spectral energy distributions or even spectral lines cannot be used in isolation to give undeniable evidence for the presence of infall. Rather, the observations must be considered together and self-consistently in the context of the chemical and dynamical models. Observational data have now been obtained for several transitions in many molecular species in B335, L1498 and L1262 (another collapse candidate) and are encouraging the development of such a modelling approach; marrying Monte-Carlo radiative transfer techniques to the chemical/dynamical models in order to provide a definitive identification of inflow as well as putting constraints on the dynamical models of the infall and the global evolution of star-forming regions. These models are not perfect; for instance they do not include the effects of non-spherical collapse or of rotation and have not been used to follow the chemical evolution beyond the isothermal stage, but with the large body of observational data that is now available for different molecules that trace different conditions in different parts of the inflow and with varying degrees of spatial resolution, the constraints on the modelling procedure should be fairly rigorous.

References

Adams, F.C., Lada, C.J., & Shu, F.H.: 1987, *Astrophysical Journal* **312**, 788.

André, P., Ward-Thompson, D., & Barsony, M.: 1993, *Astrophysical Journal* **406**, 122.

André, P.: 1995, *Astrophysics and Space Science* **224**, 29.

Bachiller, R., & Cernicharo, J.: 1986, *Astronomy and Astrophysics* **166**, 283.

Beckwith, S.V.W., Sargent, A.I., Chini, R., & Güsten, R.: 1990, *Astronomical Journal* **99**, 924.

Blake, G.A., Sutton, E.C., Masson, C.R., & Phillips, T.G.: 1987, *Astrophysical Journal* **315**, 621.

Blake, G.A., Van Dishoeck, E.F., Jansen, D.J., Groesbeck, T., & Mundy, L.G.: 1994a, *Astrophysical Journal* **428**, 680.

Blake, G.A., Sandell, G., Van Dishoeck, E.F., Groesbeck, T., Mundy, L.G., & Aspin, C.: 1994b, *Astrophysical Journal* **441**, 689.

Bonnor, W.B.: 1956, *Monthly Notices of the Royal Astronomical Society* **116**, 351.

Bontemps, S., André, P., & Ward-Thompson, D.: 1993, *Astronomy and Astrophysics* **297**, 98.

Cabrit, S., Goldsmith, P.F., & Snell, R.L.: 1988, *Astrophysical Journal* **334**, 196.

Calvet, N., Hartmann, L., Kenyon, S., & Whitney, B.: 1994, *Astrophysical Journal* **4434**, 330.

Calvet, N.: 1995, *Revista Mexicana de Astronomia y Astrofisica* In press

Caselli, P., & Myers, P.C.: 1995, *Astrophysical Journal* **446**, 665.

Choi, M., Evans II, N.J., Gregersen, E.M., & Wang, Y.: 1995, *Astrophysical Journal* **448**, 742.

Ciolek, G.E., & Mouschovias, T.Ch.: 1994, *Astrophysical Journal* **425**, 142.

Ebert, R.: 1955, *Zs. Ap.* 217.

Galli, D., & Shu, F.H.: 1993a, *Astrophysical Journal* **417**, 220.

Galli, D., & Shu, F.H.: 1993b, *Astrophysical Journal* **417**, 243.

Hartquist, T.W., Rawlings, J.M.C., Williams, D.A., & Dalgarno, A.: 1993, *Quarterly Journal of the Royal Astronomical Society* **34**, 213.

Hayashi, M., Ohashi, N., & Miyama, S.: 1993, *Astrophysical Journal* **418**, L71.

Ho, P.T.P., Martin, R.N., Myers, P.C., & Barrett, A.H.: 1977, *Astrophysical Journal* **215**, L29.

Iglesias, E.: 1977, *Astrophysical Journal* **218**, 697.

Lada, C.J.: 1985, *Annual Review of Astronomy & Astrophysics* **23**, 267.

Larson, R.B.: 1969, *Monthly Notices of the Royal Astronomical Society* **145**, 271.

Larson, R.B.: 1972, *Monthly Notices of the Royal Astronomical Society* **157**, 121.

Mauersberger, R., Wilson, T.C., Metzger, P.G., Gaume, R., & Johnston, K.J.: 1992, *Astronomy and Astrophysics* **256**, 640.

Minchin, N.R., Ward-Thompson, D., & White, G.J.: 1995, *Astronomy and Astrophysics* In press.

Mouschovias, T.Ch., & Psaltis, D.: 1995, *Astrophysical Journal* **444**, L105.

Myers, P.C., & Benson, P.C.: 1983, *Astrophysical Journal* **266**, 309.

Myers, P.C., Fuller, G.A., Goodman, A.A., & Benson, P.J.: 1991, *Astrophysical Journal* **376**, 561.

Nejad, L.A.M., Williams, D.A., & Charnley, S.B.: 1990, *Monthly Notices of the Royal Astronomical Society* **246**, 183.

Penston, M.V.: 1969, *Monthly Notices of the Royal Astronomical Society* **144**, 425.

Rawlings, J.M.C., Evans II, N.J., & Zhou, S.: 1994, *Astrophysics and Space Science* **216**, 155.

Rawlings, J.M.C., Hartquist, T.W., Menten, K.M., & Williams, D.A.: 1992, *Monthly Notices of the Royal Astronomical Society* **255**, 471.

Shu, F.H.: 1977, *Astrophysical Journal* **214**, 488.

Shu, F.H., Adams, F.C., & Lizano, S.: 1987, *Annual Review of Astronomy & Astrophysics* **25**, 23.

Strom, K.M., Strom, S.E., Edwards, S., Cabrit, S., & Skrutskie, M.F.: 1989, *Astronomical Journal* **97**, 1451.

Terebey, S., Chandler, C.J., & André, P.: 1993, *Astrophysical Journal* **414**, 759.

Ungerechts, H., Walmsley, C.M., & Winnewisser, G.: 1982, *Astronomy and Astrophysics* **111**, 339.

Walker, C.K., Lada, C.J., Young, E.T., Maloney, P.R., & Wilking, B.A.: 1986, *Astrophysical Journal* **309**, L47.

Walker, C.K., Narayanan, G., & Boss, A.P.: 1994, *Astrophysical Journal* **431**, 767.

Ward-Thompson, D., Scott, P.F., Hills, R.E., & André, P.: 1994, *Monthly Notices of the Royal Astronomical Society* **268**, 276.

Whitney, B.A., & Hartmann, L.: 1993, *Astrophysical Journal* **402**, 605.

Winkler, K-H., & Newman, M.J.: 1980, *Astrophysical Journal* **238**, 311.

Wootten, A., Snell, R., & Evans, N.J., II: 1980, *Astrophysical Journal* **240**, 532.

Wootten, A.: 1995, *Astrophysics and Space Science* **224**, 43.

Zhou, S., Evans II, N.J., Butner, H.M., Kutner, M.L., Leung, C.M., & Mundy, L.G.: 1990, *Astrophysical Journal* **363**, 168.

Zhou, S., Evans II, N.J., Kömpe, C., & Walmsley, C.M.: 1993, *Astrophysical Journal* **404**, 232.

Zhou, S., Evans II, N.J., Wang, Y., Peng, R., & Lo, K.Y.: 1994, *Astrophysical Journal* **433**, 131.

PLASMA SPECTROSCOPY AND DIAGNOSIS

ATOMIC PROCESSES IN ASTROPHYSICS

JOHN C. RAYMOND and NANCY S. BRICKHOUSE

Harvard-Smithsonian Center for Astrophysics, 60 Garden Street, Cambridge, MA 02138, USA

Abstract.
   We review the atomic processes responsible for astrophysical emission line spectra, paying particular attention to the reliability of the rates used to determine the physical conditions in the emitting gas. We discuss particular cases where often-neglected processes play important roles.

1. Introduction

The spectrum of an astrophysically interesting object can, at least in principle, be reduced to an unambiguous set of emission line intensities, continuum fluxes, and absorption line profiles. This reduction is generally limited by blending (especially at the low to moderate spectral resolution of most EUV and X-ray instruments), by uncertainty in the intensity calibration (up to 30%, though relative intensities at closely-spaced wavelengths may be more reliable), or "irrelevant" complications such as uncertain interstellar attenuation.

   Given a measured spectrum, one wishes to derive the temperature distribution, density, ionization state and elemental composition of the observed plasma. An enormous number of atomic rates is needed to interpret the spectra of celestial objects in terms of physical parameters. Recent reviews of theoretical emission spectra include Raymond (1988) and Mewe (1991) concentrating on X-ray spectra, and Brickhouse, Raymond, and Smith (1995) and Mason and Monsignori-Fossi (1994) concentrating on the EUV. Critical reviews of atomic rates include the collision strength bibliography of Pradhan and Gallagher (1992), an issue of "Atomic Data and Nuclear Data Tables" (Vol. 57, Nos. 1/2, 1994) devoted to collision rates, the ionization rate study of Kato, Masai, and Arnaud (1991), The Opacity Project book just published by The Institute of Physics Publishing and the review of nebular emission line data of Pradhan and Peng (1995).

   The above reviews are reasonably comprehensive, and progress has not been so rapid that they are entirely out of date. Therefore, we will try to complement these works by giving a brief overview of which rates are most important in common applications. We also give some opinions as to the typical level of uncertainty in the rates. These are based partly on the degree of consistency among measured or computed cross sections and partly on the scatter obtained when one attempts to fit $\sim$ 100 lines at once (Brickhouse *et al.* 1995). It is also important to maintain a healthy skepticism toward generalizations about which processes are important. The

Astrophysics and Space Science **237**: 321–340, 1996.

Universe is large enough that even the most obscure process is probably important somewhere. Therefore, we give examples of some of the lesser-known correction terms which can dominate in unusual circumstances.

We close this introduction with a brief harangue. One approach to the bewildering amount of information in a high quality spectrum is to choose one or two lines from which to derive one physical quantity, such as N_e, T_e, or an elemental abundance ratio. This approach has the considerable advantage that one can choose lines which are bright and easy to measure, and for which reliable atomic data are available. Its disadvantages are also considerable. Astrophysical plasmas are seldom isothermal or isochoric. Also, someone who selects a single diagnostic ignores a great deal of information obtained at great expense and may obtain results which are inconsistent with the intensities of other lines in the spectrum. And finally, our opinions as to which atomic rates are most reliable have frequently proven incorrect. Therefore, we believe it to be more profitable to fit a large number of lines at once. This provides a sobering perspective as to the reliability of both the observations and the theory, and it provides a feeling for the range of conditions within the object observed.

2. Basic Assumptions of Plasma Emission Models

2.1. Level Population Calculations

The radiative decay of an excited state of an atom or ion produces an emission line whose intensity depends on the population of the excited level and on the Einstein transition probability A_{kj} from upper level k to lower level j. For an optically thin plasma, all emitted photons "escape" and thus the radiated power per volume, or volume emissivity ε, is given by

$$\varepsilon = \frac{hc}{\lambda} N_k A_{kj}, \tag{1}$$

with ε in ergs cm^{-3} s^{-1} and N_k the level population density. For small optical depths (of order a few) Equation 1 may be modified by a photon escape probability factor, which depends on the scattering geometry. For larger optical depths, the assumptions of our model break down, and a full radiative transfer model is required.

The population density of level k is in general a function of time, which includes all processes that populate and depopulate the level.

$$\frac{dN_k}{dt} = \sum_i N_i R_{ik} - N_k \sum_i R_{ki} + N_k F_k, \tag{2}$$

where the rates R_{ik} and R_{ki} in s^{-1} are summed over all possible energy levels i. F_k is the net flux of ions in excited state k into (or out of) the emitting region by advection or diffusion processes. In this section we assume $F_k = 0$, and focus on the atomic rates that contribute to line emission.

Under conditions of statistical equilibrium the level populations are constant with time, and are found by the simultaneous solution of the rate equations for all energy levels. With a model of time-dependent heating or cooling, non-equilibrium problems may be solved as well, although models may be less constrained as more free parameters become available. We discuss some examples in the next section. The most general treatment of transition rates involves the computation of transitions between all levels, including transitions between levels of different ionization stages. The rate from any level m to any other level n is then given by

$$R_{mn} = R_{ioniz} + R_{recomb} + \sum_s N_s q_{s,cx} + A_{rad} \qquad + \sum_s N_s q_{s,coll}, \qquad (3)$$

where R_{ioniz} is the sum of photoionization and collisional impact ionization rates, R_{recomb} is the sum of radiative, dielectronic, and 3-body recombination rates, and the charge exchange rate is the sum of the individual charge exchange rates, where $q_{s,cx}$ are the rate coefficients and N_s is the population density of the interacting species. A_{rad} includes stimulated absorption (photo-excitation) as well as spontaneous radiative decay, and the collisional rate includes collisional excitation and de-excitation processes, with $q_{s,coll}$ the collisional rate coefficient for interaction with species s. In detail the rate from lower level j to upper level k is given by

$$R_{jk} = N_e q_{ioniz} + \overline{S}\beta_{photoioniz} + N_p q_{cxioniz} + \overline{J}B_{jk} + N_e q_{e,ex} + N_p q_{p,ex}, (4)$$

where $\overline{J}$ and $\overline{S}$ are radiative source terms for photoexcitation and photoionization, respectively. The collisional excitation processes are generally dominated by electron collisions, but protons or other species may also be important. The collision rates are proportional to the densities of the impact species through N_e and N_p. The rate coefficients q_{ioniz} for electron impact ionization from level j to level k, $q_{e,ex}$ for electron impact excitation, $q_{p,ex}$ for proton impact excitation, and $q_{p,cxioniz}$ for charge exchange have units of cm^3 s^{-1}.

$$B_{jk} = \frac{g_k}{g_j}\frac{\lambda^3}{2hc}A_{kj}, \qquad (5)$$

where g_k/g_j is the ratio of the statistical weights. For a similar level of detail, the depopulation rate from level k to level j is given by

$$R_{kj} = N_e \alpha_{rad} + N_e \alpha_{di} + N_e^2 \alpha_{3-body} \qquad + N_{H^\circ}\, q_{cxrecomb} + A_{kj}$$
$$+ N_e q_{e,de-ex} + N_p q_{p,de-ex}, \qquad (6)$$

where α_{rad}, α_{di}, and α_{3-body} are the radiative, dielectronic, and three-body recombination rate coefficients, respectively. The rate coefficients $q_{e,de-ex}$

for electron impact excitation, $q_{p,de-ex}$ for proton impact excitation, and $q_{cxrecomb}$ for charge exchange have units of cm^3 s^{-1}. By the principle of detailed balancing, the three-body recombination is the inverse of electron impact ionization, and thus

$$\alpha_{3-body} = \frac{1}{2} \frac{h^3}{(2\pi m k T_e)^{\frac{3}{2}}} \frac{g_j}{g_k} \exp(\frac{\triangle E}{kT_e}) q_{ioniz}, \tag{7}$$

where α_{3-body} has units of cm^6 s^{-1}. Similarly, $q_{cxrecomb}$ is the inverse of $q_{cxioniz}$. With typical astrophysical abundances, charge exchange with neutral hydrogen (H^o) is the most important charge exchange recombination process. Through detailed balance q_{de-ex} is related to q_{ex},

$$q_{de-ex} = \frac{g_j}{g_k} \exp(\frac{\triangle E}{kT_e}) q_{ex}, \tag{8}$$

where $\triangle E$ is the energy difference between levels j and k and T_e is the electron temperature.

The rate equations for all levels of all ionization stages may be solved simultaneously in the steady-state approximation for a given T_e, N_e, and, where collisions with hydrogen are important, N_H/N_e. Double ionization and recombination processes are relatively rare, and are ignored here, although K-shell ionization followed by Auger ionization may be significant in rapidly ionizing plasmas, such as young supernova remnants. The general case outlined here is in practice made more tractable by imposing a number of simplifying assumptions. Solving for the ionization balance may be decoupled from the line excitation problem for the low density case in which most of the ion population resides in the ground state level. First the population density of a given ionization stage is determined, then the level populations for that ion may be computed, including ionization and recombination population processes. At higher densities, the population of metastable levels and fine-structure levels within the ground state may build up substantially. For example, one might need to include a collisionally excited metastable level in the ionization balance if collisional ionization from that level is important.

The statistical equilibrium problem may also be simplified by lumping together energy levels. Two examples are the treatment of collisional ionization from subshells and the treatment of excitation and decay for multiplets (statistically weighted averages of transitions between fine-structure levels). In practice, the number of energy levels n included in the model ion is limited by the program's goals. Above some critical n collisional processes are much more important than radiative processes, and the level populations are determined by their statistical weights. The number of levels can be limited much further if the levels that emit observable lines are populated only by collisional excitations and not by radiative cascades from the decay of higher levels.

Sobelman, Vainshtein, and Yukov (1981) describe three basic types of plasma emission models: coronal, collisional-radiative (CR), and local thermodynamic equilbrium (LTE). The electron density and optical depth essentially determine the range of applicability of these models. For very high densities (LTE) or optical depths the calculation of level populations is greatly simplified since collisions dominate. At intermediate densities ($N_e \sim 10^{12} - 10^{18}$ cm^{-3}), the most stringent requirements are placed on the formulation of the CR (or non-LTE) model ion and on the coupling of ionization balance and level population. At very low densities, the problem again is simplified, as most of the population of an ion exists in the ground state, so only ground state levels determine the ionization balance. In the coronal model, heating balances the radiative power loss, and the heating mechanism need not be specified for a unique temperature determination. For many high energy astrophysical sources, photoionization by an extremely hot source may dominate the ionization of surrounding gas. For these cases the nebular approximation is analogous to the coronal case, in that most of the population of the ions is found in the ground state (cf. Osterbrock 1974). Again, for moderately high densities, such as might be expected in low mass X-ray binaries ($\sim 10^{13}$ cm^{-3}), a full CR model is necessary (Liedahl *et al.* 1990, 1992). In the nebular model the photoionizing source is often the dominant heating source, and thus heating and ionization state are coupled. Kahn and Liedahl (1995) review X-ray spectroscopy with comparisons of coronal and photoionized sources.

The coronal equilibrium model then separates into two independent sets of equations:

$$N_z N_e(\alpha_{rad,z} + \alpha_{di,z}) + N_z N_e q_{ioniz,z} = N_{z-1} N_e q_{ioniz,z-1}$$
$$+ N_{z+1} N_e(\alpha_{rad,z+1} + \alpha_{di,z+1}), \tag{9}$$

for the ionization balance, and

$$N_k N_e \sum_l q_{e,kl} + N_k \sum_j A_{kj} = \sum_l N_l A_{lk} + N_e \sum_j N_j q_{e,jk}, \tag{10}$$

for the level populations. In the very low density coronal model, and for a resonance line which decays only to the ground, the coupled level population equations reduce to the simple formulation

$$N_k A_{kg} = N_g N_e q_{e,gk}, \tag{11}$$

where g denotes the ground state. With the line emissivity given by Equation 1, coupled level population calculations are then unnecessary to produce the spectrum of strong resonance lines. On the other hand, electron densities in the solar corona, for which the coronal model is named, are not sufficiently low to ignore collisional de-excitation for multiplet ground states and low-lying metastable levels. Even strong resonance lines such as Fe IX $\lambda 171.03$

may be affected by the depopulation of the ground state at active region densities ($> 10^9$ cm^{-3}). Furthermore, for metastable levels of certain coronal ions, such as the $3p^3$ $^2P_{3/2}$ excitation of [Fe XII] $\lambda 1242$, cascades from upper levels can contribute more than half of the level population. We prefer to calculate the full set of transitions in order to solve as accurately as possible for useful N_e and T_e diagnostic lines.

2.2. ATOMIC RATE COEFFICIENTS

The rate coefficients given in these equations are of general importance and we discuss these rate coefficients in some detail now. Fitting formulas which are useful for calculating rates can be found in the many review articles listed in the Introduction. We concentrate here on the rates of dominant processes.

We expect the radiative rate coefficients to be fairly accurate for strong transitions. Radiative recombination is the inverse of photoionization, and thus related to the photoionization cross sections by detailed balance. Photoionization cross sections in the threshold region for ions of astrophysical interest are recently available as part of the Opacity Project (1995), and continue to be updated. The extrapolation to high energy of Verner and Yakovlev (1994) includes the partial photoionization cross sections for subshells (Verner *et al.* 1993). These data represent a great improvement over the last few years. Thus the accuracy might be expected to be 10%.

As mentioned above, the strong resonance lines in coronal models do not even depend on the radiative transition probabilities. While transition probabilities, or A-values, for these lines are probably accurate to a few percent, the intersystem and forbidden line A-values may be much less accurate. An example is C III] $\lambda 1909$ for which theoretical rates agree with each other to better than 10%, but the laboratory measurement is discrepant with some of the calculations by 20% (Kwong *et al.* 1993). For multi-electron ions of high Z elements, the theoretical values may disagree by 25% or more, as in the case of the well-known coronal green line, [Fe XIV] $\lambda 5303$ (Bhatia and Kastner 1993; Huang 1986; Froese-Fischer and Liu 1986). Since line ratio diagnostics for N_e ultimately derive from the trade-off between collisional de-excitation and radiative decay, relatively large uncertainties are possible in the derived densities. Furthermore, line ratios of coronal forbidden lines, which are used as temperature diagnostics, are also subject to inaccuracies of the A-values (*cf.* Esser *et al.* 1995).

Under equilibrium conditions, electron collisional rate coefficients are averages of the cross section over the electron velocity distribution. This velocity distribution f(E) is usually taken to be Maxwellian, such that,

$$f(E)dE = \frac{2}{\sqrt{\pi}} \left(\frac{E}{kT} \right)^{3/2} \frac{e^{(-E/kT)}}{\sqrt{E}} dE. \tag{12}$$

When collisions with protons are important, the proton velocity distribution may also be assumed Maxwellian, even without equal proton and electron temperatures. Since collisional processes dominate in coronal plasmas, the uncertainties in their calculated values tend to dominate the errors. Very few laboratory measurements have been made, and many assessments of accuracy are based on the relative convergence of different theoretical approaches.

Direct collisional ionization rate coefficients may be calculated as a sum over the subshells of the ionizing ion, with each term in the sum proportional to a Boltzmann factor $e^{-\triangle E/kT}$, where $\triangle E$ is the energy difference between the subshell and the ground state of the ionized ion. For highly stripped ions an additional ionization process, excitation-autoionization, is important. Here an inner subshell electron is collisionally excited to a bound state above the ionization threshold, and subsequently autoionizes. (Autoionization is also called Auger ionization. An alternative decay path for this unstable bound state is the radiative decay process called excitation fluorescence.) Experimental measurements such as those of Gregory *et al.* (1986, 1987) confirm both direct and excitation-autoionization contributions. The typical uncertainties are probably about 20%. The situation is better for ions studies with crossed beams, and worse for very complicated ions.

Dielectronic recombination generally dominates recombination for coronal plasmas. The exceptions are very cool plasmas and ions which have no low-lying excited states: bare nuclei, H-like, He-like and Ne-like ions. The process involves formation of a doubly excited level, followed by radiative decay of the captured electron. The incoming electron has an energy below the threshold for the collisional excitation, and thus is captured into a highly unstable level above the ionization threshold of the recombined ion. This electron can either autoionize, or it can radiatively decay. The radiative decay process leads to the recombination. The photons emitted contribute to dielectronic satellite lines in the emission line spectrum. While the satellite lines are important diagnostics for X-ray spectra, they are usually weak relative to the collisionally excited lines. Among the complications of calculating dielectronic recombination rate coefficients are the need for very large numbers of energy levels (several hundred for some cases) and the calculation of accurate autoionization rates for doubly excited levels. For high n, LS coupling breaks down. Furthermore, the experimental verification of dielectronic recombination cross sections is limited (*e.g.* Story, Lyons, and Gallagher 1995). For H-like and He-like ions, as well as for most low to moderate Z ions, the uncertainties may be about 20%, but discrepancies as large as factors of two exist between different theoretical calculations (*cf.* Arnaud and Raymond 1992).

The effective collisional excitation rate coefficient (i.e. the Maxwellian integral $q_{e,ex}$) depends on the product of $\frac{1}{\sqrt{T}}e^{-\triangle E/kT}$ and the collision

strength Ω_{jk}. At high temperature (relative to $\triangle E$), Ω_{jk} is a slowly varying function of energy; however, for energies near threshold, the colliding
electron may be captured during the excitation process as described for
dielectronic recombination. When autoionization occurs, rather than photon emission leading to recombination, the resonant process can dramatically increase Ω_{jk} at that energy. Thus Ω_{jk} may have a complicated structure
due to resonances near threshold. The net effect in the Maxwellian aver-
age is to increase the effective collision strength particularly for forbidden or
intersystem transitions. Collision strengths may be as accurate as 10% if resonances are properly treated, as in close coupling calculations (*e.g.* Callaway
1994). In distorted wave approximations resonances are ignored and results
are generally accurate to only about 30%. If the electron temperature is low,
the Maxwellian average is more heavily weighted by the resonant structure
and the accuracy decreases accordingly. Furthermore, ignoring resonances
systematically affects the very line ratios (forbidden-to-allowed) that are
used for electron density diagnostics. For the complex multi-electron ions,
we believe factor of two errors in $\triangle n = 0$ transition lines are likely to remain;
as $\triangle n$ increases, the simple approximations (*e.g.* Gaunt factor) still in use
in astrophysics become increasingly less accurate. Factor of five errors have
been suggested for $\triangle n = 2$ transitions of highly ionized Fe around 1.0 keV
(Liedahl, Osterheld, and Goldstein 1995). More accurate calculations for
X-ray and EUV lines are a high priority.

2.3. Spectral Emission

Once the level populations are calculated the line emission is given, in units
of ergs cm^3 s^{-1}, by

$$P_\lambda = \frac{\varepsilon}{N_e N_H} \tag{13}$$

It is necessary to relate the level population N_k to the given quantity in the
model N_e, where

$$N_k = \left(\frac{N_k}{N_z}\right)\left(\frac{N_z}{N_Z}\right)\left(\frac{N_Z}{N_H}\right)\left(\frac{N_H}{N_e}\right) N_e, \tag{14}$$

and $\frac{N_k}{N_z}$ is the solution of the level population equations for level k relative
to the entire population of that ionization stage z; $\frac{N_z}{N_Z}$ is the population of
z relative the population of element Z, derived from the ionization balance;
$\frac{N_Z}{N_H}$ is the element abundance relative to hydrogen; and $\frac{N_H}{N_e}$ is about 0.8
for fully ionized gas of cosmic abundance. A standard set of solar elemental
abundances is very widely assumed for interpreting low spectral resolution
observations, but abundances in the solar corona are observed to vary by

factors of 3, so this assumption is dangerous. The intensity of a line observed at the Earth, from a source at distance r, in units of ergs cm^{-2} s^{-1}, is then

$$I_{Earth} = \frac{P_\lambda}{4\pi r^2} \int N_e N_H dV. \tag{15}$$

The integral $\int N_e N_H dV$ is the emission measure, generally taken over a specified temperature interval.

Spectral observations which yield a large number of high signal-to-noise, unblended lines from a range of temperatures offer the opportunity to construct emission measure distributions and thus utilize the wealth of spectral information on densities, abundances, and other parameters in the context of the temperatures of line formation. Until recently, few non-solar moderate resolution spectra for temperatures above $\sim 10^6$ K existed. Since the launch of the $EUVE$ satellite in 1992, a number of high quality spectra have been obtained for stellar coronae, and analysis based on emission measure modeling is yielding insights into the conditions of these sources. While the spectral resolution is modest, and line blending compromises the interpretation for some lines, the flux calibration is remarkable ($\sim 15\%$). Fig. 1 shows three sample spectra, chosen for their different dominant temperatures, obvious from their strong spectral lines: Capella ($\sim 10^7$ K, based on Fe XVIII–XXIII), Procyon ($\sim 10^6$ K, based on Fe IX–XIV), and Epsilon Eri ($\sim 4 \times 10^6$ K, based on strong Fe XV and XVI). Analysis based on emission measure distribution modeling of strong lines for these sources may be found in Dupree et $al.$ (1993), Brickhouse et $al.$ (1995), and Brickhouse (1995) for Capella, in Drake, Laming, and Widing (1995) for Procyon, and Laming, Drake, and Widing (1995) and Schmitt et $al.$ (1995b) for Epsilon Eri.

As an example of modeling a large number of emission lines, we present in Fig. 2 models of the emission measure distribution for a solar active region (Brickhouse et $al.$ 1995). Other such analyses may be found in Raymond (1988) and Lang, Mason, and McWhirter (1990). The intensities for nearly 100 emission lines of iron over several ionization stages (Fe IX–XVII) are taken from the Solar EUV Rocket Telescope and Spectrograph ($SERTS$) active region catalogue of Thomas and Neupert (1994). Several branching ratio pairs are found in this set of lines, and the agreement for all but a few is within the expected range. The models compare the ionization balance of Arnaud and Rothenflug (1985) and Arnaud and Raymond (1992). Each line emissivity is integrated over the entire emission measure distribution, and the fit is achieved iteratively. The emission measure distribution is given in steps of 0.1 dex in T_e (K). The comparison between observed and predicted lines in Fig. 3 is for the newer ionization balance, but the agreement is comparable for both models.

This comparison gives some sense of the accuracy of the atomic rates, when the instrumental flux calibration ($\sim 25\%$ for first order lines) is taken into account. While uncertainties in the ionization and recombination rates are quite large, these data do not provide a test of the ionization balance models, since shifts in peak temperatures and relative ion population maxima may be easily accommodated by the fitting. In fact, the overall agreement of the observed lines with these models is only slightly worse than for lines taken for each ionization stage at its temperature of maximum ionization population. Thus, a factor of $\pm 35\%$ is the overall agreement of the line emissivities. While this level of agreement with theory is somewhat worse than one might expect from the collision strength uncertainties, we point out that many of the lines of Fe IX–XIV are sensitive to N_e. We also expect some spread in the forbidden-to-allowed line ratios due to omission of resonances in some of the collision strength calculations and the additional radiative terms. Furthermore, the observations may sample regions of different density along the line of sight. Of course, multiple densities would create a large spread, since the lines all have different nonlinear density dependences. While this model has been fit for all lines calculated at the best fitting density, $N_e = 10^{10}$ cm^{-3}, the spread in N_e derived from different line ratios is several orders of magnitude. Part of the scatter is undoubtedly due to measurement error as well. We include weak lines, and some of the lines may be blended. More detailed discussion of this analysis is found in Brickhouse *et al.* (1995).

Continuum emission has three major contributions. Bremsstrahlung produced by electrons which are accelerated in the electric fields of fully stripped H and He is a major continuum source throughout the high temperature region. Two-photon continua result from the decay of the H-like $2s\,^2S$ level and the He-like $1s2s\,^1S$ level. The sum of the photon energies equals the excitation energy of the level, and two-photon emission is an important contribution around 10^6 K. Radiative recombination produces a continuum spectrum which, as described above for the recombination rate coefficient, depends on the photoionization cross section. The current generation of model codes generally do not include the continua from recombination to excited levels. Under most circumstances this contribution is only a few percent. The calculations for all three sources of continuum emission should be accurate to about 10%.

3. Things That are Often Left Out

3.1. MIXED COLLISIONAL IONIZATION AND PHOTOIONIZATION

Most model calculations are based on the assumption that photoionization is either dominant or completely negligible. Generally photoabsorption is taken to dominate the heating in photoionization models. Intermediate cases where both photoionization and collisional ionization contribute are not much more difficult to compute, but they represent an unpleasant increase in

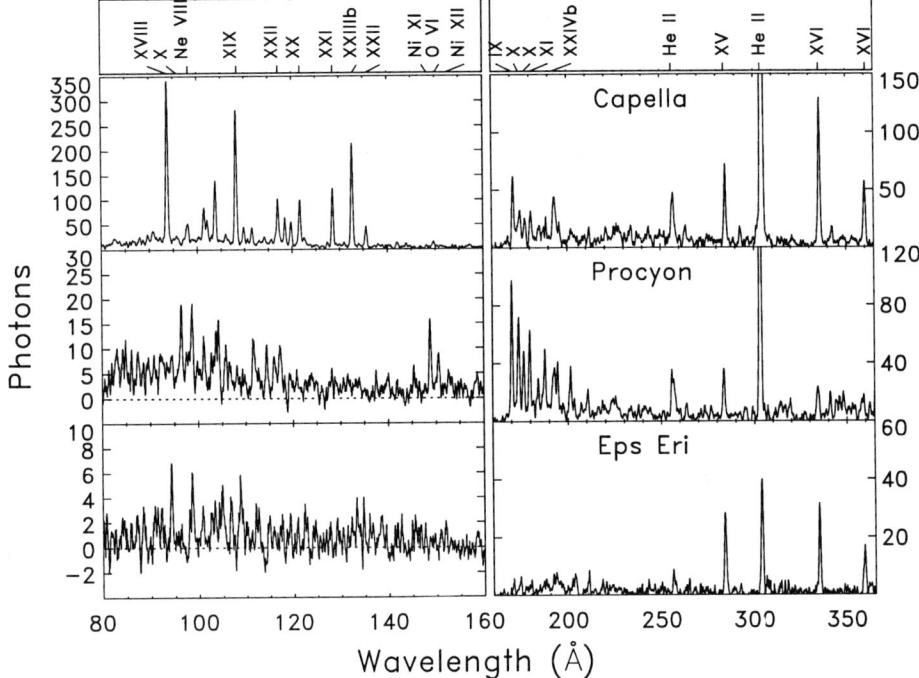

Fig. 1. *EUVE* spectra in the Short Wavelength (*left panel*) and Medium Wavelength (*right panel*) spectrometers for Capella (a binary system with G0 III and G8 III components), Procyon (F5 IV), and Epsilon Eridani (K2 V) (Courtesy of A. K. Dupree and G. J. Hanson). Species of Fe are marked by ionization stage. Fe XXIII λ132.85 is blended with Fe XX λ132.85; Fe XXIV λ192.02 is blended with Fe XII λ192.39, and may be difficult to resolve from Ca XVII λ192.86, and O V λλ192.75, 192.80, and 192.91.

the number of free parameters. Radiatively cooling gas, for instance behind a shock wave, presents a situation in which collisional ionization dominates the hotter regions, while photoionization becomes more important as the gas cools, so shock wave models always include both processes (*e.g.* Hartigan, Raymond and Hartmann 1987). Further complications arise if an external radiation source, such as an AGN or planetary nebula central star, is nearby (*e.g.* Contini and Viegas-Aldrovandi 1987). Any other interface between hot and cold plasma will present the same problem at some level. Examples include the helium lines from the Sun (*e.g.* Athay 1988), cooling flows in galaxy clusters (Voit, Donahue and Slavin 1994) or galactic halos (Sokolowski 1993), and thermal conduction boundaries (Slavin 1989).

3.2. EXCITATION BY IONS

In a plasma with equal electron and ion temperatures, electrons move faster than protons by a factor of 43. Furthermore, the transitions most often

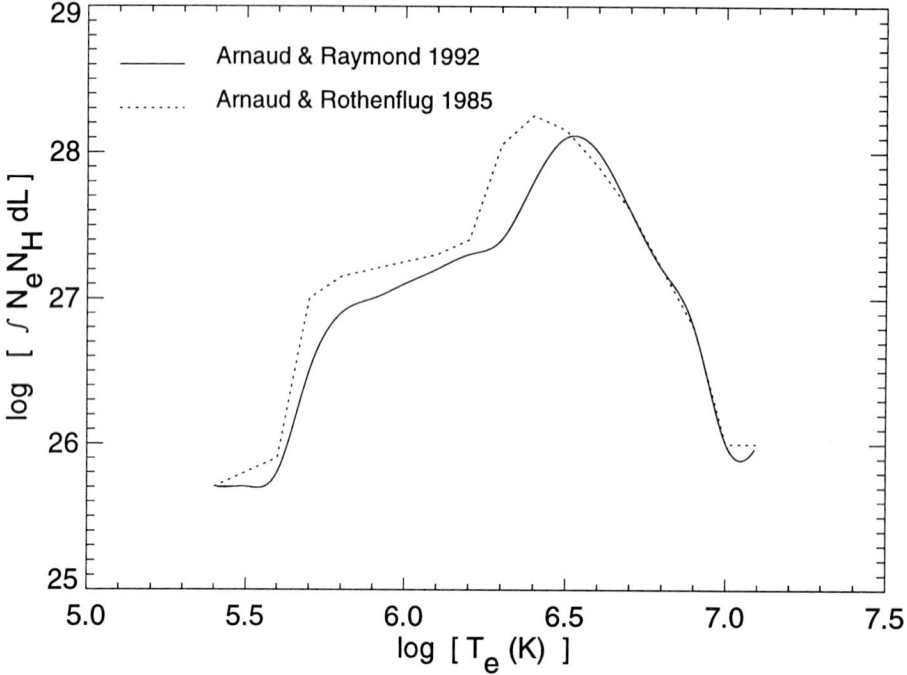

Fig. 2. Models of the emission measure distribution for the SERTS data, from Brickhouse *et al.* (1995). The solid line is the model for $N_e = 10^{10}$ cm^{-3}, assuming the ionization equilibrium of Arnaud and Raymond (1992). The dotted line is the model which assumes the ionization equilibrium of Arnaud and Rothenflug (1985).

observed have $\Delta E \sim kT$, and the cross sections at the ion speeds are small. However, there are cases where proton excitation is quite important. If $kT \gg \Delta E$, the cross section for excitation by electrons drops at least as fast as ln E/E, and a proton appears to the ion much like a lower energy electron. Most of the cases which have been studied are fine structure transitions, such as infrared lines in the ISM (Bahcall and Wolf 1968), transitions which affect density diagnostics (*e.g.* those among 2s2p^{3}P levels in Be-like ions; Doyle *et al.* 1980) and the coronal lines [Fe X] $\lambda6374$ and [Fe XIV] $\lambda5303$ (Seaton 1964). Protons can initiate 2s-2p transitions in H-like ions, affecting the Lyα doublet ratio (Zygelman and Dalgarno 1987). Protons are also important for redistributing the populations of Rydberg levels, with consequences for radio recombination lines and for the density sensitivity of dielectronic recombination. Walling and Weisheit (1988) review these applications and provide formulae for semiclassical estimates of the cross sections.

Very fast shock waves also create conditions where proton excitation must be considered. Immediately behind the shock, low ionization species are immersed in extremely hot gas, so $kT_i \gg \Delta E$ for many emission lines.

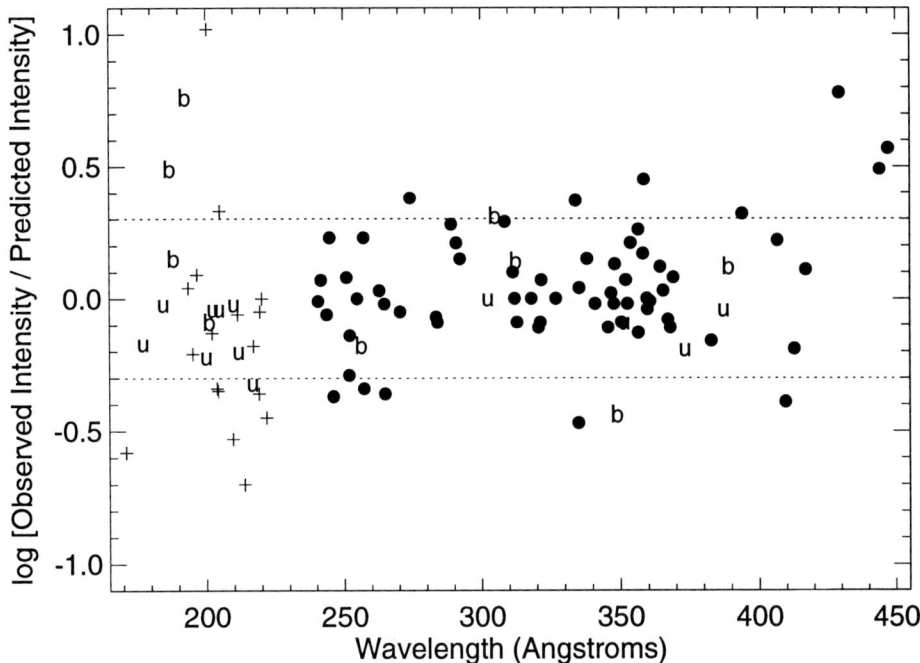

Fig. 3.   Comparison of the observed and predicted intensities for the SERTS data and the solid line model in Fig. 2, shown versus wavelength. Filled circles = first order; plus signs = second order; b = blended (with a different element or ion); i = upper limit for the observation. Between the dotted lines, agreement is within a factor of 2.

Smith, Laming and Raymond (in preparation) find proton rates for excitation of Hα and for ionization of H$^\circ$ which are comparable to the electron rates for 1000 km/s shocks. Polarization of the Hα line is a possible diagnostic (Laming 1990b). Proton excitation of the UV lines of C IV, N V and O VI observed in the 2200 km/s shock of SN 1006 (Raymond, Blair and Long 1995) are larger than the electron rates, according to the formulae of Walling and Weisheit. The cross sections for excitation by ions generally scale as the square of the charge of the impinging ion, so helium nuclei can be significant in some cases.

3.3. DENSITY SENSITIVE IONIZATION BALANCE

With the coronal approximation that all ions reside in their ground states, only the ionization and recombination rates from these states are of interest. However, Be-like and Mg-like ions have relatively low-lying metastable levels of high statistical weight ($g_m = 9$ vs. $g_g = 1$ for the ground state) so that most of the ions can be in the excited levels. Vernazza and Raymond (1979) attempted to incorporate those levels, but few of the relevant rates were available. A potentially interesting case, though we know of no specific examples where it has been demonstrated to be important, is an ion whose ground state ionization potential is just above the photon energy of a strong line, but whose metastable level can be photoionized. The ^{3}P and ^{1}D levels of C I lie on either side of the Lyα photon energy, so one might expect a substantial decline in the C I emission of photodissociation regions or slow shock waves with increasing population of ^{1}D at high density.

Dielectronic recombination rate coefficients tend to decline at high densities as the highly excited intermediate states suffer collisional ionization (Burgess and Summers 1969). For charge states z $\sim$ 4, this can reduce the rate by a factor of 2 at the densities N_e $\sim 10^{11}$ cm^{-3} found in solar active regions. The effect is strongest for ions whose dielectronic recombination goes by way of low photon energy transitions, such as Li-like ions (Summers 1976). The sensitivity scales roughly as $(N_e/Z^7)^{0.2}$. It is less likely to be important for highly charged ions or H-like, He-like or Ne-like ions, where the (relatively weak) dielectronic recombination requires $\Delta n \neq 0$ transitions. It is less important for low temperature dielectronic recombination (Storey 1981), which proceeds via much lower excited states. The dielectronic recombination is also sensitive to electric fields.

3.4. ELECTRIC FIELDS

The Stark effect can blend the higher levels of hydrogen into a quasicontinuum. A subtler effect is the enhancement of dielectronic recombination rates by Stark mixing of the ℓ levels of the doubly excited intermediate states (Reisenfeld 1992; Badnell et al. 1993; Story et al. 1995), but this effect is just beginning to be explored. Foukal, Little and Gilliam (1988) discuss the use of Stark broadening of the Paschen lines as an electric field diagnostic for the solar chromosphere. For both applications, the effects of plasma microfields (the random fields due to nearby ions) may be difficult to distinguish from macroscopic electric fields.

3.5. RESONANT SCATTERING

Resonant scattering can affect observed line intensities in several ways. At optical depths $\sim$ 1, the main effect is likely to be a geometrical redistribution of photons from optically thick directions to optically thin ones. Radiative shock waves in supernova remnants produce thin sheets of emitting gas, and the bright filaments seen in optical forbidden lines are tangencies of those sheets to the line of sight. While the forbidden lines have negligible optical depths, UV lines such as C IV λ1550 are strongly attenuated in the brightest filaments (Cornett et al. 1992). A similar effect has been invoked to explain Fe XVII line ratios in solar active regions (Schmelz, Saba, and Strong 1992). At somewhat greater optical depths, several scatterings can convert a photon to two or more lower energy photons. For an example, it takes on average eight scatterings to convert Lyβ to Hα plus a two photon pair (at low density) or Hα and Lyα (at higher density). Similar conversions ought to be observable in the X-ray emission of hydrogenic and helium-like ions if optical depths are large. Finally, at large optical depths and high densities, collisional deexcitation becomes important as the lines begin to thermalize. Schrijver, van den Oord, and Mewe (1994) invoke optical depth effects to explain discrepancies between EUVE spectra of late-type stars and model calculations, but a complete calculation is likely to show drastic effects for some lines they did not consider. Schmitt, Drake, and Stern (1995a) suggest that contributions of weak lines to what Schrijver *et al.* identified as continuum could better explain the spectrum.

3.6. RADIATIVE EXCITATION

Bowen (1935) investigated the resonance fluorescence process which converts a He II Ly α photon at 304 Å to O III photons near 3000 Å plus an EUV photon. Subsequent investigators have used improved atomic rates or extended the calculations to higher densities and line widths (Kallman and McCray 1980; Bhatia and Kastner 1990). Many other wavelength coincidences have been explored, including Ly β and O I (Athay and Judge 1995) and Ly α and H$_2$ (Jordan *et al.* 1977; Black and van Dishoeck 1987). Even an approximate coincidence can be important. Schmid (1989) showed that Raman scattering of the O VI doublet by hydrogen produces the $\lambda\lambda$ 6830, 7088 lines seen in symbiotic stars.

Continuum fluorescence can also be important. Ferguson *et al.* (1995) discuss the production of C III λ977 and other UV emission lines by continuum fluorescence in NGC 1068 as an alternative to shock heating. Continuum fluorescence of H$_2$ produces strong molecular features in photodissociation regions (Sternberg 1989).

Mason (1975) shows the effects of the photospheric radiation field on the emissivity calculations for a number of Fe and Ca optical forbidden lines in the solar corona. Calculations with updated atomic rate coefficients and photospheric radiation show that the excitation of [Fe X] λ6374 and [Fe XIV] λ5303 in low-density coronal hole regions may by dominated by this term (Esser *et al.* 1995).

3.7. POLARIZATION AND EMISSION ANISOTROPY

Particle velocity distributions, and therefore the radiation from the particles, are generally assumed to be isotropic. An exception is a directed particle beam producing X-rays in a solar flare. Laming (1990a) has shown that excitation by a beam produces different anisotropies in the Mg XII Ly α 1s-2p $^2P_{1/2}$ and 1s-2p $^2P_{3/2}$ lines. This can account for observed departures from the expected 1:2 ratio, depending upon the observer's location. Significant departure from isotropy also occurs behind a very fast shock wave. A neutral hydrogen atom can find itself immersed in very hot electrons and ions moving at 3/4 the shock speed. Laming (1990b) has shown that this can produce polarization in Hα, which could be a useful diagnostic.

3.8. NON-MAXWELLIAN VELOCITY DISTRIBUTIONS

Direct measurements of the electrons in the only accessible astrophysical plasma - the solar wind - always seem to show non-Maxwellian distributions. Nevertheless, nearly all interpretations of astronomical spectra begin with the assumption that the distribution is Maxwellian. The most common departure from Maxwellian may be a non-thermal power law tail at high energies. This will naturally affect those processes with the highest thresholds, such as ionization or excitation of high-lying levels. Recombination and excitation of lines with $\Delta E \leq kT$ will be relatively unaffected. Owocki and Scudder (1983) explored the consequences of these distributions for the ionization state of the solar wind. Anderson, Raymond and van Ballegooijen (1995) considered the ionization balance and EUV emission line intensities for several velocity distributions in the velocity filtration picture of coronal heating (Scudder 1992). Laming (1990a) considered X-ray lines in solar flares. In general, a departure from Maxwellian which sets in only far above kT will have little effect on the dominant radiation processes, though it may affect the ionization state and energy balance (e.g. cosmic rays in a molecular cloud). If the departure sets in at ~ 10 kT, the ionization state is drastically affected. It is unfortunately difficult to come up with a clean diagnostic for non-Maxwellian distributions. Line ratios of He-like ions are a recently suggested possibility (Gabriel et al. 1991).

3.9. RECOMBINATION TO EXCITED LEVELS

In a collisionally ionized plasma, recombination to excited levels contributes little to emission line intensities. This comes about because ionizations must balance recombinations in equilibrium, and ionization rates are typically slower than the excitation rates of bright emission lines by at least an order of magnitude. Recombination to an excited level may be more competitive if the excitation from the ground state is forbidden or if the gas is overionized due to photoionization or extremely rapid cooling. Of course, if one is considering an important diagnostic line ratio, even a modest recombination contribution must be considered, as in He-like and H-like ions (Mewe and Gronenschild 1981). Other important cases include the X-ray emission in photoionized coronae in Low Mass X-ray Binaries (Liedahl et al. 1992), emission lines from gas strongly cooled by high concentrations of metals or dust (e.g. nova shells - Escalante and Dalgarno 1991), and specific lines pumped by dielectronic recombination at low temperatures (e.g. C III λ2297, N IV λ1718; Harrington, Lutz and Seaton 1981) or charge transfer (e.g. [O III] λ4363; Dalgarno and Sternberg 1987).

The most important recombination lines, for both energetic and diagnostic purposes, may be the satellite lines formed during dielectronic recombination. These are caused by excitation of doubly excited levels lying below the threshold energy of the parent transition, and if the energy difference is substantial, the satellites can be relatively strong. They are resolvable from the parent transitions in highly charged ions, especially He-like and H-like ions (*e.g.* Beiersdorfer *et al.* 1993) and Fe XVII (Smith *et al.* 1985). These are included in the current generation of X-ray and UV model codes, and they are used as temperature and ionization state diagnostics in crystal spectrometer experiments on P78-1, SMM and YOHKOH (*e.g.* Doschek *et al.* 1990). Satellites to other iron lines in the 10 Å range may become important as better spectra and better rates for competing processes become available.

3.10. IONIZATION TO EXCITED LEVELS

Ionization from an inner shell or inner subshell can leave an ion in an excited state and produce a photon. Feldman *et al.* (1992) have proposed that this might account for spectral anomalies in Fe XV. It requires severe departure from ionization equilibrium, again because ionization rates are generally slower than collisional excitation. It is not clear that such extremely rapid ionization would not create worse anomalies in the spectrum, and SERTS spectra suggest that the Fe XV intensities agree with equilibrium predictions after all (Brickhouse *et al.* 1995). Ionization to excited levels may be important in Fe XXV and some of its satellite lines, along with inner shell excitation.

3.11. DIFFUSION

Diffusion in a strong temperature gradient can modify both the ionization state at a given temperature and the relative elemental abundances. It has been invoked to explain the unexpected brightness of the He II lines in the Sun (Shine, Gerola, and Linsky 1975) and it may be important in enhancing the coronal abundances of low First Ionization Potential elements and in the energy balance in the upper chromosphere (Fontenla, Avrett and Loeser 1993). The effects of shifts in the ionization balance may be difficult to distinguish from time-dependent ionization (*e.g.* impulsive heating of the solar corona as in Raymond 1990) or non-Maxwellian distributions. The abundance variations observed, for example in the solar wind (*cf.* Meyer 1985), may provide a less ambiguous signature for some type of diffusion than emission line intensities.

4. Acknowledgements

This work was supported by NASA Grant NAG-528 to the Smithsonian Astrophysical Observatory.

References

Anderson, S. W., Raymond, J. C., and van Ballegooijen, A. 1995, submitted to *Ap. J.*

Arnaud, M. and Raymond, J. 1992, *Ap. J.*, **398**, 394.

Arnaud, M. and Rothenflug, R. 1985, *Astron. & Astrophys. Supp.*, L**60**, 425.

Athay, R. G. 1988, *Ap. J.*, **329**, 482.

Athay, R. G. and Judge, P. G. 1995, *Ap. J.*, **438**, 491.

Badnell, N. R., Pindzola, M. S., Dickson, W. J., Summers, H. P., Griffin, D. C. and Lang, J. 1993, *Ap.J. Lett.*, **407**, L91.

Bahcall, J. N. and Wolf, R. A. 1968, *Ap. J.*, **152**, 701.

Beiersdorfer, P., Phillips, T., Jacobs, V. L., Hill, K. W., Bitter, M., von Goeler, S., and Kahn, S. M. 1993, *Ap. J.*, **409**, 846.

Bhatia, A. K. and Kastner, S. O. 1990, *Ap. J.*, **352**, 772.

Bhatia, A. K. and Kastner, S. O. 1993, *JQSRT*, **49**, 609.

Black, J. H., and van Dishoeck, E. 1987, *Ap. J.*, **322**, 992.

Bowen, I. 1935, *Ap. J.*, **81**, 1.

Brickhouse, N. S. 1995, *IAU Coll. 152*, in press.

Brickhouse, N. S., Raymond, J. C. and Smith, B. W. 1995, *Ap. J. Supp.*, **97**, 551.

Burgess, A. and Summers, H. P. 1969, *Ap. J.*, **157**, 1007.

Callaway, J. 1994, *At. Data Nucl. Data Tables*, **57**, 9.

Contini, M. and Viegas-Aldrovandi, S. M. 1987, *Astron. & Astrophys.*, **185**, 31.

Cornett, R. H., *et al.* 1992, *Ap. J. Lett.*, **395**, L9.

Dalgarno, A. and Sternberg, A. 1982, *Ap. J. Lett.*, **257**, 87L.

Doschek, G. A., Fludra, A., Bentley, R. D., Lang, J., Phillips, K. J. H., and Watanabe, T. 1990, *Ap. J.*, **358**, 665.

Doyle, J. G., Kingston, A. E., and Reid, R. H. G. 1980, *Astron. & Astrophys.*, **90**, 97.

Drake, J. J., Laming, J. M., and Widing, K. G. 1995, *Ap. J.*, **443**, 393.

Dupree, A. K., Brickhouse, N. S., Doschek, G. A., Green, J. C., and Raymond, J. C. 1993, *Ap. J. Lett.*, **418**, L41.

Escalante, V. and Dalgarno, A. 1991, *Ap. J.*, **369**, 213.

Esser, R., Brickhouse, N. S., Habbal, S. R., Altrock, R. C., and Hudson, H. S. 1995, *J. Geophys. Res.*, in press.

Feldman, U., Laming, J. M., Mandelbaum, P., Goldstein, W. H., and Osterheld, A. 1992, *Ap. J.*, **398**, 692.

Ferguson, J. W., Ferland, G. J., and Pradhan, A. K. 1995, *Ap. J. Lett.*, **438**, L55.

Fontenla, J. M., Avrett, E. H., and Loeser, R. 1993, *Ap. J.*, **406**, 319.

Foukal, P., Little, R., and Gilliam, L. 1988, *Sol. Phys.*, **114**, 65.

Froese-Fischer, C. and Liu, B. 1986, *At. Data Nucl. Data Tables*, **34**, 261.

Gabriel, A. H., Bely-Dubau, F., Faucher, P., and Acton, L. W. 1991, *Ap. J.*, **378**, 438.

Gregory, D. C., Meyer, F. W., Müller, A., and DeFrance, P. 1986, *Phys. Rev. A*, **34**, 3657.

Gregory, D. C., Wang, L. J., Meyer, F. W., and Rinn, K. 1987, *Phys. Rev. A*, **35**, 3256.

Harrington, J. P., Lutz, J. H., and Seaton, M. J. 1981, *M.N.R.A.S.*, **195**, 21p.

Hartigan, P., Raymond, J. C. and Hartmann, L. H. 1987, *Ap. J.*, **316**, 323.

Heil, T. G., Kirby, K. and Dalgarno, A. 1983, *Phys. Rev. A*, **27**, 2826.

Huang, K.-N. 1986, *At. Data Nucl. Data Tables*, **34**, 1.

Jordan, C., Brueckner, G. E., Bartoe, J.-D. F., Sandlin, G. D., and van Hoosier, M. E. 1977, *Nature*, **270**, 326.

Kahn, S. M. and Liedahl, D. A. 1995, "X-Ray Spectroscopy of Cosmic Sources," in *Physics with Multiply Charged Ions*, NATO Advanced Study Institute Series, in press.

Kallman, T. R. and McCray, R. 1980, *Ap. J.*, **242**, 615.

Kastner, S. O., and Bhatia, A. K. 1990, *Ap. J.*, **362**, 745.

Kato, T., Masai, K., and Arnaud, M. 1991, *National Institute for Fusion Science Report*, ISSN 0915-6364.

Kwong, V. H. S., Fang, Z., Gibbons, T. T., Parkinson, W. H., and Smith, P. L. 1993, *Ap. J.*, **411**, 431.

Laming, J. M. 1990a, *Ap. J.*, **357**, 275.

Laming, J. M. 1990b, *Ap. J.*, **362**, 219.

Laming, J. M., Drake, J. J., and Widing, K. G. 1995, submitted *Ap. J.*

Lang, J., ed. 1994, 'Electron Excitation Data for Analysis of Spectral Line Radiation from Infrared to X-ray Wavelengths: Reviews and Recommendations,' *At. Data Nucl. Data Tables*, **57**.

Lang, J., Mason, H. E., and McWhirter, R. W. P. 1990, *Sol. Phys.*, **129**, 31.

Liedahl, D. A., Kahn, S. M., Osterheld, A. L., and Goldstein, W. H. 1990, *Ap. J. Lett.*, **350**, L37.

Liedahl, D. A., Kahn, S. M., Osterheld, A. L., and Goldstein, W. H. 1991, *Ap. J.*, **391**, 306.

Liedahl, D. A., Osterheld, A. L., and Goldstein, W. H. 1995, *Ap. J. Lett.*, **438**, L115.

Mason, H. E. 1975, *M.N.R.A.S.*, **170**, 651.

Mason, H. E. and Monsignori-Fossi, B. C. M. 1994, *A & A Rev.*, **6**, 123.

Mewe, R. 1991, *A & A Rev.*, **3**, 127.

Mewe, R. and Gronenschild, E. H. B. M. 1981, *Astron. & Astrophys.*, **45**, 11.

Mewe, R., Kaastra, J. S., Schrijver, C. J., van den Oord, G. H. J., and Alkemade, F. J. M. 1995, *Astron. & Astrophys.*, **296**, 477.

Meyer, J. P. 1985, *Ap. J. Supp.*, **57**, 173.

Osterbrock, D. E. 1974, "Astrophysics of Gaseous Nebulae," W. H. Freeman and Co.: San Francisco.

Opacity Project Team, 1995, "The Opacity Project," Vol. 1, Institute of Physics Publishing: Bristol.

Owocki, S. P. and Scudder, J. D. 1983, *Ap. J.*, **270**, 758.

Pradhan, A. K. and Gallagher, J. W. 1992, *At. Data Nucl. Data Tables*, **52**, 227.

Pradhan, A. K. and Peng, J. 1995, "Atomic Data for the Analysis of Emission Lines," in *Proceedings of the Space Telescope Science Institute Symposium in Honor of the 70th Birthdays of D. E. Osterbrock and M. J. Seaton.*

Raymond, J. C. 1988, in "Hot Thin Plasmas in Astrophysics," R. Pallavicini, ed., Kluwer: Dordrecht, p. 1.

Raymond, J. C. 1990, *Ap. J.*, **365**, 387.

Raymond, J. C., Blair, W. P., and Long, K. S. 1995, *BAAS*, **27**, 854.

Reisenfeld, D.B. 1992, *Ap. J.*, **398**, 386.

Schmelz, J. T., Saba, J. L. R., and Strong, K. T. 1992, *Ap. J.*, **398**, 115.

Schmid, H. M. 1989, *Astron. & Astrophys.*, **211**, L31.

Schmitt, J. H. M. M., Drake, J. J., and Stern, R. A. 1995a, *IAU Coll. 152* abstracts.

Schmitt, J. H. M. M., Drake, J. J., Stern, R. A., Haisch, B. M. 1995b, submitted *Ap. J.*

Schrijver, C. J., van den Oord, G. H. J., and Mewe, R. 1994, *Astron. & Astrophys.*, **289**, L23.

Scudder, J. D. 1992, *Ap. J.*, **398**, 319.

Seaton, M. J. 1964, *M.N.R.A.S.*, **127**, 191.

Shine, R., Gerola, H., and Linsky, J. L. 1975, *Ap. J. Lett.*, **202**, L101.

Slavin, J. D. 1989, *Ap. J.*, **346**, 718.

Smith, B. W., Raymond, J. C., Mann, J., and Cowan, R. D. 1985, *Ap. J.*, **298**, 898.

Sobelman, I. I., Vainshtein, L. A., and Yukov, E. A. 1981, "Excitation of Atoms and Broadening of Spectral Lines," Springer-Verlag: New York.

Sokolowski, J. K. 1992, Ph.D. Thesis, Rice University.

Sternberg, A. 1989, *Ap. J.*, **347**, 863.

Storey, P. J. 1981, *M.N.R.A.S.*, **195**, 27P.

Story, J. G., Lyons, B. J., and Gallagher, T. F. 1995, *Phys. Rev. A*, **51**, 2156.

Summers, H. P. 1974, *M.N.R.A.S.*, **169**, 663.

Thomas, R. J. and Neupert, W. M. 1994, *Ap. J. Supp.*, **91**, 461.

Vernazza, J.E. and Raymond, J. C. 1979, *Ap. J. Lett.*, **228**, L89.

Verner, D. A., Yakovlev, D. G., Band, I. M., and Trzhaskovskaya, M. B. 1993, *At. Data Nucl. Data Tables*, **55**, 223.

Verner, D. A. and Yakovlev, D. G. 1995, *Astron. & Astrophys. Supp.*, **109**, 125.

Voit, G. M., Donahue, M., and Slavin, J. D. 1994, *Ap. J. Supp.*, **95**, 87.
Walling, R.S. and Weisheit, J.C., 1988, *Physics Reports*, **162**, 1.
Zygelman, B. and Dalgarno, A. 1987, *Phys. Rev. A*, **35**, 4085.

FUSION SPECTROSCOPY

NICOL J. PEACOCK

UKAEA-Government Division, FUSION (EURATOM/UKAEA Fusion Association)
Culham Laboratory, Abingdon, Oxfordshire, OX14 3DB UK

Abstract. This article traces developments in the spectroscopy of high temperature laboratory plasmas used in controlled fusion research from the early 1960's until the present. These three and a half decades have witnessed many orders of magnitude increase in accessible plasma parameters such as density and temperature as well as particle and energy confinement timescales. Driven by the need to interpret the radiation in terms of the local plasma parameters, the thrust of fusion spectroscopy has been to develop our understanding of (i) the atomic structure of highly ionised atoms, usually of impurities in the hydrogen isotope fuel; (ii) the atomic collision rates and their incorporation into ionization structure and emissivity models that take into account plasma phenomena like plasma-wall interactions, particle transport and radiation patterns; (iii) the diagnostic applications of spectroscopy aided by increasingly sophisticated characterisation of the electron fluid. These topics are discussed in relation to toroidal magnetically confined plasmas, particularly the Tokamak which appears to be the most promising approach to controlled fusion to date.

1. Introduction

"Plasma spectroscopy — the study of radiation emitted by plasmas at frequencies above the plasma frequency — is not a new subject but has been the primary observational tool of the astrophysicist for several decades. However, the growth of controlled thermonuclear research has presented spectroscopic problems that are somewhat different from from those generally treated by the astrophysicist. These problems are mainly due to the non-thermal nature of the high temperature laboratory plasmas". Even today this quote from R Wilson (1962) is an apt introduction to fusion spectroscopy, albeit there is now considerable doubt about the assumption of ionisation equilibrium in the upper solar atmosphere, Feldman, (1995) while in Tokamak fusion plasmas it is now possible to have core parameters with the product $n_e\tau > 1 \times 10^{19}$ m^{-3}s, sufficiently high for us to presume ionisation-recombination balance. The reputation of Sir Robert (Bob) Wilson's group at the Culham Laboratory stemmed from their seminal spectroscopic studies of the ZETA fusion device, Jones and Wilson (1962), Burton and Wilson (1961), Fawcett, Jones and Wilson (1961), Fawcett et al. (1963), the symbiotic VUV studies of the solar spectrum eg., Burton and Wilson (1965); the developments in models of the ionisation structure and emissivity, Wilson (1964a, 1967), McWhirter and Hearn (1963), Bates et al, (1962,a,b); and the developments in diagnostic techniques and engineering, Jones (1962), De Silva et al. (1964). This reference list is not exhaustive but gives some indication of the breadth of Bob Wilson's activities. The study of ZETA radiation was aimed specifically at (i) a spectroscopic analysis of the the

Astrophysics and Space Science **237**: 341–399, 1996.
© 1996 *Kluwer Academic Publishers. Printed in Belgium.*

plasma constituents with a view to controlling the impurity content of the thermonuclear fuel, (ii) measuring the plasma confinement and (iii) measuring the ion energies.

Burton and Wilson (1961) studied the particle confinement in ZETA by comparing the observed transient spectral signature of injected test ions with a model for the time variation of the number densities $n_z(X)$ of the species X^{z+}, where 'X' is the atomic number of an element. This model, which includes a source term, $\gamma_z(X)$, and a loss timescale, $\tau_z(X)$, in the continuity equation governing $n_z(X)$ and, as described in section 8, is essentially the same as used in present day Tokamak diagnostics. In the ZETA experiments, Ar ions were produced by edge gas injection and their loss time from the plasma was estimated at $\tau_z(\mathrm{Ar}) < 1$ msec. In contrast, in modern Tokamaks, ion loss times from the core can exceed 1 second during high confinement, 'H-mode' operation.

The study of ion energies in ZETA relied on measurements of emission line shapes, usually of the VUV lines of inert gases ions injected as trace species. The profiles were assumed initially to be determined by thermal broadening,

$$I(\delta\lambda) = \frac{I_t}{\sqrt{\pi}\Delta\lambda_D}\exp(-\frac{\delta\lambda}{\Delta\lambda_D})^2 \qquad (1)$$

with,

$$\frac{\Delta\lambda_D}{\lambda_0} = 7.7 \times 10^{-5}\sqrt{(T/A)} \qquad (2)$$

where $I(\delta\lambda)$ is the intensity $\delta\lambda$ away from the wavelength λ_0 at line centre; I_t is the total line intensity and $\Delta\lambda_D$ is the full width at half maximum. T is the ion temperature in eV and A is the atomic mass. Jones (1962) and Jones and Wilson (1962) made use of the half-intensity width of the the VUV lines as a measure of the ion energies. In the case of thermal broadening $(\lambda_0/\Delta\lambda_D)$ should be constant for all λ_0. In fact $(\lambda_0/\Delta\lambda_D)$ was found to have a non-thermal component due to fluctuating electric fields associated with plasma turbulence. If the power spectrum of the fields is $\mathcal{E}(w,t) = \mathcal{E}_0\exp(i\omega t)$ then the equation of motion for an ion X^{z+}, is

$$\mathbf{F} = Ze\left(\mathcal{E}(\omega t) + \mathbf{v} \times \mathbf{B}\right) - \nu_{e,z}m\mathbf{v} \qquad (3)$$

where $\mathbf{B}$ is the the large scale applied magnetic field, $\mathbf{v}$ is the X^{z+} ion velocity and $\nu_{e,z}$ is the 'effective' collision frequency of the X^{z+} ions due to binary encounters with all other particles and with density fluctuations. For electrostatic turbulence, the mean ion kinetic energy due to turbulence scales as $W_z = \mathcal{E}_0^2(z^2/M)$ as distinct from $W_z = \mathcal{E}_0^2 M$ expected from low frequency Alfvén turbulence at $\omega < \omega_c$, where ω_c is the ion cyclotron frequency. In fact, the total ion energy, W_{tot}, was found to be the sum of a thermal component and a component due to mass motion with $W_{\text{tot}} \sim 100$ eV. The non thermal or turbulent nature of the ion heating has drastic consequences for the achievement of steady–state thermonuclear burn in reversed field pinches, Sawyer et al. (1963); Peacock et al. (1971).

Typically, in present-day, atomic beam heated Tokamaks, the ion temperature $T_{i,z}$ can exceed 20 keV, but we are at the same time interested in the significance, for ion transport, of small line shifts due to weak fluid motion, ≤ 10 km/s, section 8. The line profile data for emission arising from a Tokamak plasma can be fitted to a pseudo Voigt function which represents a convolution of the source and instrument functions and is given by,

$$f(\lambda, \alpha) = \frac{1}{\left\{ 1 + \left(2^{\alpha^2} - 1 \right) \left[2 \frac{\lambda - \lambda_0}{\Delta \lambda_{1/2}} \right]^2 \right\}^{\alpha^{-2}}} \tag{4}$$

For $\alpha = 1$ the function is a Lorentzian while for $\alpha \to 0$, the function approximates a Doppler shape, Fraser and Suzuki (1970).

Thomson scattering as a diagnostic tool for measuring local parameters of the electron fluid was developed and demonstrated by De Silva et al.(1964) and Evans and Kazenstein (1969) within Bob Wilson's group. Figure 1 shows a group of scientists examining a rudimentary laser system for Thomson scattering at the Culham Laboratory in 1963. These early initiatives culminated at the end of the decade in the confirmation of keV temperatures and particle confinement timescales ~ 30 ms in the Soviet T-3a Tokamak, Peacock et al. (1969). Since then Thomson scattering has been thought of as an essential diagnostic in the assessment of Tokamak performance and in the interpretation of spectral intensities. Since then also, Tokamaks have been regarded as the leading confinement configuration in controlled fusion research.

In the following sections, (2) through (10), we discuss areas of plasma spectroscopy which have played an important role in the development of the Tokamaks as a confinement system for controlled fusion. The topicality of the subjects chosen here has changed with time. For example, the study of long wavelength forbidden lines has largely been replaced by charge exchange spectroscopy. While it is still the case that impurities in the core plasma can, through enhanced radiation and fuel dilution, prevent ignition and the sustainment of thermonuclear fusion, there are conditions where impurities can be beneficial. For example, the control of impurities to form an outer radiation mantle is now seen as a constructive way to avoid the worst effects of energy flow to the material walls. A relevant topic in fusion burning plasmas is the build-up, not just of metals and high Z ions in the core but of He^{2+} ash.

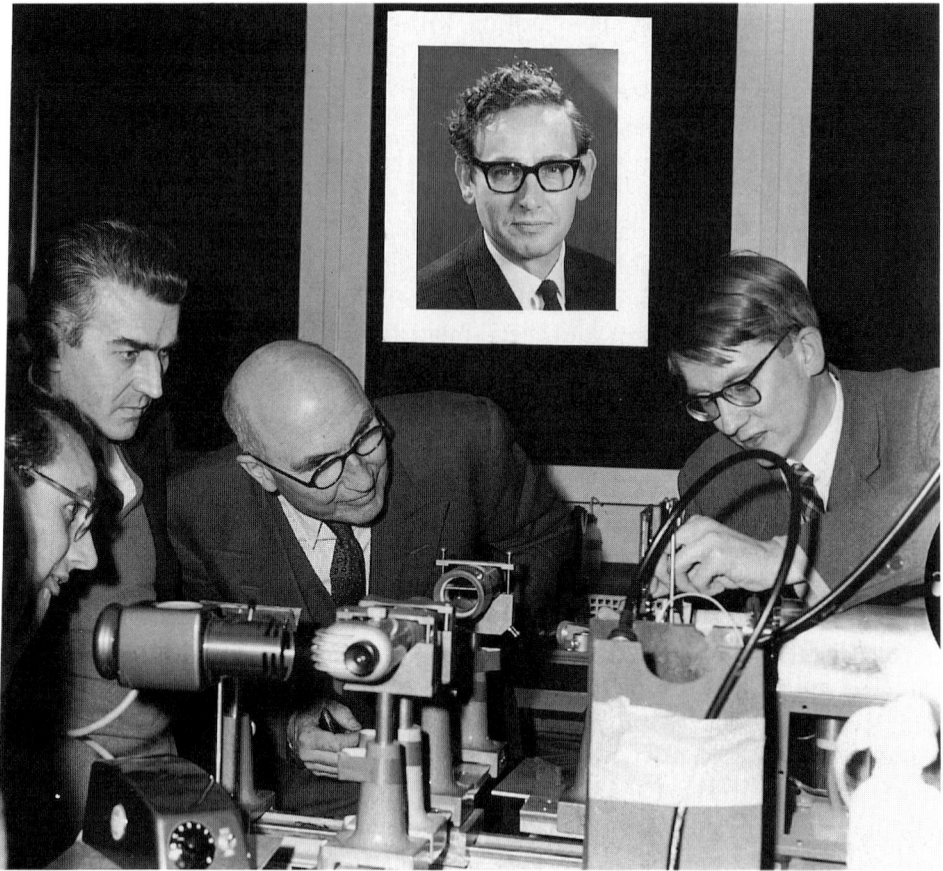

Fig. 1.   Dr D E Evans demonstrating a laser system for Thomson Scattering at the
Culham Laboratory during 1963. From right to left the photograph includes D E Evans,
Sir Roger Makins (Chairman of the UKAEA), Sir John Adams (Director of the Culham
Laboratory) and (including inset) Sir Robert Wilson.

2. Highly Ionised Atoms in Space and in the Laboratory

The accessibility to 100eV plasmas in the laboratory and space during the
1960's, stimulated VUV and XUV spectral studies of 'new' ion species, see
eg. Fawcett et al. (1963,1964), which brought an early success with the identification of a hitherto puzzling group of strong solar lines between 170–220Å.
These lines were also prominent features of the ZETA spectrum as shown
in Figure 2.   As a result of using isoelectronic extrapolation techniques,
Wilson (1964(b)), Gabriel, Fawcett and Jordan (1965) and independently,
after comparison of laboratory data with *ab initio* calculations, Cowan and
Peacock (1965) eventually attributed these lines to $3p^n 3d^m \rightarrow 3p^{n-1} 3d^{m+1}$
configurations in Fe VIII to Fe XIV. At shorter wavelengths around 20Å,
the θ-Pinch plasma, Peacock et al. (1966) provides an excellent mimic of

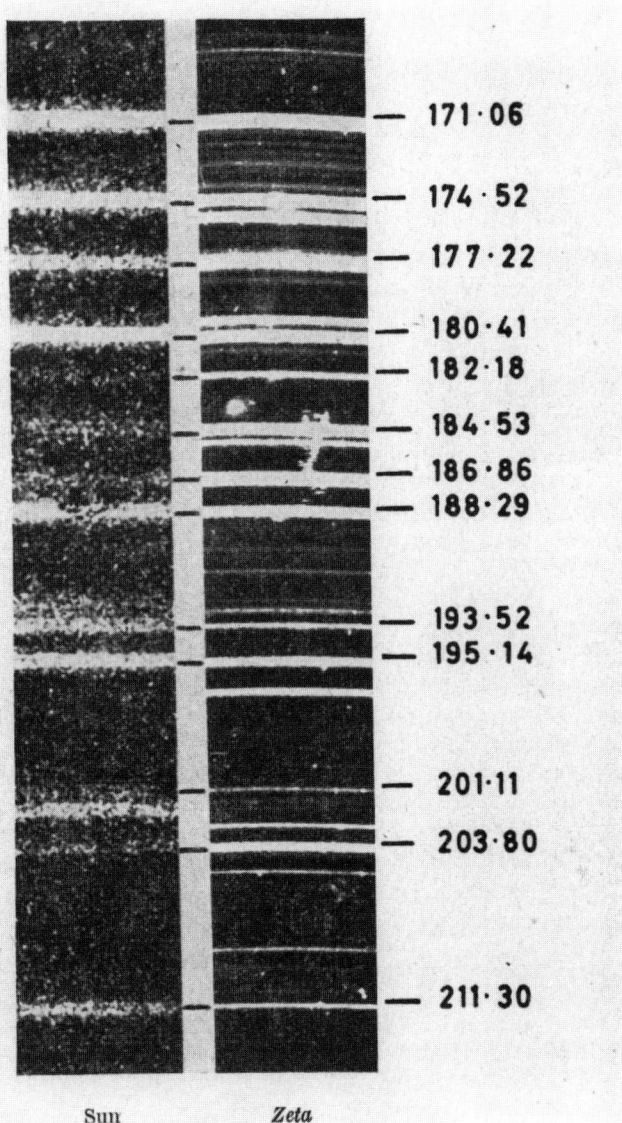

Fig. 2. Comparison of the emission spectrum from the sun and from ZETA in the wavelength region 170Å to 220Å; Fawcett et al. (1963)

the solar spectrum as illustrated in Figure 3. In passing, we should note that these more transient, plasma pinch devices with higher energy density and $T_e \leq$ 1keV, like the θ-Pinch, Bearden et al.(1961) and the Plasma Focus, Peacock et al. (1971) could be used to extend the study of hitherto unexplored ions well into the X-ray region. Since the 1960's, the study of the atomic term structure of highly ionised atoms has become something of an industry as one might judge from the extensive NBS tables produced on the subject over the last thirty years and published as Supplements to the

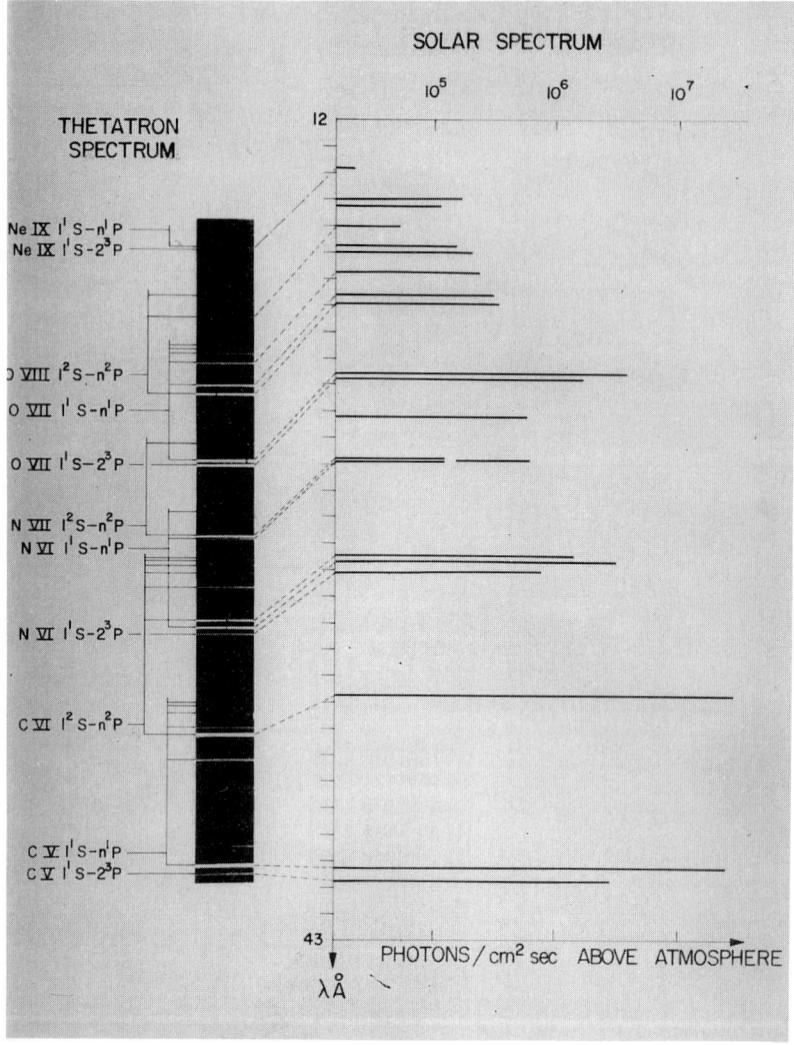

Fig. 3.   Comparison of the solar spectrum and the emission in the soft X-ray region from the MAGGI1 theta-pinch; Peacock et al. (1966)

Journal of Physical and Chemical Reference Tables.

The 1970's and later witnessed profound developments in the techniques for generating highly stripped ions in the laboratory. The use of foil or gas targets for stripping electrons from energetic ion beams produced in nuclear accelerators was perhaps one of the earliest 'non-plasma' sources. Using GeV heavy ion accelerators, it was shown that very weak currents of extreme charge states such as U^{+91} could be achieved. This area of beam foil spectroscopy was introduced by and, appropriately, has been reviewed by Bashkin (1976). The development of compact ion traps, in which the captive ions lose all or most of their electrons through collisions with electrons accelerated to $\geq$ 10 keV energies, either by cyclotron resonance heating as in the case of the ECRIS source, Geller and Jacquot (1980), or by electrostatic fields as in the case of the EBIS source, Marrs (1992), has been as rewarding as it has been technically demanding. There are now of the order of fifty ion trap sources of various designs in operation world-wide. With an upgraded 150kV 'super' EBIT, Marrs et al. (1994) have recently been able to produce U^{+91}.

The range of ions produced in high current plasma pinches has also continued to expand with the development of terrawatt pulsed transmission lines coupled to 'exploding wire' loads and low inductance vacuum sparks where spectra from ions as highly charged as Mo^{+40} have been studied, Turechek and Kunze (1975). These extremely high energy-density pinches tend to be inherently irreproducible, a problem that is not shared, however, with perhaps the most versatile and effective source of all, the laser-produced plasma. Since the development of powerful lasers for inertial confinement in the early 1970's laser-irradiated plasmas have proved to be one of the most useful sources for atomic structure studies, Lawson and Peacock (1980). Figure 4 shows an example of isoelectronic sequence mapping of the term structure of highly ionised atoms of Mo and Zr using relatively modest irradiation intensities, $I_0 \simeq 5 \times 10^{13} Wcm^{-2}$ at the target surface. Irradiation of targets with what is now an unexceptional laser power of 10J in 2ns, can readily produce, Ni-like ions of the rare earths, see eg., Zigler et al., (1980).

In large Tokamaks, electron temperatures $T_e \simeq$ 10keV can now be achieved with the result that the diffusive equilibrium of ion charge states as high as Kr^{+34} can be studied, Bitter et al. (1993).

A review of the physics interest in highly ionised ions and a description of these sources is given by Martinson (1989). In addition, there are the Proceedings of the International Conferences on the Physics of Highly Charged Ions, the most recent of which was assembled by guest editors, Winter and Aumayr (1995).

3. Infrastructure of Fusion Spectroscopy

The acquisition of a data base of energy levels, transition probabilities, collision rates, opacity coefficients and indeed, everything which makes up the infrastructure required to interpret the spectral signature of ionised atoms in plasmas, is a worldwide activity which greatly exceeds, in proportion, the actual business of taking spectra. Much of the material is published in the Atomic Data and Nuclear Data Tables (Academic Press), or in Data Centre publications such as ORNL or NIST in the USA, the NIFS Data Series and IPP-Nagoya reports in Japan and the Lebedev and Spectroscopy Institute reports in Russia. The IAEA has a coordinating role in arranging topical

348

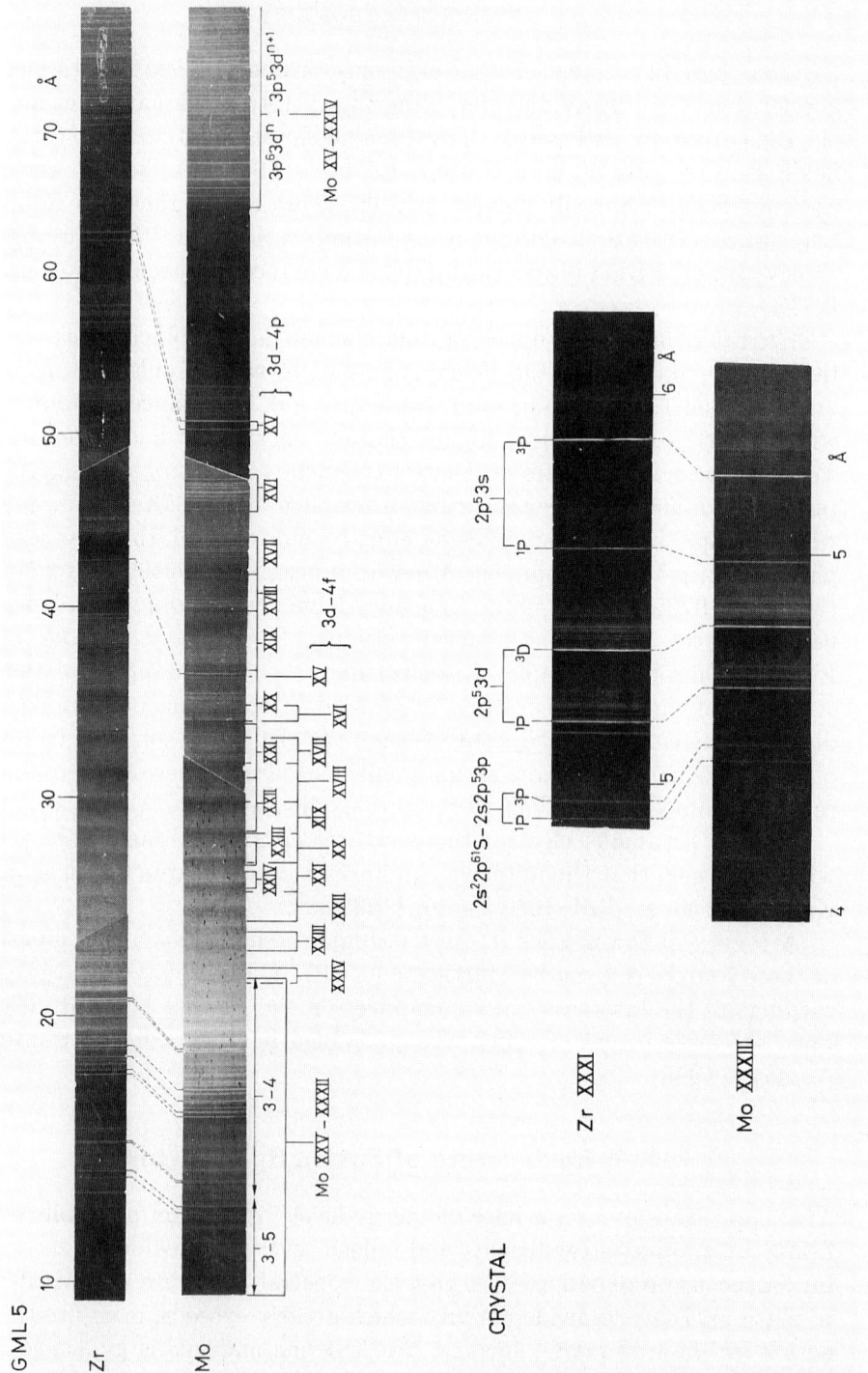

Fig. 4. Laser irradiation of solid targets allows isoelectronic mapping of multiplet structure of highly ionised Zr and Mo. The spectral region from 10Å to 70Å was recorded with a 5 meter grazing-incidence grating spectrometer (GML-5) while crystal dispersion was used in the 4Å to 6Å region.

reviews of the subject, see eg., Janev and Drawin (1993). A summary of the available documentation on wavelengths, λ_{ij}, and transition rates, A_{ij}, between states 'i' and 'j' for all ions of the elements throughout the atomic table, is given by Wiese (1993). In the case of multiple-electron configurations, direct transition energy measurements and semi-empirical extrapolations have been supplemented in recent years by a number of *ab initio* atomic structure computer packages which include relativistic effects and exchange terms, and can include configuration interactions, Cowan (1981). The outcome of these numerical techniques are energy levels and A_{ij} values. A brief overview of the different numerical methods is given by Moores (1987).

Organisation of all the atomic data on energy levels, A_{ij} values, collision rates etc, into a generalised collisional-radiative computer program and data package, ADAS, which can predict level populations, line and continuum radiances and power functions, has been developed by Summers (1994). The equilibrium power function from carbon ions with the separate contributions from each ion and from their metastable levels, is illustrated in Figure 5.

4. Tokamak Spectroscopy

A number of reviews have been written on the analyses of spectral features from Tokamak plasmas. These include articles by Isler (1984), Peacock (1980,1984), DeMichelis and Mattioli (1981), Hinnov (1983) and Bitter et al. (1993). In Tokamak plasmas the emission characteristics are determined by the relatively low electron number density, n_e, which typically satisfies $10^{19}\text{m}^{-3} \leq n_e \leq 10^{20}\text{m}^{-3}$, the high electron temperature, $0.5\text{keV} \leq T_e \leq 5\text{keV}$ and long particle confinement, $0.03\text{sec} \leq \tau_z(X) \leq 3\text{sec}$. Line features are optically thin, except perhaps for Lyman α at the walls. The continuum peaks at a photon energy of $2 \times T_e(\text{eV})$ in the X-ray region. Because of the low collision rates, the spectrum is rich in forbidden lines, (see section 5).

The presence of small quantities of impurities, of the order of a percent of light ions such as carbon, or considerably less of the common metals, can have far-reaching consequences in fusion plasmas. At low concentrations $\sim 10^{-3}n_e$, radiation from metal ions can cool the plasma thus, as shown in Figure 6, fatally increasing the $n\tau$ threshold for fusion ignition, where n is the fuel density and τ is the particle energy confinement time, Jensen et al. (1977). The main effects of light ion impurities with $X \leq 10$, occur at levels of $\geq 10^{-2}n_e$. At these concentrations, light impurity ions increase the effective ion charge, Z_{eff}, while diluting the fuel and thus reducing the fusion reaction rate as indicated in Figure 6. Z_{eff} is a useful parameter to quantify the impurity content in a plasma since it is the average ion charge 'seen' by

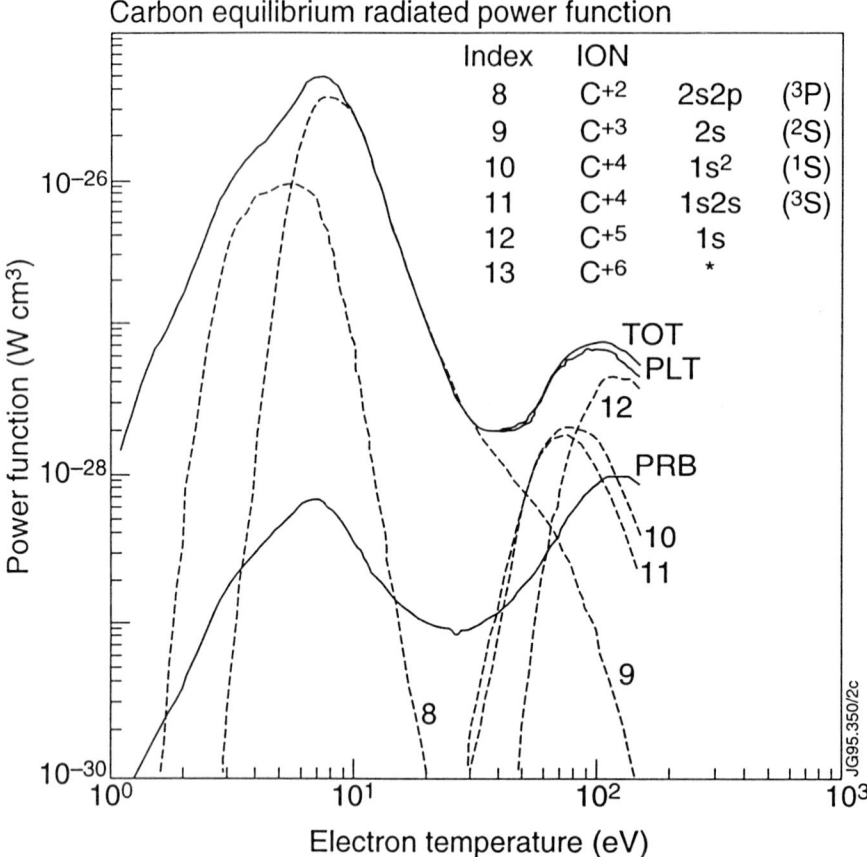

Fig. 5.   Radiated power function for carbon and its variation with the equilibrium electron temperature. The contributions from the separate ground and metastable states are
indexed. 'PLT' and 'PRB' indicate the total line radiation summed over all of the ion
charge states and the sum of the bremsstrahlung continua, respectively, Summers (1994).

the electrons during collisions with the ions,

$$Z_{\text{eff}} = \sum_{X,Z} \frac{n_z(X)Z^2}{n_z(X)Z} = \sum_{X,Z} \frac{n_z(X)Z^2}{n_e} = 1 + \sum_Z q_z Z(Z-1) \qquad (5)$$

where

$$q_z \equiv \sum_X \frac{n_z(X)}{n_e} \qquad (6)$$

In the calculation of the bremsstrahlung continuum intensity, Z_{eff} is equal
to the factor by which the free-free radiation exceeds that of hydrogen. The

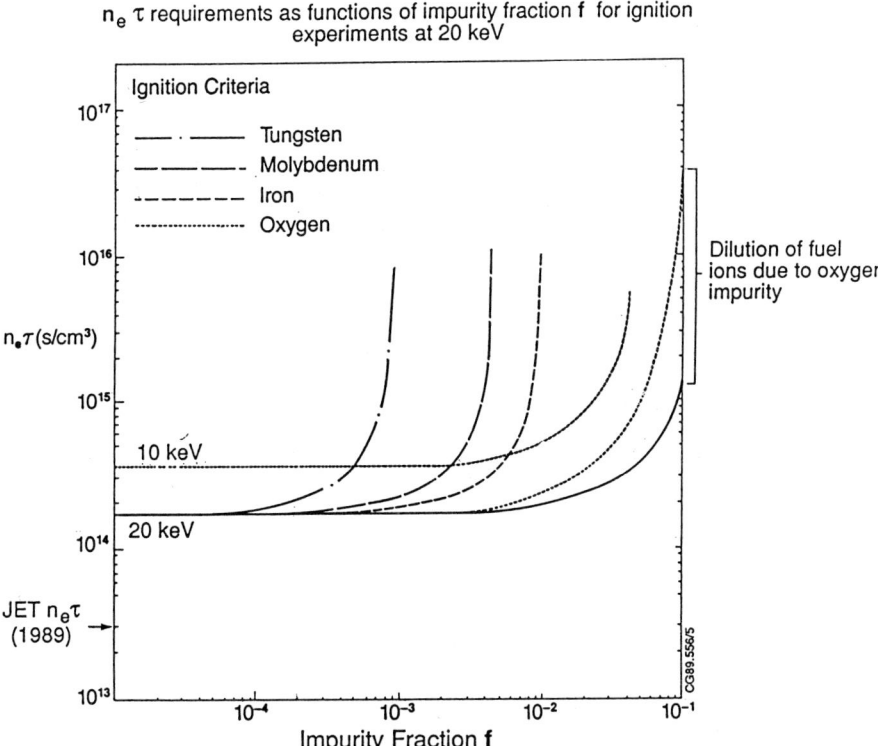

Fig. 6. The product of hydrogen isotope fuel density and confinement time, $n\tau$, required for thermonuclear ignition versus impurity fraction. The increase in $n\tau$ following metal contamination is due to increased radiation losses over hydrogen bremsstrahlung. Relatively large fractions of low Z contaminants such as oxygen can be tolerated but have the effect, eventually, of diluting the fuel. This poisoning effect of the reaction rate by low Z ions is most serious for fusion reaction rate products such as He^{2+} ash.

power radiated per unit volume due to bremsstrahlung is, with T_e in keV

$$P_{\mathrm{Br}} = 5.35 \times 10^{-37} n_e^2 Z_{\mathrm{eff}} T_e^{1/2} \quad \mathrm{Wm}^{-3} \tag{7}$$

The electrical resistivity, η, due to electron-ion collisions also depends on

Z_{eff} and is given by

$$\eta^{-1} = \sigma = \frac{n_e e^2}{m_e \nu_{ee}} \tag{8}$$

where σ is the electrical conductivity, ν_{ee} is the stopping frequency of the electrons with thermal speed v_{th} and e and m_e are their elementary charge and mass.

$$\nu_{e,z} \propto \frac{(n_e z_\Omega)}{v_{\text{th}}^3} \tag{9}$$

and z_Ω, apart from the inclusion of electron-electron collisions, is equivalent to Z_{eff}. In practical units,

$$\sigma = 1.9 \times 10^4 T_e^{3/2} / z_\Omega \ln \Lambda \quad \Omega^{-1} \text{m}^{-1} \tag{10}$$

$\ln \Lambda$ being the Coulomb logarithm, Wesson (1987). In the absence of anomalous, ie. collective, scattering of the electron flow, impurities will determine the current distribution and therefore the plasma equilibrium in the Tokamak. Typically, Z_{eff} in Tokamaks will have a value between about 1.5 and 3. The total radiated power, P_{rad}, is usually in the range $0.1 P_{\text{in}} \leq P_{\text{rad}} \leq P_{\text{in}}$ where the total input power, P_{in}, is due to a variety of heating mechanisms, including ohmic dissipation, P_Ω, wave resonance heating, $P_{\text{ECRH}} + P_{\text{ICRH}}$, at the electron and ion cyclotron frequencies, or neutral beam injection, P_{NBI}, ie.,

$$P_{\text{in}} = \sum P_\Omega + P_{\text{ECRH}} + P_{\text{ICRH}} + P_{\text{NBI}} + \cdots \tag{11}$$

Tomographic reconstruction of the bolometric signals from a Tokamak, usually show that the radiation intensity profile is saddle-shaped with the continuum from the fully stripped ions in the core being relatively weak compared to the line radiation from the partially stripped ions at or near the plasma boundary, Reichle et al. (1995); Fuchs et al. (1995). In the case of Tokamaks where the outer flux surfaces form a 'divertor' as illustrated in Figure 7, line emission from the low temperature divertor region dominates. The spectra recorded depend on the line-of-sight (LOS) as illustrated in the example shown in Figure 8, where the vertical view is dominated by VUV emission from low temperature ions, CIV, HeII etc. in the divertor region. In contrast, the same VUV spectral region when viewed along the horizontal major axis LOS is dominated by ionised metals in this particular JET discharge. The problem is to interpret this raw data in terms of elemental components of P_{rad} and Z_{eff}. The obvious route is to compare the absolute line radiances with those predicted from a transport model of the ions (cf section 8), coupled with the ADAS data base with q_z as a floating parameter. This turns out to be a hazardous exercise because of the accumulation of errors involved at each stage of the calculation. A semi-empirical

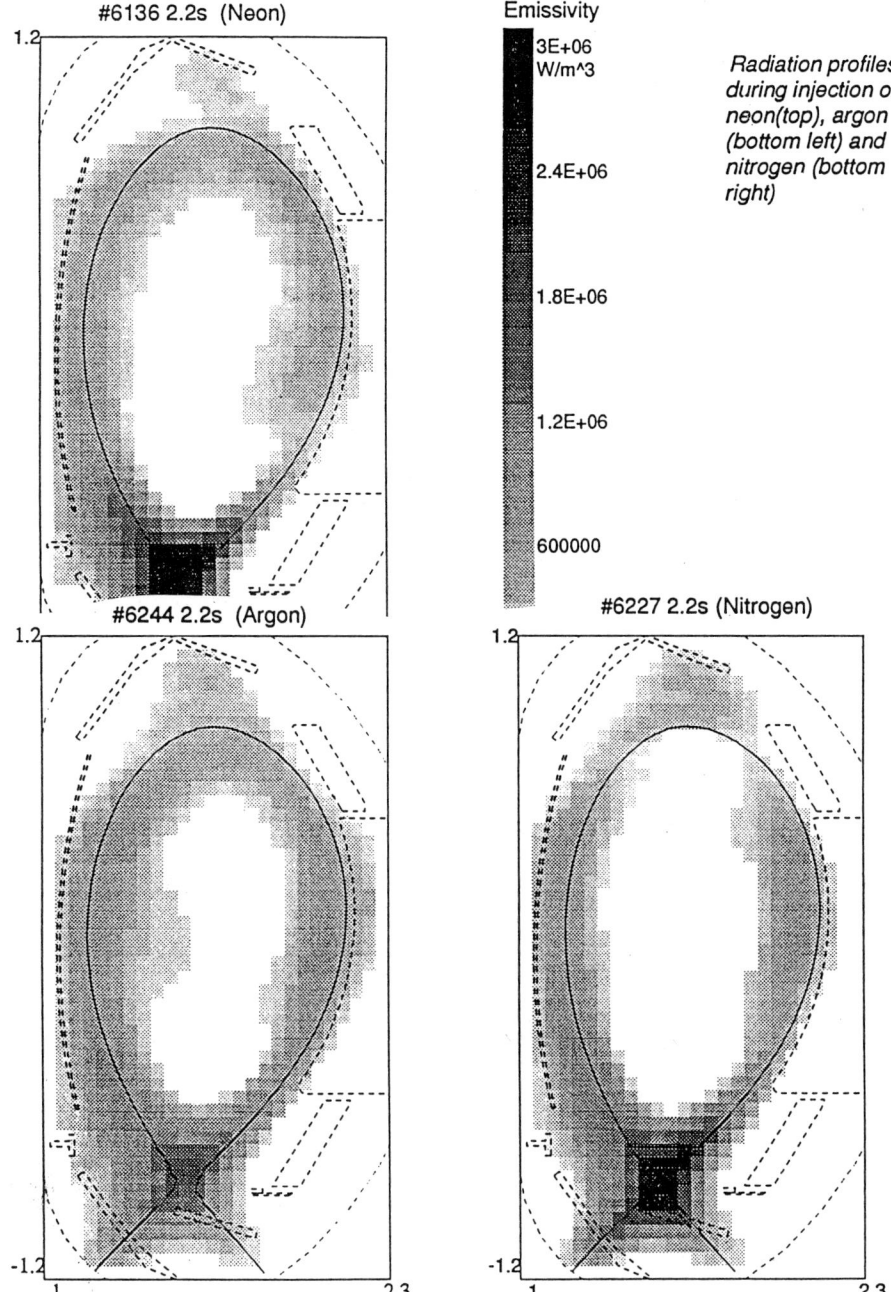

Fig. 7. Two-dimensional reconstructions of the radiation distributions in a poloidal plane of ASDEX-U using data from bolometer arrays with many, crossed, lines-of-sight. The bolometers integrate the emission over a spectral range of 1Å to 2000Å. Controlled gas puffing in these discharges emphasise the changing position of the radiation mantle relative to the outer poloidal field contours which define the 'X point' separating the main plasma from the bottom 'divertor' region, Fuchs et al. (1995).

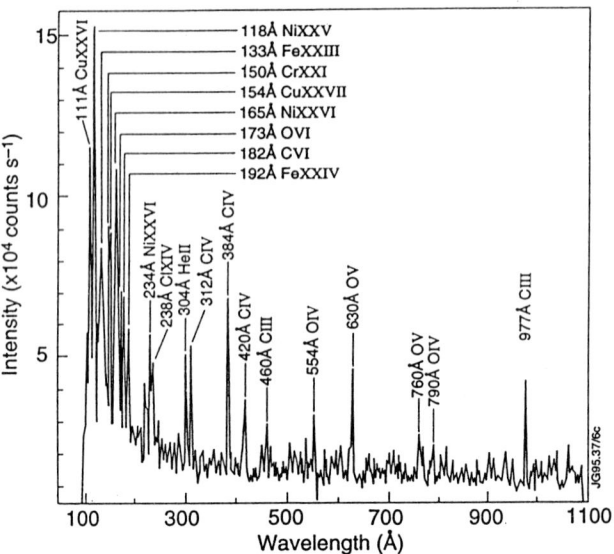

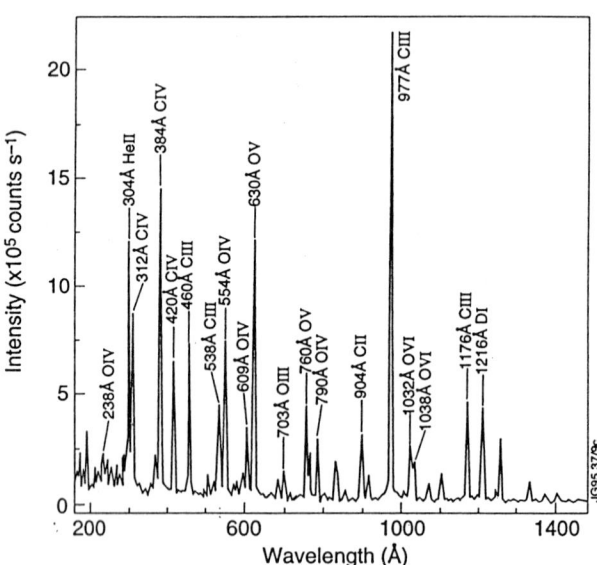

Fig. 8.   Emission spectrum in the VUV region from JET. The upper spectrum along a horizontal line-of-sight indicates the presence of highly ionised metals, in contrast to the spectrum along a vertical chord which is dominated by lowly charged ions from the low temperature 'divertor' region.

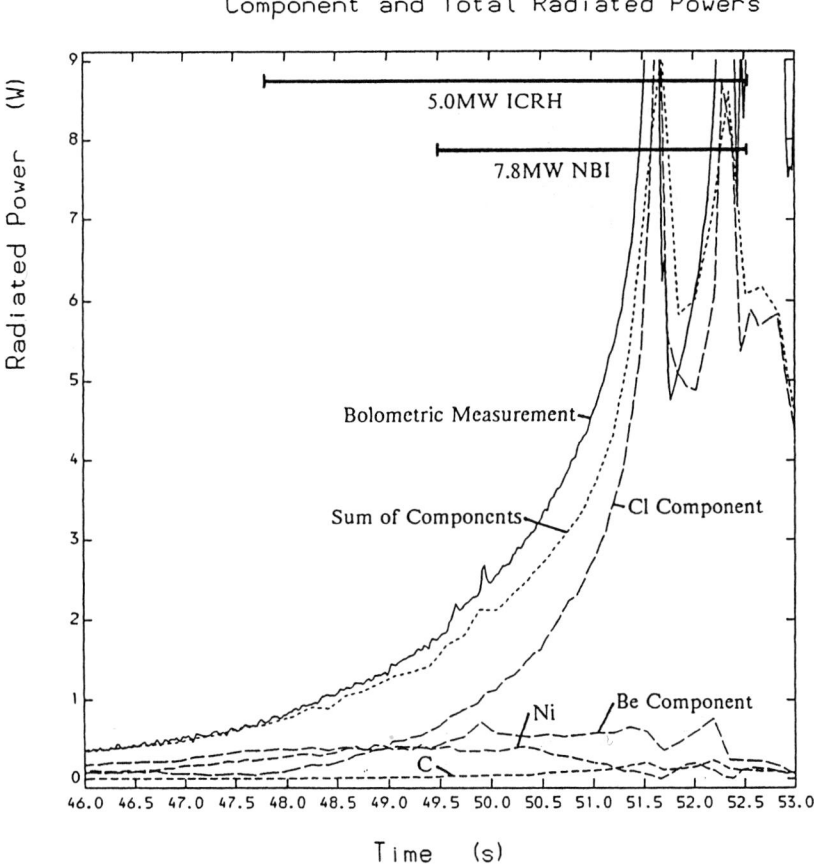

Component and Total Radiated Powers

Fig. 9.   Elemental contributions to the total radiated power based on a semi-empirical analyses of the VUV and XUV emission line intensities in JET, Lawson et al. (1995).

approach, Lawson (1995), where selected line radiances are assumed to have a proportionality to the bolometric signals, seems to account quite well for the separate $P_{rad}(Z)$ components, at least for discharges where the bulk of the radiation emanates from within the closed flux surfaces. An interesting example of this type of analysis is shown in Figure 9, where the accumulation of chlorine leads to $P_{rad} \geq P_{in}$ and a disruption follows. A necessary check on the viability of the semi-empirical method is the determination of whether P_{rad}, as measured by the bolometry, is equal to $\sum P_{rad}(Z)$, ie. the sum of all the elemental impurity contributions, Figure 9.

5. Forbidden Lines

Since forbidden lines have played an important role in establishing the properties of the solar corona, Edlén (1969) it is not surprising that in Tokamak plasmas, non electric dipole (E1) transitions, usually in metallic ions, have attracted a deal of attention due to their potential use as a diagnostic of the high temperature region, Peacock (1980), Peacock and Burgess (1981), Feldman et al., (1980), Hinnov (1983). Apart from interpretating their line profiles in terms of ion motion, the forbidden line intensities can be used as a measure of n_e over limited regions of density. In a three level (i, j, k) system with ground state i for example, the intensity ratio between the forbidden I_{ij} line and the allowed I_{ik}, coronal line is

$$\frac{I_{ij}}{I_{ik}} = \frac{C_{ij}}{C_{ik}} \frac{A_{ij}}{(A_{ij} + n_e C_{jk})} \tag{12}$$

where C_{ij} is the collision rate coefficient for transitions between states i and j. In the range where $A_{ij} \simeq n_e C_{jk}$, then one should expect a density dependent line ratio.

Forbidden transitions can, in principle, occur throughout the Tokamak spectrum from the visible to the X-ray region. Inspection of the plots in Figure 10, however, suggest that forbidden lines at $\lambda < 100\text{Å}$ with $A_{ij} > 10^4 \text{s}^{-1}$ should always be readily observable and indeed the (E2) resonance decay of the $3d4s(J = 2)$ level in the first excited configuration of Ni-like Mo XV contributes the strongest XUV line feature in the DITE Tokamak with Mo limiters, Mansfield et al. (1978). The resonance (M1) decay of the $1s2s\,^3S_1$ metastables in He-like ions, (see section 7), is an important example in the X-ray region.

Much of the interest has been focused on the long wavelength, VUV or near visible lines, since these are amenable to high resolution techniques. In particular, (M1) transitions within the $2s^n 2p^m$ ground configurations of ionised metals have been studied, Suckewer et al. (1980), Suckewer and Hinnov (1978, 1979), Lawson et al. (1981), Finkenthal et al. (1984). In Tokamaks with trace metals like Zr and Mo, (M1) transitions within the $n = 3, 4$ shell ground configurations have also been identified, Suckewer et al.(1982), Hinnov (1983). An extensive compilation of the forbidden lines in the $ns^2 np^k$ ground and $nsnp$ excited configurations has been published by Kaufmann and Sugar (1986).

The dominance of the collision rates over the A_{ij} values of forbidden lines at low transition energies, see Figure 10, would lead us to expect that only those forbidden lines with $A_{ij} > 10^4 \text{s}^{-1}$ and transition energies $\epsilon \geq 10\text{eV}$ would be present in the Tokamak spectrum. Emission of forbidden lines with A_{ij} as low as $10^2 - 10^3 \text{s}^{-1}$ have been observed however as illustrated for TiXIV and TiXVIII in Figure 11. The near constancy of the

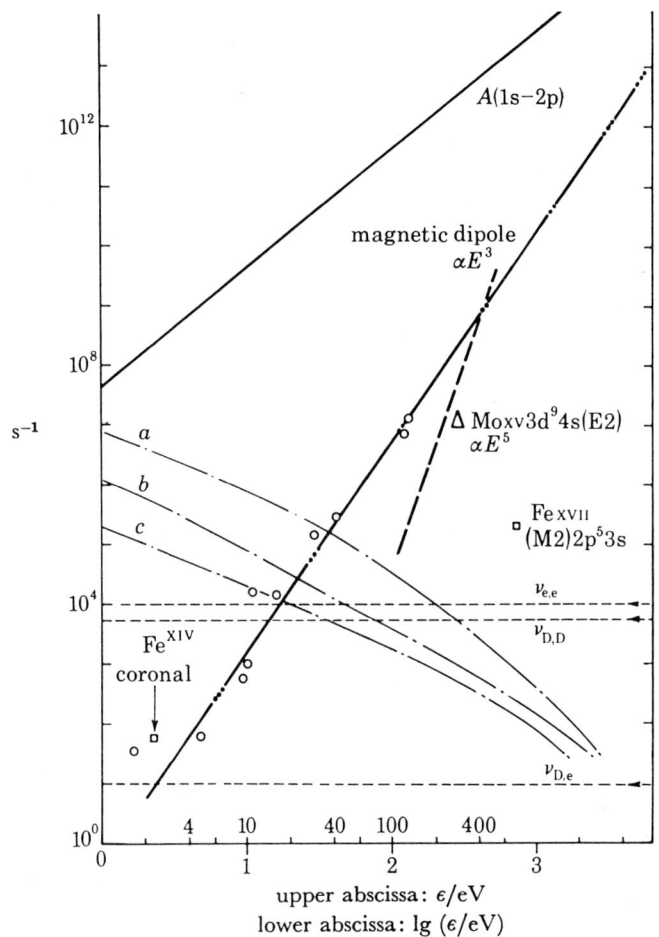

Fig. 10. Spontaneous decay rates for various types of transitions (solid line, allowed $1s - 2p$; dot-dash and $\bigcirc$, magnetic dipole; $\square$, magnetic quadrupole; dashed line and $\triangle$, electric quadrupole) against the transition energy ϵ/eV for $T_e = T_i = 1\mathrm{keV}$ and $n_e = n_i = 10^{19}\mathrm{m}^{-3}$. The collisional decay rates $\langle\sigma v\rangle n_e$ (Lotz 1967) $(a, \langle\sigma v\rangle^1_{\mathrm{Lotz}} n_e; b, \langle\sigma v\rangle^{\mathrm{d.e.-ex}}_{i-k} n_e; c, \langle\sigma v\rangle^{\mathrm{ex}}_{i-k} n_e)$ and the thermalization rates $\nu_{ee}, \nu_{\mathrm{D}e}$ and ν_{DD} for typical tokamak conditions are plotted for comparison.

quasi-Boltzmann populations of the states within the ground configuration as density is increased above about $10^{19}\mathrm{m}^{-3}$, Feldman et al., (1980), contrasts with the linear rise in the populations of the upper excited levels. Although the resonance decays from these two configurations will show a density dependence, the increasing intensity of the allowed lines throughout the spectrum and their subsequent masking effect are primarily responsible for explaining the failure to observe long wavelength forbidden lines for

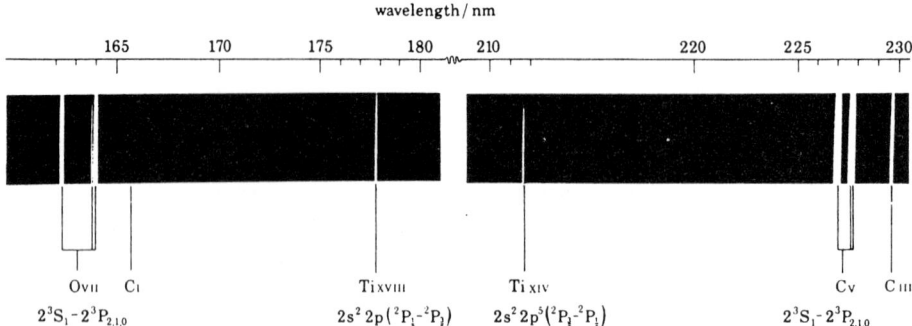

Fig. 11.   Forbidden transitions (spontaneous decay rate $A \simeq 10^3 \mathrm{s}^{-1}$ within the $1s^2 2p^n$ ground configuration of ionized titanium impurities in the DITE tokamak ($n_e \simeq 2 \times 10^{19} \mathrm{m}^{-3}, T_e \simeq 800\mathrm{eV}$). Reference emission lines from the $1s2s\,^3S_1 - 1s2p\,^3 P_{2,1,0}$ triplets of oxygen and carbon impurities ($A \simeq 10^8 \mathrm{s}^{-1}$ and $N(0) \simeq 10 \times N(T_i)$) are seen to have intensities comparable with those of the metal lines.

$n_e \geq 10^{20} \mathrm{m}^{-3}$. Despite this, suggestions for using near visible forbidden lines in very highly stripped ions continue to be considered as a diagnostic of very high temperature Tokamaks, see eg Feldman et al. (1991).

The unrelenting progress in technology has made a huge impact on spectroscopy as on other diagnostic techniques. Perhaps the most striking advance has been in the shear volume of spectroscopic data generated, for example, tens of megabytes of spatial and spectral data is a normal acquisition from one JET tokamak discharge. Another fundamental change relates to advances in the electron fluid diagnostics based on refractive index measurements and laser light scattering, Soltwisch et al. (1986). Multi-spatial channel interferometry, reflectometry, polarimetry, and multi-point Thomson laser light scattering define the characteristics of the electron fluid with such precision that it is now possible to use spectroscopy to check on the basic atomic physics models and the atomic cross sections, Keenan et al. (1989). The effort on plasma diagnostics in the JET Tokamak, Keilhacker (1989), can be judged by the systems identification shown in Figure 12.

The use of tomographic reconstruction from individual LOS views recorded on non-dispersive bolometer and *PIN* diode arrays, gives immediate meaning to $P_{\mathrm{rad}}(r)$ and soft X-ray images. We have seen in Figure 7 the quality of the information derived from arrays of bolometers. X-ray images of radiation above about 2keV photon energy have proved extremely informative in the detailed study of instability phenomena on time scales as short as $10\mu\mathrm{s}$. Using *PIN* diode arrays, the evolution of instabilities in the high

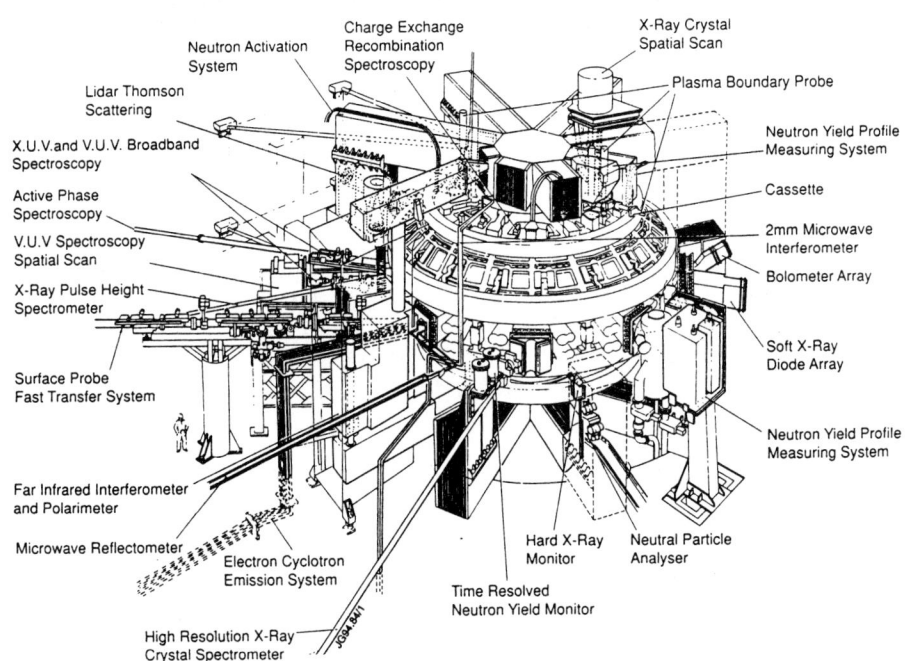

Fig. 12.   Schematic of diagnostic systems deployed on JET.

temperature core, like 'sawteeth' or 'snakes', can be followed, as illustrated in Figures 13 and 14.   More surprisingly, X-ray tomography can monitor instabilities, such as localised edge modes (ELM's) in the relatively cool, outer region of the plasma, as indicated in Figure 15.

Advances in spectroscopic instrumentation has been most notable in the field of multi-channel detectors such as channeltrons, linear Si-diode arrays, CCD detector arrays, Barnsley (1993). These detectors are often used in flat-field and Rowland circle grating spectrometers which can survey the spectrum down to about 10Å. In the visible region, the effects of wall conditioning, for example by gettering the oxygen with boron, beryllium etc., are most obvious. The information which can be derived during a getter-

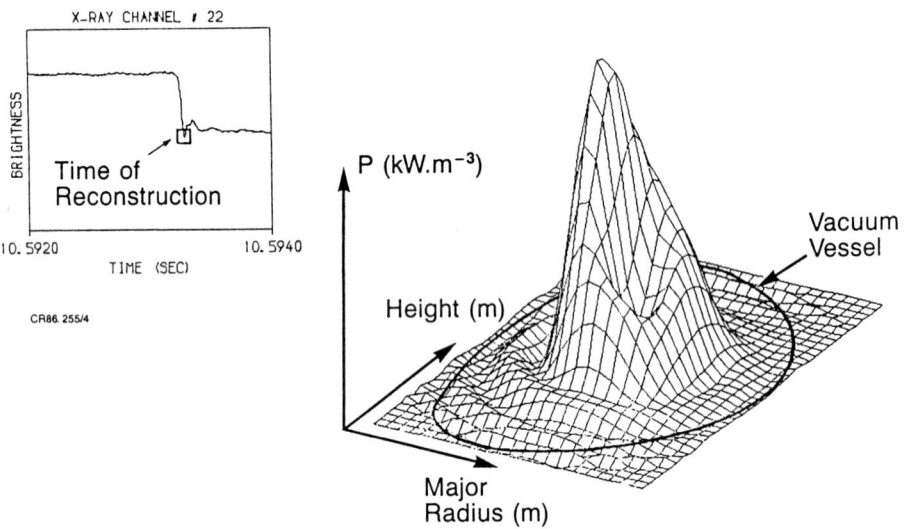

SOFT X-RAY EMISSIVITY
JET PULSE ¡ 9588
TIME = 10.593280 S
MAX = 3.48 KW/M3

X-RAY CHANNEL ¡ 22

BRIGHTNESS

Time of
Reconstruction

10.5920   10.5940
TIME (SEC)

CR86.255/4

P (kW.m⁻³)

Vacuum
Vessel

Height (m)

Major
Radius (m)

Fig. 13. X-ray image reconstruction of thermal X-ray emission during periodic 'sawteeth', MHD instabilities of the core plasma in JET. A single frame, exposure time < 0.1 ms, illustrates the redistribution of the core emission following a 'sawtooth' collapse, Edwards (1988)

ing experiment is illustrated in Figure 16. In this spectral region, also, the appearance of molecular bands attributed to CD, BeH etc., are often noted at the plasma-wall interaction surface. We have discussed typical VUV spectra from a flat-field survey instrument in Figure 8. At shorter wavelengths, in the XUV, improved resolution over a narrower spectral range can be achieved by employing diode arrays in higher dispersion, grazing incidence spectrometers as illustrated by the Be spectrum in Figure 17. Sensitivity calibration

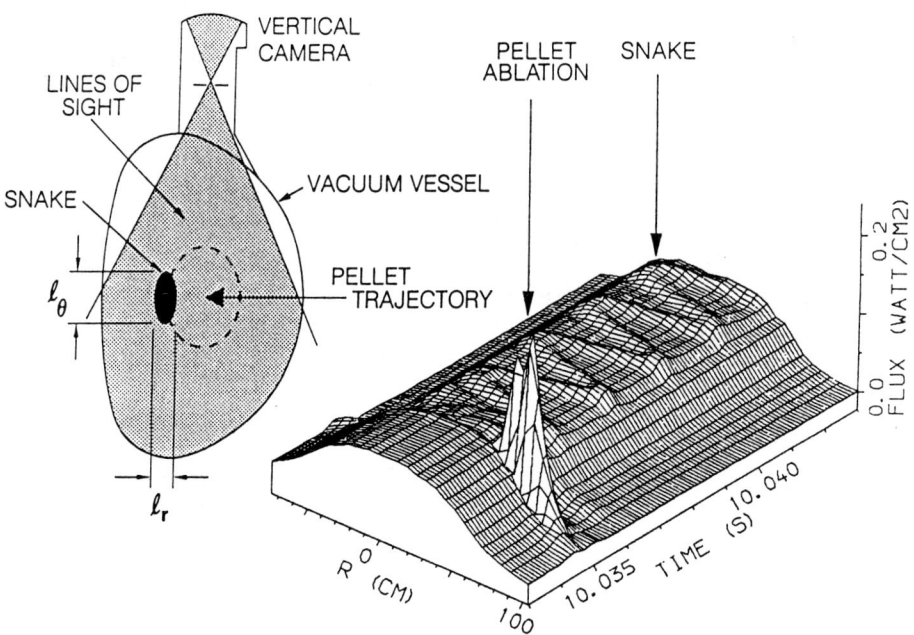

Fig. 14. Isolated regions of relatively high X-ray emission, dimensions l_r and l_θ, following solid fuel pellet injection into JET. The poloidal rotation of the region about a rational flux surface appears as a snake-like disturbance after the diode array data is reconstructed in time, Gill et al. (1992).

in the XUV/VUV region is a problem, however, but using a combination of branching ratios as transfer reference standards and charge-exchange cascades (see section 6) for the relative line intensities, $S^{-1}(\lambda)$ curves have been successfully constructed, Hawkes et al. (1993). In the X-ray region, $\lambda < 40\text{Å}$, CCD detectors have been used directly in high resolution crystal spectrometers as illustrated in Figure 18, to give what is effectively X-ray streak spectroscopy, Abbey et al. (1993).

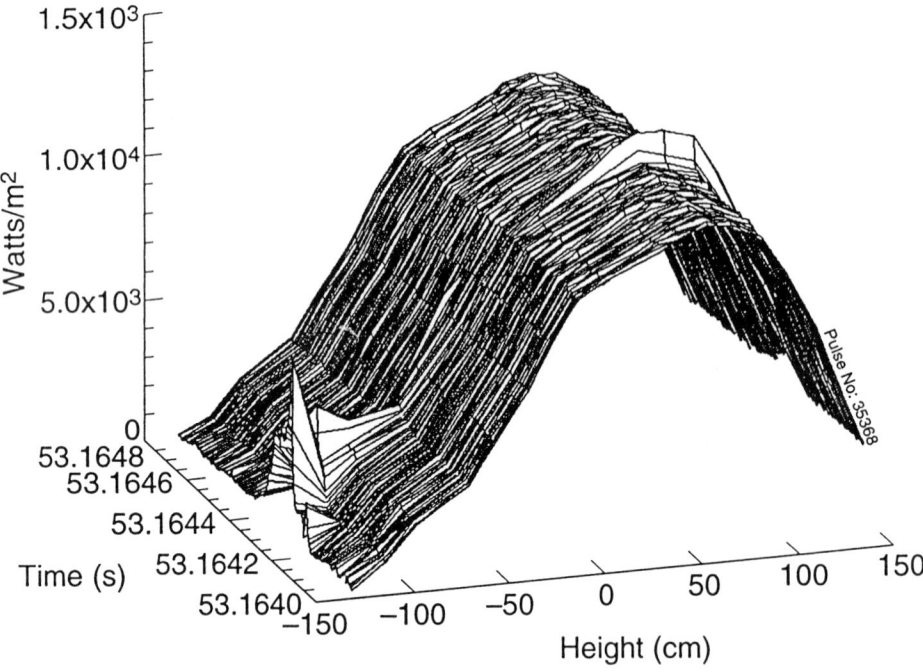

Fig. 15. An example of an Edge Localised Mode (ELM) instability of the plasma boundary in JET as it appears in the tomographic reconstruction of the data from the X-ray diode arrays, Alper (1995).

In the soft X-ray region, $15\text{Å} < \lambda < 150\text{Å}$, there have been considerable developments in diffraction systems, Barnsley (1993), where organic crystals, such as OHM, give 'grating-equivalent' resolution and where synthetic multi-layers, giving high reflectivity over a narrow bandwidth at near normal incidence, have been proposed, Underwood and Barbee (1981). Both of these types of diffractors and others, are used in a versatile Bragg rotor survey spectrometer, Barnsley et al. (1993), which allows detection of all the impurities with $Z > 3$, likely to be found in a Tokamak. The use of the Bragg rotor on JET is illustrated in Figure 19.

6. Charge Exchange Spectroscopy and Atomic Beam Diagnostics

Resonance charge transfer between confined fuel ions and refuelling atoms

$$H^0 + H^+ = H^+ + H^0 \tag{13}$$

was recognised as one of the main energy loss mechanisms in early mirror machine research. In Tokamaks, one might expect atom-ion collisions to be

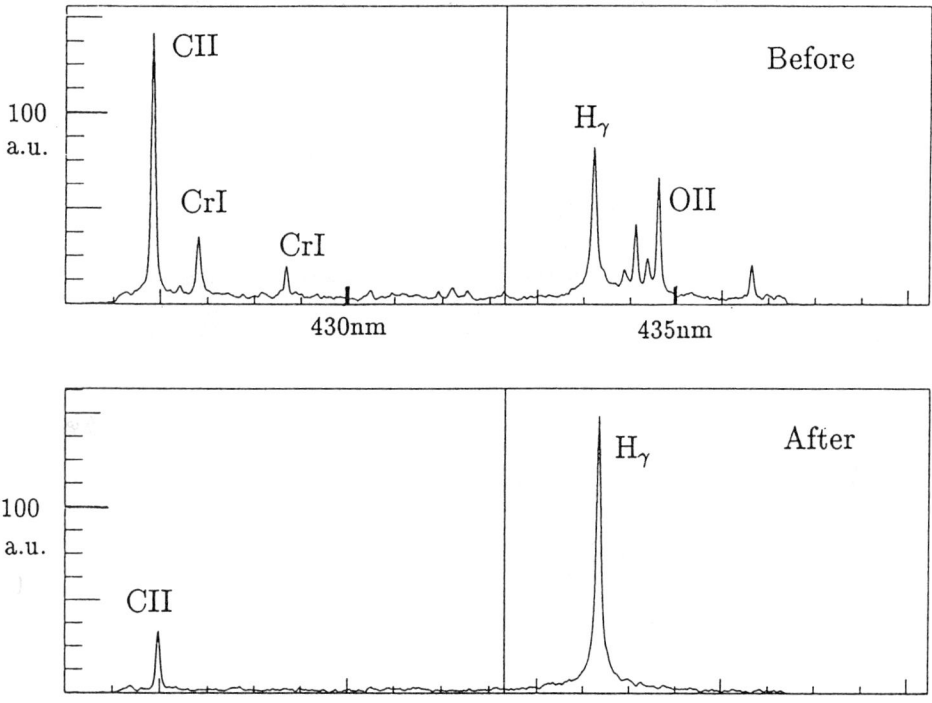

Fig. 16.   Comparison of visible line emission, centered at 4325Å, in standard (plasma current, $I_p = 100$kA; toroidal field, $B_T = 1.1$T; $n_e = 1.8 \times 10^{19}$ m^{-3}) hydrogen discharges in Compass, (a) before and (b) after boronisation; Esser et al. (1992).

important at the plasma boundary where n_e(edge) $\leq 10^{19}$m^{-3}. The penetrating distance, given by the atomic velocity divided by the ionisation rate, is about 1cm for thermal (0.05eV) hydrogen from the walls and several cms for H^0 with a few eV energy due to Frank-Condon, molecular dissociation. The neutral atom products of the charge exchange collisions have, of course, the plasma thermal energy and will penetrate further, to more than 10cm and can build up in the bulk plasma to a number density $n(\text{H}^0)$ of about 10^{15}m^{-3}. At this fractional abundance $10^{-4} - 10^{-5}$ relative to n_e, charge exchange (CX) recombination is already competitive with electronic recombination, Hulse et al., (1980) and so the ionisation balance and level populations will be affected by the presence of background neutrals. In fact, the appearance near the plasma boundary of high quantum level ($n = 27$) distortions to the resonance spectrum of Ar^{16+} in the ALCATOR-C Tokamak, Rice et al (1986) has been attributed to CX collisions between excited states of thermal H^0 and Ar^{17+} ions which have diffused outwards from the core. Distortions to the CVI Lyman decrement in the XUV spectrum from

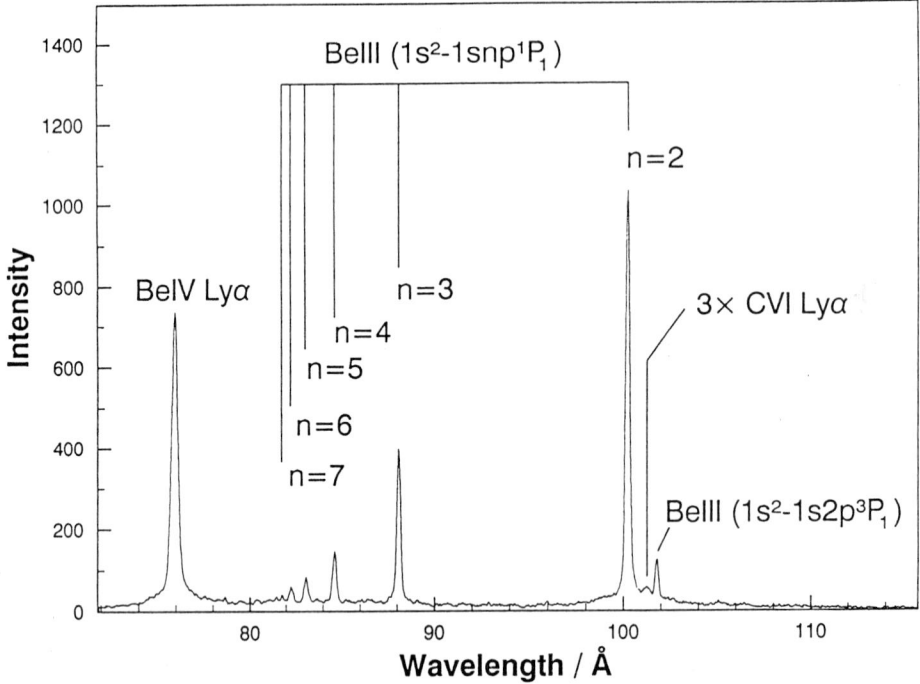

Fig. 17. Survey of part of the XUV spectrum of JET showing the Be resonance line emission, including the $1s^2 - 1s2p\,^3P_1$ intercombination line at 101.69Å which can be used together with the $1s^2s\,^3S_1 - 1s2p\,^3P_1$ line at 3724Å for branching ratios calibration, Hawkes et al. (1993).

the JET plasma edge, shown in Figure 20, have been successfully modelled taking into account contributions of CX from all the levels in H^0 up to the thermal limit, $n \simeq 5$, Mattioli, Peacock et al. (1989).

The increasing availability of atomic beams with sufficient energy, $\epsilon \gg$ 10keV, to penetrate and heat the core fuel ions and with sufficient current, $\sim 10\,\mathrm{mAcm}^{-2}$, to refuel the tokamak, has expedited the development of an unique CX diagnostic of the local plasma conditions, Isler (1977); Afrosimov, Gordeev et al. (1979). The uniqueness, which has been extensively exploited in beam-heated Tokamaks over the last decade, stems from the space- and time-resolved spectral signature of the atom-ion collisions. Since almost all of the ions in the bulk of the plasma, apart from a fraction of a percent of

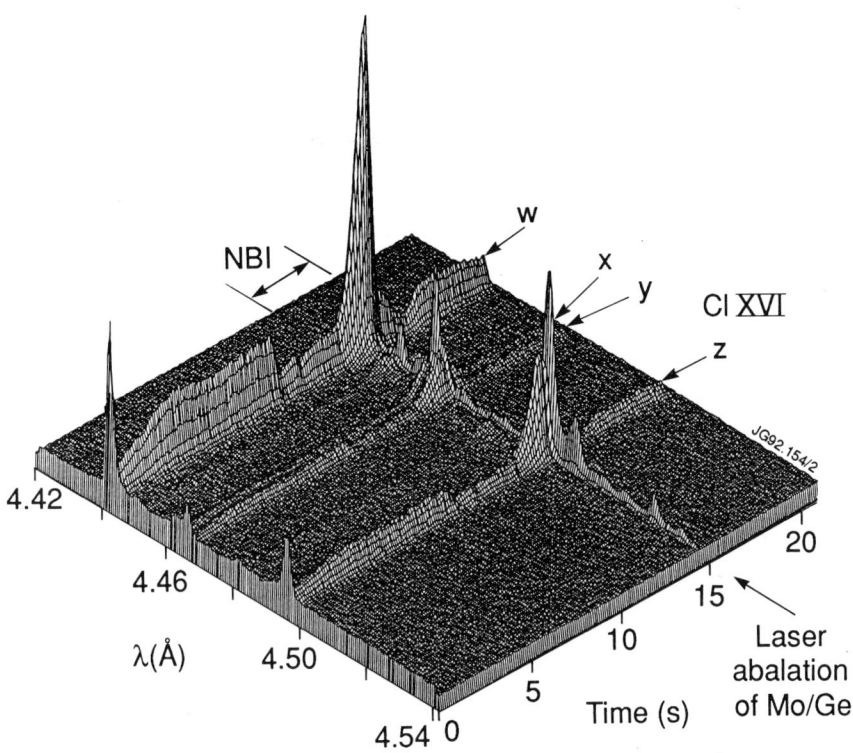

Fig. 18.   Isometric plot of the resonance 'w','x','y','z', lines (using the Gabriel (1972) notation) from the n=2 levels of Cl XVI during a JET discharge with a period of neutral beam injection (NBI) and with a transient injection of Mo and Ge. Each time slice, results from a compression of the the 2-D image onto a single row of pixels in the frame store region of the CCD detector, giving, in this case, a time resolution of 40ms; Abbey et al. (1993).

$Z > 10$ impurities, exist as bare nuclei, then the most common CX collisions are of the type,

$$X^{z+} + H^0 = (X^{(z-1)+})^* + H^+ = X^{(z-1)+} + H^+ + nh\nu \qquad (14)$$

where $nh\nu$ represents the photon cascade from the majority state, $\hat{n}$, selected in the charge transfer collision. For collisions between hydrogenic ions and H^0 atoms, $\hat{n} = Z^{0.77}$. For CX collisions with O^{8+}, for example, $\hat{n} = 5$, while for C^{6+}, $\hat{n} = 4$. The charge transfer cross section peaks at at an impact velocity equal to the orbit velocity of the exchanging electron which for H^0 ground state is,

$$v_0 = e^2/\hbar = 2.2 \times 10^6 \mathrm{ms}^{-1} \qquad (15)$$

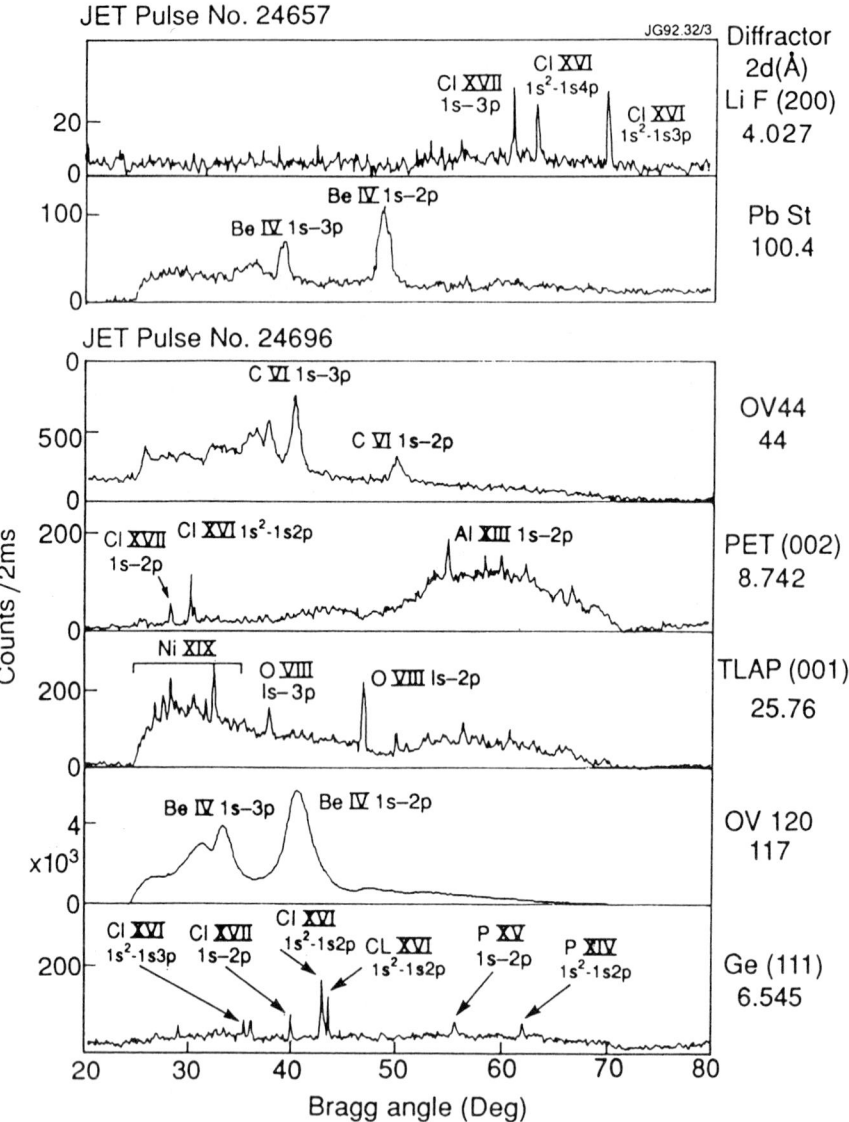

Fig. 19.   Typical survey spectra from JET, using a range of crystals and multilayers on a hexagonal rotor. Wavelength coverage is almost complete between 2Å and 100Å, thus monitoring all the main impurity elements.

where $\epsilon_0(H^0) \equiv 25\text{keV/AMU}$ is the kinetic energy per AMU corresponding to V_0. At the equivalent thermal electron energy ie., $\epsilon_0(H^0)(m_e/m_i)$, electronic recombination is about 10^{-5} of the charge transfer recombina-

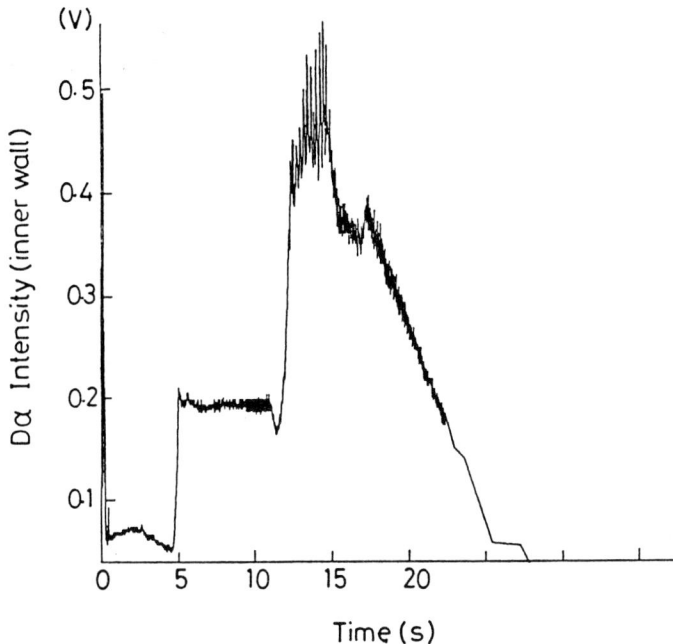

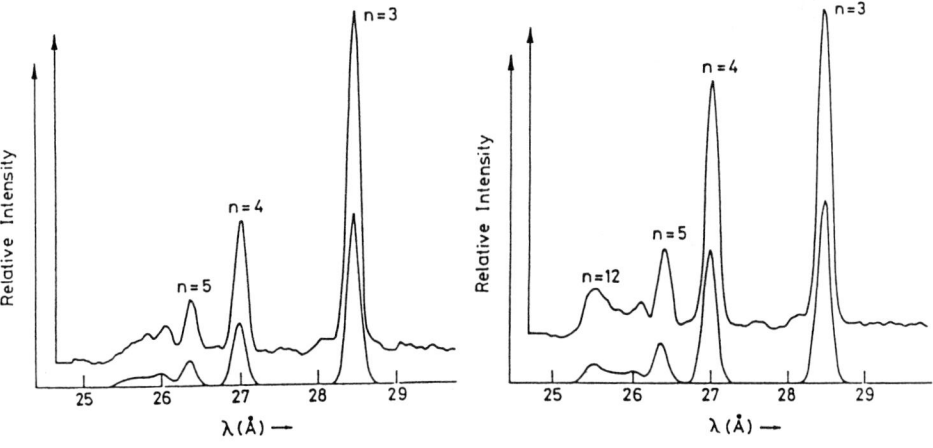

CVI LYMAN SERIES LINE INTENSITIES

Fig. 20. Upper curve shows Balmer-α intensity from inner wall of JET discharge. The time-dependent intensity of D_α is related to the local influx of D^0 at the plasma boundary. The steps in the emission result from a shift of the plasma boundary into contact with the inside wall at just before 5s followed by application of an RF heating pulse at 11s. The lower curves illustrate the effect on the CVI Lyman series line intensities of CX between C^{6+} and D^0, due to the increased influx of D^0 from 3.5s (on left) to 13s (on right). Comparison is made between the JET data (upper spectrum) and model calculations (lower spectrum) for each of the two time frames.

tion coefficient for say oxygen. This very large total cross section for CX, Janev et al. (1980,1981), coupled with the fact that relaxation to ground of the exchanged electron involves relatively few quantum jumps, mainly Y-RAST, $\Delta n = 1$ lines in the VUV and XUV regions of the spectrum, means that charge exchange lines can dominate the VUV spectrum of beam-heated Tokamaks as illustrated in Figure 21. Examples of the space-time and of the line width signatures of the CX emission are shown in Figure 22.

The diagnostic capability, unique to the CX process, of being able to 'illuminate' bare nuclei, measure their density and temperature and follow their transport in space and time, has been developed by Isler et al. (1981), Fonck et al. (1984), Duval et al. (1985) and Carolan et al. (1987), among others. Analysis of the line intensity requires, of course, not the total cross section as is required for calculations of ionisation balance, but the *partial* $\sigma_{CX}(n, l)$ cross sections into all the states contributing to the observed line intensity through cascades. Collisional smearing of the l-states and possibly Stark and Zeeman splitting have to be taken into account, Fonck et al. (1984). A great deal of effort, both theoretical and experimental has gone into the measurement of these 'effective' cross sections, Gilbody (1981). Of particular interest are the $\sigma_{CX}(n, l)$ partial cross sections for CX into $n > \hat{n}$ which give rise to transitions in the visible region. While, for a given collision energy, $\sigma_{CX}(n)$ scales as n^{-3}, these 'Rydberg' levels are sufficiently well populated for beam energies, $\epsilon(H^0) > \epsilon_0(H^0)$, that quantitative visible spectroscopy of bare nuclei, Carolan et al (1987), becomes an attractive proposition in Tokamaks with atomic beam injection and where the nuclei have an abundance $\geq 10^{-3}/n_e$. The CX cross-section $\sigma_{CX}(n)$ falls off approximately as $\epsilon(H^0)^{-3}$, but the Rydberg transitions and visible CX spectroscopy are still useable up to impacting energies of ~ 120 keV/AMU, von Hellermann and Summers (1993). Indeed, CX collisions between beam atoms and quasi-thermal alphas, which are products of the fusion reactions in a burning fusion plasma and have slowed down into this energy range, should just be detectable above the noise in the bremsstrahlung as enhanced intensity in the wings of the 4686Å line, von Hellermann and Summers (1993), von Hellermann, Mandl et al. (1993).

Visible emission, resulting from charge transfer collisions between beam atoms and light ions such as H^+, He^{2+}, C^{6+} etc., can present a problem of interpretation arising from severe blending of the CX light with background edge emission caused by electron impact excitation of the ground state of the H-like ions and indeed, other ions which mimic the Rydberg levels, Boileau, von Hellermann, Horton et al. (1989). These residual blends can, in principle be corrected for since they will, of course, be present even in the absence of atomic beams. Another problem, shared by the VUV lines, is that secondary CX collisions involving the H^0 thermal atoms, the 'halo' neutrals, which are products of the beam attenuation, can also contribute to the CX line

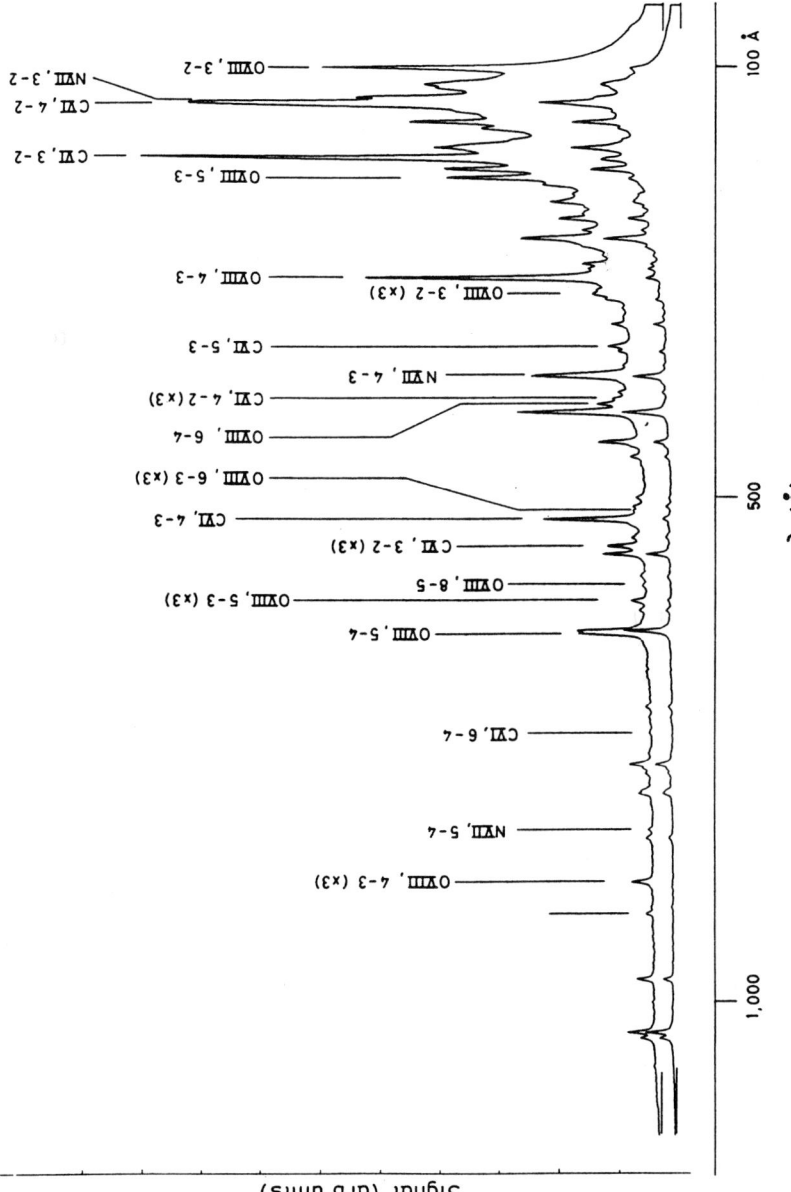

Fig. 21. The VUV spectrum at two different times during an ASDEX discharge. The lower spectrum is taken at 1.4s just after the switch-off of the neutral beams. The upper spectrum is taken at 1.370s during the beam injection. Exposure times for both spectra are 16ms. All the features annotated are charge-exchange recombination lines which appear only during the beam-injection period.

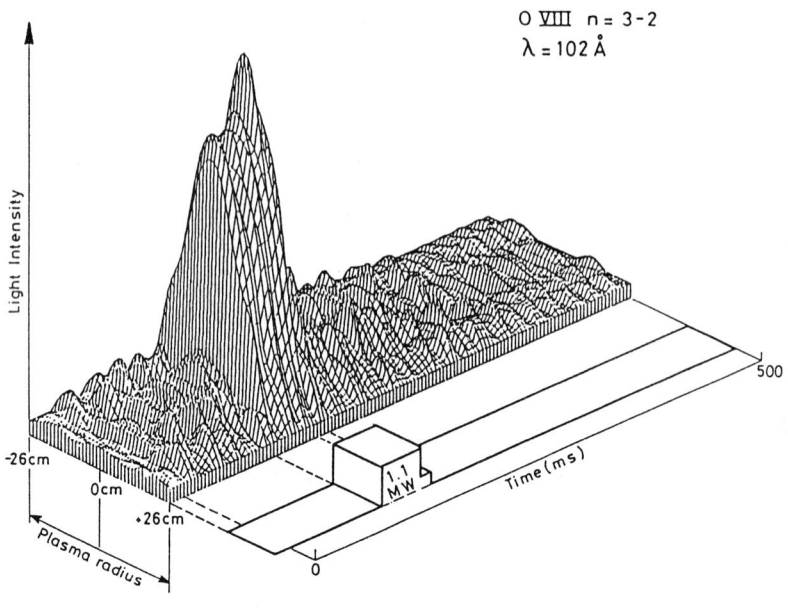

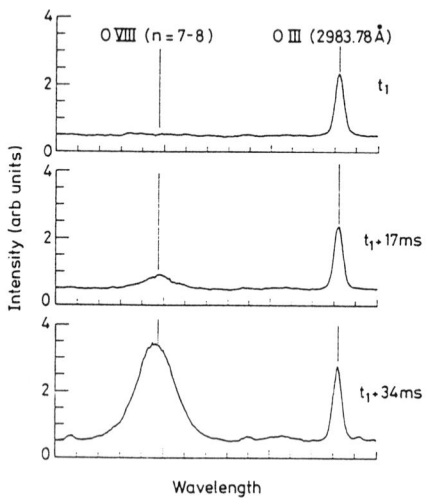

Fig. 22. (a)Radial emission profiles of OVIII (n=3-2) at 102Å in the DITE tokamak with a period of atomic beam injection. The CX signature viewed along a chord intersecting one of the heating beams is synchronous with the beam current. (b)Time sequence of OVIII ($n = 8-7$) emission prior to (t_1) and during beam injection (t_1+17ms and t_1+34ms). The breadth of the CX feature, determined by the local ion plasma temperature, contrasts with the widths of the recycling edge lines, eg., OIII. Line widths as well as temporal variations are important features of the CX signature.

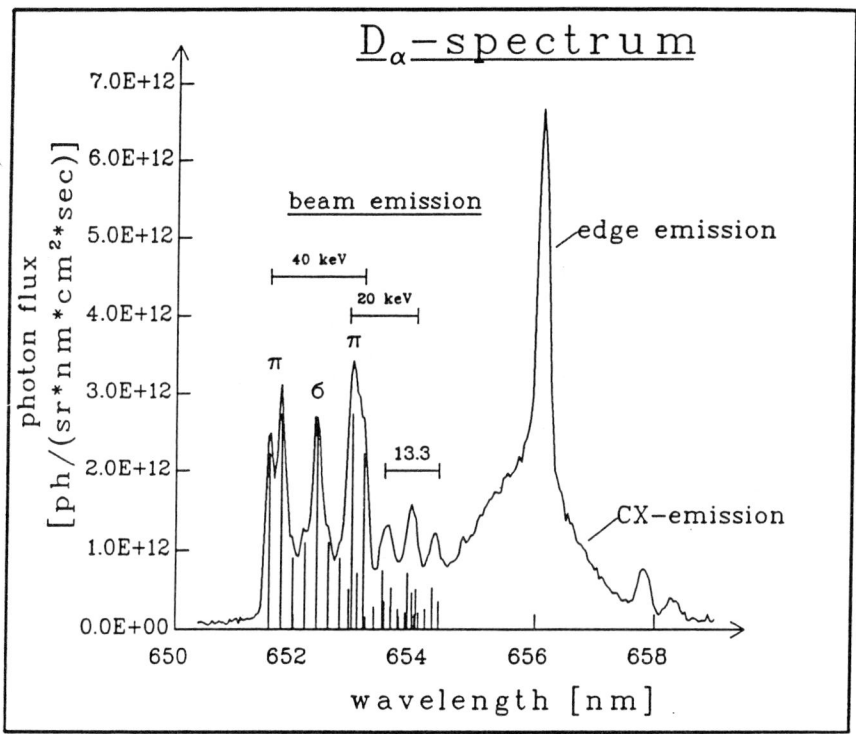

Fig. 23. D_α light observed along a LOS through a beam of fast D^0 neutrals injected into a deuterium plasma in JET. The feature shows the CX emission from the plasma as well as the beam particle emission which is Stark split due to the $\mathbf{v} \times \mathbf{B}$ Lorentz field. Radiation from each of the three different energy components of the beam can be distinguished by their appropriate doppler shifts. The beam emission consists of 27 Stark split lines, each being indicated by a vertical bar, Mandl (1991).

profile, as can 'plume' effects due to thermal electron collisions with the recombined ion, Fonck et al. (1984). The complexity of the visible emission is best illustrated as in Figure 23 by the Balmer alpha emission from JET when heated with atomic hydrogen beams. The complicated spectral structure can be unravelled on realising that emission features from the beams themselves contribute to the spectrum and are displaced many Angstrom units from the thermal plasma emission by the projection of the beam velocity along the LOS. There are typically three energy components in the beam. Each of these components are split into Stark sub-components by the $\mathbf{v}_b \times \mathbf{B}_{(\phi+\theta)}$ Lorentz field, where $\mathbf{v}_b$ is the beam velocity and $\mathbf{B}_{(\phi+\theta)} = \mathbf{B}_\theta(r) + \mathbf{B}_\phi(R)$ is the vector sum of the poloidal and toroidal components of the magnetic field. Typically in JET, with B $\sim$3.5 Tesla and $\mathbf{v}_b \sim 3\times10^6 \mathrm{ms}^{-1}$, then the motional Stark field strength is $E_s = 10^5 \mathrm{V/cm}$, which is about $\times10^2$ more than the plasma microfield. This realisation, Boileau, A., von Hellermann, Mandl, et al. (1989), has opened up many new diagnostic possibilities based on beam emission spectroscopy (BES). The information content in the Balmer alpha light is worth a study on its own, Mandl (1991). The CX line intensity is given by

$$I_{CX}(\lambda) = n_Z(X) \sum_{k=1}^{3} j\left(\frac{\epsilon_0}{k}\right) \sigma_{CX}\left(\lambda; \frac{\epsilon_0}{k}\right) \tag{16}$$

where $\sigma_{CX}(\lambda; \epsilon_0/k)$ is the effective cross section for CX induced emission of a photon of wavelength λ. The different energy components in the atomic beam are denoted by ϵ_0/k, $k = 1, 2, 3$ and have associated atomic fluxes or 'currents' denoted by $j(\epsilon_0/k)$. For the deuterium fuel ions, we replace $n_Z(X)$ by $n(D) \simeq n_e$, after some correction factor for the impurity dilution of the fuel. The deuterium beam emission signal I_{BES}, on the other hand, is given by,

$$I_{BES}(\lambda) = \sum_k j\left(\frac{\epsilon_0}{k}\right) \left[\sum_z n_z(X) Z^2 \sigma_{i,X}\left(\frac{\epsilon_0}{k}\right) + \frac{n_e \langle \sigma_{e,X} \nu_e \rangle_{ex}}{v_b}\right] \tag{17}$$

where $\sigma_{i,X}$ is the ion impact excitation, which for beam energies of several 10's of keV/AMU, well exceeds excitation of the beam atoms due to the electron impacts. The ratio of the Balmer emission from the deuterium beam atoms to the Balmer emission due to CX with the plasma deuterium ions, $I_{BES}(D_\alpha)/I_{CX}(D_\alpha)$, is then independent of the beam attenuation and is a function only of the relevant beam energy dependent cross-sections and the local value of the effective charge state of the ions in the plasma, viz, $Z_{eff} f(\epsilon_0/k; \sigma_{i,X}/\sigma_{e,X})$. The beam emission and CX features which are major components of the Balmer spectrum as shown in Figure 23 can therefore yield $Z_{eff}(r)$ directly.

The next feature of interest in the H-alpha emission is the polarisation pattern of the Stark multiplet. The wavelength separation at 6561Å between the adjacent sub-levels is $\Delta\lambda_{MSE}(\overset{\circ}{A}) = 2.75 \times 10^{-7} E_s$ (Volts/m) and the σ and π components are easily resolvable as seen in Figure 23. Levington et al.(1989) have developed a novel technique for measuring the local pitch angle of the magnetic field $\tan^{-1}(B_\theta(r)/B_\phi(r))$, which is related to the Tokamak safety factor $q = rB_\phi(r)/RB_\theta(r)$ and to the current distribution, Wesson (1987), by using polarimetry on the plane-polarised, central σ components of the MSE complex. This has proved to be the most accurate technique for measuring $q(r)$ which has so far been developed.

A further development by Durst, Fonck et al. (1992), Fonck, Bretz et al. (1992) related the BES, Balmer-α line intensity fluctuations, which are proportional to the local ion density fluctuations ie., $\sqrt{\tilde{I}(H_\alpha)} \propto \tilde{n}_j$, to the level and frequency spectrum, $S(k, \omega)$, of turbulence in the plasma. Radial scale lengths $k_r^{-1} \sim$ 1cm, with frequencies in the 1–100 kHz region are typical in Tokamaks. The fluctuation levels $\tilde{n}_i$ are considerably reduced during good confinement operation.

A summary of the diagnostic information which can be derived from atomic beam spectroscopy makes remarkable reading. That information includes the fuel and impurity ion densities, $n_i(r)$ and $n_Z(X, r)$ respectively; the effective charge state, Z_{eff}; the impurity ion temperature and ion velocity distributions, $T_{i,Z}$ and $f_Z(v_{i,Z})$, including thermalized alphas $f_{Z=2}(v_{i,Z=2})$; the ion fluctuation level and power function, $\tilde{n}_i$ and $S(k, \omega)$ and finally the magnetic field components, $B_\theta(r) + B_\phi(R)$.

7. K-shell Spectra from Tokamaks

Satellite lines, lying to the long wavelength side of the H- and He-like resonance lines in the soft X-ray region, had been documented, prior to 1970, in spectral studies of the Vacuum Spark, the θ-Pinch, and of the Plasma Focus discharges where, for example, Ly α of Ar XVIII could be readily produced, (Peacock et al. 1969). Following the identification by Gabriel and Jordan (1969) of the M1 forbidden transition, $1s^2\,^1S_0 - 1s2s\,^3S_1$ in the solar spectrum of O VII, the diagnostic potential of the satellite lines in laboratory plasmas was quickly realised, (Gabriel and Jordan, 1972). In the spectra from K-shell ions with $Z > 12$, the $1s^2\,^1S_0 - 1s2p\,^1P_0, 1s2p\,^3P_{1,2}, 1s2s\,^3S_1$, transitions become interspersed with a rich spectrum of dielectronic lines and inner-shell excited lines of the type $1s^2nl - 1s^2pnl$, where the non-participating or 'spectator' electron, or electrons, have quantum states $nl \geq 2$. The theory determining the wavelengths and intensities of these lines was outlined by Gabriel (1972) but has subsequently been refined and improved over a sequence of papers, eg., Bhalla et al. (1975), Merts et al. (1976), Dubau and Volonte (1980), Bely-Dubau et al. (1982). More precise energy levels and A values have also been compiled using, for example, relativistic Z-expansion techniques, as favoured, among others, by Vainshtein and Safronova (1978, 1980) or using multi-configuration Dirac-Fock models such as used by Chen (1986).

Satellite spectra were observed in a wide range of laboratory plasmas extending from Tokamaks with densities $n_e \sim 10^{19}\text{m}^{-3}$, ie., near solar flare conditions to transient high density ($\geq 10^{27}\text{m}^{-3}$) plasmas such as are produced by laser-irradiation of solid targets, as illustrated for an Al target in Figure 24, Peacock et al., (1973), or in vacuum sparks, (Golts et al., 1974). In Tokamaks, where forbidden lines with $A_{ij} \geq 10^3\text{s}^{-1}$ are prominent features of the satellite spectra, and where the electron parameters can be independently measured, it has been possible to assess in some detail the diagnostic potential of the K-shell emission. The studies referenced below, while not exhaustive, are illustrative of the activity in the field of Tokamak

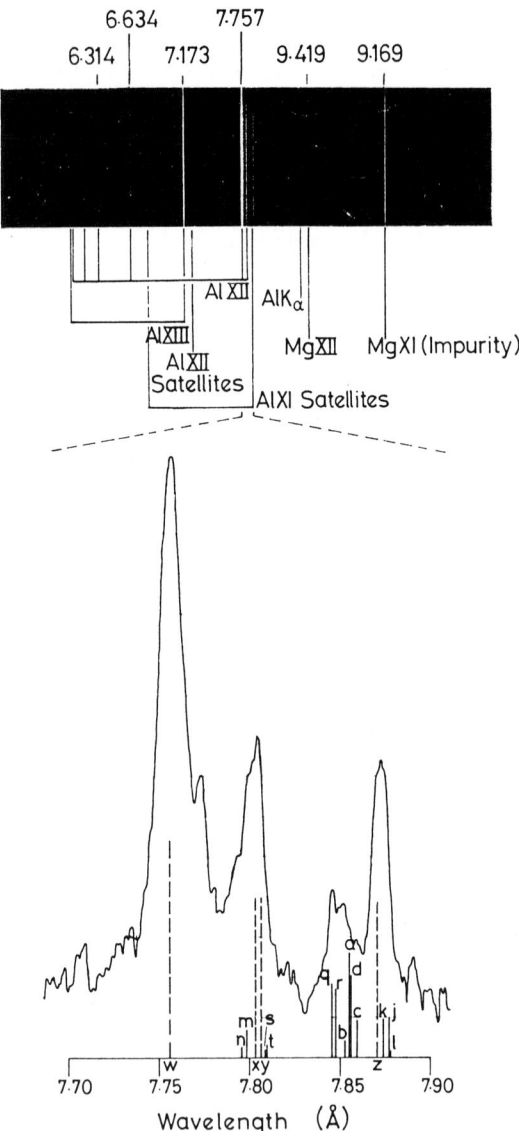

Fig. 24.   X-ray emission from Al target irradiated by a laser pulse of $10^{14}\,\mathrm{Wcm}^{-2}$ peak intensity. The spectrum was recorded on film using a mica dispersion element in a de Broglie spectrograph. The line notation follows Gabriel (1972).

spectroscopy. Analyses of the Fe XXV spectrum and its associated satellites have been carried out by Bitter et al., (1979) while S XV and Cl XVI have been studied by Källne et al. (1983) and A XVII by Källne et al. (1984), by the TFR group, Bombarda et al. (1985) and by Phillips et al. (1994). Ti XXI has been studied by Bitter et al. (1984; 1985) and by Bely-Dubau et al. (1982). Ni XXVI and its satellites have been analysed by Hsuan et al. (1987) while the TFR group, (TFR, Cornille et al. 1985) have studied the progressive changes in satellite positions and intensities throughout the isoelectronic sequences from Ar XVII through Mn XXIV. Reviews of these Tokamak studies are given by Bartiromo (1986) and by Bitter et al. (1993).

We illustrate the diagnostic potential of the satellite spectra by reference to the Cl XVI emission from the JET and COMPASS Tokamaks in Figure 25. The energy diagram and the line features, often blended, are labelled 'w', 'x', 'y', 'z' etc. after Gabriel (1972). A self-consistent derivation of the core parameters T_e, T_i, n_e, $V_{\theta,\phi}$ where the subscripts, θ and ϕ, refer to the poloidal and toroidal components of the ion fluid velocity, can in principle be derived from the Cl emission line intensities and line shapes within the narrow spectral region from $4.44 \rightarrow 4.50$ Å. The G-ratio $(\frac{x+y+z}{w})$ of line intensities is nearly density-independent for Cl XVI in Tokamaks but is temperature sensitive due to the different behaviour of the singlet and triplet excitation rates. The R-ratio $(\frac{z}{x+y})$ is of limited interest in Tokamaks since it will only show density dependence, due to collisional de-excitation of the metastable $1s2s\,^3S_1$, above 10^{20}m^{-3}. The intensity ratios of the dielectronic satellites to the resonance line 'w', for example (k/w) and (d_{13}/w) are conveniently independent of ionisation balance and of electron density (n_e) but are steeply dependent on temperature (T_e) with the form

$$\frac{I(k)}{I(w)} \sim \frac{1}{T_e} \exp[(E_w - E_s)/kT_e] \tag{18}$$

where E_w and E_s refer to excited state energies above the ground state of Cl XVI. On the other hand, the satellite lines such as 'q', produced by inner-shell excitation from the ground states of Cl XV and the resonance line 'w' of Cl XVI have similar excitation functions and so the ratio

$$\frac{q}{w} \simeq \frac{N(\text{ClXV})\langle\sigma(q)v\rangle_{\text{ex}}}{N(\text{ClXVI})\langle\sigma(w)v\rangle_{\text{ex}}} \frac{A_r}{\sum(A_r + A_a)} \tag{19}$$

is nearly independent of T_e and measures the ionisation balance or departures from it.

In JET with a minor plasma radius, $a = 1$m, the Cl XVI emission shell is located on the flanks of the $T_e(r)$ profile at $r/a \simeq 0.4$ where, typically, T_e could be a few keV. In COMPASS, with $a = 0.2$m, $T_e(0) \simeq 0.8$keV and Cl XVI is a core ion. In each device, the location of the ions is determined by diffusive equilibrium. In JET, the satellite lines are weak relative to the (w, x, y, z) resonance lines. Figure 26 shows the G-ratio as a function of T_e in both devices. At $T_e > 1.5$keV, the data are exclusively JET and the decrease in the G ratio is due to diffusion and spreading inwards to higher

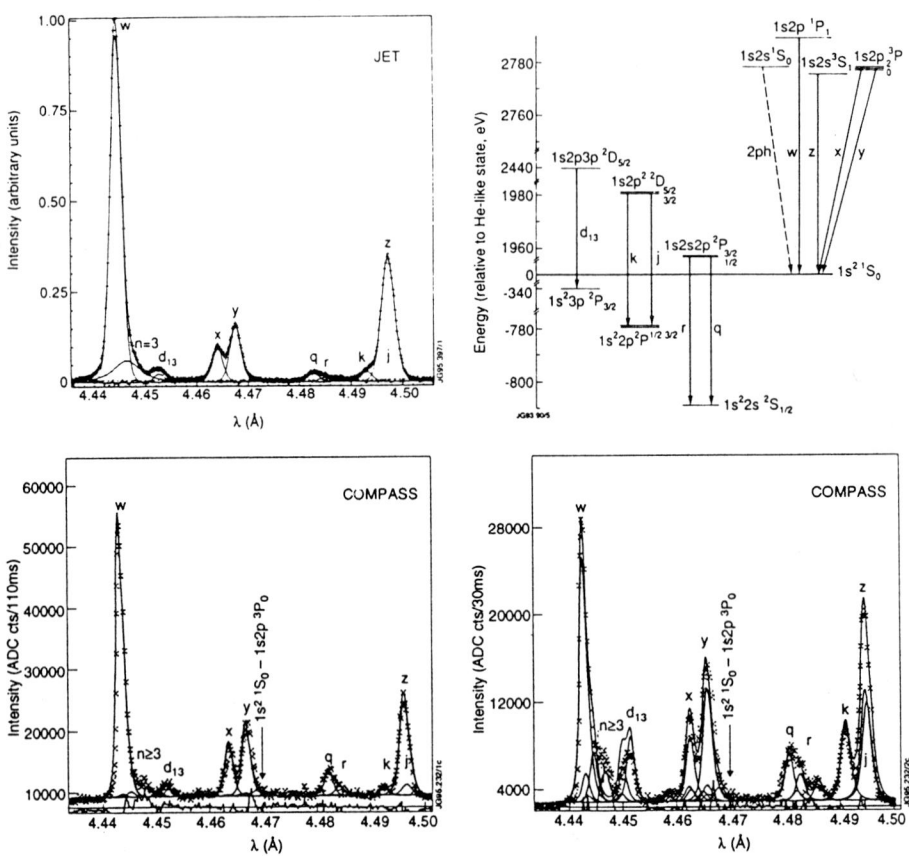

Fig. 25. Spectrum of Cl XVI from JET (above left) compared to those from COMPASS-D (below) using the same high resolution ($\lambda_0/\Delta\lambda = 3000$), curved crystal spectrometer. The energy level diagram and identification of the transitions are shown (upper right). The dielectronic satellite intensities of Cl XVI are relatively weak in the high, ($T_e \geq 1.5$keV) JET spectrum and the forbidden 'z' line is almost unblended. In the lower temperature COMPASS-D tokamak, ($T_e < 1.5$keV) the Cl XVI lines ('w','x', 'y','z') are heavily overlaid with the dielectronic and inner-shell excited satellites. The satellite lines are most prominent in the lowest temperature plasma ($T_e \simeq 0.5$keV, lower right), Coffey et al.(1995).

temperatures than would be predicted by coronal ionisation balance. This process is almost equivalent to reducing the recombination rate, Figure 26. At the lower end of the T_e scale, the COMPASS data lies above coronal values. This is not fully understood but may be due to inner-shell ionisation such as $1s^2 2p\,^2S_{3/2,1/2} \rightarrow 1s2p\,^3P_{1,2}$, thus enhancing the triplet states. It is a common and so far unexplained phenomenon in the COMPASS studies

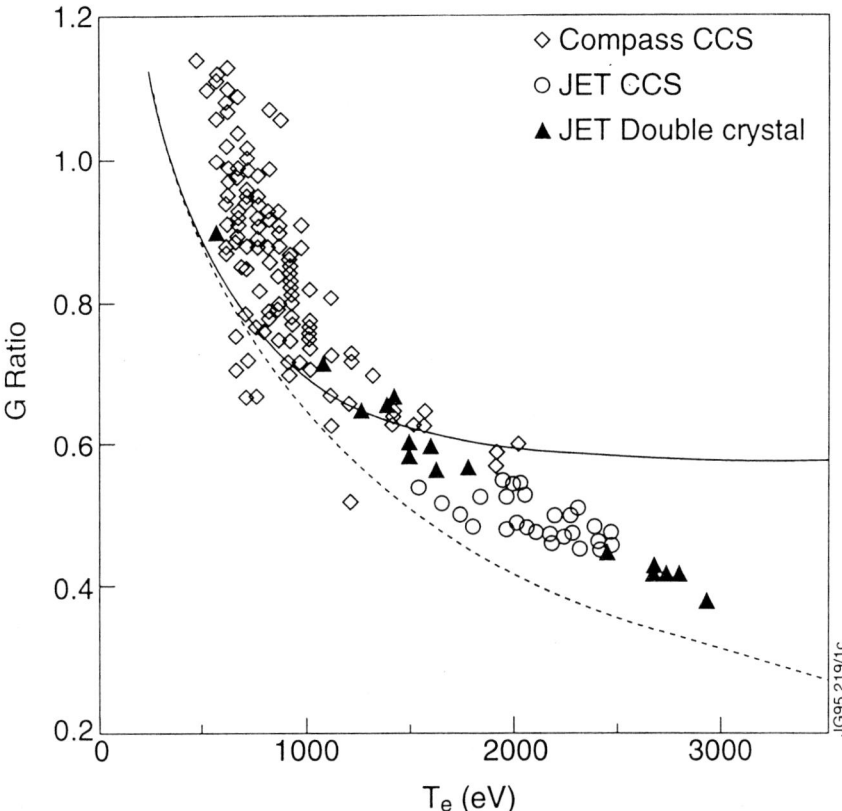

Fig. 26. The theoretical ratio of $(x + y + z)/z$ line intensities, ie., the G ratio, in the Cl XVI spectrum as a function of T_e at an electron density of $n_e = 3.2 \times 10^{19}\,\mathrm{m}^{-3}$, with dielectronic recombination included (solid line) or excluded (dashed line). Experimental points from JET (○) and COMPASS-D (◇) are shown and, in addition, JET results with a double crystal spectrometer (△) are displayed to extend the T_e range, Coffey et al. (1995).

that (x, y) line intensities are enhanced by $\leq 30\%$, relative to the (w, z) lines as calculated with existing theoretical models.

We now turn our attention to analyses of the Cl XV satellites. Ion diffusion should also alter the (Li-like / He-like) ion abundance ratio as indicated by the (q/w) line ratio, Figure 27, both lines being excited from the ground states of the respective ions. Figure 27 shows that coronal balance is a lower bound to the data set and that the largest departure from coronal equilibrium is to be found at the highest T_e, and consequently lowest n_e, where diffusive processes are relatively rapid. Section 9 treats the spectroscopic derivation of ion transport in more detail.

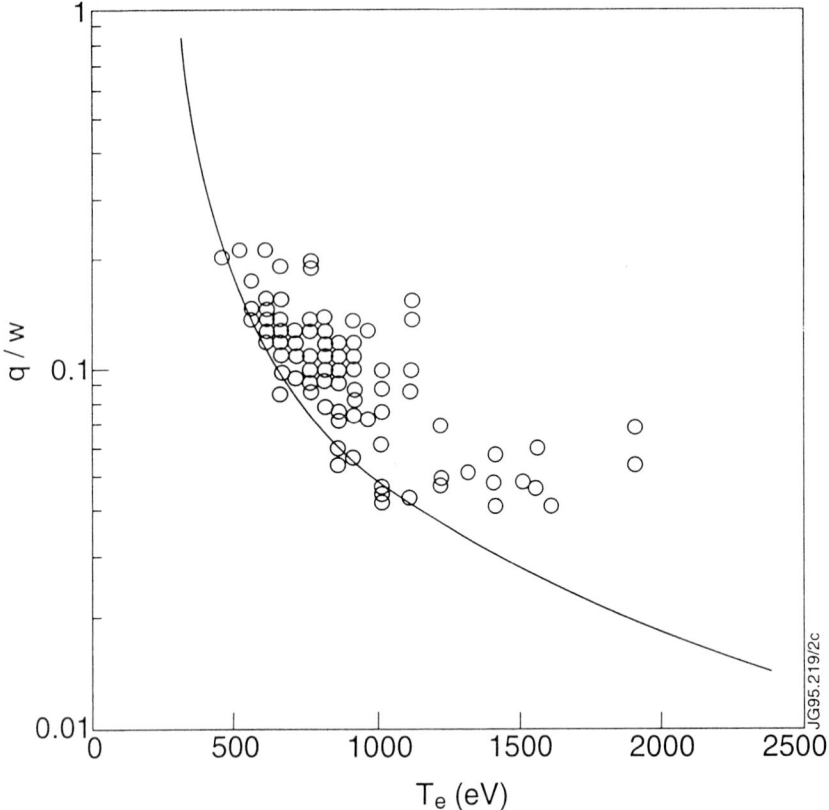

Fig. 27. The theoretical ratio of q/w line intensities in the Cl XVI spectrum plotted as a function of T_e. Equilibrium values for $N(\mathrm{Li-like})/N(\mathrm{He-like})$ have been used in the calculations. Experimental points from COMPASS-D are shown (circles).

The ability to measure T_e (from the (k/w) ratio) and T_i (from the pseudo-Voigt function applied to unblended lines), simultaneously on a 10ms timescale in COMPASS, allows an investigation of the heat transport, in particular the electron-ion transfer rate. Figure 28 shows the measured values of T_e against the corresponding value of T_i in COMPASS. The T_i measurements are compared with scaling relations involving the parameter $2.8\times10^{-6}(IB_\theta R^2 n_e)^{1/3}\sqrt{A}$, suggested by Artsimovich (1972). The measured values are in general slightly lower than the predicted ones but increase with the parameter values, particularly n_e, as expected. The T_e values, which are in general agreement with Thomson scattering results apply to all data for $T > 450\mathrm{eV}$. The tendency for the temperatures of the ion and electron fluids to come together at higher values of the Artsimovich parameter can be

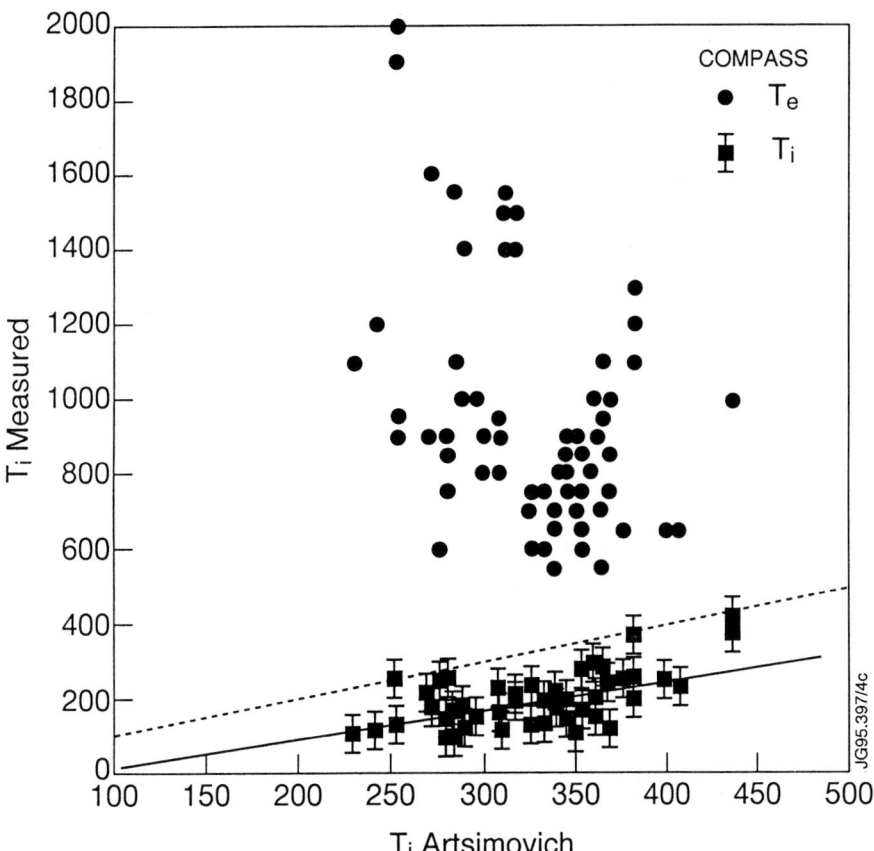

Fig. 28. Electron and ion temperatures derived from analyses of the COMPASS-D spectrum of Cl XVI. Line intensity ratios (d_{13}/w) and (k/w) were used for T_e while line profiles of 'z' were used for T_i. The lower group of points $(< 450\text{eV})$ refer to T_i while the upper group $(> 450\text{ eV})$ refer to T_e. T_i scaling due to Artsimovich (1972) is shown broken, while the solid line is the best fit to the T_i data.

understood in terms of the electron power balance,

$$\frac{dw_e}{dt} = \int^{\text{vol}} P_{\text{OH}}\, dv + \int^{\text{vol}} P_{\text{ECRH}}\, dv - \int^{\text{vol}} P_{e,i} + P_{\text{RAD}}\, dv - \frac{w_e}{\tau_{\text{Diff}}} \quad (20)$$

where the LHS is the rate of change of the electron energy. The terms on the RHS are in sequence, the input ohmic power, the applied electron cyclotron heating power, the heat loss to the ions and by radiation and finally the characteristic cross-field loss rate due to heat conduction (coefficient: χ_e) and particle diffusion (coefficient: $D_\perp$), where,

$$\frac{1}{\tau_{\text{Diff}}} = \frac{1}{\tau_e(\chi)} + \frac{1}{\tau_e(D_\perp)} \quad (21)$$

and

$$\frac{w_e}{\tau_{\text{Diff}}} = \int^{\text{vol}} (\nabla \cdot n_e \chi_e \nabla T_e + \nabla \cdot \frac{3}{2} T_e D_\perp \nabla n_e)\, dv \qquad (22)$$

Similar equations can be written for the ion fluid. When applied to the range of data, Figure 28, we find that an estimate of the heat conduction loss time for the lowest n_e, ECR-heated plasmas ($\langle n_e \rangle = 6.4 \times 10^{18}\text{m}^{-3}$, $T_e = 1500\text{eV}$, $T_i = 180\text{eV}$) is $\tau_e(\chi) = a^2/(3.8\chi) \sim 5$ ms for the electrons and ~ 30 ms for the ions. Since, at these low densities, the time for collisional transfer of the electron energy to the ion fluid much exceeds the ion heat conduction loss time, ie., $\tau_{ei} \gg \tau_i(\chi)$, this accounts for the maintenance of the large difference in the fluid temperatures at the lower bound of n_e. At the highest densities ($\langle n_e \rangle = 1.2 \times 10^{20}\text{m}^{-3}$, $T_e = 800\text{eV}$, $T_i = 380\text{eV}$) almost all the power to the electrons is transferred directly to the ions through what is then the main loss channel for the electron energy.

A quite different application of the 1- and 2-electron spectra in Tokamaks has been the study of atomic structure. These ions are the subject of precise *ab initio* calculations, including relativistic and QED corrections to high order in Z, see eg., Mohr, 1982 and Drake, 1988 for H- and He-like ions respectively. A review comparing theory and experiment in H-like high Z ions is given by Briand (1993). Precision crystal measurements of Ly $\alpha_{1,2}$ for Cl XVII at an accuracy approaching the 10ppm level have been reported by Källne, Källne, Patrick and Stöckli (1983) who used the ALCATOR Tokamak. An alternative approach followed by Stamp et al. (1981) and Peacock et al. (1984) using the DITE and TCA Tokamaks has been to measure the $1s2s\,^3S_1 - 1s2s\,^3P_{2,1,0}$ transitions in the VUV region, where high resolution is still possible with a diffraction grating at normal incidence. The QED contribution to the these $\Delta n = 0$ transitions at $Z = 20$ is at the level of two percent or so. So far, no systematic discrepancies between the most sophisticated theory and the Tokamak (and other) experiments have been established.

A somewhat related observation is the Ly $\alpha_{1,2}$ intensity or 'β' ratio which, in Tokamaks, is generally observed to exceed slightly the statistical ratio of 0.5 (Dunn, 1990; Källne, Källne, Patrick and Stöckli (1983). For $Z \sim 16$ or 17 as in these studies, it would appear that collisional and satellite effects contribute about equally to the departure of β from 0.5. At higher $Z > 20$, blending with the M1 decay of the $2s\,^2S_{1/2}$ excited state is expected to make a significant apparent increase in β.

8. Ion Transport in Tokamaks

Background fuel ions and impurities are strongly linked by collisions so that a study of the spectral emission from impurities directly yields information on the overall particle transport and often on the total plasma energy. Particle and energy transport in Tokamaks often behave similarly in many plasma operating conditions.

According to theory, Rutherford (1974, 1976), Connor (1973), Hirshman and Sigmar (1981), neoclassical diffusion (classical density and temperature gradient driven fluxes in cylindrical geometry enhanced by the Tokamak curvature) should be a slow process with diffusion coefficients $\sim 0.01 \mathrm{m}^2\mathrm{s}^{-1}$. Furthermore, if the temperature gradient terms are neglected, ambipolarity of the ions and electrons should cause a Z-dependent accumulation of impurities relative to the fuel ions, n_i, at the peak n_e on axis,

$$\frac{n_z(r=0)}{n_z(r=a)} = \left(\frac{n_z(r=0)}{n_z(r=a)}\right)^Z \tag{23}$$

While neoclassical ion transport is not commonly achieved in Tokamak experiments, in some operating scenarios, such as fuelling the core by solid fuel pellet injection, Behringer, Denne et al. (1978), Lauro-Taroni, Alper, et al. (1994), such impurity accumulation has been observed.

The time evolution of ions of charge state z is given by the 1-D radial continuity equation,

$$\frac{\partial n_z}{\partial t} = -\nabla \cdot \Gamma_z + \Delta_z \tag{24}$$

where Δ_z contains the impurity source and sink terms and all the ionisation and recombination processes. The ion flux is given by

$$\Gamma_z = -D_\perp \frac{\partial n_z}{\partial r} \tag{25}$$

The set of equations for all ionisation stages, $n_z(r)$ throughout the plasma radius, can be summed to give the total elemental impurity density, $\sum_z n_z(X)$, in which case ionisation and recombination cancel out and $\Delta_z = 0$ in the continuity equation. It is almost the rule in Tokamak experiments that both particle and energy transport are anomalously fast, by factors of up to 100, relative to neoclassical predictions. We define therefore, an anomalous coefficient given by

$$\Gamma_z = -D_{\mathrm{an}} \frac{\partial n_z}{\partial r} \tag{26}$$

with values of $D_{\mathrm{an}} \sim 0.1 \to 1 \mathrm{m}^2\mathrm{s}^{-1}$, depending on the Tokamak operating conditions.

Since the diffusion term by itself does not account for steady-state particle profiles in Tokamaks, the inclusion of convection has been suggested, Coppi and Sharky (1981). It is common to include the additional convection term in the form

$$\Gamma_z = -D(r)\nabla n_z(r) + V(r)n_z(r) \tag{27}$$

where

$$V(r) = -2\frac{D(r)r\mathcal{S}}{a^2} \tag{28}$$

and $\mathcal{S}$ is the dimensionless 'pinch' parameter, Seguin et al. (1983). In many cases $0 < \mathcal{S} < 3$, Barnsley (1993) and the ion confinement time, τ_z is then approximately,

$$\tau_z = \frac{a^2}{(2.4^2 - 1.4\mathcal{S})D_z} \quad (0 < \mathcal{S} < 3) \tag{29}$$

The continuity equation then becomes

$$\frac{\partial n_z}{\partial t} = -\nabla \cdot \Gamma_z + n_e[-n_{z-1}S_{z-1} - n_zS_z + n_{z+1}\alpha_{z+1} - n_z\alpha_z] - \frac{1}{\tau_z} \tag{30}$$

where S_z and α_z are the ionisation and recombination coefficients, respectively.

Figure 29 illustrates the difference between the calculated steady-state iron ion distributions, Hulse (1983), when diffusion is appreciable and constant, and the coronal distribution which prevails when diffusion is switched off. In practice, however, both D and V are found to vary with radius. Appropriate 1-D numerical codes have been developed to solve the coupled set of continuity equations describing each ion charge, Hulse (1993). These include the STRAHL code, Behringer (1987) and the SANCO code, Lauro-Taroni (1994), the latter code being able to cope with time dependent profiles of $T_e(r)$ and $n_e(r)$ and is supported with ionisation, recombination and line emission measures from the ADAS atomic physics database, Summers (1994). Figures 30 and 31, for example, indicate the steady-state volume emissivities of Al and the radial ion distributions of Cl in the COMPASS-D tokamak using the SANCO and ADAS codes.

In ohmically heated plasmas τ_e (energy confinement time) and τ_p (particle confinement) are usually not very different. Auxiliary heating mechanisms, such as electron cyclotron resonance heating (ECRH) or neutral beam NBI etc. tend to increase transport and cause a reduction in τ, typically by a factor of two. This factor and more can be retrieved by gaining access to improved confinement modes (IOC), such as described by Barnsley et al., (1987); modes produced by atomic beam injection in the counter ion flow direction, Isler et. al.(1985) and 'high' confinement modes produced in Tokamaks with divertors, the so-called H-modes, Wagner (1982). In all of these modes, particle confinement can well exceed energy confinement. In the best JET H-modes for example, particle confinement can be ≥ 4 s while the energy confinement is about 1s.

Improved particle confinement in H-mode is thought to be due to the generation of electric field shear associated with a narrow region of steepened

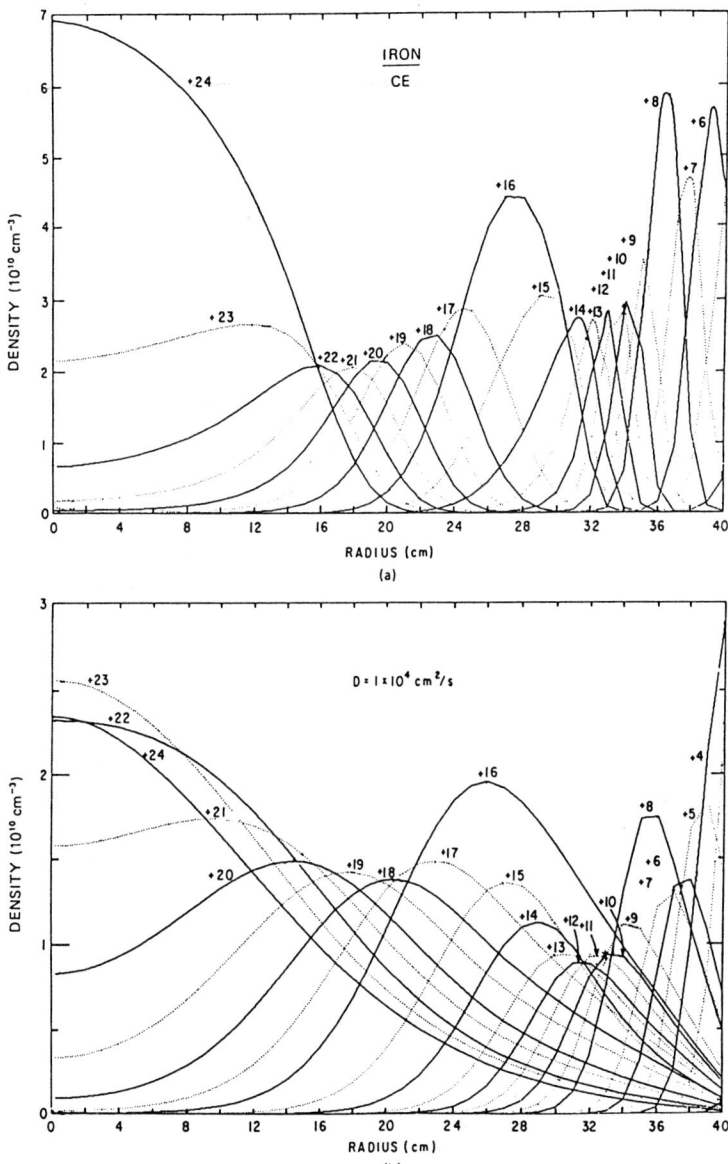

Fig. 29.   1-D calculations of the radial charge-state density profiles for iron impurity ions in a plasma with central electron parameters of $n_e = 3 \times 10^{13} \mathrm{cm}^{-3}$ and $T_e = 2\mathrm{keV}$ and an iron density $n(\mathrm{Fe}) = 1 \times 10^{11} \mathrm{cm}^{-3}$. Coronal equilibrium has been assumed in the calculation of the upper profiles, while in the lower set of profiles, the plasma is in diffusive equilibrium with a constant D of $1 \times 10^4 \mathrm{cm}^2\mathrm{s}^{-1}$; Hulse (1983).

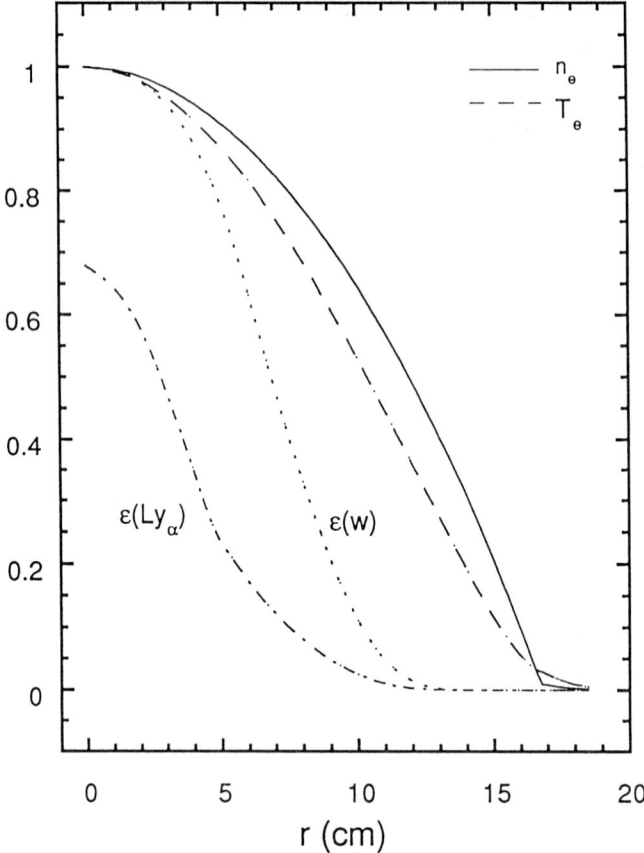

Fig. 30.   SANCO ion diffusion model of Al ion emission assuming diffusion coefficients appropriate to L-mode conditions in COMPASS-D tokamak with $T_e(r = 0) = 1\text{keV}$ and $n_e(r = 0) = 6.6 \times 10^{13}\,\text{cm}^{-3}$. The volume emissivity $\epsilon(\text{Ly}\alpha)$ of Al XIII is normalised to the $\epsilon(w)$ emissivity of the 'w', $1s^2 - 1s2p$, transition in Al XII.

pressure gradient close to the plasma boundary. In this 'transport barrier' region, the ratio of V/D is relatively high. It is not clear whether the confining E_r-fields arise from $\mathbf{v} \times \mathbf{B}$ terms due to fluid motion or from pressure gradient, ∇P terms. Impurity ion fluid velocities of the order of 20kms^{-1} have been reported during the change to H-mode confinement from L-mode in Tokamaks. The zero order radial force balance on any one charge state, neglecting T_e gradients, is given by

$$\nabla P_z = eZN_z \left(\mathbf{E}_r + \mathbf{v} \times \mathbf{B} - \mathbf{v}\nu_{z,i} \right) \tag{31}$$

where the last term which accounts for friction due to ion collisions is usually ignored.

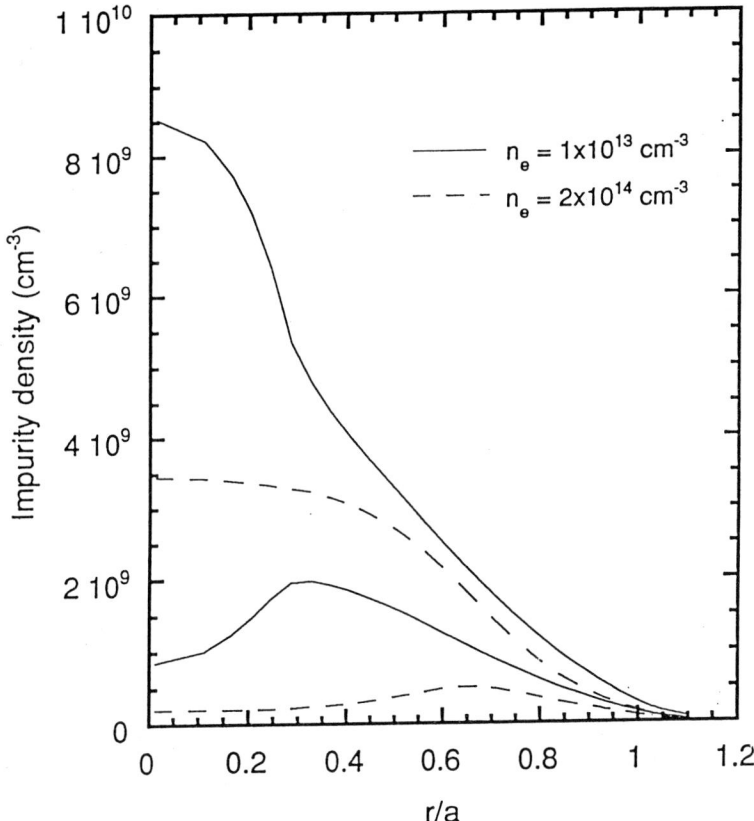

Fig. 31. SANCO ion diffusion model of Cl XVI (upper) and Cl XV (lower) ion abundances assuming diffusion coefficients appropriate to L-mode conditions in COMPASS-D tokamak, with $T_e(r = 0) = 1\text{keV}$ as in Figure 30, and $n_e(r = 0)$ values as shown.

There is conflicting evidence from the various Tokamak experiments as to the relative importance of the terms in this equation in determining the radial electric field, Hawkes (1995). The whole subject of impurity transport is complex. It remains, nonetheless, of considerable importance in the design of a Tokamak-based fusion reactor.

Spectroscopic methods for measuring particle transport can be divided into 'passive' studies of indigenous impurity ions which eventually adopt steady-state radial distributions, or 'active' injection of test atoms into the plasma boundary with a few eV energy, using laser 'blow-off', Marmar et al. (1975), or similar techniques. The advantage of measuring the transient rise and decay of the test ions is that D and V can be measured individually and indeed approximate solutions for unfolding the inflow time τ_1 and the decay time τ_0 in terms of the $\mathcal{S}$ parameter are given by Seguin et al. (1983). In the COMPASS Tokamak, Peacock et al. (1994), Fielding et al. (1994), the transport coefficients are derived, Figures 32(a) and (b), by fitting the transient spectral signature of injected Al ions to the predictions of the SANCO code with $D(r/a)$ and $V(r/a)$ functions somewhat similar to those suggested by Pasini et al. (1990). In the H-mode, after taking into account any density increase, confinement is sometimes essentially infinite (on the H-mode period time-scale). The monotonically increasing intensity of Al XIII Ly α after accessing the H-mode in Figure 32(b), is in part due to increasing electron density and possibly also to an increasing X-ray background continuum due to neoclassical peaking of impurities in the core, Peacock et al. (1995). In the larger Tokamaks such as JET, particle confinement in H-mode can extend to several seconds. The time variation of cobalt ions injected into the H-mode in JET and shown in Figure 33(a), indicates a decay time constant $\tau_0(\text{Co}) \sim 4\text{sec}$. The relatively rapid decay of the test ions following the end of the H-mode is successfully modelled as shown in Figure 33(b), assuming changes in $D(r/a)$ and $V(r/a)$ similar to that indicated in Figure 32.

In passive spectroscopy the ions of interest such as C, O, Cl etc. are indigenous, there being a steady state loss and influx, to and from the plasma boundary. Profiles of the D and V coefficients are selected that give computed emission profiles in agreement with the radial distributions of the ions. In practice, this technique cannot lead to independent determination of D and V. Ion transport has been derived from the (q/w) ratio of the emission lines, shown in Figure 25, of Cl XVI from the COMPASS-D tokamak, by comparing model emissivities as predicted by the time-independent ion transport code STRAHL, Behringer (1986). The D and $\mathcal{S}$ parameters are varied until a fit with the data is obtained. Less time consuming is to use a zero dimensional approach and solve the full set of coupled ionisation balance equations which include a source term γ_z and a loss term τ_z. These are written as,

$$\frac{\partial n_z}{\partial t} = -\frac{n_z}{\tau_z} + \gamma_z + n_e[n_{z-1}S_{z-1} - n_z S_z + n_{z+1}\alpha_{z+1} - n_z\alpha_z] \qquad (32)$$

A solution is shown in Figure 34. There is general agreement between the these 'passive' measurements of τ_z and the ion injection results in L-mode. For $n_e \geq 3 \times 10^{19}\text{m}^{-3}$, however, the ratio approaches the coronal value for all sensible values of τ_z in COMPASS and the (q/w) ratio in CL XVI is no longer a viable method in this region of parameter space.

In passing, we should note the close similarity of these experiments, both in concept and in technique, with the early studies, discussed in the introduction, of impurity confinement in ZETA by Bob Wilson and his co-workers.

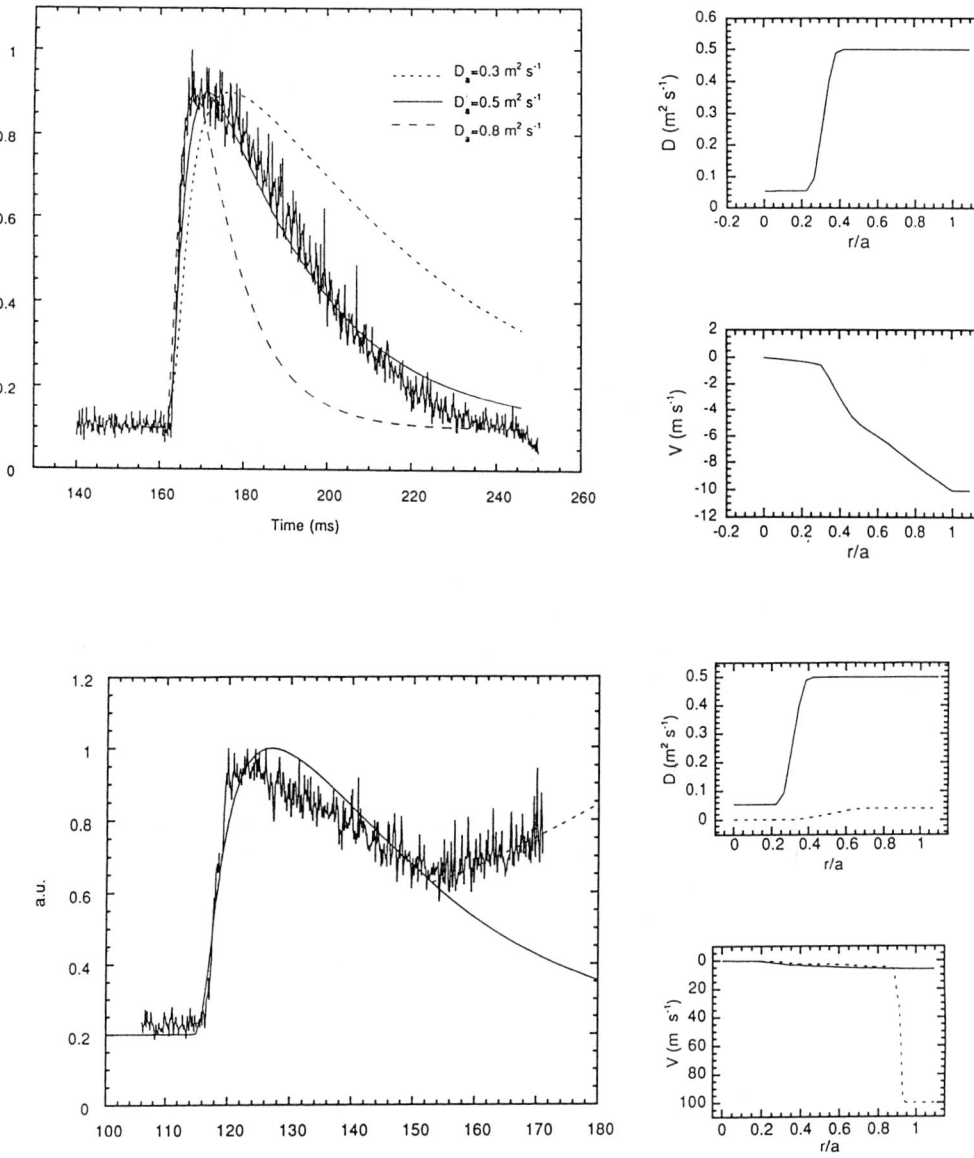

Fig. 32.   (a)SANCO model simulation of temporal variation of Al XIII, Ly α, intensity following injection of pulse of Al atoms at the plasma boundary during L-mode (#12016) in the COMPASS-D tokamak. The best fit diffusive and convective coefficients are shown. (b)SANCO model simulation of Al XIII, Ly α, intensity following injection of pulse of Al atoms at the plasma boundary (#12210) in the COMPASS-D tokamak. Injection occurs at 150ms into L-mode. The arrest in the decay of the signal at 152ms is due to accessing an H-mode. The best fit diffusive and convective coefficients for both confinement modes are shown.

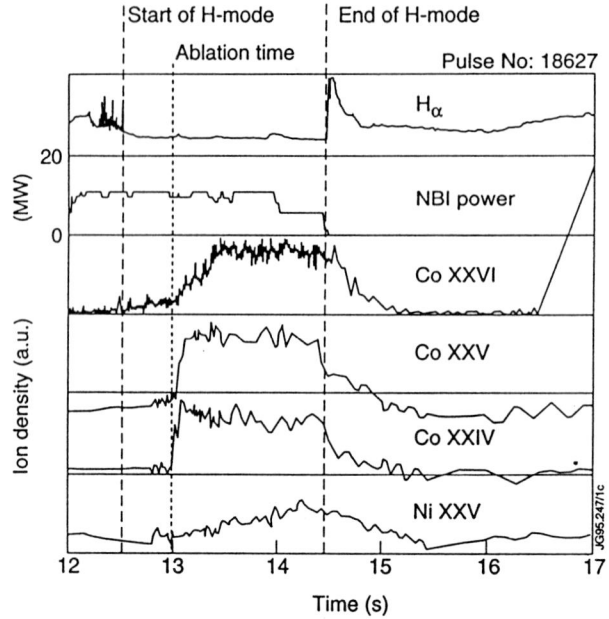

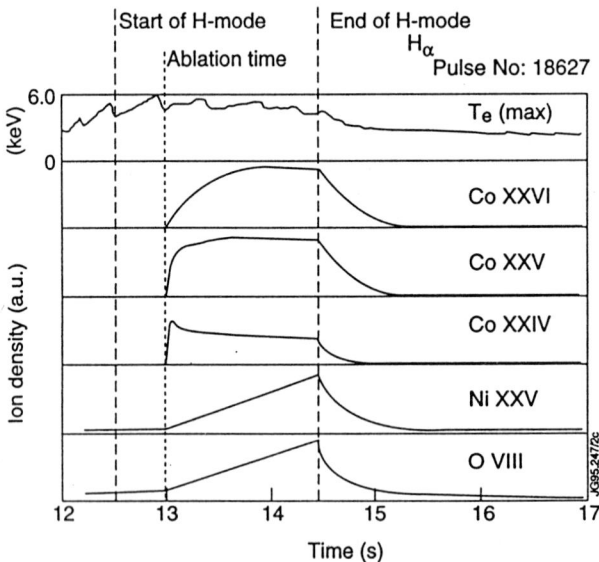

Fig. 33. (a)Temporal variation of cobalt ion densities following injection of Co atoms into H-mode in JET discharge (#18627). Characteristic confinement time τ_z(Co) of the highly charged ions such as Co XXVI in the core plasma in JET is several seconds in H-mode. (b)STRAHL model of temporal variation of the Co and Ni ion abundances for same JET discharge (#18627) as shown in Figure 33(a). The diffusion and convection coefficients were adopted from Pasini et al. (1990).

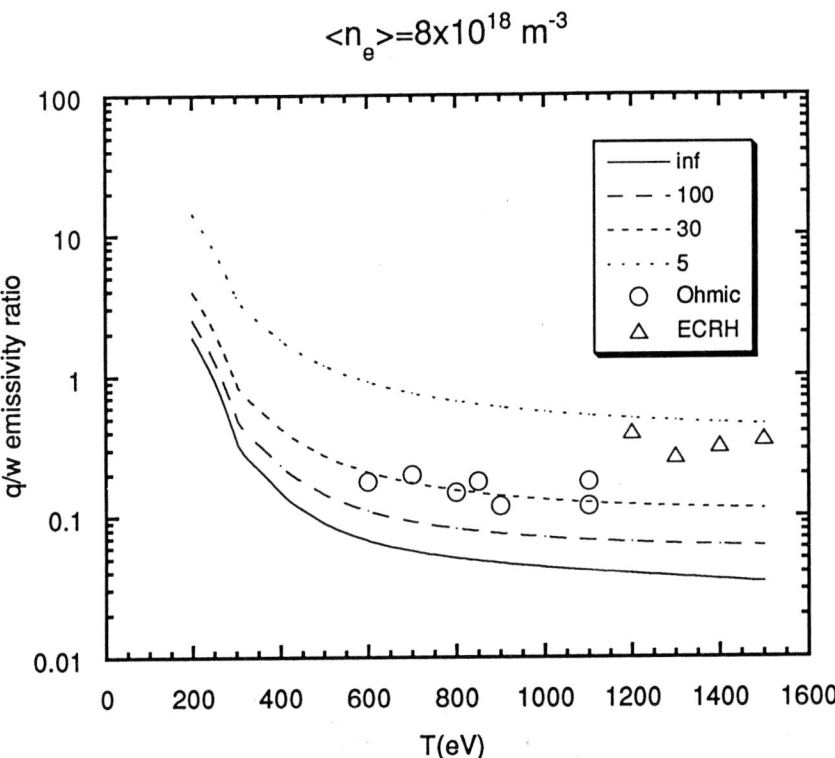

$<n_e> = 8 \times 10^{18} \; m^{-3}$

Fig. 34. Analyses of the K-shell emission from Cl XVI and its satellites in terms of ion confinement in the COMPASS-D tokamak. The line intensity ratio q/w is a measure of the relative ion abundance of *Li-like Cl XV/ He-like Cl XVI*, as a function of T_e with ion confinement time (τ_z) as a parameter.

9. Plasma Condensations

The concept of thermal instabilities is a familiar one in astrophysics and has been proposed for example by Parker (1953) to explain solar prominences, but it is also thought to have wider applicability to the formation of condensations in planetary nebulae and in the interstellar medium, Field (1965). In the laboratory, radiation-cooling instabilities have been credited for the formation of 'hot spots' in the Vacuum Spark, Negus and Peacock (1979). In Tokamaks, plasma condensations were discovered about ten years ago in the ALCATOR Tokamak, Lipschultz et al., (1984), Lipschultz, (1987) and were given the acronym of MARFE, Multifaceted Asymmetric Radiation From the Edge. While MARFE's are indeed asymmetric in the poloidal plane, it is equally true that they are toroidally *axisymmetric*. MARFEs have the unusual property of not following the helical field lines in Tokamaks but locate themselves typically in the edge plasma, close to the inner wall. Since there is pressure equilibrium in the Tokamak, radiation induced cooling is accompanied by an equivalent increase in density, usually by a factor of up to ten, making condensation densities at the edge $\sim 1 \times 10^{20} \mathrm{m}^{-3}$. Like a 'curtain ring', MARFEs can 'slide' up and down the inner wall, but their preferred position above or below the horizontal mid-plane is determined by the $\mathbf{B} \times \nabla \mathbf{B}$ ion drift. Models for MARFE formation and stability, eg., Stringer (1985); Wesson and Hender (1993), are based on the balance between radiation loss and heat conduction from the surrounding plasma,

$$\nabla \cdot \kappa \nabla T_e = n_e n_z P_{\mathrm{rad}}(T_e) \tag{33}$$

where κ is the heat transport coefficient, ($\kappa_\parallel$ is proportional to $T_e^{5/2}$) and $P_{\mathrm{rad}}(T_e)$ is the radiation power loss as a function of T_e. Clearly, when the radiation loss term exceeds the heat inflow ie., at high density and in the presence of impurities, and when $\frac{\partial P_{\mathrm{rad}}(T_e)}{\partial T_e}$ is a minimum, MARFE formation is a possibility. The ionisation balance radiation loss curves for the most common impurity, namely carbon, Figure 5, indicate MARFE formation at about 15eV.

Stringer's (1985) model, however, based on the balance between radiation loss and heat flow, predicts a condition for MARFE onset given by

$$n_e^2 \sum q_z \left(\frac{2P_z}{T_e} - \frac{dP_z}{dT_e} \right) > \frac{1.72 \times 10^{22} T_e^{5/2}}{q^2 R^2} \tag{34}$$

In this inequality, P_z represents radiation power loss by ions of charge state z, q_z is the ion abundance relative to n_e and q is the Tokamak safety factor, Wesson (1987). The expression within the brackets is plotted in Figure 36 for various confinement times of carbon ions transiting the MARFE region. Clearly, the shapes of the condensation criterion shown in Figure 36 will depend not only on the transient nature of the passage of ions through the MARFE region, but on the details of the atomic processes, perhaps including charge exchange collisions if there is an appreciable influx of atomic H^0 into the MARFE region.

Experimentally, MARFEs are recognised by their radiation signature in the bremsstrahlung continuum, in CIII line emission or in $\mathrm{H}\alpha$ light as illustrated in the photograph of the ASDEX Tokamak, Figure 37. Since the power radiated from only about 5%-10% of the plasma volume can amount to as much as 40%-100% of the input power, MARFEs are readily identified by multi-chord bolometery as displayed in Figure 38. Evidently, in JET, MARFEs last a considerable time ~ 2 s but spend most of this period oscillating vertically along the inboard edge of the plasma boundary.

Unfortunately, direct spectroscopic measurements and indeed, direct measurements of even the electron parameters in MARFEs are few and far

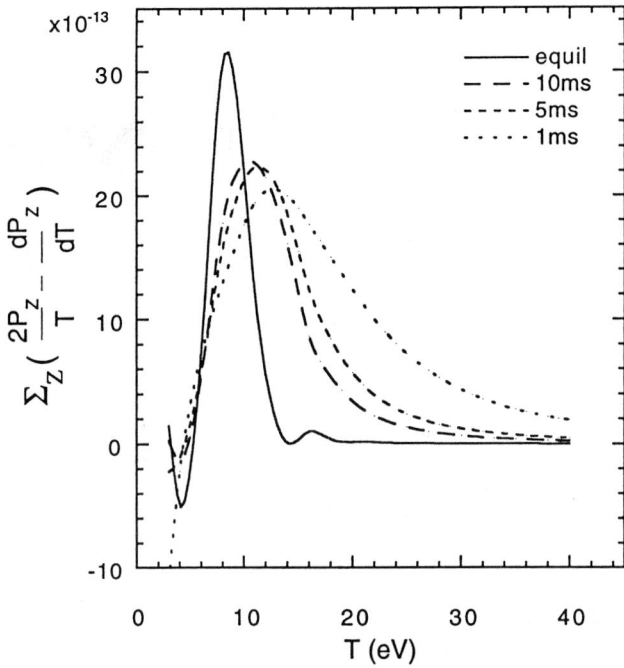

Fig. 35. Criterion for MARFE condensation (Stringer, 1985) as a function of electron temperature in a hydrogen plasma with carbon impurity. The radiated power coefficient, P_z for each charge state z, and consequently the criterion for MARFE formation varies with the ion confinement time as shown.

between. O'Mullane et al. (1995) have considered the spectral signature of a MARFE. Ions in low charge states are generally the causal impurities in MARFE formation due to the relatively low T_e near the edge and the shape of the cooling curves, Figure 5. Two impurity sources give rise to MARFE cooling, viz., fresh radial influxes of atoms and ions from outside the plasma boundary or, alternatively, existing more highly ionised impurities which flow through the condensation region from the surrounding bulk plasma. The initial distribution of the ionisation stages and their spectral signature reflects the origin of the sources. Influxing ions are neutral or weakly ionised with velocities corresponding to their sputtering energies of a few eV while streaming ions with $T_e \geq 100$ eV reflect the local diffusive ionisation balance. Globally the ionisation balance in the plasma is not greatly affected by the the presence of the condensation since the re-ionisation time of ions leaving the MARFE is shorter than the transit time of ions around the torus. Within the MARFE, impurities are subject to parallel and perpendicular transport. Over a range of MARFE temperatures, frictional slowing down $\tau_{||}$ time and

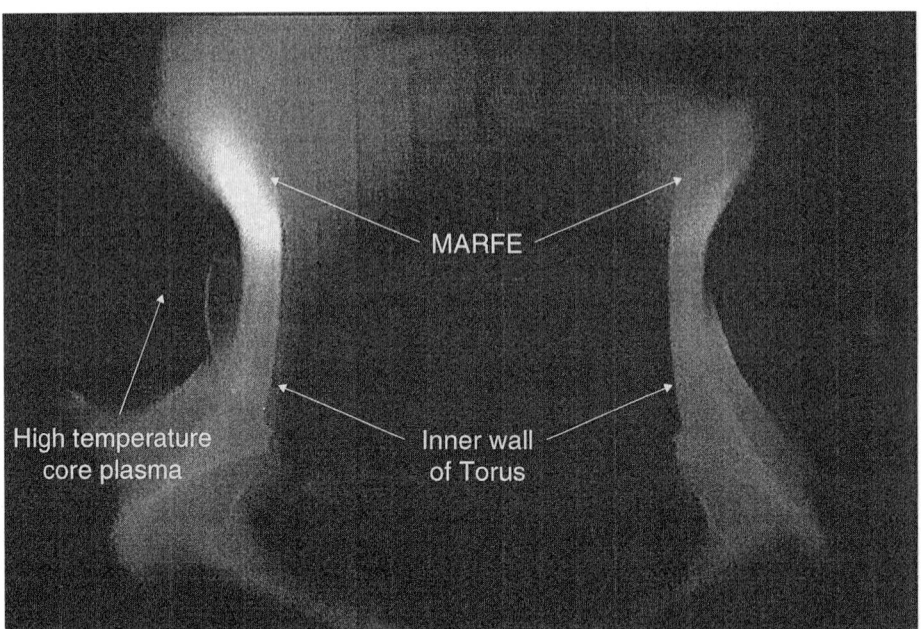

Fig. 36.   CCD camera picture of H_α emission from MARFE near the inside wall in ASDEX-U Tokamak. The ion drift due to $\mathbf{B} \times \nabla \mathbf{B}$ fields in this discharge (#4489: $t = 2.055$ s) is upwards; Büchl and Junker (1994).

the diffusive deflection time $\tau_\perp$ are comparable and of the order of a few milliseconds. The establishment of ionisation equilibrium occurs on about this same time scale, as indicted in Figure 38 by the calculations for carbon ions passing through the MARFE from the surrounding plasma. Prior to ions adopting a new ionisation balance in the MARFE, a recombination feature should appear initially as the background ions first enter the MARFE region as indicated in Figure 39. This feature should serve to distinguish between the two sources of ions. Dielectronic recombination cascade lines as predicted in the VUV have yet to be seen from MARFEs. However, Y-RAST lines arising from charge exchange recombination of $C^{+6} + H^0$ have been reported, Hess et. al.,(1993) and O'Mullane et al.(1995). The latter authors reach the conclusion that the concentration of intrinsic impurity charge states, in diffusive equilibrium, can be sufficient to form the MARFEs. Values for $T_e \simeq 15$ eV have been measured from line ratios. Ideally, we should like to

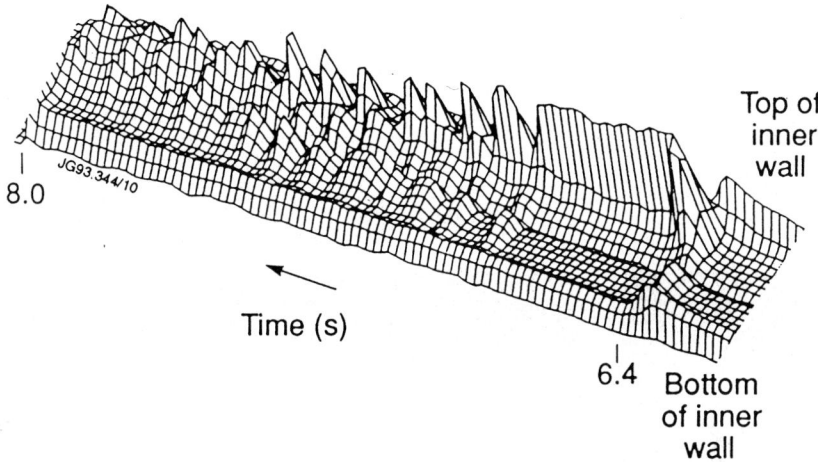

Fig. 37. Total radiation from a MARFE condensation near the inner wall of JET as a function of time. The tomographic reconstruction uses the data from a bolometric camera viewing the inside wall of the torus. The MARFE, after periods, of the order of a second, of positional stability in the upper half of the torus, breaks into oscillatory motion up and down the inside wall.

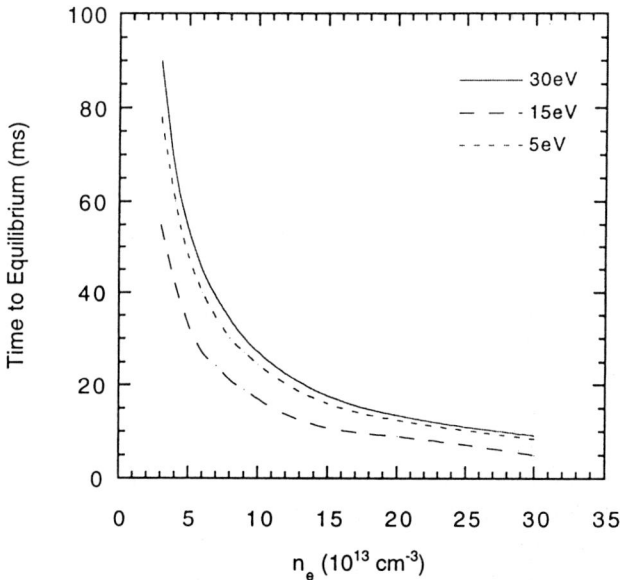

Fig. 38. Time taken for ions of carbon, from the surrounding plasma at a temperature of 300eV, to attain 95% of the equilibrium ion abundances in the MARFE as a function of MARFE temperature and density.

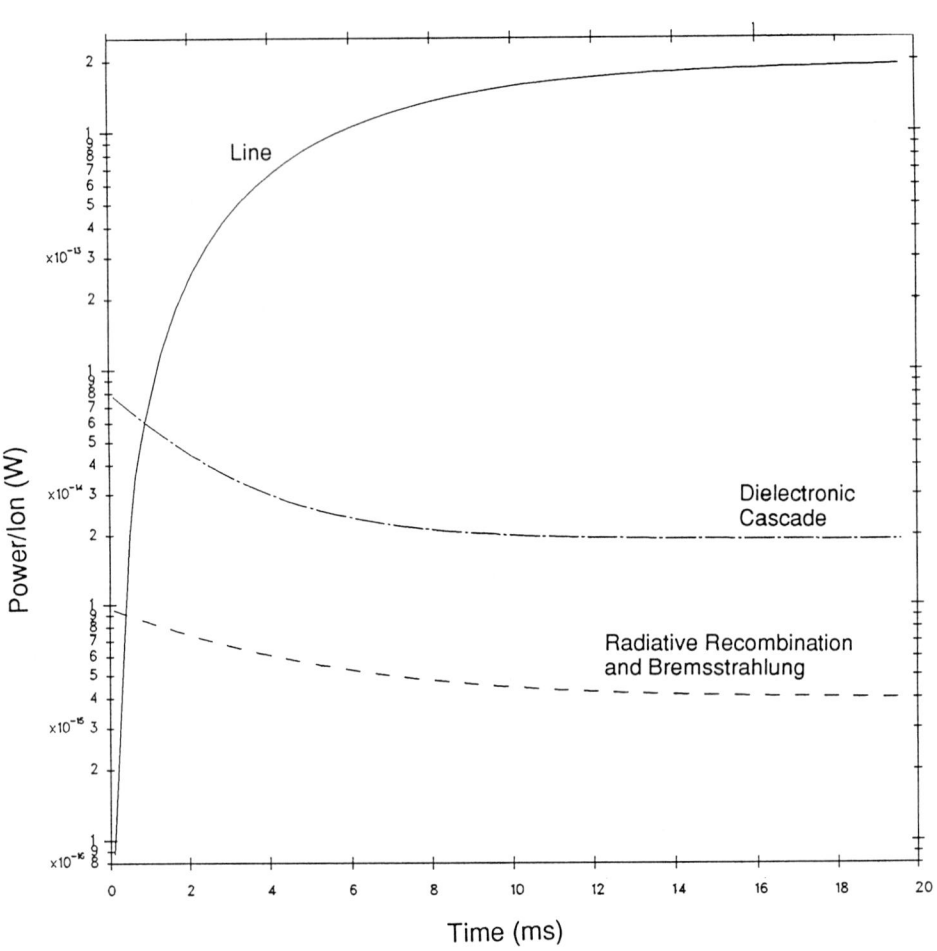

Fig. 39.   The power loss components of the spectrum from carbon ions streaming into a MARFE region with a temperature of 5eV and density of $8 \times 10^{19}\,\mathrm{m}^{-3}$. The continuum loss is initially high relative to the line emission. This 'recombination' signature is lost within about ten milliseconds as the ion charge states assume near equilibrium abundances. The ions at time zero are assumed to have a characteristic of a 300eV and $2 \times 10^{19}\,\mathrm{m}^{-3}$.

measure, simultaneously, the electron and ion density components in order to confirm pressure balance between the MARFE and its surrounding plasma. The possibility of using MARFE condensations to control power loss from a fusion reactor has a certain appeal.

10. Summary

The reader has been introduced to a number of topics, taken from Tokamak research, in order to trace the the development of applications of spectroscopy in controlled fusion research over the last 35 years, from the early toroidal devices like ZETA to present-day Tokamaks. The subject of plasma spectroscopy has grown in sophistication in terms of the expansion of the atomic processes which have to be considered and their associated data base, the complexity of the experimental techniques and the wide range of diagnostic applications. Plasma spectroscopy has increased our appreciation of the subtle role of impurities in determining much of the plasma behaviour. Control of impurities, by techniques such as wall conditioning, magnetic divertors, pellet or atomic beam injection and radiation mantles, offers a wealth of future investigations.

Acknowledgements

The author would like to acknowledge the help and inspiration he has derived from his students past and present in writing this article. In particular he is indebted to M O'Mullane for his technical help in preparing the manuscript and whose research work is featured in the sections on MARFEs and ion transport.

References

Abbey, A. F., Barnsley, R., Dunn, J., Lea, S. N. and Peacock, N. J.: 1993, *UV and X-ray Spectroscopy of Laboratory and Astrophysical Plasmas. (editors, E Silver and S. Khan) Cambridge University Press*, 493.

Afrosimov, V. V., Gordeev, Y.S. et al.: 1979, *J.E.T.P. Lett.* **28**, 501.

Alper, B.: 1995, *private communication, JET*.

Artsimovich, L. A.: 1972, *Nuclear Fusion* **12**, 215.

Barnsley, R., Fielding, S. J., Hawkes, N. C. et al.: 1987, *Culham Laboratory Report (UKAEA) CLM P792*.

Barnsley, R., Lea, S., Patel, A. and Peacock, N. J.: 1993, *UV and X-ray Spectroscopy of Laboratory and Astrophysical Plasmas. (editors, E Silver and S. Khan) Cambridge University Press*, 513.

Barnsley, R.: 1993, 'X-ray Spectroscopic Diagnostics of Magnetically Confined Plasmas. Instrumentation and Techniques.', *PhD thesis, Dept. of Physics, Leicester University, UK*.

Bartiromo, R.: 1986, *Proc. of International School of Plasma Physics, Varenna, Italy, (Course and Workshop on Basic and Advanced Diagnostic Techniques for Fusion Plasmas) EUR 10797 EN* **1**, 227.

Bashkin, S., (editor): 1976, 'Beam Foil Spectroscopy', (Berlin: Springer)

Bates, D. R., Kingston, A. E. and McWhirter, R. W. P.: 1962, *Proc. Roy. Soc.* **A267**, 297.

Bearden, A. J., Ribe, F. L., Sawyer, G. A, and Stratton, T. F.: 1961, *Phys. Rev. Lett.* **6**, 257.

Behringer, K. H. 1987, *Joint European Torus JET Report — R(87)08*.

Behringer, K. H., Denne, B. et al.: 1988, *Proc. 15th EPS Conf. on Plasma Physics and Controlled Nuclear Fusion Research (Dubrovnik)* **12B**, 338.

Bely-Dubau, F., Dubau, J., Faucher, P. and Gabriel, A. H.: 1982, *Mon. Not. Roy. Astr. Soc.* **198**, 239.

Bely-Dubau, F., Faucher, P., Steenman-Clark, L. et al.: 1982, *Phys. Rev. A* **26**, 3459.

Bhalla, C. P., Gabriel, A. H. and Presnyakov. L. P.: 1975, *Mon. Not. Roy. Astr. Soc.* **172**, 359.

Bitter, M. et al.: 1979, *Phys. Rev. Lett.* **42**, 304.

Bitter, M., Hill, K. W., von Goeler, S., Hill, K. W., Sesnic S. et al.: 1984, *Phys. Rev. A* **29**, 661.

Bitter, M., Hsuan, H., Hill, K.W. and Zarnstroff, M., von Goeler, S. et al.: 1985, *Phys. Rev. A* **32**, 3011.

Bitter, M., Hsuan, H., Hill, K.W. and Zarnstorff, M.: 1993, *Physica Scripta* **T47**, 87.

Bitter, M., Hsuan, H. et al.: 1993, *Phys. Rev. Lett.* **71**, 1007.

Boileau, A., von Hellermann, M. G., Mandl, W. et al.: 1989, *J. Phys. B: At.Molec. Phys.* **22**, L145.

Boileau, A., von Hellermann, M. G., Horton, L. D. et al.: 1989, *Plasma Phys. and Controlled Fusion* **31**, 779.

Briand, J. P.: 1993, *Physica Scripta* **46**, 157.

Büchl, K. and Junker, W. 1994, *MARFE emission ASDEX-U, private communication.*

Burton, W. and Wilson, R.: 1961, 'Spectroscopic Measurements of Plasma Containment in ZETA' *UKAEA Culham Laboratory preprint CLM-P1, Proc. Phys. Soc.* **78**, 1416.

Burton, W. and Wilson, R.: 1965, *Nature* **207**

Carolan, P. G., Duval, B. P. et al.: 1987, *Phys. Rev. A* **35**, 3454.

Chen, M. S. : 1986, *Atomic Data and Nuclear Data Tables* **34**, 301.

Coffey, I. H. et al.: 1995, *Proc. 11*[th] *Colloqu. on UV and X-ray Spectroscopy of Astrophysical and Laboratory Plasmas. (Nagoya, 1995), to be publ. Universal Academy Press.*

Connor J.W.: 1973, *Plasma Physics* **15**, 765. Hirshman, S. P. and Sigmar, D. J.: 1981, *Nuclear Fusion* **21**, 1079.

Coppi, B. and Sharky, N.: 1981, *Nuclear Fusion* **21**, 1363.

Cowan, R. D. and Peacock, N. J.: 1965, *The Astrophysical J.* **142**, 390.

Cowan, R. D.,: 1981, 'The theory of Atomic Structure and Spectra', *Univ. of California Press.*

De Michelis, C. and Mattioli M.: 1981, *Nuclear Fusion* **21**, 677.

De Silva, Evans, D. E. and Forrest, M. J.: 1963, *Nature* **4952**, 1321.

Drake, G. W.: 1988, *Can. Jnl. Phys.* **66**, 586.

Dubau, J. and Volonte, S.: 1980, *Rep. Prog. Phys.* **43**, 199.

Dunn, J.: 1990, 'High Resolution X-Ray Spectroscopy of Laboratory Sources', *PhD Thesis, Dept. of Physics, Leicester University, UK.*

Durst R., Fonck, R. J., Cosby, G. and Evensen, H.: 1992, *Rev. Sci. Instr.* **63**, 4907.

Duval, B., P.,Hawkes, N. C. et al.: 1985, *Nuclear Instruments and Methods in Physics Research* **B9**, 689.

Edlén, B.: 1969, *Solar Physics* **9**, 439.

Edwards, A. W.: 1988, *Joint European Torus, private communication.*

Evans, D. E. and Katzenstein, J.: 1969, *Rep. Prog. Phys.* **32**, 207.

Fawcett, B. C., Jones, B. B. and Wilson, R.: 1961, *Proc. Phys. Soc.* **78**, 1223.

Fawcett, B. C., Gabriel, A. H., Griffin, W. G., Jones, B. B. and Wilson, R.: 1963, *Nature* **200**, 1303.

Fawcett, B. C., Gabriel, A. H., Jones, B. B. and Peacock, N. J.: 1964, *Proc. Phys. Soc.* **84**, 257.

Feldman, U., Doschek, G. A. et al.: 1980, *J. Appl. Phys.* **51**, 190.

Feldman, U., Indelicato P and Sugar J.: 1991, J. Opt. Soc. Am. B **8**, 3.

Feldman, U.: 1995, *Comments on Atomic and Molecular Physics* **31**, 11.

Fielding, S. J., Carolan, P. G. et. al., 1994, *Proc. 21*[st] *EPS Conf. on Plasma Physics and Controlled Nuclear Fusion Research (Montpellier)* **18B**, 322.

Field, G. B.: 1965, *Astrophysical J.* **142**, 531.

Finkenthal, M. et al.: 1984, *J. Appl. Phys.* **56**, 2012.

Fonck, R. J., Darrow, D. S. and Jaehnig et al.: 1984, Phys. Rev. 29,

Fonck, R. J., Bretz, N., Cosby, G., Durst R. et al.: 1992, *Plasma Physics and Controlled Fusion* **34**, 1993.

Freeman, R. L. and Jones, E. M. : 1974, *Culham Laboratory Report CLM -R137.*

Fuchs, J. C. et al.: 1995, *Proc. 22nd EPS Conf on Plasma Physics and Controlled Nuclear Fusion Research (Bournemouth) to be published.*

Gabriel, A. H., Fawcett, B.C. and Jordan, C.: 1965, *Nature* **206**, 390.

Gabriel, A. H. and Jordan, C.: 1969, *Nature* **221**, 947.

Gabriel, A. H. and Jordan, C.: 1972, *Case Studies in Atomic Collision Physics, chapter 4, (editors, McDaniel and McDowell, Publ. North Holland)* **2** 209.

Gabriel, A. H.: 1972, *Mon. Not. Roy. Astr. Soc.* **160**, 99

Geller, R. and Jacquot, B.: 1983, *Physica Scripta* **T3**, 19.

Gilbody, H. B.: 1981, *Physica Scripta* **23**, 143.

Gill, R. D. et al.: 1992, *Nuclear Fusion* **32**, 723.

Golts, E. Ya., Zhitnik, I.A., Kononov, E. Ya., Mandelstam, S. L. and Sedilnikov, Y. V.: 1974, *Report No.4 of the Inst. of Spectroscopy, Troitsk, USSR Academy of Sciences.*

Hawkes, N. C., Lawson, K. D. and Peacock, N. J.: 1993, *UV and X-ray Spectroscopy of Laboratory and Astrophysical Plasmas. (editors, E Silver and S. Khan) Cambridge University Press*, 324.

Hawkes, N. C.: 1995, 'Plasma Rotation and Electric Field measurements during the H-mode in JET.', *PhD thesis, Dept. of Physics, Imperial College, University of London.*

Hender, T. C. et al.: 1995, *Proc. 22nd EPS Conf. on Plasma Physics and Controlled Nuclear Fusion Research (Bournemouth) — to be published.*

Hess, W. R., Mattioli, M. et.al., 1993, 20th EPS Conf on Plasma Physics and Controlled Nuclear Fusion Research (Lisbon), 17 C, Part 3, 1079.

Hinnov, E.: 1983, *in Atomic Physics of Highly ionised Atoms, (editor, R Marrus), Plenum Publishing Corp.*, 49.

Hsuan, H., Bitter, M., Hill, K. W., von Goeler, S., et al.: 1987, *Phys. Rev. A* **35**, 4280.

Hulse, R. A., Post, D. E. and Mikkelsen D. R.: 1980, *J. Phys. B: At. Molec. Phys.* **13**, 3895.

Hulse, R. A.: 1983, *Nuclear Technology/ Fusion (Computational Methods)* **3**, 259.

Hulse, R. A.: 1993, 'Modeling of Impurity Transport in the Core Plasma', *Atomic and Plasma-Material Interaction Processes in Controlled Thermonuclear Fusion, publ. Elsevier*, 165.

Isler, R. C.: 1977, *Phys. Rev. Lett.* **38**, 1359.

Isler R.C. et al.: 1981, *Phys. Rev. A* **24**, 2701.

Isler, R. C.: 1984, *Nuclear Fusion* **24**, 1599.

Isler R. C., Morgan, P. D. and Peacock, N.J.: 1985, *Nuclear Fusion* **25**, 386.

Janev, R. K. and Grozdanov, T. P.: 1980, *Physics of Ionised Gases (M Matic, editor) Publ. Boris Kidrič Inst. Nucl. Sc., Belgrade*, 181.

Janev, R. K. and Presnyakov, L. P.: 1981, *Phys. Reports* **70**, 1.

Janev, R. K. and Drawin H W., (editors): 1993, 'Atomic and Plasma-Material Interaction Processes in Controlled Thermonucleaar Fusion' *publ. Elsevier.*

Jensen, R. V. et al.: 1977, *Nuclear Fusion* **176**, 1187.

Jones, B. B.: 1962, *Applied Optics* **1**, 239.

Jones, B. B. and Wilson, R.: 1962, *Nucl. Fusion: Suppl. Part 3*, 889

Källne, E. and Källne, J. and Pradhan, A.K.: 1983, *Phys. Rev. A* **27**, 1476.

Källne, E. and Källne, J., Richard, P. and Stöckli, M.: 1984, *J. Phys. B: At. Mol. Phys.* **17**, L115.

Källne, E. and Källne et al.: 1984, *Phys. Rev. Lett.* **52**, 2245.

Kaufmann, V. and Sugar, J.: 1986, *J. of Physical and Chemical Reference Data* **15**, 321.

Keenan, F. P., McCann, S. M. et al.: 1989, *Phys. Rev. A* **39**, 4092.

Keilhacker, M.: 1989, 'Diagnostics for Magnetically Confined Plasmas', *JET report JET-IR(89)09.*

Lauro-Taroni, L.: 1994, *Joint European Torus, private communication.*

Lauro-Taroni, L., Alper, B.: 1994, *Proc. 21st EPS Conf. on Plasma Physics and Controlled Nuclear Fusion Research (Montpellier)* 1, 102.

Lawson, K. D. and Peacock, N. J.: 1980, *J. Phys. B: Atom. Molec. Phys.* 13, 3313.

Lawson, K. D. Peacock, N. J. and Stamp, M. F.: 1981, *J. Phys. B: At. Molec.Phys.* 14, 1929.

Lawson, K. D., et al.: 1995, 'Derivation of Elemental Radiated Power Components and Impurity Concentrations', *JET report to be published.*

Levington, F. M., Fonck, R. J. et al.: 1989, *Phys. Rev. Lett.* 63, 2060.

Lipschultz, B., LaBombard, B., Marmar, E. S. et al.: 1984, *Nucl. Fusion* 24, 977.

Lipschultz, B.: 1987, *J. Nucl. Mater.* 145-147, 15.

Lotz, W.: 1967, *Z. Phys.* 206, 205.

Mandl, W.: 1991, 'Development of active Balmer-alpha Spectroscopy at JET', *PhD Thesis, University of Munich and JET report JET-IR(92)05.*

Mansfield, M. W. D., Peacock, N. J. et al.: 1978, *J. Phys. B.* 11, 152.

Marmar, E. S., Cecchi, J. L. and Cohen, S. A.: 1975, *Rev. Sci. Instr.* 46 1149.

Marrs, R. E. (editor): 1992, 'Selected publications from the Electron Beam Ion Trap program at Lawerence Livermore National Laboratory", *UCRL-ID-110491.*

Marrs, R. E., Elliot, S. R. and Knapp, D. A.: 1994, *Phys. Rev. Lett.* 72, 4082.

Martinson, I.: 1989, *Rep. Prog. Phys.* 52, 157.

Mattioli, M., Peacock, N. J. et al.: 1989, *Phys. Rev. A*, 40, 3886.

McWhirter, R. W. P. and Hearn, A. G.: 1963, *Proc. Phys. Soc.* 82, 641.

Merts, A. L., Cowan R. D. and Magee, N. H.: 1976, *Los Alamos Sci. Laboratory, report No. LA 6220 MS*

Mohr, P. J.: 1983, *At. Data and Nucl. Data Tables* 29, 453.

Moores, D. L.: 1987, 'Atomic Structure Theory', *Astrophysical and Laboratory Spectroscopy, Proc. 33rd Scottish Summer School in Physics*, pp27.

Negus, C. R. and Peacock, N. J.: 1979, *J. Phys. D: Appl. Phys.* 12, 91.

O'Mullane M. G., Coffey, I. H. and Peacock, N. J.: 1995, *Proc. 11th Colloq. on UV and X-Ray Spectroscopy of Astrophysical and Labratory plasmas. (Nagoya) — to be published by Universal Academy Press.*

Parker, E. N.: 1953, *Astrophysical J.* 117, 431

Pasini, D., Mattioli, M. et. al.: 1990, *Nuclear Fusion* 30, 2049.

Peacock, N. J., Cowan, R. D. and Sawyer, G. A.: 1966, *Proc. 7th Internat. Conf. on Phenomena in Ionised Gases, (Belgrade)* 2, 599.

Peacock, N. J., Speer R. J. and Hobby, M. G.: 1969, *J. Phys. B: At. Mol. Phys.* 2, 798.

Peacock, N. J.: 1971, *Plasma physics and Controlled Nuclear Fusion Research, (Publ. IAEA, Vienna)* 1, 537.

Peacock, N. J., Robinson, D. C., Forrest, M. J., Wilcock, P. D. and Sannikov, V. V.: 1969, *Nature* 224, 488.

Peacock, N. J., Hobby, M. G. and Galanti, M.: 1973, *J. Phys. B. (Atomic and Molecular Physics)* 6, L298.

Peacock, N. J.: 1980, 'Topical Spectroscopic features of the Emission from Highly ionised Atoms in tokamak Discharges', *Culham Laboratory Report CLM-P619*

Peacock, N. J.: 1980, *in The Physics of Ionised gases, Publ. Boris Kidriç Inst. of Nuclear Sciences, Belgrade(editor M Matic)*, 687.

Peacock, N. J. and Burgess, D. D.: 1981, *Phil. Trans. Roy. Soc. Lond.* 300, 665.

Peacock, N. J., Stamp, M. F. and Silver, J. D.: 1984, *Physica Scripta* T8, 10.

Peacock, N. J., 1984, 'Diagnostics based on Emission Spectra', *Applied Atomic Collision Physics (Academic Press Inc.)* 2, 143.

Peacock, N. J., Barnsley, R. et. al., 1994, *Proc. 21st EPS Conf. on Plasma Physics and Controlled Nuclear Fusion Research (Montpellier)* 18B, 134.

Peacock, N. J. et al. 1995, Proc. 22nd EPS Conf. on Plasma Physics and Controlled Nuclear Fusion Research (Bournemouth) — to be published.

Phillips, K. J. H., Keenan, F. P., Harra, L. K. et al., 1994, *J. Phys. B: At. Mol. Opt.Phys.* 27, 1939.

Reichle, R. et al.: 1995, *Proc. 22nd EPS Conf on Plasma Physics and Controlled Nuclear Fusion Research (Bournemouth) to be published*.

Rice, J. E., et al.: 1986, *Phys. Rev. Lett.* **56**, 50.

Rutherford, P.: 1974, *Physics of Fluids* **17**, 1782.

Rutherford, P. et al.: 1976, *Proc. Int. Symp. on Plasma Wall Interactions, KFA Jülich FRG*.

Sawyer, G. A. et al.: 1963, *Phys. Rev.* **131**, 1891.

Seguin, F. H., Petrasso, R. and Marmar, E. S.: 1983, *Phys. Rev. Lett.* **51**, 455.

Soltwisch, H., Costley, A. E., Salzmann, H. et al.: 1986, 'Diagnostic Methods for magnetic confinement Experiments', *Basic and Advanced Diagnostic Tecques for Fusion Plasmas. International School of Plasma Physics 'Piero Caldirola'*, *Varenna, Italy, Publ. by Commission of the European Communities (editors Stott P E et al.)* **2**.

Stamp, M. F., Armour, I. A., Peacock, N. J., and Silver, J.D.: 1981, *J. Phys. B: At. Mol. Phys.* **14**, 3551.

Stringer, T. E.: 1985, *Proc. 12th EPS Conf on Plasma Physics and Controlled Nuclear Fusion Research (Budapest)* **1**, 86.

Suckewer, S. and Hinnov, E.: 1978, *Phy. Rev. Lett.* **41**, 756.

Suckewer, S. and Hinnov, E.: 1979, *Phys. Rev. A* **20**, 478.

Suckewer, S. Fonck, R. and Hinnov, E.: 1980, *Phys. Rev.. A* **21**, 924.

Suckewer, S., Hinnov, E., Cohen, S., et al.: 1982, *Princeton Plasma Physics Report PPPL-1899*.

Summers, H. P.: 1994, 'Atomic Data and Analysis Structure – User Manual', *JET Joint Undertaking Report, EURATOM*.

TFR Group, Cornille, M. et al.: 1985, *Phys. Rev. A* **32**, 3000.

TFR Group, Bombarda, F. et al.: 1985, *Phys. Rev. A* **32**, 2374.

Turechek, J.J. and Kunze, H.J.: 1975, *Zeit. Phys. A* **273**, 111.

Underwood, J. H. and Barbee, T. W.: 1981, *AIP Conference on Low Energy X-ray Diagnostics, No 75, (editors, Attwood and Henke)*, 170.

Vainshtein, L. A. and Safranova, U. I.: 1978, *Atomic Data and Nuclear Data Tables* **21**, 49.

Vainshtein, L. A. and Safranova, U. I.: 1980, *Atomic Data and Nuclear Data Tables* **25**, 311.

von Hellermann, M. G., Mandl, W. et al.: 1991, *Plasma Phys. and Controlled Fusion* **33**, 1805. VIIIK-shell Spectra from Tokamaks: References:

von Hellerman, M. G. and Summers, H. P.: 1993, 'Active Beam spectroscopy at JET', in *Atomic and Plasma-Material Interaction Processes in Controlled Thermonuclear Fusion, publ. Elsevier*, 135.

Wagner, F., et al.: 1982, *Phys. Rev. Lett.* **49**, 1408.

Wesson, J., 1987, 'Tokamaks', *Oxford Science Publications*.

Wesson, J. A. and Hender, T. C.: 1993, *Nucl. Fusion* **33**, 1019.

Wiese, W. L.: 1993, 'Atomic Spectroscopic Data Base, *Atomic and Plasma-Material Interaction Processes in Controlled Thermonucleaar Fusion publ. Elsevier*, 355.

Wilson, R.: 1964, 'Proc. Symp. on Atomic Collision Processes in Plasmas. Culham Laboratory' *AERE Report 4818*, 16.

Wilson, R.: 1964, *Annales d'Astrophysique* **27**, 771.

Wilson, R.: 1967, 'Proc. ESRIN Study Group on Plasmas in Space and the Laboratory, Frascati May/June 1966', *ESRO Report SP-20*

Wilson, R.: 1992, 'Spectroscopy of Non-Thermal Plasmas' *J. Quant. Spect. Radiat. Transfer* **2**, 477.

Winter, H. P. and Aumayr, F., (editors): 1995, 'Proc. 7th International Conference on the Physics of Highly Charged Ions, (Vienna, Sept. 1994)', *to be publ. Nucl. Instr. and Methods in Physics Research, B*.

Zigler, A. et al.: 1980, *J. Opt. Soc. Am.* **70**, 129.